RS**Means**

Illustrated

Construction

Dictionary

RSMeans

ILLUSTRATED

CONSTRUCTION

DICTIONARY

Student Edition

WILEY

John Wiley & Sons, Inc.   RSMeans

For general information on our other products and services, or technical support, please contact our Customer Care Department within the United States at 800-762-2974, outside the United States at 317-572-3993 or fax 317-572-4002.

Wiley publishes in a variety of print and electronic formats and by print-on-demand. Some material included with standard print versions of this book may not be included in e-books or in print-on-demand. If this book refers to media such as a CD or DVD that is not included in the version you purchased, you may download this material at http://booksupport.wiley.com.

For more information about Wiley products, visit our Web site at www.wiley.com.

Library of Congress Cataloging-in-Publication Data:

RSMeans illustrated construction dictionary. — Student ed.

    p. cm.

 ISBN 978-1-118-13352-1 (pbk.); 978-1-118-35170-3 (ebk.); 978-1-118-35171-0 (ebk.); 978-1-118-35315-8 (ebk.); 978-1-118-35316-5 (ebk.); 978-1-118-35318-9 (ebk.)
 1. Rev. ed. of Means illustrated construction dictionary. 2003. 2. Building—Dictionaries. 3. Construction industry—Dictionaries. I. R.S. Means Company. II. Means illustrated construction dictionary. III. Title: Illustrated construction dictionary.

 TH9.M42 2013
 624.03—dc23

                                                                              2012009782

Printed in the United States of America
SKY10076950_060624

Contents

Preface

RSMeans Illustrated Construction Dictionary, Student Edition is a must-have companion to any construction-related curriculum. The engineering staff at RSMeans has edited the original *RSMeans Illustrated Construction Dictionary* to apply specifically for the student. Terms are defined in easy-to-understand language, and supplemented by over 1,400 illustrations. In addition to a higher percentage of illustrations, many illustrations new to this Student Edition have been added to make learning easier. Whenever possible, words or phrases are explained in non-technical language. When technical terms are used, they are also defined as separate entries. Very old architectural terms are purposely omitted, as this is intended to be a current construction dictionary with up-to-date terminology. A useful appendix illustrates and identifies the symbols that a building professional is likely to encounter.

Terms for new construction trends, such as BIM, building automation, energy and environmental conservation, "green" construction practice, engineered lumber products, the newest seismic technologies, and historic preservation are also covered. The editors obtained the assistance of industry groups, associations, societies, and manufacturers, as well as published authors who are nationally recognized authorities, to achieve this comprehensive construction trades, practices, and equipment dictionary.

Acknowledgments

This edition would not have been possible without the work of the RSMeans engineering staff. The editors also thank John Schaufelberger at the University of Washington and Clark Cory of Purdue University for their review of the material and recommended revisions for this student edition. Finally, a special thanks to the American Association of Cost Estimators (AACE) for allowing us to reprint some of their invaluable cost engineering terminology.

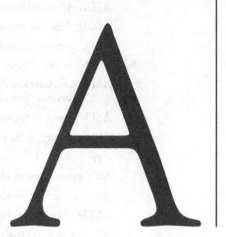

Abbreviations

a acre, ampere

A area, area square feet, ampere

A&E architect-engineer

AAMA Architectural Aluminum Manufacturers Association

AASHTO American Association of State Highway and Transportation Officials

ABC aggregate base course, Associated Builders and Contractors

ABS acrylonitrile butadiene styrene, asbestos-bonded steel

ABT air blast transformer, about

ac, a-c, a.c. alternating current

a.c. asphaltic concrete (a.c. paving)

AC air conditioning, alternating current (on drawings), armored cable (on drawings), asbestos cement

ACB asbestos-cement board, air circuit breaker

ACC accumulator

Access. accessory

ACD automatic closing device

ACEC American Consulting Engineers Council

ACGIH American Conference of Governmental Industrial Hygienists

ACI American Concrete Institute, Association of Construction Inspectors

ACM asbestos-containing material, asbestos-covered metal

ACS American Ceramic Society

ACSR aluminum cable steel reinforced, aluminum conductor steel reinforced

Acst acoustic

Actl actual

a.d. air-dried

AD access door, air-dried, area drain, as drawn

ADA Americans with Disabilities Act

ADAAG Americans with Disabilities Act Accessibility Guidelines

ADD addendum (on drawings), addition (on drawings)

Addit additional

ADF after deducting freight (used in lumber industry)

ADH adhesive

ADI after date of invoice

adj adjacent, adjoining, adjust, adjustable

ADS automatic door seal

af audio frequency

AFE Association for Facilities Engineering (formerly the American Institute of Plant Engineers)

AFL-CIO American Federation of Labor and the Committee for Industrial Organization

AFUE Annual fuel utilization efficiency

AG above grade

AGA American Gas Association

AGC Associated General Contractors

Agg, Aggr aggregate

AGL above ground level

AH, A HR, amp hr ampere-hour

AHERA Asbestos Hazard Emergency Response Act

AHU air-handling unit

AI Asphalt Institute

A

AIA American Institute of Architects, Asbestos Information Association

AIC ampere interrupting capacity

AIEE American Institute of Electrical Engineers

AIMA Acoustical and Insulating Materials Association

AISC American Institute of Steel Construction

AISE Association of Iron and Steel Engineers

AISI American Iron and Steel Institute

AITC American Institute of Timber Construction

AL, alum aluminum

Allow, ALLOW allowance

ALM alarm

ALS American Lumber Standards

alt altitude

ALT alternate

ALTN alteration

ALY alloy

AM ante meridiem

AMB asbestos millboard

AMD air-moving device

amp, Amp ampere

ANFO ammonium nitrate fuel oil mix

ANL anneal

Anod anodized

ANSI American National Standards Institute

AP access panel

APA The Engineered Wood Association (formerly the American Plywood Association)

APC acoustical plaster ceiling, American Plastics Council

APF acid-proof floor

API American Petroleum Institute

APPA The Association of Higher Education Facilities Officers (formerly the Association of Physical Plant Administrators)

Appd approved

Approx approximate

APR air-purifying respirator

Apt apartment

APW Architectural Projected Window

AR as required, as rolled

ARC W, ARC/W arc weld

ARS asbestos roof shingles

ART artificial

AS automatic sprinkler

ASA American Standards Association

asb asbestos

ASBC American Standard Building Code

asbe asbestos worker

ASC asphalt surface course

ASCE American Society of Civil Engineers

ASCII American Standard Code for Information Interchange

ASEC American Standard Elevator Codes

ASES American Solar Energy Society

ASHRAE American Society of Heating, Refrigeration, and Air-Conditioning Engineers

ASIS American Society for Industrial Security

ASME American Society of Mechanical Engineers

asph asphalt

ASR automatic sprinkler riser

ASSE American Society of Sanitary Engineering

ASTM American Society for Testing and Materials

AT asphalt tile, airtight

ATB asphalt-tile base

ATC acoustical tile ceiling; architectural terra-cotta, automatic temperature control

ATF asphalt-tile floor

atm atmosphere, atmospheric

aux auxiliary

av, ave, avg average

A/W all-weather

AW actual weight

AWEA American Wind Energy Association

AW&L all widths and lengths

AWG American wire gauge

AWI Architectural Woodwork Institute

AWPI American Wood Preservers Institute

AWS American Welding Society

AWWI American Wood Window Institute

Definitions

abaciscus, abaculus 1. A tessara or small square stone used in mosaic tile. 2. A small abacus.

abamurus A masonry buttress for the support of a wall.

abandonment 1. To surrender the right or claim of interest without specifically transferring it. 2. The act of deserting one's obligations under a contract frequently manifested by removing personnel, materials, and equipment from the job site.

abate 1. To cut away in stone or to beat down on metal in order to create figures or a pattern in relief. 2. To reduce or decrease concentrations of pollutants.

abatement 1. The encapsulation or removal of building materials containing pollutants (such as lead or asbestos) to prevent the release of or exposure to fibers. 2. In lumber industry, the amount of wood lost as waste during the process of sawing or planning.

abatvent A wall louver that restricts wind from entering a building, but admits light and air.

abatvoix An acoustical reflector for a single voice, such as behind and over a church pulpit.

ABC extinguisher A fire extinguisher suitable for use on type A, B, and C fires.

A-block A hollow masonry unit with one closed end commonly used at wall openings.

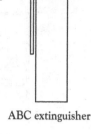

ABC extinguisher

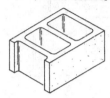

A-block

Abney level A handheld level used for measuring elevations and vertical angles.

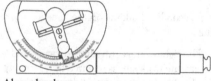

Abney level

above finished floor Datum or benchmark for measuring the height above a floor, abbreviated as AFF.

above-grade subfloors A floor above ground level, but with no head-room below.

abrade To scrape or wear away a surface by friction or striking.

Abrams' law The rule stating that with given materials, curing, and testing conditions, concrete strength is inversely related to the ratio of water to cement. Low water-to-cement ratios produce high strengths.

abrasion resistance index A comparison of the abrasion resistance of a given material to that of rubber. The index is applied principally to aggregate handling equipment.

abrasive 1. A hard material used for wearing away or polishing a surface by friction. 2. The material that is adhered to or embedded in a surface such as sandpaper or a whetstone.

abrasive blasting A method of cleaning surfaces with a high-pressure stream of air and an abrasive material such as sand or steel grit.

abrasive floor A floor with an abrasive adhered to or embedded in the surface to provide traction and prevent slipping.

abrasive floor tile Floor tile with an abrasive adhered to the surface.

abrasive nosing A strip of anti-skid abrasive adhered to or attached to the nosing of a stair tread.

abrasive stair tread A stair tread with an abrasive surface.

abrasive terrazzo A terrazzo floor with an abrasive surface rather than a high polish.

abreuvoir The mortar joint between masonry units.

absorbed moisture Moisture that has been absorbed by a solid such as masonry.

absorbent 1. A material that has an affinity for certain substances and attracts these substances from a liquid or gas with which it is in contact, thus changing the physical and/or chemical

properties of the material. 2. A substance that attracts and holds large quantities of liquid.

absorber 1. A device containing liquid for the absorption of vapors. 2. In a refrigeration system, the component on the low-pressure side used for absorbing refrigerant vapors.

absorber plate That part of a solar energy system that collects the solar energy.

absorber plate

absorption 1. The process by which a liquid is drawn into and fills permeable pores in a solid body, increasing its weight. 2. The process by which solar energy is collected on a surface. 3. The increase in weight of a porous object resulting from immersion in water for a given time, expressed as a percent of the dry weight.

absorption air conditioning An air cooling and dehumidifying system powered by solar or other energy collected on absorbing plates.

absorption bed or field (disposal field, drain field) A network of trenches that may contain coarse aggregate and distribution pipe and is used to distribute septic tank effluent into the surrounding soil.

absorption chiller Heat-operated refrigeration unit that uses an absorbent (lithium bromide) as a secondary fluid to absorb the primary fluid (water), which is a gaseous refrigerant in the evaporator. The evaporative process absorbs heat, thereby cooling the refrigerant (water), which in turn cools the chilled water circulating through the heat exchanger.

absorption loss 1. Water losses that occur until soil particles are sufficiently saturated, such as in filling a reservoir for the first time. 2. Water losses that occur until the aggregate in a concrete mix is saturated.

absorption rate (initial rate of absorption) 1. The weight of water absorbed by a brick or concrete masonry unit that is partially immersed in water for one minute, expressed in grams or ounces per minute. 2. The annual rate at which new housing or leasable space is being sold or leased. The absorption rate of a prior year often is used to predict the needs for next year. (A gross absorption rate measures the consumption of new housing/space only.)

absorption-type liquid chiller A system using an absorber, condenser, and associated accessories to cool a secondary liquid.

ABS plastic pipe Acrylonitrile-butadiene-styrene plastic pipe, which is resistant to heat, impact, and chemicals.

abstract of bids A list of the bidders for a sealed bid procurement indicating the significant portions of their bids.

abstract of title A deed for a parcel of land showing encumbrances and a history of ownership.

abut To join or touch at one edge or end without overlapping.

abutment (butment) 1. The structure that supports the end of a bridge or arch or that anchors the cables of a suspension bridge. 2. The surface at which one member meets another.

abutment piece In structural framing, the horizontal member that distributes the load of vertical members and is thus the sole plate of a partition.

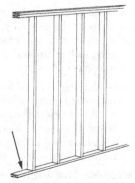

abutment piece

A

abuttals The properties adjacent to a parcel of land or body of water and which mark the boundaries of that land or water body.

abutter A property owner contiguous or within a specified distance from a parcel of land.

Accelerated Cost Recovery System (ACRS) A method of calculating depreciation of assets placed in service from 1980 to 1986 for federal income tax purposes. Replaced by the Modified Accelerated Cost Recovery System (MACRS) by the Tax Reform Act of 1986.

accelerated depreciation Asset depreciation at a faster rate than the straight-line method. Examples include the 200% or 150% declining balance methods.

accelerating admixture An admixture for hydraulic concrete that shortens setting time and inhibits early strength development.

accelerator An additive that, when added to paint, concrete, mortar, or grout mix, speeds the rate of hydration and thereby causes it to set or harden sooner.

accelograph An instrument used to measure displacement during an earthquake. Often installed in buildings to measure movement.

accent lighting Fixtures or directional beams of light arranged so as to bring attention to an object or area.

acceptance Compliance by an offeree with the terms and conditions of an offer.

acceptance, final (partial) The formal action by the owner accepting the work (or a specified part thereof), following written notice from the engineer that the work (or specified part thereof) has been completed and is acceptable subject to the provisions of the contract regarding acceptance.*

acceptance certificate A dated and signed document issued to a contractor by an owner certifying that all the work of a construction project is complete and in accordance with all provisions of the contract.*

accepted bid The proposal or bid a contractor and an owner or owner's representative use as the basis for entering into a construction contract.

access 1. The means of entry into a building, area, or room. 2. A port or opening through which equipment may be inspected or repaired.

access connection A ramp or roadway for entering or exiting an arterial highway.

access control system Computerized building security equipment, such as badge readers, designed to protect against unauthorized entry into buildings or building zones.

access door or panel A means of access for the inspection, repair, or service of concealed systems, such as air-conditioning equipment.

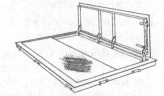
access door or panel

access flooring A raised flooring system with removable panels to allow access to the area below. This type of flooring is frequently used in computer rooms because it provides easy access to cables.

accessible That which is easily removed, repaired, or serviced without damaging the finish of a building.

accessible route A continuously unobstructed path connecting all accessible elements and spaces of a building or facility.

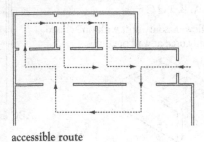
accessible route

accessories In the placing of concrete, the items used to assemble scaffolding, shoring, and forms, other than the wales, frames, and the forms themselves.

accessory building A secondary building on the same lot adjacent to the main building.

access to the work The right of the contractor to ingress and egress, and to occupy the work site as required to reasonably perform the work described in the contract documents. An example of denial of access to the work would be on the segment of a sewer installation project where no easements or work limits are indicated, but the contractor is ordered, after contract award, to conduct operations within a narrow work corridor necessitating different or unanticipated construction methods (e.g., use of sheeting).*

accolade Ornamental treatment over an arch, doorway, or window formed by two ogee curves meeting in the middle.

accordion door A retractable door, usually fabric-faced, hung from an overhead track and folding like the bellows of an accordion.

accordion partition A retractable partition having the same features as an accordion door.

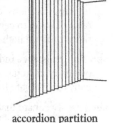
accordion partition

accouplement 1. In architecture, the pairing of pilasters or columns, as in a colonnade or buttress. 2. In carpentry, a tie or brace between timbers.

accumulator (surge drum, surge header) 1. A pressure vessel whose volume is used to maintain a constant pressure. 2. In refrigeration, a storage chamber for low-side refrigerant.

acetone A highly flammable organic solvent used with lacquers, paint thinners, paint removers, and resins.

acetylene A carbon gas which, when combined with pure oxygen and ignited, produces an extremely hot flame used in gas welding and metal cutting.

acetylene torch The torch used for welding and cutting. Contains compressed acetylene and oxygen.

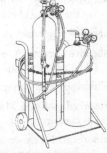
acetylene torch

A

AC generator A generator that produces alternating current.

achromatic color White, colorless light.

acid- and alkali-resistant grout or mortar A grout or mortar that is highly resistant to prolonged exposure to alkaline compounds, acid liquids, or gases.

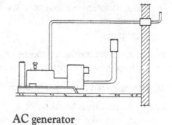

AC generator

acid etch (aciding) A method of cleaning the latence from concrete by washing it with an acid solution and rinsing with water.

acid-proof floor A floor that resists deterioration when exposed to acid.

acid resistance A measurement of a surface's ability to resist the corrosive effect of acids.

acid-resistant brick Brick that resists deterioration caused by exposure to acid. This type of brick should be laid with acid-resistant mortar.

acid soil Soil with a pH value of less than 6.6.

acid steel Steel made with a silica flux or in a silica-lined furnace.

acorn nut Nut with hexagonal base and rounded top that encases the end of the screw.

acoustical A term used to define systems incorporating sound control.

acoustical barrier A building system that restricts sound transmission.

acoustical block (acoustic block) A masonry block with sound-absorbing qualities, usually defined in terms of its NRC (noise reduction coefficient) rating.

acoustical board A construction material in board form that restricts or controls the transmission of sound.

acoustical ceiling A ceiling system constructed of sound-control materials. The system may include lighting fixtures and air diffusers.

acoustical door A door constructed of sound-absorbing materials and installed with gaskets around the edges.

acoustical enclosure (acoustical booth, acoustical room) An enclosure constructed of acoustical materials for privacy in speaking, listening, and recording, as in a recording studio or a telephone booth.

acoustical lining Insulating material secured to the inside of ducts to limit sound and provide thermal insulation.

acoustical materials Materials that absorb and isolate sound and reduce reverberation, including felts, tiles, boards, and plasters.

acoustical metal deck A metal decking that includes a sound-absorbing material installed at a small additional cost per square foot.

acoustical panel Modular units composed of a variety of sound-absorbing materials for ceiling or wall mounting.

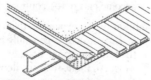

acoustical metal deck

acoustical reduction factor A value, expressed in decibels, that defines the reduction in sound intensity that occurs when sound passes through a material.

acoustical sprayed-on material A fibrous material with acoustical properties applied to a surface by spraying through a nozzle.

acoustical tile A term applied to modular ceiling panels in board form with sound-absorbing properties. This type of tile is sometimes adapted for use on walls.

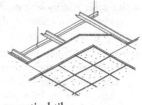

acoustical tile

acoustical transmission factor The reciprocal of the sound reduction factor. A measure of sound intensity as it passes through a material, expressed in decibels.

acoustical wallboard Wallboard with sound-absorbing properties.

acoustical window wall Double-glazed window walls with acoustical framing. This type of wall system is used particularly at airports.

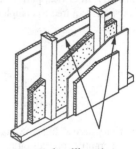

acoustical wallboard

acoustic lining Insulating material secured to the inside of ducts to attenuate sound and provide thermal insulation.

acoustics 1. The science of sound transmission, absorption, generation, and reflection. 2. In construction, the effects of these properties on the acoustical characteristics of an enclosure.

acquiescence A term frequently used when owners of adjacent properties agree on a boundary between their properties, if the original boundary is difficult or impossible to establish.

acre A common unit of land-area measurement equal to 160 square rods, or 43,560 square feet.

acre foot A unit of volume measurement equal to one acre times one foot thick. The acre foot is used to measure the volume of water or ore deposits.

acrylic See **acrylic resin.**

acrylic fiber Fiber produced from polymerized acrylonitrile, a liquid derivative of natural gas. A tough economical fiber commonly used in commercial and residential carpets and draperies.

acrylic plastic glaze A clear plastic sheet that is bonded to glass and that increases the ability of the glass to resist breaking and shattering.

acrylic resin (acrylate resin) In construction, clear, tough, thermoplastic resin manufactured in sheet and corrugated form, used as an adhesive, and as the main ingredient in some caulking and sealing compounds.

action item An element of work, design, research, or other task to be competed before a specific date or time, such as the before a subsequent meeting of involved parties.

action level The point when a concentration of hazardous materials reaches a level where OSHA regulations dictate protective steps be taken.

activated sludge Sludge that has settled out of oxygenated sewage.

active earth pressure The horizontal component of pressure exerted on a wall by earth.

active leaf In a double-leaf door, the leaf to which the latching or locking mechanism is attached.

5

active solar energy system A system that primarily collects and transfers solar energy using mechanical means that are not powered by solar energy.

active walls Building walls that act as a generator or collector of energy. An example is a double glass wall that collects solar energy and reflects excess heat when the desired interior envelope temperature has been reached. This combination reduces a facility's net heating and cooling load.

activity In critical path method (CPM) scheduling, a task or item of work required to complete a project. Also called *task*.

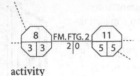

activity

activity arrow In arrow diagrams in critical path method scheduling, a graphic representation of an activity.

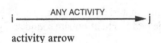

activity arrow

activity duration In critical path method scheduling, the estimated time required to complete an activity in time units (weeks, days, hours, etc.). There are three types of duration: original (or planned) duration, actual duration, and remaining duration. For in-progress activities, the completion duration is calculated by combining actual duration (so far) and remaining duration.

act of God An unforeseeable, inevitable event caused by natural forces over which an insurance policyholder has little or no control.

actual costs The actual expenditures incurred by a program or project.*

actual cost records Contemporaneous construction and accounting records detailing actual costs from a constructed project, including invoices, contracts, subcontracts, change orders, and applications for payment.*

actual damages Damages that can be assessed against an owner or contractor if either or both fail to perform their respective responsibilities and obligations as contained in the construction contract. Actual damages are considered economic (monetary) damages that can be clearly determined and proven, typically awarded by a court as the result of a lawsuit brought by one of the parties to the construction contract.*

actual dimension The real dimensional measurement of a piece of lumber, masonry unit, or other construction material.

actual finish date Date when work on an activity is substantially complete. Activity substantial completion is when only minor or remedial work remains and successor activities may proceed without hindrance from the predecessor's remaining work. It is not necessarily the last day work will be performed on that activity. The remaining duration of this activity is zero.*

actual start date Date when work on an activity actually started with intention of completing activity within the planned duration. The actual start date is not necessarily the first date work was performed on that activity. Interim starts and stops for an activity may show the need for splitting the activity into component parts.*

actuator In hydraulics, a motor or cylinder designed to convert hydraulic energy into mechanical energy.

acute angle An angle less than 90°.

adapt To modify a building or space to make it suitable for new requirements or purposes.

adaptable building A building that can be easily updated or modified to meet changing needs or requirements.

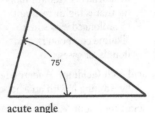

acute angle

adapter Any device designed to match the size or characteristics of one item to those of another, particularly in the plumbing, air-conditioning, and electrical trades.

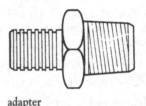

adapter

ADA Standards for Accessible Design As an adjunct to the Americans with Disabilities Act, a set of standards that establish minimum technical requirements for the design and construction of buildings and facilities. Their intent is to increase the level of accessibility in the built environment, in existing facilities as well as new construction and alterations.

ADA Standards for Accessible Design

addendum A document describing an addition, change, correction, or modification to contract documents. An addendum is issued by the design professional during the bidding period or prior to the award of contract, and is the primary method of informing bidders of modifications to the work during the bidding process. Addenda become part of the contract documents.

addition 1. An expansion to an existing structure, generally in the form of a room, floor, or wing. An increase in the floor area or volume of a structure. **2.** A chemical added to cement at the time of its manufacture to help the process or to alter the cement's characteristics.

addition

additive A substance that is added to a material to enhance or modify its characteristics, such as curing time, plasticity, color, or volatility.

additive alternate A specific alternate option for construction specifications or plans that results in a net increase in the base bid.

address system An electronic audio system with a microphone and speakers installed for either fixed (permanent) or mobile use. Wiring for a permanent system should be done prior to any finish work.

addressable system An advanced fire alarm or security system that provides for easy monitoring, remote testing, and quick location of an alarm condition.

adhesion The binding together of two surfaces by an adhesive.

adhesion-type ceramic veneer Ceramic tile or veneer attached to a backing by mortar, grout, or adhesive only. No anchors are used.

adhesive Generally, any substance that binds two surfaces together. In construction, the term is used principally in the wallboard and roofing trades.

adiabatic process A thermodynamic process occurring in the absence of heat gain or heat loss.

adit 1. The entrance or approach to a building. 2. The entrance to a mine.

adjustable base anchor An attachment to the base of a door frame above a finished floor.

adjustable clamp A temporary clamping device that can be adjusted for position or size.

adjustable square (double square) A carpenter's tool used for marking and scribing lumber. An adjustable square usually incorporates a level bubble.

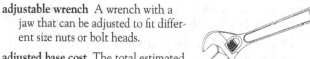

adjustable square (double square)

adjustable wrench A wrench with a jaw that can be adjusted to fit different size nuts or bolt heads.

adjustable wrench

adjusted base cost The total estimated cost of a project after adding or deducting addenda or alternatives.

adjuster A representative of the insurance company who negotiates with all parties involved in a loss in order to settle the claim equitably. An adjuster deals with the policyholder, repair contractor(s), witnesses, and police (if necessary), and acts as a middleman between these parties and the insurance company.

adjusting nut A threaded nut used for alignment of an object. Often coupled with a locking nut to secure it in position.

adjusting nut

adjusting screw A screw used for alignment of an object. Often coupled with a locking nut to secure it in position.

adjustment The determination of: (a) the cause of a loss, (b) whether it is covered by the policy, (c) the dollar value of

the loss, and (d) the amount of money to which the claimant is entitled after all allowances and deductions have been made.

admixture An ingredient other than cement, aggregate, or water that is added to a concrete or mortar mix to affect the physical or chemical characteristics of the concrete or mortar. The most common admixtures affect plasticity, air entrainment, and curing time.

adobe Earthen, sun-cured brick. A relatively labor-intensive, but low-embodied energy material, adobe absorbs excess heat during hot days and releases it during cool nights, thereby moderating a building's internal temperature.

adobe brick A large, roughly formed, unfired brick made from adobe and straw.

adsorbed water Water that is held on the surface of materials by electrochemical forces. This water, such as that on the surfaces of aggregate in a concrete mix, has a higher density and thus different physical properties from those of the free water in the mix.

adsorbent A material that has the ability to extract certain substances from gases, liquids, or solids by causing them to adhere to its surface without changing the physical properties of the adsorbent. Activated carbon, silica gel, and activated alumina are materials frequently used for this application.

adsorption The process of extracting specific substances from the atmosphere or from gases, liquids, or solids by causing them to adhere to the surface of an adsorbent without changing the physical properties of the adsorbent.

ad valorem Latin for based on value. Real property taxes, as they are based on the value of real property, are an ad valorem tax. An ad valorem tax is levied in proportion to value.

advance payment A partial payment to a contractor made shortly after the contract is signed. Similar to a down payment.

advance payment bond The generic term for the assurance of performance provided by a contractor to an owner that any money advanced to the contractor will be properly used to pay for project costs.

advance slope method A method of placing concrete in which the sloped face of the fresh concrete moves forward as the concrete is placed.

advance waiver of liens A waiver of all the contractor's rights to file mechanic's or materialmen's liens against the owner for nonpayment for work performed. Such advance waiver may be a condition of the owner's contract.

advertisement for bids Published notice of an owner's intention to award a contract for construction to a constructor who submits a proposal according to instructions to bidders.

adz A long-handled tool with a curved blade set perpendicular to the handle. Used for dressing lumber.

adz

adz-eye hammer A claw hammer with a long eye for receiving the handle.

aerate To introduce air into soil or water, for example, by natural or mechanical means.

aeration The process of introducing air into a substance or area by natural or mechanical means.

aeration plant A sewage treatment plant in which air is introduced into the sewage to accelerate the decomposition process.

aerator A mechanical device that introduces air into a material such as soil, water, or sewage.

aerator fitting A pipe fitting used to introduce air into a flow of water.

aerial Pertaining to, caused by, or present in the air.

aerial ladder An extension ladder capable of reaching high places and often mounted on a vehicle such as a fire truck.

aerial lift A term commonly applied to mobile working platforms that are elevated hydraulically or mechanically.

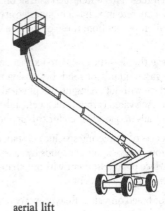

aerial lift

aerial survey A survey of the earth's surface based on aerial photographs and ground control points.

aerodynamic instability A harmonic motion occurring in a structure during high winds and endangering structural integrity. The term was used to define the failure of the Tacoma Narrows Bridge.

aerofilter A bed of coarse aggregate used for filtering sewage.

aerofilter

affidavit of noncollusion A sworn statement by the bidders on a project that the prices on their proposals were arrived at independently without consultation between or among them.

affinity A tendency for two substances to unite chemically or physically.

A-frame 1. A structural system or hoisting system with three members erected in the shape of an upright capital letter "A." **2.** A building with a steep gable roof that extends to the ground.

A-frame

afterfilter (final filter) In air conditioning, a filter located at the outlet end of the system.

age hardening A term used to describe a hardening process of metals at room temperature.

agent Under agency law, an agent is authorized by the principal to act on the principal's behalf. Generally, an agent's acts bind the principal as though the principal had acted directly.

aggregate Granular material such as sand, gravel, crushed gravel, crushed stone, slag, and cinders. Aggregate is used in construction for the manufacturing of concrete, mortar, grout, asphaltic concrete, and roofing shingles. It is also used in leaching fields, drainage systems, roof ballast, landscaping, and as a base course for pavement and grade slabs. Aggregate is classified by size and gradation.

aggregate

aggregate, abrasive An antiskid aggregate worked into the surface of a concrete floor.

aggregate bonding capacity The maximum total contract value that a bonding company will cover (in performance bonds) for all of a construction company's current contracts.

aggregate, coarse Aggregate that is larger than ⅛" and is retained on the No. 8 sieve.

aggregate, coarse-graded Aggregate with a continuous grading from coarse to fine, with a predominance of coarse particles.

aggregrate, concrete The fine and course aggregate used in manufacturing concrete. Both are usually washed and graded.

aggregrate, exposed A concrete surface with the aggregate exposed, formed by applying a retarder to the surface before the concrete has set, and subsequently removing the cement paste to the desired depth.

aggregrate, fine Aggregate smaller than ⅛". Fine aggregate passes through the No. 8 sieve.

aggregate, heavyweight The aggregate produced from materials with high specific gravity, such as limonite, iron ore tailings, and magnetite.

aggregate interlock The term applied to a situation in which the aggregate from one side of a concrete joint projects between the aggregate of the other side of the joint, thus resisting shear.

aggregate, lightweight One of several materials used to decrease the unit weight of concrete, thereby reducing the structural load and the cost of the building. The materials most commonly used are perlite and vermiculite. The use of lightweight aggregate is costly, but sometimes necessary in construction.

aggregate limit The maximum amount an insurance policy will pay for the sum of all personal injury and property damage claims that may arise during the term of the policy as the result of multiple occurrences. Legal defense costs may be excluded from this limit.

aggregate, masonry Washed sand used in a mortar mix.

aggregate, open-graded An aggregate in which a skip between the sieve gradations has been deliberately achieved so that the voids are not filled with intermediate-size particles.

aggregate panel A precast concrete panel with exposed aggregate.

aggregate, plaster Natural or manufactured washed sand used in a plaster mixture.

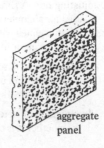

aggregate panel

aggregate, roof 1. The aggregate used for a tar-and-gravel application. 2. The ballast used for membrane-type roofing.

aggregate spreader A piece of equipment used for placing aggregate to a desired depth on a roadway or parking lot.

aggregrate testing Any of a number of tests performed to determine the physical and chemical characteristics of an aggregate. Common tests are for abrasion, absorption, specific gravity, and soundness.

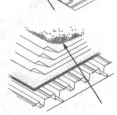

aggregate, roof

aggregate, well-graded An aggregate that incorporates sizes from the maximum to the minimum specified so as to fill most of the voids. This type of aggregate is used for asphaltic concrete mixes and for base courses.

aggressive sampling During removal of hazardous materials, the agitation of air to test success of remediation effort.

aging 1. A method of classifying individual receivables by age groups, according to the time elapsed from the date due. 2. A process used to make building materials appear old or ancient. 3. The chemical and physical changes in a material incurred by the passage of time.

agitation The rotation of, or moving of blades through, a drum containing concrete or mortar to prevent segregation or setting of mixture.

agitator A mechanical device used to maintain plasticity and to prevent segregation, particularly in concrete and mortar.

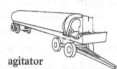

agitator

A-grade wood 1. A plywood surface that is smooth and paintable, and considered the best standard veneer. May be composed of more than one piece well jointed together. 2. Plywood designation A-face, best veneer grade.

agreement 1. A promise to perform, made between signatories to a document. 2. In construction, the specific documents setting forth the terms of the contracts among architect, owner, engineer, construction manager, contractor, and others.

agreement form A standard printed form used by the signatories to an agreement, with blank spaces to fill in information pertinent to a particular contract.

agricultural lime A granular hydrated lime used for soil conditioning.

air admittance valve A one-way valve that allows air to enter into a plumbing drainage system if negative pressure in the piping occurs.

air balancing The process of adjusting a heating or air-conditioning duct system to provide equal distribution to all areas.

air barrier A component of the building envelope system that prohibits air leakage into a building and reduces the risk of condensation buildup. House wrap and fluid-applied air barriers are examples.

airborne transmission A term that refers to sound traveling through air in a structure.

air break A piping arrangement in which a drain from a fixture appliance or device discharges through an open connection into a receptacle or interceptor. Used to prevent backflow or back siphonage.

airbrush A device with a nozzle for applying paint with compressed air.

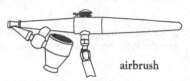

airbrush

air chamber In water piping, a vertical pipe containing entrapped air to absorb the pressure shock when a valve is closed suddenly.

air change The volume of air in an enclosure that is being replaced by new air. The number of air changes per hour is a measure of ventilation.

air circuit breaker A breaker that discontinues current flow in air.

air cleaner A device, often hung from the ceiling, for removing impurities from the air. The device may have a mechanical or electrostatic filter.

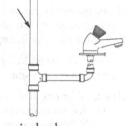

air chamber

air compressor A machine that extracts air from the atmosphere and compresses it into a holding chamber. The most common use of compressed air is for the operation of pneumatic tools. Air compressors are classified by the number of CFM (cubic feet per minute) of compressed air they can produce.

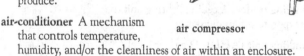

air-conditioner A mechanism that controls temperature, humidity, and/or the cleanliness of air within an enclosure.

air compressor

air-conditioning system An air treatment system designed to control the temperature, humidity, and cleanliness of air and to provide for its distribution throughout the structure.

air content The volume of air present in a concrete or mortar mix, expressed as a percentage of the total volume. A controlled air content prevents concrete from cracking during the freeze/thaw cycle.

air curtain (air wall) A narrow stream of air directed across an opening to deter the transfer of hot or cold air, contaminants, and insects from one side to the other.

air density The weight per unit volume of air, expressed in pounds per cubic foot.

air diffuser An outlet in an air-supply duct for distributing and blending air in an enclosure. Usually, a round, square, or rectangular unit mounted in a suspended ceiling.

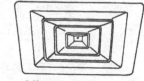

air diffuser

air-distributing ceiling A suspended ceiling system with small perforations in the tiles for controlled distribution of the air from a pressurized plenum above.

air door An invisible barrier of high-velocity air that separates different environments. Sometimes called *air walls*, air doors are typically used for garage-type or larger doors to reduce infiltration and ex-filtration.

air drain An empty space left between a foundation wall and a parallel wall to prevent the fill from lying directly against the foundation wall.

air-entraining agent An admixture for concrete or mortar mixes that causes minute air bubbles to form within the mix. Air entrainment is desirable for workability of the mix and prevention of cracking in the freeze/thaw cycle.

air-entraining hydraulic cement Hydraulic cement containing an air-entraining addition in such amount to cause the product to entrain air in mortar within specified limits.

air escape In plumbing, a valve for automatically discharging excess air from a water line.

air exchange rate Refers to the rate at which outside air replaces indoor air in a building, expressed in either air changes per hour (ACH) cubic feet per minute (CFM).

air gap In plumbing, the distance between the outlet of a faucet and the overflow level of the fixture.

air grating A fixed metal grating, particularly in masonry foundation walls, for ventilation.

air grating

air hammer A portable, pneumatic percussion tool used for breaking and hammering.

air-handling troffer A ceiling lighting unit that incorporates an air diffuser.

air-handling unit (AHU) The traditional method of heating, cooling, and ventilating a building by which single- or variable-speed fans push air over hot or cold coils, then through dampers and ducts and into one or more rooms.

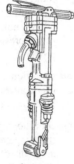

air hammer

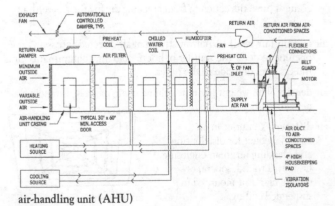

air-handling unit (AHU)

air leakage The air that escapes from a system or enclosure through cracks, joints, and couplings.

air lift A device that uses compressed air to lift slurry or dry powder through piping.

airlock 1. An airtight chamber such as that used in tunnel and caisson excavation. **2.** A system of double doorways permitting entry and exit while preventing airflow from one area to another, as from a contaminated area to an uncontaminated area. **3.** An entrance room between areas of different pressures, such as the entrance to an air-supported structure. **4.** In plumbing, air trapped in a system and preventing flow.

air makeup unit A system for introducing fresh, conditioned air into an enclosure from which air is being exhausted.

air-mixing plenum In an air-conditioning system, a chamber in which fresh air is mixed with recirculated air.

air monitoring In asbestos abatement, a procedure used to determine the fiber content in a volume of air over a measurable period of time.

air permeability test A procedure for determining the fineness of powdered material such as cement.

air pocket A void filled with air, such as in a water piping system or in a concrete form when placing concrete.

air purge valve A device for eliminating trapped air from a piping system.

air-purifying respirator A device that removes pollutants from a contaminated atmosphere as a person breathes.

air-purifying respirator

air receiver The air storage tank on a compressor.

air regulator An instrument for regulating the flow or pressure of air in a system.

air release valve A valve that releases air from a water pipe or fitting.

air rights The exclusive right of real property owners to possess the airspace above their land, as long as they comply with building and zoning laws.

air separator A pneumatic device that uses air to sort materials by size.

air shaft (air well) A roofless enclosed area within a building, admitting light and ventilation.

air splitter Device inside an air duct that divides a single air stream into several streams.

air-supported structure A nonrigid structure supported by atmospheric pressure that is slightly higher inside the tank than outside. The difference in pressure is created by fans.

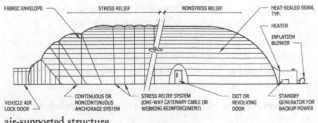

air-supported structure

air terminal The top of a lightning protection system on a building.

air test A test for leaks in ductwork and in drainage and pipe systems where compressed air is forced into a sealed system and leaks are detected with a pressure gauge.

airtight Refers to the inability to permit air passage.

air tube system A tubular conveying system that uses air pressure to move capsules containing paperwork from one station to another.

air washer A water spraying mechanism for cleaning and humidifying air in a ventilation system.

airway The air space between the thermal insulation and sheathing on a roof.

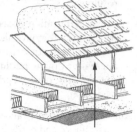
airway

aisleway Any open passageway permitting access and traffic flow between sections within a building.

alarm system An installed electrical system devised to protect against unauthorized entry or fire by giving off an audible and/or visual signal.

alclad A product having an aluminum or aluminum alloy coating metallurgically bonded to the surface. The coating is anodic to the core, thus protecting it physically and electrolytically against corrosion.

alcove A recess or partly enclosed extension opening into a larger room.

algorithm A set of mathematical instructions, or a computer program, used to produce a control output.

alidade A sighting apparatus often used with a plane table for determining and plotting horizontal and/or vertical angles.

aligning punch A tool used for aligning holes in structural steel. Often referred to as a spud wrench.

alignment 1. The adjustment of elements in a plane such as structural steel. **2.** The plane or horizontal orientation of a structure or roadway.

aliphatic resin glue Thermoplastic adhesive used to bond wood and other porous materials.

alite The primary constituent of Portland cement clinker. Alite is composed of tricalcium silicate and small amounts of magnesium oxide, aluminum oxide, ferric oxide, and other materials.

alkali 1. A liquid that has a pH greater than 7.0. **2.** Water-soluble salts of alkali metals, such as sodium and potassium, which occur in concrete and mortar mixes. The presence of alkaline substances may cause expansion and subsequent cracking.

alkali resistance The ability, particularly of paint, to resist attack by alkaline materials.

alkali soil Soil that has a pH value of 8.5 or higher and is thus harmful to some plant life.

alkyd paint A paint, with an alkyd resin base, that produces a quick-drying, hard surface.

alkyd plastics Thermoset plastics with good heat and electrical insulation properties. Commonly used in paints, lacquers, and molded electrical parts where temperatures will not exceed 400°F.

alkyd resin A synthetic resin used as a binder in lacquers, adhesives, paints, and varnishes.

Allen wrench A section of hexagonal stock used to turn an Allen head screw or bolt.

alligatoring Rough cracking of a painted surface, often caused by applying another coat before the first is dry or by exposing a painted surface to extreme heat.

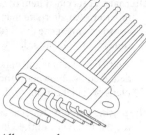
Allen wrench

alligator shears (lever shears) A shop tool used for shearing sheet metal.

allocable cost A cost that is assignable to a particular contract or other cost objective.

allocated cost The total cost of an asset's individual components, summarized, including incorrect costs, and classified into one of four categories.

allowable bearing value (allowable soil pressure) The bearing capacity of a soil, in pounds per square foot (psf), determined by its characteristics, such as shear, compressibility, water content, and cohesion. The higher the allowable bearing value of a soil, the smaller the footing required to support a structural member.

allowable cost Any reasonable cost that may be recovered under the contract to which it is allocable.

allowable load The ultimate load divided by a safety factor.

allowable pile-bearing load The allowable load used to design a pile cluster to support a structure.

allowable stress The maximum stress allowed by code for members of a structure, depending upon the material and the anticipated use of the structure.

allowance 1. A stated requirement of the contract documents whereby a specified sum of money is incorporated, or allowed, into the contract sum in order to sustain the cost of a stipulated material, assembly, piece of equipment, or other part of a construction contract. This device is convenient in cases where the particular item cannot be fully described in the contract documents. **2.** In bidding, an amount budgeted for an item for which no exact dollar amount is available. **3.** A contingency for unforeseen costs. **4.** The classification of connected parts or members according to their tightness or looseness.

alloy A homogeneous mixture of two or more metals developed and used because of its lower cost and/or the certain desirable properties it exhibits.

alpha gypsum A specially processed calcined gypsum with an extremely high compressive strength.

ALTA survey Short for American Land Title Association, a land survey that identifies the title commitments of the parcel in addition to the normal as-built conditions.

alteration Construction within a structure or to its exterior closure that does not change the overall dimensions of the structure. Alteration includes remodeling and retrofitting.

alternate A specified item of construction that is set apart by a separate sum. An alternate may or may not be incorporated into the contract sum at the discretion and approval of the owner at the time of contract award.

alternate bid An amount stated in a bid that can be added or deducted by an owner if the defined changes are made to the plans or specifications of the base bid.

alternating current An electric current that reverses direction at regular intervals. In the United States, most current for domestic use reverses direction at 60 cycles per second.

alternative dispute resolution (ADR) A confidential method of settling a dispute without going to court, typically negotiation, mediation, or arbitration.

alternator A machine that develops alternating current by mechanical rotation of its rotor.

altitude In surveying or astronomy, the angular distance of a celestial body above the horizon.

alum A double sulfate added to plaster as a hardener and accelerator.

alumina Aluminum oxide found in the clay used to make brick and clay tile.

aluminum A silver-colored, nonmagnetic, lightweight metal used extensively in the construction industry. It is used in sheets, extrusions, foils, and castings. Sheets are often anodized for greater corrosion resistance and surface hardness. Because of its light weight and good electrical conductivity, aluminum is used extensively for electrical cables. Aluminum is usually used in alloy form for greater strength.

aluminum-clad window A factory-finished and sealed window whose wooden construction is enclosed with aluminum sheeting.

aluminum-coated steel Steel coated with aluminum to inhibit corrosion.

aluminum door A glazed door with aluminum stiles and rails.

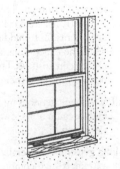
aluminum door

aluminum foil A very thin aluminum sheet used extensively for thermal reflection and moisture protection.

aluminum nitrate fuel oil mix (ANFO) An inexpensive explosive used in blasting and mining operations.

aluminum paint A paint containing aluminum paste, which gives the paint good heat-, light-, and corrosion-resistant properties.

aluminum window A glazed window with an aluminum sash and muntins.

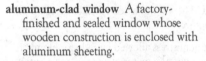
aluminum window

ambient lighting The general background lighting, whether natural or artificial, of an area.

ambient noise The total noise level from all sources in a given area, either within a building or in an outside environment.

ambient temperature The temperature of the environment surrounding an object.

amendment A modification of the contract by a subsequent agreement. This does not change the entire existing contract but does alter the terms of the affected provisions or requirements.*

American Arbitration Association (AAA) A private nonprofit organization that provides education, training, and administrative assistance to parties who use nonjudicial methods, such as alternative dispute resolution (ADR) for resolving disputes. The AAA is involved primarily with binding arbitration and mediation.

American basement (walk-out basement) The floor of a building partly above and partly below grade.

American Conference of Governmental Industrial Hygienists (ACGIH) An organization of professionals skilled in the science of industrial hygiene.

American Federation of Labor (AFL) A labor organization or union formed in the United States under the leadership of Samuel Gompers in 18 86. The American Federation of Labor provided an "umbrella" organization, the purpose of which was to represent to management the interests of workers in various trades, crafts, and other skilled disciplines related to manufacturing and construction.

American Federation of Labor and the Committee for Industrial Organizations (AFL-CIO) A major union formed by the merger of the two organizations listed above under the leadership of John L. Lewis in 1955. The AFL-CIO represents the interests of various types of member workers in industry and other endeavors (including construction) for the purpose of negotiating with management for acceptable wages, benefits, and other material interests of worker-employees.

American Lumber Standards Committee (ALSC) The committee that maintains standards for hardwood and softwood grading, including those for size and nomenclature. Rules for grading are established by the U.S. Department of Commerce and enforced by regional organizations.

American National Standards Institute (ANSI) Publisher of the American National Standards, a reference book outlining the approved standards and specifications for all facets of building construction.

American standard beam A hot-rolled steel I-beam designated by the prefix S before the size and weight.

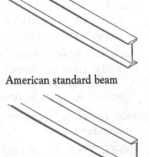
American standard beam

American standard channel

American standard channel A hot-rolled steel channel designated by the prefix C before the size and weight.

American Standard Code for Information Interchange (ASCII) An accepted standard for computerized data transmission.

American standard pipe threads (Briggs standard) The thread size and pitch commonly used in the United States for connecting pipe and fittings.

American wire gauge (American standard wire gauge, Brown and Sharpe gauge) The standard in the United States for specifying and manufacturing wire and sheet metal sizes, particularly electrical wire and metal flashing.

Americans with Disabilities Act (ADA) A federal civil rights act prohibiting discrimination against people with disabilities. There are five sections that cover different aspects of discrimination: employment, state and local government, public accommodations and commercial facilities, telecommunications, and miscellaneous provisions.

Americans with Disabilities Act Accessibility Guidelines (ADAAG) The minimum guidelines that must be followed to meet ADA Standards for Accessible Design. *See also* **ADA Standards for Accessible Design**.

ammeter An instrument for measuring the rate of ampere flow through an electric circuit.

amorphous A type of rock that has no crystalline structure.

amortization The process of paying off stock, bonds, a mortgage, or other indebtedness through installments or by a sinking fund.

amount of mixing The mixing action employed to combine the ingredients of concrete or mortar, measured in time or number of revolutions.

ampacity A designation of the current-carrying capacity of an electrical wire, expressed in amperes.

ampere The electromotive force required to move one volt of electricity across one ohm of resistance. A measure of electrical current.

amplitude In sound or vibration, the maximum variation from the mean position.

analog point In Building Automation Systems, a sensor, such as a damper or temperature sensor, that has a continuous range of settings that can be monitored or controlled by the system.

analog signal A signal in the form of a fluctuating quantity (such as voltage or current strength) that reflects variations, such as loudness. It is not limited to discrete units.

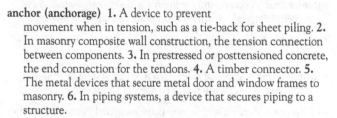

anchor (anchorage)

anchor (anchorage) 1. A device to prevent movement when in tension, such as a tie-back for sheet piling. 2. In masonry composite wall construction, the tension connection between components. 3. In prestressed or posttensioned concrete, the end connection for the tendons. 4. A timber connector. 5. The metal devices that secure metal door and window frames to masonry. 6. In piping systems, a device that secures piping to a structure.

anchorage bond stress (development bond stress) The forces on a deformed reinforcing steel bar divided by the product of the perimeter times the embedded length.

anchorage deformation (anchorage loss, anchorage slip) In prestressing concrete members, the deformation of an anchor or slippage of tendons when the prestressing device is released.

anchorage zone 1. In pretensioning, the area of the member in which the stresses in the tendon anchor are developed. 2. In posttensioning, the area adjacent to the anchorage that develops secondary stresses.

anchor block A block of wood in a masonry wall that provides a means of attaching other wood members.

anchor bolt (foundation bolt, hold-down bolt) A threaded bolt, usually embedded in a foundation, for securing a sill, framework, or machinery.

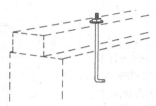

anchor bolt (foundation bolt, hold-down bolt)

anchor bolt plan A plan view showing size and location of all anchor bolts for a building's systems components. May be included in structural steel and shop drawings.

anchor plate A plate attached to an object to which accessories or structural members may be attached by welding, screwing, nailing, or bolting.

anchor rod A threaded metal rod attached to hangers and used to support pipe and ductwork.

anchor strip A wooden, plastic, or metal board surrounding a window and nailed to the building's framing to serve as a windbreak.

anemometer An instrument that measures the velocity of airflow.

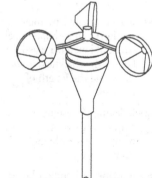

anemometer

angle 1. The figure or measurement of a figure formed when two planes diverge from a common line. 2. In construction, a common name for an L-shaped metal member.

angle bead (angle staff, staff angle) A metal or wood strip set at the corner of a wallboard or plaster wall to serve as a guide and to provide protection. Angle beads are most commonly made of nonferrous or galvanized perforated sheet metal.

angle block (glue block) A small block of wood used to fasten or stiffen the joint of two adjacent wood members, usually at right angles.

angle bond A metal tie that projects into each wall at a corner and is used to bond masonry.

angle brace (angle tie) A piece of material temporarily or permanently secured across an angle to make it rigid, such as a strip of wood nailed across the corners of a window frame to keep it square during installation.

angle brick A brick cast with an oblique angle on one of its corners.

angle cleat (angle clip) A short section of angle iron used to attach structural members, such as precast panels, to structural steel.

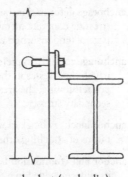

angle cleat (angle clip)

angle closer A special brick or a portion of a brick used to close the bond on the outside corner of a brick wall.

angle collar (bevel collar) A cast-iron pipe angle fitting with a bell-type connection at each end.

angle float (angle trowel) A trowel with two surfaces meeting at right angles. An angle float is used for finishing plaster or concrete in an inside corner.

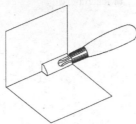

angle float (angle trowel)

angle framing Light-gauge framing with an angle iron.

angle gauge A template used to set or maintain an angle during construction.

angle iron (angle bar, angle section) An L-shaped steel structural member classified by the thickness of the stock and the length of the legs.

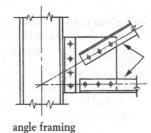

angle framing

angle lacing A system of connecting two structural components with angle irons.

angle iron (angle bar, angle section)

angle plane A hand tool used to remove projections and smooth inside corners on a plaster brown coat covering once it has set.

angle strut An angle iron erected to carry a compression load.

anglet A groove in a material or structure, most often containing a right angle.

angle valve A valve with the inlet at right angles to the outlet for controlling flow in a pipe.

angle valve

angular aggregate An aggregate made of crushed material with sharp edges, as opposed to screened gravel with rounded edges.

angular measure The deviation between two lines that meet at a point, measured in degrees, minutes, and seconds.

anhydrite An additive used in the manufacture of Portland cement to control the set.

anhydrous calcium sulfate (dead-burnt gypsum) Gypsum from which all the water of crystallization has been removed.

anhydrous gypsum plaster A high-grade finish plaster with most of the water of crystallization removed.

animal glue A strong adhesive with poor water resistance made from bones and hides of animals. Often used in furniture manufacture.

anionic surfactant A negatively charged adjuvant with limited compatibility used in asbestos abatement.

annealed wire A pliable wire used in construction primarily for reinforcing steel tie wires.

annealing The process of subjecting a material, particularly glass or metal, to heat and then slow cooling to relieve internal stress. This process reduces brittleness and increases toughness.

annex A secondary structure either near or adjoining a primary structure.

annual fuel utilization efficiency (AFUE) A seasonal efficiency rating that is an accurate estimation of fuel used for furnaces and direct-fired forced hot air systems. It measures the system efficiency and accounts for start-up, cool-down, and other operating losses.

annular ring nail A nail with a series of threadlike rings on its shank to give it good holding power. This type of nail is used for attaching gypsum board to wood studs.

annular ring nail

annunciator An electrical signaling device that identifies when a circuit is engaged.

anode The conductor rod used in an electrical system to protect underground tanks and pipes from electrochemical action.

anodize The process of creating electrolytically a hard, noncorrosive film of aluminum oxide on the surface of a metal. This film can be either clear or colored.

antechamber An entrance, vestibule, or foyer.

anticorrosive paint A paint containing corrosive-resistant pigments such as zinc chromate, lead chromate, or red lead. This type of paint is used as a primer on iron and steel products.

antiflotation pads Concrete pads secured to underground tanks to add sufficient weight to the tank to overcome buoyancy when empty.

antimicrobial A compound commonly added to other products to prevent bacterial growth on the surface of a finished product.

antioxidant Any substance that inhibits oxidation, which deteriorates plastics and other materials.

antisiphon trap (deep-seal trap) In a drainage system, a plumbing trap that provides a water seat to prevent siphonage.

antislip paint A paint with coarse particles mixed in to roughen the surface to which it is applied. This type of paint is used on steps, ramps, walkways, and porches.

antistatic agent An additive that reduces the development of static electricity on the surface of plastics or on carpeting.

anvil The part of a pile hammer that transmits the driving force to a pile.

aperture In construction, any opening left in a wall for a door, window, or for ventilation.

apex The peak, or highest point, of any structure.

apex stone (keystone, saddle stone) The highest stone or block in an arch, gable, dome, or vault. Apex stones are often decorative.

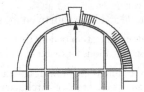

apex stone (keystone, saddle stone)

apparent density The mass per unit volume (or the weight per unit volume) of a material, taking into consideration any voids.

appliance An electric or gas device used to provide occupant comfort or convenience such as by producing light, heat, air conditioning, refrigeration, and so forth.

appliance panel An electrical service panel with circuit breakers or fuses specifically designed for service to appliances.

application bond The measurement of the strength of adhesion between two adhered surfaces.

application butyl An adhesive with a butyl base used in caulking and sealants.

application cement A common term for mastics used in flooring and roofing applications.

application failure The separation by chemical or physical means of two adhered surfaces.

application for payment A formal written request for payment by a contractor for work completed on a contract and, if allowed for in the contract, materials stored on the job site or in a warehouse.

application mortar A mixture with an adhesive additive used for affixing ceramic wall or ceiling tile.

application neoprene A liquid neoprene compound applied to concrete foundation walls for waterproofing.

application spreader A trowel with notched edges used for applying adhesive.

applied trim Strips or moldings applied to, as opposed to manufactured with, door and window frames and wood paneling.

applied trim

appraisal A dollar estimate of the value of a certain item of property, or the assessment of the value of a loss. The estimate is developed from market value, replacement cost, income produced, or a combination of these factors. Appraisals are usually made by qualified professional appraisers.

apprentice A person who works with a skilled craftsman for a number of years in order to learn the trade. An apprentice is generally rated by the number of years served.

approach ramp 1. An access for vehicles to a highway. 2. A sloped access for the handicapped to a building, in lieu of stairs.

approved In construction, materials, equipment, and workmanship in a system, or a measurable portion thereof, which have been accepted by an authority having jurisdiction. Usually, the term refers to approval for payment, approval for continuation of work, or approval for occupancy.

approved equal Material, equipment, or method of construction that has been approved by the owner or the owner's representative as an acceptable alternative to that specified in the contract documents.

appurtenance 1. Something added on to a main structure or system. 2. A condition added to a property deed, such as a right-of-way.

apron 1. A piece of finished trim placed under a window stool. 2. A slab of concrete extending beyond the entrance to a building, particularly at an entrance for vehicular traffic. 3. The piece of flat wood under the base of a cabinet. 4. Weather protection paneling on the exterior of a building. 5. A splashboard at the back of a sink. 6. Bowl-front closing device for a scraper bowl. 7. At an airport, the pavement adjacent to hangars and appurtenant buildings.

apron

apron flashing 1. The flashing that diverts water from a vertical surface on a building to a sloped roof, such as that around a chimney. 2. Flashing that leads water from a roof into a gutter.

apron molding The piece of flat wood under the base of a cabinet.

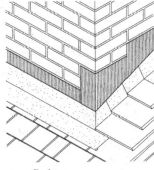

apron flashing

apron piece (pitching piece) A piece of lumber protruding from a wall to support the rough stringers at the top or at a landing of a wooden staircase.

apron wall A distinct exterior wall panel extending from a windowsill to the window below.

aquastat An electrical control activated by changes in water temperature.

arbitration The process by which parties agree to submit their disputes to the determination of a third, impartial party (referred to as the arbitrator), rather than pursuing their claims before a judge and jury in a court of law.

arbor 1. An enclosure of closely planted trees, vines, or shrubs that are either self-supporting or supported on a framework. 2. The rotating shaft of a circular saw or shaper.

arc **1.** The electrical discharge between two electrodes. When the electrodes are surrounded by gas in a lamp, they become a bright, economical light source. **2.** Any portion of a circle or the angle that it makes.

arcade A covered passageway between buildings, often with shops and offices on one or both sides.

arc cutting A method of cutting metal with an electric welding machine. The metal melts from the heat produced by the arc between the electrode and the metal.

arch A curved or flat structure spanning an opening. The shape and size of arches are limited by the materials used and the support provided.

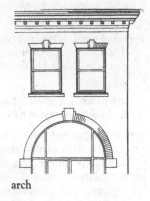

arch

archaic materials Historical components and assemblies that are essential to the integrity of a historic structure, but that are not in common use for new construction. Examples include traditional solid masonry construction, historic reinforced concrete assemblies, terra cotta masonry, lead-coated copper sheet metal assemblies, and hollow clay tile interior partition walls.

arch brick (compass brick, feather edge brick, radial brick, radiating brick, radius brick, voussoir brick) **1.** One of a number of types of brick manufactured to construct curved surfaces such as arches and round manholes. **2.** Extremely hard-burned brick from an arch of a scove kiln.

arching The bridging of shear stresses in a soil mass across an area of low shear strength to adjacent areas of higher shear strength.

architect A professionally qualified and licensed person who prepares plans and specifications for a building or structure. Architectural services include such duties as project analysis, development of the project design, and the preparation of construction documents (including drawings, specifications, bidding requirements, and general administration of the construction contract).

architect-engineer A person or company providing services as both architect and engineer.

architect's approval Permission granted by the architect, acting as the owner's representative, for actions and decisions involving materials, equipment, installation, change orders, substitution of materials, or payment for completed work.

architect's scale A draftsman's tool with proportionate, graduated spaces. May be flat, like a ruler, or three sided. The three-sided scale has 10 separate scales: ⅛″ and ¼″, 1″ and ½″, and ⅜″. ³/₁₆″, and ³/₃₂″, and 1½″ and 3″.

architectural Pertaining to a class of construction, particularly in home building, of higher-than-average quality. The term often pertains to the ornamental features of a structure.

architectural area of buildings The total of all stories of a building, after adjustments, computed according to AIA standards, measured from the exterior faces of exterior walls and from the center line of walls between buildings.

architectural barrier An architectural feature that is not compliant with accessibility for disabled users or prohibits usage or access to a building.

architectural concrete Structural or nonstructural concrete that will be permanently exposed to view and therefore requires special attention to uniformity of materials, forming, placing, and finishing. This type of concrete is frequently cast in a mold and has a pattern on the surface.

architectural concrete

architectural door A grade classification of door that designates higher-than-standard specifications for material and appearance.

architectural drawings Also called core drawings, these show the layout of the building and its use of space. Architectural drawings convey the structure's aesthetic value and show the dimensions and placement of all key features.

architectural fee The cost of architectural services to an owner. The fee varies according to the services provided and the complexity of the project.

architectural glass Glass with a configured surface to obscure vision or diffuse light.

architectural floor plan The most common plan view that shows doors, windows, walls, and partitions.

architectural millwork (custom millwork) Millwork manufactured to meet the specifications of a particular job, as distinguished from stock millwork.

architectural millwork (custom millwork)

architectural precast concrete Precast concrete that, through application, finish, shape, color, or texture, contributes to a building's architectural form and finished effect.

architectural programming A process that identifies a structure's proposed use, code, agency review, and approval requirements and identifies the necessary rehabilitation and restoration of existing components, and any other needed improvements.

architecture The art and science of designing and building structures.

architrave In classical architecture, **1.** the bottom-most beam that spans from column to column resting directly upon the capitals; **2.** ornamental moldings around door or window openings.

arch stone The wedge-shaped masonry units used in building an arch.

---

arch truss A roof truss having a curved upper chord and a straight lower chord.

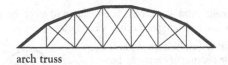
arch truss

arc voltage The reaction of a circuit's inductance to the rate of current change in the circuit.

arc welding The joining of metal parts by fusion. Heat is produced by the electricity passing between an electrode and the metal, and is usually accompanied by a filler metal and/or pressure.

area 1. A measurement of a given planar region or of the surface of a solid. **2.** A particular part of a building that has been set aside for a specific purpose.

area drain A catch basin or other device designed to collect surface water.

area light A light source used to illuminate a significant area, either indoors or outdoors.

area method A construction cost estimating system employing unit square foot costs multiplied by the adjusted gross floor area of a building.

area wall A masonry wall surrounding or partly surrounding an open area, particularly one below grade, such as an areaway at the entrance to a basement.

areaway An open area located below grade and adjacent to a building to provide light, air, or access to a basement or crawl space.

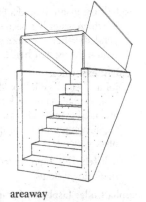
areaway

areaway grating A steel or cast-iron grating placed over an areaway, usually at grade level.

armature The rotating part of a motor or generator consisting of copper wire wound around an iron core.

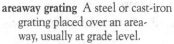
armature

arm conveyor A belt with protruding arms or angles to carry materials into a building.

armor coat Durable pavement comprised of two or more thin layers of aggregate and asphalt.

armored cable (metal-clad cable) An electrical conduit of flexible steel cable wrapped around insulated wires.

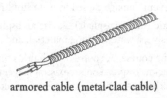
armored cable (metal-clad cable)

armored concrete Concrete with a surface treatment containing steel or iron and used in areas with heavy, steel-wheeled traffic.

armored faceplate A metal faceplate mortised into the edge of a door to protect the lock mechanism.

armored front A tamperproof metal plate that covers the set screws of a mortise lock.

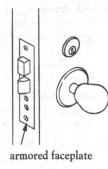
armored faceplate

armored plywood Plywood that is faced on one or both sides with metal cladding.

armor plate (kick plate) A metal plate that is installed on the lower part of a door to protect it from kicks and scratches.

arrester 1. A wire screen at the top of a chimney or incinerator to prevent burning material from flying out. **2.** In electrical equipment, a protective device that limits surge voltages by diverting current.

arrow diagram A CPM (critical path method) diagram in which arrows represent activities in a project.

articles Also referred to as *clauses*, these separate and numbered paragraphs within a construction contract state the rights, duties, responsibilities, and obligations of the parties (e.g., the owner and the contractor) to the contract.

artificial intelligence Computer systems that solve problems symbolically rather than algorithmically. Similar to the warning, decision-making, and problem-solving process in the human brain.

artificial stone A material containing stone chips and cement, mortar, or plaster that is seasoned for several months, then polished for a finish that simulates stone.

artificial turf A synthetic material designed to simulate a natural surface, and used to form playing surfaces for indoor or outdoor sports arenas, such as football fields.

asbestos (asbestos fiber) A flexible, noncombustible, inorganic fiber used primarily in construction as a fireproofing and insulating material.

asbestos encapsulation An airtight enclosure of asbestos fibers with sealant or film that prevent fibers from becoming airborne and creating a potential health hazard.

asbestos removal A special trade that has developed since the health hazards of airborne asbestos have been revealed. Applies principally to ceiling tile, fireproofing, and pipe insulation.

asbestos work A classification system designed by OSHA that rates the level of training needed to perform asbestos-related tasks. Class I involves the removal or abatement of thermal insulation or surfacing asbestos-containing materials (ACM); Class II involves removing asbestos floor or ceiling tiles, siding, roofing, or piping; Class III involves repair and maintenance operations where employees may disturb ACM; and Class IV involves custodial activities during which employees contact ACM.

A

as-built drawings Record drawings made during construction. As-built drawings record the locations, sizes, and nature of concealed items such as structural elements, accessories, equipment, devices, plumbing lines, valves, mechanical equipment, and the like. These records (with dimensions) form a permanent record for future reference.

as-built schedule A time-scaled graphic depiction of the historical record of events, activities, and progress of a given project.

ash A sturdy, long-grained hardwood with excellent bending qualities. This wood is used in veneers, trim, and flooring.

ash dump An opening in the bottom of a firebox or fireplace into which ashes are swept, falling into an ashpit below.

ashlar 1. Any squared building stone. The term usually refers to thin stone used as facing. If the horizontal courses are level, it is called coursed ashlar, if they are broken, it is called random ashlar. **2.** Short vertical studs between the ceiling joists and the rafters.

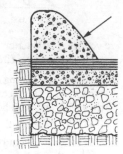
ash dump

ashlar brick (rock-faced brick) A brick with a broken face resembling stone.

ashlar line A horizontal line on the exterior face of a masonry wall.

ashlar masonry A stone masonry wall or veneer composed of rectangular units bonded with mortar.

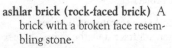
ashlar masonry

ashlar veneer A nonstructural wall facing composed of ashlar masonry.

ashpit A cleanout under a fireplace, usually at the base of a chimney, where ashes are removed.

as-late-as-possible (ALAP) An activity for which the scheduling application sets the early dates as late as possible without delaying the early dates of any successor.*

aspect The orientation of a building with respect to the points of a compass.

aspect ratio 1. In any configuration, the ratio of the long dimension to the short dimension. **2.** The ratio of the width of a duct to its height.

aspen A smooth-grained, white hardwood used for trim and veneer.

asphalt A dark brown to black bitumen pitch that melts readily. It appears in nature in asphalt beds and is also produced as a by-product of the petroleum industry.

asphalt, blown Asphalt that has had air blown through it at high temperatures to give it workability for roofing, pipe coating, foundation waterproofing, and other purposes.

asphalt base course A bottom paving course consisting of coarse aggregate and asphalt.

asphalt block A manufactured paving block made from asphaltic

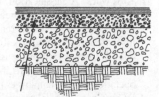
asphalt base course

concrete and aggregate. The block is typically manufactured in squares, rectangles, and hexagons, and comes in dark gray or black colors.

asphalt cement Asphalt that has been refined to meet the specifications for use in paving and other special uses.

asphalt coating (asphalt-lined pipe) The asphaltic coating of corrugated metal pipe. Coatings can be inside, outside, or just on the invert.

asphalt color coat An asphalt surface treatment that has been impregnated with aggregate of a specified color.

asphalt curb An extruded or hand-formed berm made from asphaltic concrete.

asphalt cutback An asphalt that has been liquefied by an additive for a specific use.

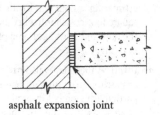
asphalt curb

asphalt cutter Any of a variety of machines designed to cut asphalt pavement.

asphalt dampproofing The application of asphalt to the surface of a concrete or masonry wall to prevent passage of absorption of water or moisture.

asphalt emulsion Liquid asphalt in which water has been suspended. When the water evaporates, the asphalt hardens.

asphalt expansion joint Premolded felt or fiberboard impregnated with asphalt and used extensively as an expansion joint for cast-in-place concrete.

asphalt expansion joint

asphalt felt Felt impregnated with asphalt and used in roofing and sheathing systems.

asphalt filler (asphalt joint filler) A liquid asphalt used for filling joints and cracks in pavement and floors.

asphalt flashing cement A semisolid asphaltic material used to apply flashing.

asphaltic A term used to describe materials containing asphalt. While sometimes used interchangeably with asphalt in the construction industry, it is usually more correct to use asphaltic.

asphaltic concrete (asphalt paving, bituminous concrete, blacktop) A mixture of liquid asphalt and graded aggregate used as a paving material for roadways and parking lots. It is usually spread and compacted in layers over a prepared base while still hot.

asphaltic macadam A term generally referring to a penetration method of paving whereby the aggregate is placed first, then liquid asphalt is sprayed into the voids, followed by the addition of a finer-graded aggregate. Penetration macadam usually needs a seal coat to prevent damage caused by water infiltration.

asphaltic mastic (mastic asphalt) A viscous asphaltic material used as an adhesive, a waterproofing material, and a joint sealant.

asphalt leveling course A course of asphaltic concrete pavement of varying thickness spread on an existing pavement to compensate for irregularities prior to placing the next course.

asphalt, liquid An asphaltic material having a fluid consistency at normal temperatures. The common types specified for pavements are cutback, rapid curing (RC), medium curing (MC), and slow curing (SC), which are blended with petroleum solvents and emulsion, which is blended with water.

asphalt overlay One or more courses of asphaltic concrete placed over existing pavement. The process of overlaying usually includes cleaning, and application of a tack coat, followed by a leveling course.

asphalt paint An economical, liquid-asphaltic product used principally for weatherproofing.

asphalt paper A paper that has been coated or saturated with asphalt for use as a moisture barrier.

asphalt pavement Any pavement made from one or more layers of asphaltic concrete.

asphalt pavement sealer A material applied to asphalt pavement after compaction to protect it from deterioration caused by exposure to weather or petroleum products.

asphalt penetration A measure of the hardness or consistency of asphalt, expressed as the distance a needle of standard diameter will penetrate a sample under given time, load, and temperature conditions.

asphalt-prepared roofing (asphaltic felt, bituminous felt, cold-process roofing, prepared roofing, rolled roofing, rolled strip roofing, roofing felt, sanded bituminous felt, saturated felt, self-finished roofing felt) A roof covering manufactured in rolls and made from asphalt-impregnated felt with a harder layer of asphalt applied to the surface of the felt. All or part of the "weather" side may be covered with aggregate of various sizes and colors.

asphalt prime coat A tack coat, usually an emulsion, to increase the adhesion of one course to another in pavement construction.

asphalt primer A liquid asphalt of low viscosity that is applied to a nonbituminous surface such as concrete to prepare the surface for an asphalt course.

asphalt seal coat A thin asphalt surface treatment used to waterproof and improve the wearing surface texture of pavement, particularly that of an asphaltic macadam. Depending on the intended purpose for the pavement, a seal coat may or may not include aggregate.

asphalt shingles (composition shingles, strip slates) Roofing felt saturated with asphalt, coated on the weather side with a harder asphalt and aggregate particles, and cut into shingles for application to a sloped roof.

asphalt shingles (composition shingles, strip slates)

asphalt surface course The top or wearing course of asphaltic concrete pavement.

asphalt surface treatment The application of liquid asphalt to any asphaltic pavement, with or without adding aggregate.

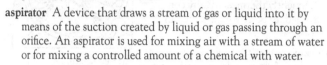

asphalt surface course

asphalt tack coat A light coat of asphalt, usually an emulsion, added to an existing pavement to create a bond between the pavement and another course.

aspirator A device that draws a stream of gas or liquid into it by means of the suction created by liquid or gas passing through an orifice. An aspirator is used for mixing air with a stream of water or for mixing a controlled amount of a chemical with water.

as-planned schedule A project schedule prepared by the contractor to indicate the intended progress and method of performance. Frequently used as the baseline schedule for calculating delay.

assembled occupancy For design purposes, the maximum number of people who will occupy a room or hall at one time.

assessed valuation The value of a property assigned by a municipality for real estate tax purposes. The valuation may be higher or lower than the market value of the property.

assessment 1. A tax on property. 2. A charge for specific services, such as sewer or water, by a government agency.

assessment ratio The ratio between the market value and assessed valuation of a property, expressed as a percent.

asset An item of monetary value, which can include real, personal, or financial property, that is expected to have some value in a future period.

assignment 1. A transfer of rights, frequently involving rights arising under a contract. 2. With respect to a contract, a document stating that payment for work completed or materials delivered must be made to someone other than the company or person specified in the contract.

associate dimensioning A feature of a computer-aided design and drafting system that dynamically recalculates all dimensions affected by a change the operator has made.

astragal 1. A molding attached to one of a pair of doors or casement windows to cover up the joint between two stiles. 2. A bead molding, most often half-round and ornate, with a narrow flat band, or fillet, on at least one side.

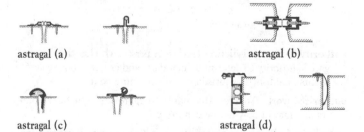

astragal (a)   astragal (b)

astragal (c)   astragal (d)

atmospheric pressure The pressure (14.7 psi) exerted by the earth's atmosphere at sea level under standard conditions.

attenuation The sound reduction process utilizing a sound-absorbing system.

Atterberg limits Terms defining the properties of soils at different water contents.

Atterberg test Laboratory tests to determine the Atterberg limits.

attic Unfinished space between ceiling joists of the top story of a building and the roof rafters.

attic tank A domestic water storage tank installed above the highest plumbing fixture in a building to provide water pressure by gravity.

attic ventilator An electric fan, frequently thermostatically controlled, to push hot air out of an attic.

attorney-in-fact One who holds a power of attorney from another to execute documents on behalf of the grantor of the power.

attic ventilator

attribute In the context of asset or project planning, a characteristic or property that is appraised in terms of whether it does or does not exist (e.g., go or not-go) with respect to a given requirement.*

audio frequencies Frequencies between 15 and 20,000 cycles per second (Hz), which is within hearing range of the human ear.

audio masking system Reducing distracting sounds and increasing speech privacy through the use of sound-masking equipment or software. Some systems provide protection from laser beams and other high-tech sound detection devices.

audit The examination of records, documents, and other evidence for the purpose of determining the propriety of transactions and assessing fiscal compliance with relevant cost and accounting requirements.

auger 1. A carpenter's hand tool used for boring holes in wood. 2. A handheld or rotary-powered tool with a helical cutting edge used for drilling holes in soil. Augers are used for taking soil samples, drilling for caissons, or drilling for cast-in-place piles.

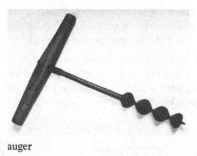

auger

authority having jurisdiction (AHJ) A person who has the delegated authority to determine, mandate, and enforce code requirements established by jurisdictional governing bodies.

authorized work An effort that has been approved by higher authority and may or may not be definitive.*

autoclave A chamber in which steam at high pressure is used to cure precast concrete members.

automatic door A power-operated door that opens at the approach of a person or vehicle and closes when the person or vehicle has passed.

automatic fire pump A pump in a standpipe or sprinkler system that turns on when the water pressure drops below a predetermined level.

automatic fire vent (automatic smoke vent) A device in the roof of a building that operates automatically to control fire or smoke.

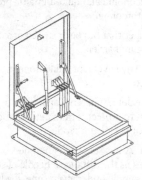

automatic fire vent (automatic smoke vent)

automatic iris An optical device, much like the iris in the eye, containing a diaphragm that expands and contracts to control the amount of light that passes through the lens.

automatic operator A remote-control operating device. The term usually refers to the opening and closing of doors by electronically actuated switches.

automatic sprinkler system A fire safety system designed to provide instant and continuous spraying of water over large areas in the case of fire.

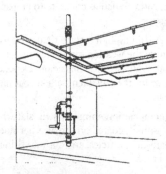

automatic sprinkler system

automatic transfer switch In an electrical system, a switch that automatically transfers the load to another circuit when the voltage drops below a predetermined level.

auxiliary contact An additional contact not normally associated with operation of a switching device such as a contact or circuit breaker.

A

auxiliary rafter (cushion rafter) A rafter used to strengthen the main rafter, usually at the area of greatest load.

auxiliary reinforcement In a prestressed concrete member, refers to all reinforcing steel other than the prestressing steel.

average annual cost The conversion, by an interest rate and present worth technique, of all capital and operating costs to a series of equivalent equal annual costs. As a system for comparing proposal investments, it requires assumption of a specific minimum acceptable interest rate.*

average bond stress The force exerted on a steel reinforcing bar divided by the product of the perimeter multiplied by the embedded length.

average grade The average of ground surface elevations within a building site.

average haul The average distance material is transported from where it originates to where it is deposited, such as from cut to fill in roadway construction.

awl A hand tool used for piercing holes, particularly in leather. Often fitted with a needle for sewing heavy materials.

awning A projection over a door or window, often retractable, for protection against rain and sun.

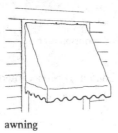

awning

awning window A window that is hinged at the top.

ax (axe) A sharp-edged hand tool for splitting wood and hewing timber.

axed brick (rough axed brick) Brick shaped by an ax so as to create rough surfaces.

awning window

ax hammer A hand tool for dressing stone.

axial fan A fan that produces pressure from the velocity of gas passing through the impeller, with no pressure being produced by centrifugal force.

ax hammer

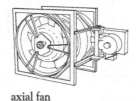

axial fan

axial force diagram In statics, a graphic representation of the axial loads acting at each section of a structural member.

axial load (axial force) The longitudinal force acting on a structural member.

axis A straight line representing the center of symmetry of a plane or solid object.

azimuth The horizontal angle measured clockwise from north to an object.

B

Abbreviations

B1S banded one side, bead one side

B2E banded two ends

B2S banded two sides, bead two sides, bright two sides

B2S1E banded two sides and one end

B3E beveled on three edges

B4E beveled on four edges

B beam, boron, brightness

BA bright annealed

BAS building automation system

bat. batten

B&B in the lumber industry, grade B and better; balled and burlapped

bbl, Bbl., brl barrel

BC building code

b&cb beaded on the edge and center

bcc body-centered cubic

BCM broken cubic meter

BCY broken cubic yard

bd in the lumber industry, board

bdl bundle

BE in the piping industry, beveled end

BET between

Beth B Bethlehem beam

bev beveled

bev sid beveled siding

BF, bd ft board foot

BFP backflow preventer

bg bag

BG below ground or below grade

Bg cem bag of cement

Bh Brinell hardness

Bhn Brinell hardness number

BHP brake horsepower, boiler horsepower

BI black iron

BICSI Building Industry Consulting Service International

BIM Building Information Modeling

Bit, Bitum. bituminous

Bk backed

Bkrs breakers

B/L bill of lading

BL building line

bldg, Bldg building

blk block, black

BLKG blocking

BLO blower

BLR boiler

blt built, borrowed light

bm board measure

BM beam, bench mark

B/M, BOM bill of materials

BMEP brake mean effective measure

B&O back-out punch

BOCA Building Officials and Code Administrators International

b of b back of board

boil boilermaker

BP blueprint, baseplate, bearing pile, building paper

bpd barrels per day

BPG beveled plate glass

BPM blows per minute

BR bedroom

brc brace

brcg bracing

BRG bearing

BRI building-related illness

BRK brick

BRKT, bkt bracket

BRS brass

Br Std, BS British Standard

BRZ bronze

BRZG brazing

B&S beams and stringers; bell and spigot; Brown and Sharpe gauge

BSMT basement

BSR building space requirements

BTB bituminous treated base

btr, Btr better

Btu British thermal unit

BTUH Btu per hour

but buttress

BW butt weld

B&W black and white

BX interlocked armored cable

Definitions

babbitt An antifriction alloy composed of tin and lesser amounts of copper and antimony. Babbitt is used in bushings and bearings.

back 1. That part, area, or surface that is farthest from the front. 2. The portion behind or opposite that is intended for use or view. 3. The reverse (scale). 4. That portion that offers strength or support from the rear. 5. The extrados of an arch or vault sometimes concealed in the surrounding masonry. 6. In slate or tile, the side opposite the bed. 7. The surface of wallboard that receives the plaster. 8. The side of a piece of lumber or plywood opposite the face. The back is the side with the lower overall quality or appearance.

back arch A concealed arch that supports the inner part (or backing) of a wall where a lintel carries the exterior facing.

backband A rabbeted molding used to surround the outside edge of a casing in an opening such as a door or window.

backbone subsystem In a premises distribution system, the cable that runs from the equipment room to the various floors in a building. In a single-floor building, the subsystem is the main trunk of the communications system.

back boxing Thin boards used in construction of double-hung windows to enclose the channel in which the sash weights hang, and to keep the channel free of mortar

back-brush To paint over a freshly painted surface with a finishing return stroke.

backcharge A charge against a contract in the form of a credit change order to a contractor for the cost of having others perform portions of their contract.

back check The mechanism in a hydraulic door closer or door check that reduces the speed with which the door can be opened.

back clip A special clip used on the hydraulic door closer or door check that reduces the speed with which the door can be opened.

back coating Asphalt coating applied to the back of shingles or rolled roofing.

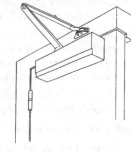
back check

back-draft damper A damper, the blades of which are gravity-controlled and allow the passage of air in one direction only.

back edging A process by which glazed ceramic pipe is cut by first chipping away the glaze and then chipping the underlying pipe until it is cut through.

backer rod foam Foam rope used to seal wide gaps and joints before caulk is applied, reducing unwanted air leakage.

backfill Earth, soil, or other material used to replace previously excavated material, as around a newly constructed foundation wall.

backfilling 1. The process of placing backfill. 2. Rough masonry laid behind a facing or between two faces. 3. Brickwork laid in spaces between structural timbers

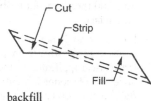
backfill

backflow 1. The unintentional reversal of the normal and intended direction of flow. Backflow is sometimes caused by back siphonage. 2. The flow of water or other liquids, mixtures, or substances into the distributing pipes of a potable water supply system from a source other than the intended source.

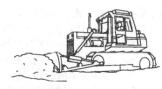
backfilling

backflow preventer A device or means to prevent backflow into the potable water system.

backhoe A powered excavating machine used to cut trenches by drawing a boom-mounted bucket through the ground toward the machine. The bucket is raised and swung to either side to deposit the excavated material.

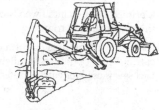
backflow preventer

backing 1. The bevel applied to the upper edge of a hip rafter. 2. Positioning furring onto joists to create a level surface on which to lay floorboards. 3. Furring applied to the inside angles of walls or partitions to provide solid corners for securing wallboard. 4. The first coat of plaster on lath. 5. The unseen or unfinished inner face of a wall. 6. Coursed masonry applied over an extrados of an arch. 7. Interior wall bricks concealed by the facing bricks. 8. The wainscoting

backhoe

between a floor and a window. **9.** The material under the pile or facing of a carpet. **10.** The stone used for random rubble walls.

backing board **1.** In a suspended acoustical ceiling, gypsum board to which acoustical tiles are secured. **2.** Gypsum wallboard or other material secured to wall studs prior to paneling to provide rigidity, sound insulation, and fire resistance.

backing brick A lower quality of brick used in places where it will be concealed by face brick or other masonry.

back iron Reinforcing steel plate on a wood plane.

backjoint A rabbet in masonry such as that over a fireplace to receive a wood nailer.

backlight To illuminate from behind (and often above) an object.

back lining (back jamb) **1.** In a weighted sash window, the thin wood strip that closes the jamb of a cased frame to provide a smooth surface for the operation of the sash and, where applicable, prevents abrasion of brickwork by the sash weights. **2.** The framing piece that constitutes the back recess for box shutters.

back lintel A lintel used to support the backing of a masonry wall, and therefore, not visible on the face.

back-mop To apply hot bituminous material, either by mop or mechanical applicator, to the underside of roofing felt during the construction of a built-up roof.

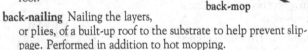

back-mop

back-nailing Nailing the layers, or plies, of a built-up roof to the substrate to help prevent slippage. Performed in addition to hot mopping.

back nut **1.** A threaded nut that helps to create a watertight joint, as on the long thread of a pipe connector, and whose one dished side accepts a grommet. **2.** A locknut.

back-paint To apply paint to the reverse or unseen side of an object, not for appearance but for protection against weather.

backplastering Plaster applied to one face of a lath system following the application and subsequent hardening of plaster that has been applied to the opposite face.

backplate A wood or metal plate that functions as backing for a structural member.

back pressure Hydraulic or pneumatic pressure in a direction opposite the normal and intended direction of flow through a pipe, conduit, or duct. Usually caused by a restriction to the flow.

back primed Back-painted woodwork. Used primarily for exterior shingles, siding, or trim.

backsaw A handsaw used in finish carpentry work. The back (noncutting) edge of the saw is stiffened with a steel or brass strip. A backsaw is used for cutting mitered joints and other joinery work.

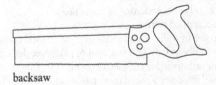

backsaw

backset The offset or horizontal distance between the face or front of a door back to the center of the keyhole or central axis of the knob.

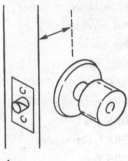

backshoring Reinsertion of shores beneath a stripped concrete slab after the original formwork and shoring has been removed from a small section. Unlike reshoring, backshoring keeps the slab from supporting its own weight or the weight of existing loads above it until the slab attains full strength.

backset

back sight In surveying, a sight on an established survey point or line.

backsplash A protective panel, apron, or sheet of waterproof material positioned on a wall behind a sink, counter, or lavatory.

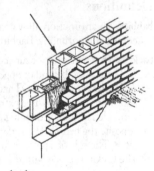

backsplash

backup **1.** That part of a masonry wall behind the exterior facing. **2.** Any substance placed into a joint to seal the joint and reduce its depth, and/or to inhibit sagging of the sealant. **3.** Overflow due to blockage in a piping system. **4.** Supporting documents for an estimate or schedule including detailed calculations, descriptions of data sources, and comments on the quality of the data.*

backup

backup figures The detailed calculations that produce unit prices.

backup strip (lathing board) A narrow strip of wood secured to the corner of a wall or partition to provide a base on which to nail the ends of the lathing.

back veneer The ply on veneer plywood opposite the face veneer and usually of lesser quality.

back vent In plumbing, a venting device installed on the downstream side of a trap to protect it from siphonage.

backwater valve (backflow valve) A check valve in a drainage pipe that prevents reversal of flow.

badger plane A large, wooden hand plane for rabbeting, whose mouth is fashioned obliquely from side to side to allow its use close to corners.

baffle **1.** A tray or partition employed on conveying equipment to direct or change the direction of flow. **2.** An opaque or translucent plate-like protective shield used against direct observation of a light source; a light baffle. **3.** A plate-like device for reducing sound transmission. **4.** Any construction intended to change the direction of flow of a liquid.

bag (sack) A quantity of Portland cement; 94 pounds in the United States, 87.5 pounds in Canada, 112 pounds in the United Kingdom, and 50 kilograms in most other countries. Different weights per bag are commonly used for other types of cement.

bag plug An inflatable drain stopper usually placed at the lowest point in a piping system; used during testing of the system's integrity.

bag trap A plumbing trap, shaped like an S, whose inlet and outlet are in alignment.

bakelite A plastic developed for use in electrical fittings, door handles, pulls, etc. Bakelite has high chemical and electrical resistance.

balance arm A supporting arm at the side of a projected window that allows the sash to be opened without an appreciable change in its center of gravity.

balance beam (balance bar) A counterbalance consisting of a long beam attached to a movable structure, such as a drawbridge or gate, whose weight it offsets during opening and closing.

balanced circuit A three-wire electric power circuit whose main conductors all carry substantially equal currents, either alternating or direct, and in which there exists substantially equal voltages between each main conductor and neutral.

balanced door A door that is installed using double-pivoted hardware, which allows it to swing open in a semi-counterbalanced manner.

balanced earthwork Cutting and filling in which the amount of one is equal to the amount of the other after swelling and compaction factors are applied.

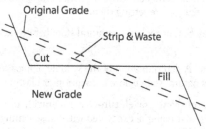

balanced earthwork

balanced load **1.** In an electric circuit, such as a three-wire system, a load connected such that the currents on each side are equal. **2.** In reinforced concrete beam design, a load that would cause crushing of concrete and yielding of tensile steel simultaneously.

balanced sash A sash in a double-hung window that requires very little effort to raise or lower because its weight is counterbalanced with weights or pretensioned springs.

balance spring A window mechanism for counterbalancing a sliding sash to keep it open as needed.

balanced sash

balanced step (dancing step, dancing winder) One of a series of winders that are balanced (as opposed to radiating from a common center). Their narrow ends are nearly equal in width to that of the straight portion of the adjacent stair flight to provide the line of traffic with a relatively even tread width.

balance point The temperature outside at which the heat lost from a building equals the heat gained from the occupants and equipment inside.

balancing **1.** Adjusting the mass distribution of a rotor to diminish journal vibrations and control the forces on the bearings from eccentric loading. **2.** In an HVAC system, adjusting the system to produce the desired level of heating and cooling in each area of a building.

balancing damper A plate or adjustable vane installed in a duct branch to regulate the flow of air in the duct.

balancing valve (balancing plug cock) A pipe valve used to control the flow rather than to shut it off.

balancing valve (balancing plug cock)

balcony **1.** A platform that protrudes from a building. It can be cantilevered or supported from below, and is usually protected by a railing or balustrade. **2.** A gallery protruding over the main floor of an auditorium; usually provides additional seating. **3.** In a theater, an elevated platform used as part of a permanent stage setting.

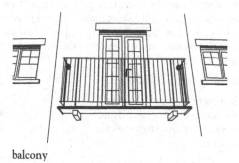

balcony

ballast **1.** A layer of coarse stone, gravel, slag, etc., over which concrete is placed. **2.** The crushed rock or gravel of a railroad bed on which ties are set. **3.** The transformer-like device that limits the electric current flowing through the gas within a fluorescent lamp. **4.** A high-intensity discharge that provides a lamp with the proper starting voltage. **5.** Material placed in a vessel to provide temporary stability.

ballast factor The ratio of the luminous output of a lamp when functioning on a ballast to its luminous output when functioning under standardized rating conditions.

ballast noise rating The degree of noise created by a fluorescent lamp ballast, represented by the letters A (the quietest) to F (the loudest).

ball-bearing hinge A butt hinge having ball bearings positioned between the knuckles to reduce friction.

ball catch A door fastener in which a spring-tensioned metal ball engages the striking plate to keep the door closed until force is applied.

ball-check valve A device used to stop the flow of liquid in one direction while allowing flow in an opposite direction. The pressure against a spring-loaded ball opens the valve in one direction of flow. Pressure from the other direction forces the ball against a seat, closing the valve and preventing flow.

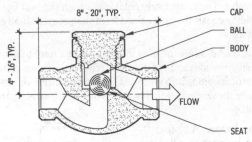

8" - 20", TYP.

4" - 16", TYP.

CAP
BALL
BODY
FLOW
SEAT

ball-check valve

ball cock A float valve incorporating a spherical float; used to control the height of water, as in a toilet tank. *See also* **float valve**.

ball float The somewhat spherical floating device by which a ball valve is controlled.

ball joint A flexible mechanical joint that allows the axis of one part to be set at an angle to the other by virtue of the design of the two components. One possesses a fixed spherical shell to accommodate the ball-shaped end of the other.

balloon framing (balloon frame) A style of wood framing in which the vertical structural members (the posts and studs) are single, continuous pieces from sill to roof plate. The intermediate floor joists are supported by ledger boards spiked to or let into the studs. The elimination of cross grains in the studding reduces differential shrinkage.

ballpark figure A rough estimate.

ball peen hammer A hammer having a hemispherical peen on one end and used by metal workers, stonemasons, and mechanics.

ball peen hammer

ball valve A spherically shaped gate valve that provides a very tight shut-off for fluids in a high-pressure piping system.

baluster (banister) **1.** One of a series of short, vertical supporting elements for a handrail or a coping. **2.** Any vase-shaped supporting member or column. **3.** The roll on the side of an Ionic capital.

balustrade A complete railing system, including a top rail, balusters, and sometimes a bottom rail.

balustrade

band **1.** A group of small bars or the wire encircling the main reinforcement in a concrete structural member to form a peripheral tie. A band is also a group of bars distributed in a slab, wall, or footing. **2.** A horizontal ornamental feature of a wall, such as a flat frieze or fascia, usually having some kind of projecting molding at its upper and lower edges. **3.** Frequencies that occur in a range between two set limits.

band clamp A metal clamp consisting of two pieces that are bolted at their ends to hold riser pipes.

banding Wood strips or veneer attached to the exposed edges of plywood or particleboard in the construction of furniture or shelves.

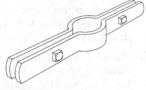

band clamp

band joist A vertical member that forms the perimeter of a floor system.

band saw A power saw consisting of a continuous piece of flexible steel that runs around two pulleys and has teeth on one or both sides. A band saw is used to cut logs into cants, to rip lumber, and to cut curved shapes.

bandwidth A range of frequencies expressed in cycles per second (Hertz). The greater the bandwidth, the more information that can be transmitted.

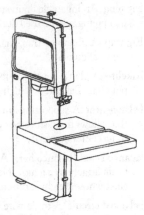

band saw

bank A mass of soil rising above a digging or trucking level. Excavation and loading are done at the face of a bank.

bank cubic yard A unit designating one cubic yard of earth or rock, measured or calculated before removal from the bank.

banker A table or bench on which stonemasons or bricklayers shape their materials before setting them.

bank material Soil or rock in its natural state before excavation or blasting.

bank measure A determination of the volume of a mass of soil or rock in its natural state before excavation or blasting.

bank-run gravel (bank gravel, run-of-bank gravel, all-in aggregate) Granular material excavated without screening, scalping, or crushing. This type of gravel is a naturally occurring aggregate comprised of cobbles, gravel, sand, and fines.

bank sand Sand that is unlike lake sand in that it has sharp edges that provide a better bond and more strength when used in plastering.

bar **1.** A deformed steel member used to reinforce concrete. **2.** A solid piece of metal whose length is substantially greater than its width.

barbed wire (barbwire) Two or more wires twisted together with intermittent barbs incorporated during manufacture. Used for security and livestock fencing.

bar bending The process of bending reinforcing steel into shapes required for reinforced concrete construction.

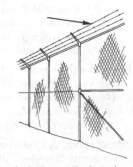

barbed wire (barbwire)

bar chart (Gantt chart) A chart that graphically describes activities on a work-versus-time scale, illustrating planned start and completion dates for the various project activities.

bar clamp A carpenter's clamping device consisting of a long bar with adjustable clamping jaws. A bar clamp is used to hold joinery components during gluing.

bare cost The estimated cost of an item of work or a project before the bidder's markup for overhead and profit.

barefaced tenon (bareface tenon) A tenon that has a "shoulder" on one side only, and is used in the construction of wood doors.

bargain and sale deed A deed in which the grantor admits that he has some interest in, though not necessarily a clear and unencumbered title to, the property being conveyed. This kind of deed often contains a warranty that the grantor did not encumber the property or convey away any part of the title during his period of ownership.

bargeboard A board that hangs from the projecting end of a gable roof, often ornamental.

barge spike (boat spike) A long, square spike with a chisel point, used primarily in heavy timber construction.

bar graph A graphical representation of simultaneous events charted with reference to time. A bar graph is a simplified method of charting events, such as the processes of building construction. The horizontal axis of the chart is scaled to increments of time; the various events are charted vertically. The duration of the event is charted by a horizontal line or bar beginning at the time the event is scheduled to begin and ending at the time the event is scheduled to be completed. At any point in time, the reader can observe the number of events that are occurring simultaneously.

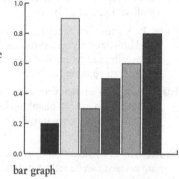

bar graph

bar joist A light steel joist of open-web construction with a single zigzag bar welded to upper and lower chords at the points of contact. Used as floor and roof supports.

bar joist

bar mat An assembly of steel reinforcement composed of two or more layers of bars placed at right angles to each other and tied together by welding or wire ties.

barometer A device that measures atmospheric pressure.

barrel 1. A unit of weight measure for Portland cement, equivalent to four bags or 376 pounds. 2. A standard cylindrical vessel with a liquid capacity of 31½ gallons. 3. That part of a pipe where the bore and wall thickness remain uniform.

barrel bolt (tower bolt) A cylindrical bolt mounted on a plate that has a case projecting from its surface to contain and guide the bolt.

barrel bolt (tower bolt)

barricade An obstruction used to deter the passage of persons or vehicles. Any of several devices used to detour or restrict passage.

barrier-free environment A building or area that is fully accessible and usable by disabled people.

barrow 1. A wheelbarrow. 2. A large mound of earth or pile of rocks intentionally placed on top of an ancient burial site for protection.

bar sash lift A handle on the bottom rail of a sash for raising and lowering the sash.

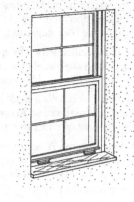

bar sash lift

bar spacing The distance between parallel reinforcing bars, measured center to center.

bar strainer 1. A screening device, fabricated from parallel bars or rods, used over a drain to prevent the entrance of foreign objects. 2. A bar screen.

bar support (bar chair) A rigid device of formed wire, plastic, or concrete, used to support or hold reinforcing bars in proper position during concrete operations.

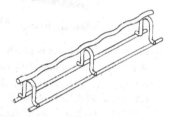

bar support (bar chair)

bar-type grating An open grate with parallel bearing bars evenly spaced and attached to a frame. The grating may be cast or welded and may have crossbars.

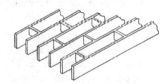

bar-type grating

barway A gate with one or more sliding bars that act as large latch bolts.

BAS *See* **Building Automation System.**

bascule A structure that rotates like a saw around a horizontal axis, and that has a counterbalance at one end. The most common use is for a bascule bridge.

base 1. The lowest part of anything upon which the whole rests. 2. A subfloor slab or "working mat," either previously placed and hardened or freshly placed, on which floor topping is placed. 3. The underlying stratum on which a concrete slab, such as pavement, is placed. 4. A board or molding used against the bottom of walls to cover their joint with the floor and to protect them from kicks and scuffs. 5. The protection covering the unfinished edge of plaster or gypsum board.

base anchor A fixed or adjustable metal device attached to the base of a door frame to secure it to the floor.

base angle Angle iron stock attached to the perimeter of a foundation for supporting and aligning tilt-up wall panels.

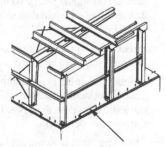

base angle

base bid The amount of money stated in the bid as the sum

for which the bidder offers to perform the work described in the bidding documents, prior to the adjustments for alternate bids that have been submitted.

base bid specifications The specifications that list or describe those specific materials, equipment, and construction methods that comprise the base bid exclusive of any alternate bids.

base block A usually unadorned, squared block that terminates a molded baseboard at an opening or serves as a base when attached to the foot of a door or the bottom of window trim.

baseboard heater A heating system in which the heating elements are housed in special panels placed horizontally along the baseboard of a wall. The heat source is commonly from hot water, steam, or electricity.

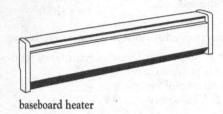

baseboard heater

base coat 1. The plaster beneath the finish coat. **2.** The initial coat of paint or stain applied to a surface.

base course 1. A layer of material of specified thickness constructed on the subgrade or subbase of a pavement to serve one or more functions, such as distributing loads, providing drainage, or minimizing frost action. **2.** The lowest course of masonry in a wall, pier, foundation, or footing course.

base flashing 1. In roofing, the flashing supplied by the upturned edges of a watertight membrane. **2.** The metal or composition flashing used with any roofing material at the joint between the roofing surface and a vertical surface, such as a parapet or wall.

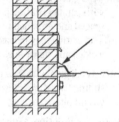

base flashing

base line 1. The meticulously established reference line used in surveying or timber cruising. **2.** In construction, the center or reference line of location of a highway, railway, building, or bridge.

baseline schedule 1. A fixed project schedule that is the standard by which project performance is measured. The current schedule is copied into the baseline schedule that remains frozen until it is reset. Resetting the baseline is done when the scope of the project has been changed significantly, for example after a negotiated change. At that point, the original or current baseline becomes invalid and should not be compared with the current schedule. **2.** Version of schedule that reflects all formally authorized scope and schedule changes.*

basement The bottom full story of a building below the first floor. A basement may be partially or completely below grade.

base metal In joining two metal pieces, the parent metal that is actually welded, brazed, or soldered, and remains unmelted after the joining process, as opposed to the filler metal deposited during the joining operation.

base molding Trim molding applied to the upper edge of interior baseboard.

baseplate A plate used to distribute vertical loads from structural columns or machinery.

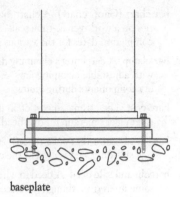

baseplate

base screed A galvanized metal screed with perforated or expanded flanges to provide ground for plaster and to separate areas of dissimilar materials.

base sheet The saturated and/or coated felt sheeting laid as the first ply in a built-up roof system.

base shoe (base shoe molding, carpet strip, carpet stripmolding, floor molding, shoe molding) The molding or carpet strip covering the joint between a floor and a baseboard, often a quarter-round bead.

base shoe corner A block or a piece of molding installed in the corner of a room so as to eliminate the need to miter the base shoe.

base tee A pipe tee that has an attached supporting baseplate.

base tile The bottom course of tile in a tiled wall.

base trim Any decorative molding at the base of a wall, column, or pedestal.

basin 1. A somewhat circular natural or excavated hollow or depression having sloping sides and usually used for holding water. **2.** A similarly shaped plumbing fixture, such as a sink.

basin wrench A wrench with a long shank and ratcheted jaws used in plumbing for difficult-to-reach areas, as when installing a faucet behind a sink.

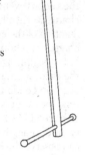

basin wrench

basket crib A construction of interlocking timbers that can be arranged to function as a shaft liner, a protective device around a concrete pier in water, or a temporary floating foundation.

basket weave A pattern of bricks placed flat or on edge and arranged in a checkerboard layout.

bas-relief (basso-riviero, basso rilievo) Sculpture, carving, or embossing that protrudes slightly from its background.

basswood (American linden) A fine-textured softwood used for carving, cabinet work, and paneling. Basswood is also the primary source of excelsior.

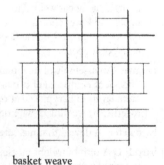

basket weave

bastard Any nonstandard item deviating from normal size, slope, fabrication, or quality.

bastard file A flat file whose grain is somewhat less than coarse and that is primarily used to smooth metal surfaces.

bastard hip Hip roof design in which the angles of the hip rafter are not equal (or 45 degrees).

B

bastard granite A gneissic rock whose formation resembles that of true granite. Bastard granite is used mostly in wall construction but is not a true granite.

bastard masonry **1.** Thin blocks of facing stones used to face brick or rubble walls. Bastard masonry is dressed and built to resemble ashlar. **2.** Rough, quarry-dressed ashlar stones.

bastard pointing In masonry, a type of pointing that emphasizes the joint by forming a small ridge projecting along its center.

bastard-sawn **1.** Lumber that has been sawn so that the annual rings make angles of 30°–60° to the surface of the piece. **2.** Any lumber that has been flat-sawn, plain-sawn, or slash-sawn.

bat **1.** A burned brick or shape that, because it is broken, has only one good end. **2.** A piece of brick. **3.** A single unit of batt insulation. **4.** A piece of wood that serves as a brace.

bat bolt A bolt whose butt or tang has been provided with barbs or similar protrusions to increase its grip.

batch The quantity produced as the result of one mixing operation, as in a batch of concrete.

batch box A container of known volume used to measure the constituents of concrete or mortar in proper proportions.

batch mixer A machine that mixes concrete, grout, or mortar in batches in accordance to a design mix. Each batch is used completely before a second batch is started.

batch mixer

batch plant An installation of equipment including batchers and mixers as required for batching and mixing concrete materials. Called a *mixing plant* when mixing equipment is included.

bathroom A room usually equipped with a water closet, a lavatory, a bathtub and/or a shower.

batted work (broad tooled) Stone whose surface has been scored downward from the top by administering 8 to 10 narrow, parallel strikes per inch with a batting tool.

batten **1.** A narrow strip of wood used as siding to cover the joints of parallel boards or plywood. The resultant pattern is referred to as *board and batten.* **2.** A strip of wood placed perpendicular to several parallel pieces of wood to hold them together. **3.** A furring strip fastened to a wall to provide a base for lathing or plastering. **4.** In roofing, a strip of wood placed over boards or roof structural members to provide a base for the application of wood or slate shingles, or clay tiles. **5.** The steel strip that fastens the metal flooring on a fire escape.

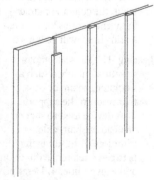

batten

batten door (ledged door, unframed door) A door in which stiles are absent and that consists of vertical boards or sheathing secured on the back side by horizontal battens.

batten roll (conical roll) In metal roofing, a roll joint fabricated over a triangular wood piece.

batten seam In metal roofing, a seam fabricated around a wood strip.

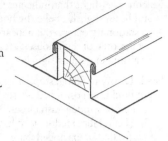

batten seam

batter **1.** To incline from the vertical. **2.** A slope, such as that of the outer side of a wall, that is wider at the bottom than at the top.

batter boards Pairs of horizontal boards nailed to wood stakes adjoining an excavation. Used with strings as a guide to elevation and to outline a proposed building. The strings strung between boards can be left in place during excavation.

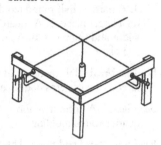

batter boards

batter brace (batter post) **1.** A bracing member positioned diagonally so as to reinforce an end of a truss. **2.** An inclined timber that forms a side support to a tunnel roof.

battered wall A wall that slopes backward, as by recessing or sloping masonry in successive courses.

batter level An instrument used to measure the inclination of a slope.

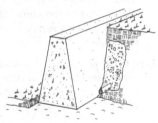

battered wall

batter pile (brace pile, raking pile, spur pile) **1.** A pile driven somewhat diagonally so as to resist horizontal forces. **2.** Any pile installed at an angle to the vertical.

batter rule An instrument used to adjust the inclination of a battered wall during its construction. A batter rule incorporates a rule or frame and a plumb line and bob.

batter stick A tapered board hung vertically by its wide end or used in conjunction with a level to check the batter of a wall surface.

batting tool A mason's dressing chisel used to apply a striated surface to stone.

batt insulation Thermal or sound-insulating material, such as fiberglass or expanded shale, that has been fashioned into a flexible, blanket-like form, often with a vapor barrier on one side. Batt insulation is manufactured in dimensions that facilitate its installation between the studs or joists of a frame construction.

bay **1.** In construction, the space between two main trusses or beams. **2.** The space between two adjacent piers or mullions, or between two adjacent lines of columns. **3.** A small, well-defined area of concrete laid in the course of placing larger areas, such as floors, pavements, or runways. **4.** The projecting structure of a bay window.

bay window A usually large window or group of windows that projects from a wall of a building forming a recess within the building.

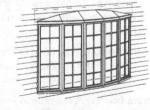

bay window

B

beacon 1. A light that indicates a location by directing its powerful beam slightly above the horizontal and rotating it so that to a stationary observer, it appears to be flashing. A beacon is used at airports, on lighthouses, etc. **2.** A transmitter that broadcasts warning or guiding signals.

bead 1. Any molding, stop, or caulking used around a glass panel to hold it in position. **2.** A stop or strip of wood against which a door or window sash closes. **3.** A strip of sheet metal that has been fabricated so as to have a projecting nosing and two perforated or expanded flanges. A bead is used as a stop at the perimeter of a plastered surface or as reinforcement at the corners. **4.** A narrow, half-round molding, either attached to or milled on a larger piece. **5.** A square or rectangular trim less than 1″ in width and thickness. **6.** A choker ferrule; the knob on the end of a choker.

bead-and-real Half-round molding into which are incorporated alternating patterns of discs and elongated beads.

bead board A decorative form of wainscoting that consists of a quarter-round molding.

bead butt (bead and butt) Thick, framed panelwork, as on a door, having one face flush with the frame and decorated with moldings (beads) on the adjoining edges, running with the grain and butting against the rail. The other side is recessed and without moldings.

bead molding Small, half-round, convex molding, either continuous or divided, that resembles a succession of beads.

beam 1. A horizontal structural member, such as a girder, rafter, or purlin, that transversely supports a load and transfers the load to vertical members, such as columns and walls. **2.** The graduated horizontal bar of a weighing scale.

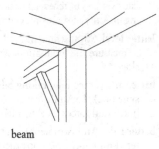

beam

beam anchor (joist anchor, wall anchor) A metal tie used to secure a beam, joist, or floor firmly to a wall.

beam-and-girder construction A type of floor construction in which slabs are used to distribute the load to evenly spaced beams and girders.

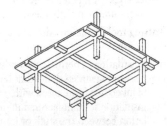

beam-and-girder construction

beam bearing plate A metal plate positioned under the end of a beam to distribute the reaction load over a larger area. When used under a column, it is called a *loose plate*.

beam blocking Covering or enclosing a beam, joist, or girder to make it appear larger than it really is.

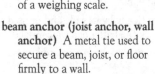

beam bolster

beam bolster A rod or heavy wire device used to support steel reinforcement in the formwork for a reinforced concrete beam.

beam brick A lintel made of brick courses, often held together by metal straps.

beam ceiling A type of construction in which the structural and/or ornamental overhead beams are left exposed to view from the room below.

beam column A structural member that transmits an axial load as well as a transverse load.

beam compass An instrument comprising a small horizontal bar with two vertical components that can slide along it. One carries a sharp-pointed tip and is held by the user in a stationary position. The other carries a pencil tip and is moved around the first to draw circles or arcs of circles for full-size working drawings.

beam fill (beam filling) Masonry, brickwork, or concrete placed between floor or ceiling joists to stiffen the joists and provide fire resistance.

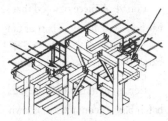

beam form

beam form A retainer or mold constructed to give the necessary shape, support, and finish to a concrete beam.

beam hanger (beam saddle) 1. A wire, strap, or other hardware device used to hang beam forms from another structural member. **2.** In timber construction, a strap, wire, or stirrup used to support a beam.

beam haunch A poured concrete section that continues beyond a beam to support the sill.

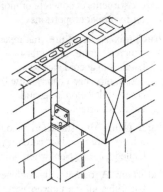

beam pocket

beam pocket 1. A space left open in a vertical structural member to receive a beam. **2.** An opening in the column or girder form where forms for an intersecting beam will be framed.

bearing 1. That section of a structural member, such as a column, beam, or truss, that rests on the supports. **2.** A device used to support or steady a shaft, axle, or trunnion. **3.** In surveying, the horizontal angle between a reference direction, such as true north, and a given line. **4.** Descriptive of any wall that provides support to the floor and/or roof of a building.

bearing

bearing bar 1. A wrought-iron bar used on masonry to offer a level support for floor joists. **2.** A supporting bar for a grating.

bearing pile A pile that supports a vertical load.

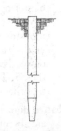

bearing pile

bearing plate A steel plate positioned under a beam, column, girder, or truss to distribute a load to a support member.

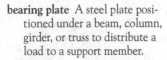
bearing plate

bearing stratum The soil or rock stratum on which a footing or mat bears or carries the load transferred to it by a pile, caisson, or similar deep foundation unit.

bearing wall Any wall that supports a vertical load as well as its own weight.

bed 1. The mortar into which masonry units are set. 2. Sand or other aggregate on which pipe or conduit is laid in a trench. 3. To set in place with putty or similar compound, as might be performed in glazing. 4. A supporting base for engines or machinery. 5. To level or smooth a path onto which a tree will be felled. 6. To set glass in place using putty.

bedding 1. A prepared base for masonry or concrete. 2. The lath or other support(s) on which pipe is laid.

bedding coat (bed coat) Ordinarily, the initial coat of joint compound on gypsum board applied over tape, bead, and other fastener heads.

bedding dot A small area of plaster built out of the face of a finished wall or ceiling that acts as a screed for leveling and plumbing in a plastering operation.

bedding stone A flat slab of marble with which bricklayers and masons check the flatness of rubbed bricks.

bed dowel A dowel placed in the mortar bed for a stone to prevent settlement before the mortar has set.

bed joint 1. The horizontal layer of mortar on which a masonry unit is laid. 2. In an arch, a horizontal joint or one that radiates between adjacent voussoirs. 3. A horizontal fault in a rock formation.

bed molding 1. A molding placed at the angle between a vertical surface and an overhanging horizontal surface, such as between a side wall and the eaves of a building. 2. The lowest molding in a band of moldings. 3. In classical architecture, a molding of a cornice of an entablature, located between the corona and the frieze.

bedplate A baseplate, frame, or platform that supports a structural element, furnace, or heavy machine.

bedrock Solid rock that underlies the earth's surface soil and that can provide, by its very existence, the foundation on which a heavy structure may be erected.

bed timber A large wood member set perpendicular to trusses that serves as a foundation or support element.

BEES (Building for Environmental and Economic Sustainability) A methodology that considers multiple environmental and economic impacts over the building product's entire life cycle to develop a rational decision-making scoring system.

Belgian block A paving stone shaped like a truncated pyramid and laid with the largest face down.

bell (hub) A portion of a pipe that, for a short distance, is sufficiently enlarged to receive the end of another pipe of the same diameter for the purpose of making a joint.

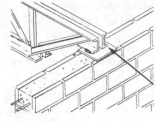
bell (hub)

bell-and-spigot joint (bell joint, spigot-and-socket joint) The most commonly used joint in cast-iron pipe. Each length is made with an enlarged diameter or bell at one end into which the plain or spigot end of another piece is inserted. The joint is sealed by cement, oakum, lead, or rubber caulked into the bell around the spigot.

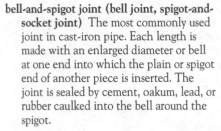
bell-and-spigot joint (bell joint, spigot-and-socket joint)

belled caisson A caisson with an enlarged, bell-shaped base.

belled excavation The bell-shaped lower portion of a shaft or footing excavation frequently used in caisson construction.

belling In pier, caisson, or pile construction, the process of enlarging the base of a foundation element at the bearing stratum to provide more bearing area.

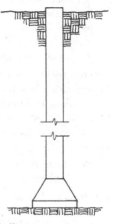
belling

bellows expansion joint In a run of piping, a joint accomplished with flexible metal bellows that can expand and/or contract linearly to allow for thermally induced linear fluctuations of the run itself.

bell transformer A very small transformer that supplies low voltage power on demand to doorbells or similar devices.

belt 1. A flexible continuous loop that conveys power (or materials) between the pulleys or rollers around which it passes. 2. A course of brick or stone that protrudes from a wall of similar material and is usually positioned in line with the windowsills.

belt course 1. In masonry, a continuous layer of stone or brick that protrudes from the face of a stone or brick wall. 2. In carpentry, a horizontal band around a building, usually made of a flat board member and molding.

belt loader An excavating machine comprising an auger or other cutting edge that digs away the earth, and a conveyer belt that elevates the excavated material for loading onto a truck or depositing elsewhere.

belt sander An electrically powered, portable sanding tool with a continuous abrasive belt driven in one direction only. A belt sander is used to smooth surfaces, usually wood.

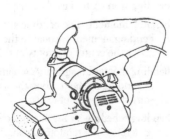
belt sander

belvedere A rooftop pavilion or a small structure, like a gazebo, for enjoying a vista.

bench brake A large, heavy, bench-mounted device for bending sheet metal.

bench dog A peg or pin partially inserted into a hole at an edge or end of a workbench to help secure a piece of work or prevent it from sliding off the bench.

benched foundation (stepped foundation) A foundation cut as a series of horizontal steps in an inclined bearing stratum to prevent sliding when loaded.

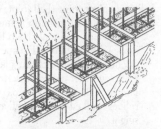

benched foundation (stepped foundation)

bench hook (side hook) In carpentry, any device used to protect the top surface of a workbench from being scarred or damaged by any slipping or sliding of the work. It usually keeps the work positioned toward the front of the bench.

benching 1. A half-round channel cast in the concrete in the bottom of a manhole to direct discharge when the flow is low. 2. Concrete placed in horizontal steps on steeply sloping fill to prevent sliding. 3. Concrete laid on the side slopes of drainage channels where the slopes are interrupted by manholes, etc. 4. Concrete laid in a pipeline trench to provide firmer support.

bench mark 1. A marked reference point on a permanent object, such as a metal disc set in concrete, whose elevation as referenced to a datum is known. 2. A mark made by a surveyor or general contractor to be used as a reference point when measuring the elevation or location of other points.

bench stop A usually notched, adjustable metal apparatus fastened close to an end of a workbench to hold a piece of wood securely during planing.

bench table A course of masonry, wide enough to form a seat, that protrudes from the foot of an interior wall or column.

bench vise An ordinary vise attached to a workbench to hold a material or component in place while it is being worked on.

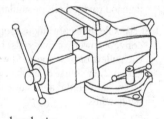

bench vise

bend 1. An elbow fitting or other short length of bent conduit used to join two lengths of straight adjacent conduit. 2. A pipe fitting used to achieve a change in direction.

bending extra A charge (an extra) by the fabricator for bending reinforcing bars. Extras are charged by the ton according to bar sizes.

bending iron A tool used to shape or expand pipe.

bending moment The bending effect at any section of a beam. The bending moment is equal to the algebraic sum of moments taken about the center of gravity of that section.

bending moment diagram A graphic representation of the variation of bending moment along the length of the member for a given stationary system of loads.

bending schedule A list of reinforcement prepared by the designer or detailer of a reinforced concrete structure that shows the shapes, dimensions, and location of every bar, and the number of bars required.

bend test Subjecting a flat bar to a 180° cold bend in order to test its weld or steel and to check its ductility, which is verified if no cracking occurs during the test.

benefit cost analysis A method of evaluating projects or investments by comparing the present value or annual value of expected benefits to the present value or annual value of expected costs.*

benefit-to-cost ratio (BTC) Benefits divided by costs, where both are discounted to a present value or equivalent uniform annual value.*

bent 1. A structural framework, transverse to the length of a structure, designed to carry lateral as well as vertical loads. 2. Any of several grasses of the genus *agrostis* that are used where a resilient, velvety texture is required.

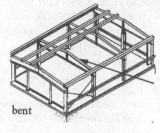

bent

bent bar A reinforcing bar bent to a prescribed shape, such as a truss, straight bar with hook, stirrup, or column tie. A bent bar is bent to pass from one face of a member to the other.

bentonite A clay composed principally of minerals in the montmorillonite group and characterized by high absorption and very large volume change with wetting or drying.

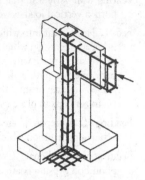

bent bar

berm 1. An artificially placed continuous ridge or bank of earth, usually along a roadside. Also called a *shoulder*. 2. A ridge or bank of earth placed against a masonry wall. 3. A ledge or strip of earth placed so as to support pipes or beams. 4. Earthen dikes or embankments constructed to retain water on land that will be flood-irrigated. 5. Earthen or paved dike-like embankments for diverting runoff water. 6. A raised wall enclosing a liquid waste storage or spill area. 7. An asphaltic concrete or concrete curb.

best practices Practical techniques gained from experience that have been shown to produce best results.*

bevel 1. Any angle (except a right angle) or inclination of any line or surface that joins another. 2. An adjustable instrument used for determining, measuring, or reproducing angles. 3. In welding, preparation on the edges to be welded.

bevel angle In welding, the angle created by the prepared edge of a member and a plane perpendicular to the surface of the member.

bevel board (pitch board) A board that has been cut to a predetermined desired, or required, angle and employed in any angular wood construction, including roof and stair framing.

bevel chisel A wood-cutting chisel whose cutting edge is angled between its sides.

bevel chisel

bevel cut Any cut made at an angle other than a right angle.

beveled edge 1. A vertical front edge on a door, cut so as to have a slope of $\frac{1}{8}$" in 2" from a plane perpendicular to the face of the door. 2. The factory-applied angle on the edge of gypsum board that creates a "vee" grooved joint when two pieces are installed together. 3. A chamfer strip incorporated into concrete forms for columns or beams so as to eliminate sharp corners on the finished product.

beveled end The end of a pipe or fitting that has been prepared for welding.

beveled end

beveled washer A washer with a bevel on one side, and used mostly in structural steel work to provide a flat surface for the nut wherever a threaded rod or bolt passes through a beam at an angle.

bevel joint A carpentry joint in which two wood pieces meet at any angle except a right angle.

B-grade wood The classification of a somewhat inferior grade of solid surface veneer that contains visible repair plugs and tight knots.

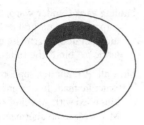

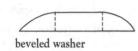
beveled washer

bias The fixed voltage applied to an electrode.

bibcock (bib, bibb, bib tap) 1. Any faucet or stopcock that has its nozzle directed downward. **2.** Any tap supplied by a horizontal pipe.

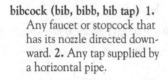

bib valve Any standard bibcock or faucet equipped with a
bibcock (bib, bibb, bib tap)
handle that is turned in one direction to initiate the flow of water, and turned in the opposite direction (screwed down) to shut off the flow by closing a washer disk onto a seating within the valve.

bid A complete signed proposal to perform work (or a designated portion) for a stipulated sum. A bid is submitted in accordance with the bidding documents.

bid abstract (summary) On a given project, a compilation of bidders and their respective bids, usually separated into individual items.

bid bond A form of security executed by the bidder or principal in conjunction with a surety to guarantee that the bidder will enter into a contract within a specified period of time and will furnish the required bonds for performance and labor and materials payment.

bid call A published announcement that bids for a specific construction project will be accepted at a designated time and place.

bid date A predetermined date for the receipt of bids, usually set by the architect and owner.

bidder An entity or person who submits a bid for a prime contract with the owner. A bidder is not a contractor on a specific project until a contract is signed between the bidder and the owner.

bidding documents Documents usually including advertisement or invitation to bidders, instructions to bidders, bid form, form of contract, forms of bonds, conditions of contract, specifications, drawings, addenda, and any other information needed to completely describe the work so that constructors can adequately prepare proposals or bids for the owner's consideration.

bidet A low, basin-like bathroom fixture used for washing the lower part of the body.

bid form A form, furnished to the bidder, on which to submit his bid.

bid opening (bid letting) A formal meeting held at a specified time and place at which sealed bids are opened, tabulated, and read aloud.

bid package All drawings, specifications, documents, estimates, paperwork, bid forms, and bid bonds relevant to a construction project. A contract is based on the bid package.

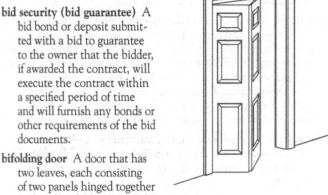
bidet

bid protest A challenge by a disappointed bidder. On a government contract, this is submitted to the contracting officer or the GAO.

bid results The display of all the bids on a project.

bid security (bid guarantee) A bid bond or deposit submitted with a bid to guarantee to the owner that the bidder, if awarded the contract, will execute the contract within a specified period of time and will furnish any bonds or other requirements of the bid documents.

bifolding door A door that has two leaves, each consisting of two panels hinged together so that they fold on each other when the door is opened. One
bifolding door
free edge of each leaf, or pair of panels, is hinged at a doorjamb; the other edge is supported from and guided by an overhead track.

bilateral contract A contract in which both contracting parties are bound to fulfill obligations toward each other.

bill of materials (BOM) 1. Set of physical elements required to build a project. **2.** Hierarchical view of the physical assemblies, subassemblies, and components needed to fabricate a manufacturing product. **3.** Descriptive and quantitative list of materials, supplies, parts, and components required to produce a designated complete end item of materials, assembly, or subassembly.*

BIM application A very broad category of any software that can be used with a BIM platform or BIM environment to support Building Information Modeling. Thus, traditional applications such as drafting, rendering, specification writing, and engineering analysis tools are all potentially BIM applications, if workflows and/or data exchange integrates them in Building Information Modeling.

The term can be further qualified to denote specific application areas. For example, "BIM Architectural Design Application" is often used to refer to applications used primarily for architectural design, such as Revit®Architecture, Bentley Architecture, DigitalProject® and ArchiCAD®, or "BIM 4D application," that supports animation of a BIM model according to an associated construction schedule.

BIM environment A BIM environment is the functional capability embedded in a BIM server. It encompasses the data management information, and software for enforcing policies and practices that

integrate the applications (tools or platforms) within an organization. Often the BIM environment is not conceptualized explicitly, but grows in ad hoc manner, driven by needs within the firm. Integration and support across multiple BIM platforms is its critical raison d'être, as well as managing communication with external systems. A BIM environment is supported by a set of policies and practices that facilitates management of BIM project data.

BIM platform A BIM design application that generates data for multiple uses and incorporates multiple tools directly or through interfaces with varying levels of integration. Most BIM design applications serve not only a tool function, such as 3D parametric object modeling, but also other functions, such as drawing production and application interface, making them also platforms.

BIM process A process that relies on the information generated by a BIM design tool for analysis, fabrication detailing, cost estimation, scheduling or other use.

BIM server A BIM server is a database system whose schema is based on an object-based format. It is different from existing project data management (PDM) systems and web-based project management systems in that PDM systems are file-based systems, and carry CAD and analysis package project files. BIM servers are object-based, allowing query, transfer, updating and management of individual project objects from a potentially heterogeneous set of applications. BIM servers are targeted to support BIM environments.

BIM system A software system that incorporates a BIM design application and other applications that utilize the BIM data. The system may be connected through a local area network or the Internet.

BIM tool A task-specific software application that manipulates a building model for some defined purpose and produces a specific outcome. Examples of tools include those used for drawing production, specification writing, cost estimation, clash and error detection, energy analysis, rendering and visualization.

bin A storage container, usually for loose materials such as sand, stone, or crushed rock.

binary code In computer technology, a system of representing numbers or letters using the base-2 (binary) number system.

binder **1.** Almost any cementing material, either hydrated cement or a product of cement or lime and reactive siliceous materials. The kinds of cement and the curing conditions determine the general type of binder formed. **2.** Any material, such as asphalt or resin, that forms the matrix of concretes, mortars, and sanded grouts. **3.** That ingredient of an adhesive composition that is principally responsible for the adhesive properties that actually hold the two bodies together. **4.** In paint, that nonvolatile ingredient, such as oil, varnish, protein, or size, that serves to hold the pigment particles together in a coherent film. **5.** A stirrup or other similar contrivance, usually of small-diameter rod, that functions to hold together the main steel in a reinforced concrete beam or column.

binder course (binding course) **1.** A succession of masonry units between an inner and outer wall that serve to bind them. **2.** In asphaltic concrete paving, an intermediate course between the base and the surfacing material composed of bituminously bound aggregate of intermediate size.

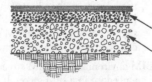

binder course (binding course)

binding post **1.** A set screw that holds a conductor against the terminal of a device or on equipment. **2.** A post attached to an electric wire, cable, or apparatus to facilitate a connection to it.

binding post

binding yarn (binder warp, crimp warp) In carpeting, the natural or synthetic yarn incorporated lengthwise into the woven fabric to "bind" the tufts of pile securely.

bin-wall A retaining, supporting, or protective structure made from a group of connected bins filled with gravel or sand. May serve as an abutment, a pier, a retaining wall, or as a shield against gunfire or explosion.

biological remediation The process of degrading a contaminant through a bioremediation process.

biological wastewater management Using natural or simulated wetlands to purify wastewater through biological processes.

bin-wall

biomass All of the living material in a given area; most often refers to vegetation.

biometrics Electronic methods of identifying and authenticating a person's identity through the confirmation of his/her physiological or sometimes behavioral characteristics. Techniques include fingerprint matching, iris and retinal scanning, facial recognition, hand geometry matching, voice recognition, and vein matching.

biparting door A sliding door with two leaves that slide in the same plane and meet at the door opening.

biparting door

birch A strong, fine-grained hardwood commonly used in veneer, furniture, flooring, and turned wood products.

bird screen Wire mesh used to cover chimneys, ventilators, and louvers to prevent birds from entering buildings through these accesses.

bird's mouth An L-shaped cut made in a diagonal timber, such as a rafter, so that it snugly fits against another timber, such as a wall plate.

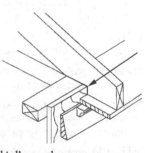

bird's mouth

bit **1.** The tool that fits into a brace or drill and is rotated to bore a hole. **2.** On a soldering iron, the (usually copper) tip that heats the joint and melts the solder. **3.** That part of a key that is inserted into a lock and engages the tumblers or bolt. **4.** A binary digit; the smallest unit of computerized data.

bite In glazing, the overlap between the innermost edge of the stop or frame and the outer edge of the light or panel.

bit extension A length of rod held at one end by the chuck of a brace and equipped at the other end for holding a bit. A bit extension permits the drilling of holes whose required depth is greater than the length of an ordinary bit.

bit key (wing key) Any key that has a bit.

bitumen Any of several mixtures of naturally occurring or synthetically rendered hydrocarbons and other substances obtained from coal or petroleum by distillation. Bitumen is incorporated in asphalt and tar and is used in road surfacing and waterproofing operations.

bituminous Composed of, similar to, derived from, relating to, or containing bitumen. The term bituminous is descriptive of asphalt and tar products.

bituminous base course (black base) Bituminously bound aggregate serving as a foundation for binder courses and surface courses in asphalt paving operations.

bituminous cement A class of dark substances composed of intermediate hydrocarbons. Bituminous cement is available in solid, semisolid, or liquid states at normal temperatures.

bituminous coating Any waterproof or protective coating whose base is a compound of asphalt or tar.

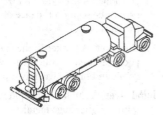

bituminous coating

bituminous distributor A tank truck equipped with a perforated spray bar through which heated bituminous material, such as tar or road oil, is pumped onto the surface of a road.

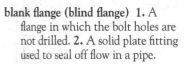

bituminous distributor

bituminous emulsion 1. A suspension of any globules of a bituminous substance in water or an aqueous solution. **2.** An invert emulsion of the above, i.e., a suspension of tiny globules of water or an aqueous solution in a liquid bituminous substance. This type of bituminous emulsion is applied to surfaces to provide a weatherproof coating.

bituminous grout A mixture of bituminous material and fine sand or other aggregate that, when heated, becomes liquid enough to flow into place without mechanical assistance. Bituminous grout will air-cure after being poured into cracks or joints as a filler and/or sealer.

bituminous leveling course In paving operations, a course consisting of a mixture of asphalt and sand and used to level or crown a base course or existing deteriorating pavement prior to the application of a surface.

bituminous macadam A paving material comprising bituminously coated coarse aggregate.

bituminous paint A thick black waterproofing paint containing substantial amounts of coal tar or asphalt.

B-labeled door A door carrying a certification from Underwriters' Laboratories that it is of a construction that will pass the standard fire door test required for a Class B opening,

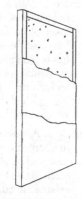

B-labeled door

and that it has been prepared (with cuts and reinforcement) to receive the hardware required for a Class B opening.

black japan A black, high-quality bituminous paint or varnish.

black plate Uncoated cold-rolled steel in 12″ to 32″ wide sheets.

black steel pipe Steel pipe that has not been galvanized.

black water Wastewater from toilets and other plumbing fixtures that may be contaminated with bacteria or other harmful organisms.

Blaine apparatus Air-permeability apparatus for measuring the surface area of a finely ground cement.

Blaine test A method for determining the fineness of cement or other material based on the permeability to air of a sample prepared under specified conditions.

blank door 1. A recess in a wall fitted with a fixed door, and used for architectural effect. **2.** A door that has been fixed in position to seal an opening.

blanket 1. Soil or pieces of rock remaining or intentionally placed over a blast area to contain or direct the throw of fragments. **2.** Insulation sandwiched between sheets of fabric, plaster, or paper facing, used for protecting fresh concrete during curing.

blanket insulation Faced or unfaced thermal or sound insulation, usually made of fiberglass, available in densities and thicknesses that allow it to conform to the various-shaped spaces it encounters in its many applications. Blanket insulation is the same as batt insulation, except that it is supplied in continuous rolls instead of sheets. *See also* **insulation blanket**.

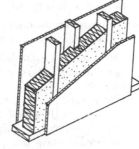

blanket insulation

blank flange (blind flange) 1. A flange in which the bolt holes are not drilled. **2.** A solid plate fitting used to seal off flow in a pipe.

blank flange (blind flange)

blank jamb A vertical component of a door frame as installed without any preparation for hardware installation.

blank-off A blank plate sealing off a sector of a diffuser to prevent airflow in the direction of the blank-off plate.

blast To loosen, crack, or move rock or hard-packed soil by the detonation of explosives.

blast cleaning Any cleaning method in which air, liquid, abrasive, or some combination of these is applied under pressure.

blast-furnace slag The nonmetallic waste that develops simultaneously with iron in a blast furnace. Consists essentially of silicates and aluminosilicates of calcium and other bases.

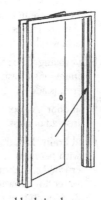

blank jamb

blast hole A vertical hole with a diameter of at least 4″ and drilled to accept a charge of blasting explosives.

blasting The process of loosening rock or hard-packed material with explosives.

blasting cap A metallic tube closed at one end, containing a charge of one or more detonating compounds, designed for and capable of detonation from the sparks or flame of a safety fuse inserted and crimped into the open end.

blasting mat A blanket of interwoven steel cable or interlocking steel rings placed over a blast to contain the resultant fragments.

BLCC A National Institute of Standards and Technology (NIST) software program that performs life-cycle analysis of buildings and components, useful for comparing alternate designs that have higher initial costs, but lower operating costs over the life of the building.

bleaching Lightening, whitening, or removing color, either by chemical means, such as chlorine or oxalic acid, or by exposure to certain kinds of light.

bleeder A small valve used to drain fluid from a pipe, radiator, or small tank.

bleeding 1. The autogenous flow of mixing water within, or its emergence from, freshly placed concrete or mortar. Bleeding is caused by the settlement of the solid materials within the mass. Also called *water grain*. 2. In painting, seepage of resin or an undercoating of paint through the finish coat. 3. In lumber, the exuding of sap or resin.

bleed-through (strike-through) Discoloration in the face plies of wood veneer caused by seepage of cement through the veneer.

blended cement A hydraulic cement consisting of an intimate and uniform blend of (a) granulated blast-furnace slag and hydrated lime, (b) Portland cement and granulated blast-furnace slag, (c) Portland cement and pozzolan, or (d) Portland-blast-furnace slag, cement, and pozzolan. Blended cement is produced by intergrinding Portland cement clinker with the other materials or by a combination of intergrinding and blending.

blending valve A three-way valve that permits the mixing required to obtain a desired liquid temperature.

blind 1. Any panel, shade, screen, or similar contrivance used to block light or inhibit viewing. 2. An assembly of wood stiles, rail and wood slats, or louvers used in conjunction with doors and windows.

blind

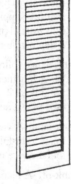

blind casing The rough window frame or subcasing to which the trim is added.

blind door A door with a louver instead of glass.

blind drain 1. A trench filled with pervious materials such as crushed stone through which water flows toward an outlet. 2. A drain that is not connected to a sewage system.

blind header 1. A brick or other masonry unit that, when laid, gives the appearance of a header while in reality is somewhat less than a whole unit. A blind

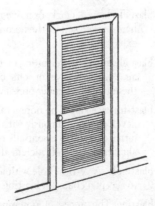

blind door

header can also be a half-brick laid where only one end is visible. 2. A header brick concealed within a wall and functioning to bond adjacent tiers of bricks.

blinding 1. Applying a layer of weak concrete or other suitable material to reduce surface voids or to provide a clean, dry working surface. 2. The clogging of the openings in a screen or sieve by the material being separated. 3. Applying small chips of stone to a freshly tarred surface.

blind joint (bastard joint) 1. A masonry joint, such as that of a *blind header*, that is entirely concealed. 2. In a double Flemish masonry bond, a thin joint between the adjacent ends of two stretchers. This joint bisects a header in the course directly below it.

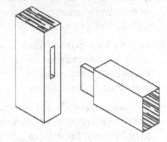

blind mortise (stopped mortise) A mortise in, but not through, a member.

blind mortise and tenon joint A joint between a blind mortise and a stub tenon.

blind mortise and tenon joint

blind nailing (concealed nailing, secret nailing) Nailing performed so that the nailhead cannot be seen on the face of the work.

blind nailing (concealed nailing, secret nailing)

blind pocket A pocket in the ceiling at a window head. A blind pocket is used to conceal an object when not in use, such as a venetian blind in the raised position.

blind rivet A small-headed pin with an expandable shank used for joining light gauge metal.

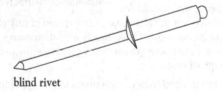

blind rivet

blind stop A rectangular molding nailed between the outside trim and the outside sash of a window frame, which serves as a stop for storm sashes or screens.

blister Usually an undesirable moisture and/or air-induced bubble or bulge that often indicates that some kind of delamination has taken place. Blisters can occur between finish plaster and the base coat, between paint or varnish and the surface to which it has been applied, between roofing membranes or between membrane and substrate, between reinforcing tape and the gypsum board to which it has been adhered, etc.

block 1. A usually hollow concrete masonry unit or other building unit, such as *glass block*. 2. A solid, often squared, piece of wood or other material. 3. A piece of wood nailed between joists to stiffen

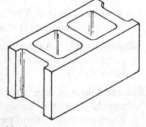

block

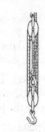

a floor. **4.** Any small piece of wood secured to the interior angle joint to strengthen and stiffen it. **5.** A pulley and its enclosure. **6.** A solid piece of wood or other material used to fill spaces between formwork members.

block and tackle A mechanical device comprised of pulley blocks and ropes or cables, and used to hoist or move heavy objects or loads.

block and tackle

block beam In structural design, a flexural member composed of individual blocks joined together by prestressing.

block bridging (solid bridging, solid strutting) Short boards fixed between floor joists to stiffen the joists and distribute the load.

block-in-course A kind of masonry employed in heavy construction that consists of squared, hammer-dressed stones laid so that the joints are close and the course is not higher than 12".

blocking **1.** Small pieces of wood used to secure, join, or reinforce members, or to fill spaces between members. **2.** Small wood blocks used for shimming. **3.** A method of bonding two parallel or intersecting walls built at different times by means of offsets whose vertical dimensions are not less than 8" (20 centimeters). **4.** The sticking together of two painted surfaces when pressed together. **5.** An undesired adhesion between touching layers of material, such as occurs during storage.

blocking

blocking course In masonry, a finishing course, usually of stones, placed on top of a cornice.

blockwork Masonry of concrete block and mortar.

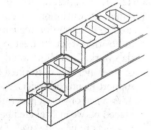

blockwork

bloom **1.** A thin, hazy film on old paint, usually caused by weathering. **2.** A similar film on glass, resulting from general atmospheric deposition of impurities, or caused more by smoke, vapor, etc. **3.** Efflorescence on brickwork. **4.** A hazy or other kind of discoloration that sometimes occurs on the surface of rubber products. **5.** The term given to steel that has been reduced from an ingot by being rolled in a blooming mill to a dimension of at least 6" square; if further reduced, it becomes a *billet*.

blowback The difference between the pressure at which a safety valve opens and the pressure at which it closes automatically after the release of excess pressure has occurred.

blow count **1.** The number of times that an object must be struck to be driven into the soil to a desired or specified depth.

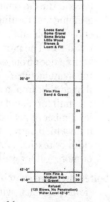

blow count

2. In soil borings, the number of times a sample spoon must be struck to be driven 6" or 12". **3.** In pile driving operations, the number of times a pile must be struck to be driven 12". **4.** The number of blows per unit distance of advance.

blower A fan, especially one for heavy-duty use such as forcing air through ducts to an underground excavation.

blowhole (gas pocket) In concrete, a bug hole or small regular or irregular cavity, not exceeding 15 mm in diameter, resulting from entrapment of air bubbles in the surface of formed concrete during placement and compaction.

blown-in insulation Insulation that is pumped or injected into walls, roofs, and other areas.

blown joint (blow joint) A plumbing joint sealed with the use of a blowtorch.

blow-off **1.** A discharge outlet on a boiler to allow for expulsion of undesirable accumulations of sediment and/or for drainage. **2.** In a sewer system, an outlet pipe for expelling sediment or water or for draining a low sewer.

blows (boils) In dewatering, small springs that bubble up on the floor of an excavated pit caused by unrelieved aquifers. Blows can cause large soil movement if not attended to properly.

blowtorch A small, portable, gas-fired burner that generates a flame hot enough to melt soft metals. A blowtorch is used to melt lead in plumbing operations, heat soldering irons, and burn off paint.

blowtorch

blow up (blowout) Localized buckling or breaking up of rigid pavement as a result of excessive longitudinal pressure.

blueboard A type of wallboard used as a base for and finished with veneer plaster.

blueprint A negative image reproduction having white lines on a blue background and made either from an original or from a positive intermediate print. Today, the term almost always refers to *photocopied* prints, which are architectural or working drawings having blue or black lines on a white background.

bluestone A hard, fine-grained sandstone or siltstone of dark green to bluish-gray color that splits readily to form thin slabs. A type of flagstone, bluestone is commonly used for paving walkways.

blue top A grade stake driven into the ground that indicates the finished grade level.

board Lumber ranging from 4" to 12" (10 × 30 cm) wide and less than 2" (5 cm) thick.

board and batten A type of siding in which the joints between vertically placed boards or plywood are covered by narrow strips of wood.

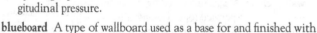

board and batten

board and brace A type of carpentry work consisting of boards grooved on both edges with thinner boards inserted between them and fitted into the grooves.

B

board foot The basic unit of measurement for lumber. One board foot is equal to a 1″ thick board, 12″ in width and 1′ in length. Thus, a 10″ long, 12″ wide, 1″ thick piece contains 10 board feet. When calculating board feet, nominal sizes are assumed.

board insulation (insulating board, insulation board) Lightweight thermal insulation, such as polystyrene, manufactured in rigid or semi-rigid form, whose thickness is very small relative to its other dimensions. Board insulation offers little structural strength and is usually applied under a finish material, although some types are surface-finished on one side.

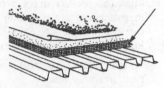

board insulation (insulating board, insulation board)

board measure A system of measuring quantities of lumber, using one board foot as the basic unit of measurement.

board sheathing A sheathing made of boards, usually tightly spaced, although open spacing may be used in some roofs.

boardwalk A walkway made of boards or planks, often used as a promenade along a beach or shore.

boasted work A dressed stone surface with roughly parallel narrow chisel grooves of varying widths that do not extend across the entire face of the stone.

boatswain's chair A seat supported by slings and attached to a suspended rope. Designed to accommodate one worker in a sitting position.

boatswain's chair

bodily injury Any form of physical injury, sickness, or disease that is experienced by a person.

body 1. The principal volume of a building, such as the nave of a church. **2.** The load-carrying part of a truck.

body coat In painting, an intermediate coat of paint applied after the priming coat but before the finishing coat.

boil 1. A wet run of material at the bottom of an excavation. **2.** A swelling in the bottom of an excavation due to seepage.

boiled linseed oil Linseed oil to which manganese, lead, or cobalt salts have been added so that it will harden rapidly when applied in thin layers.

boiler A closed vessel in which a liquid is heated or vaporized, either by application of heat to the outside of the vessel, by circulation of heat through tubes within the vessel, or by circulating heat around liquid-filled tubes in the vessel.

boiler

boiler economizer The last pass of boiler tubes or a heat exchanger located in the flue pipe that extracts some of the heat from the flue gases before they are vented to the atmosphere.

boiler horsepower 1. Boiler horsepower is equal to the evaporation of 34.5 lbs of water per hour from and at 212° F, 33, 475 Btu/hour. **2.** The largest rating obtained by dividing the square feet of a boiler's surface by 10.

boiler jacket The thermal insulation surrounding a boiler. A more aesthetically acceptable enclosure, often made of metal, usually covers the jacket.

boiler plate 1. Medium-hard steel from which boilers are fabricated. The steel is rolled into plates whose thicknesses may vary from 0.25″ to 1.5″. **2.** Standard text used in documents such as contract agreements.

boiler rating The heating capacity of a boiler, expressed in Btus per hour. The boiler rating should not be confused with the horsepower rating.

boiler scale Disintegrated bits of steel plate from a boiler's interior lining, which flake and fall to the boiler floor and should be removed periodically.

boiler steel A medium-hard steel that is rolled to become boiler plate.

bollard 1. A series of short posts set to prevent vehicular access or to protect property from damage by vehicular encroachment. A bollard is sometimes used to direct traffic. **2.** A post on a slip or wharf for securing mooring or docking lines.

bollard

bolster 1. In concrete, an individual or continuous support used to hold reinforcing bars in position. Usually used in slab work. **2.** A short wood or steel member positioned horizontally on top of a column to support beams or girders. **3.** A mason's blocking chisel. **4.** A piece of wood, generally a nominal 4″ in cross section, placed between stickered packages of lumber or other wood products to provide space for the entry and exit of the forks of a lift truck.

bolster

bolt 1. An externally threaded, cylindrical fastening device fabricated from a rod, pin, or wire, with a round, square, or hexagonal head that projects beyond the circumference of the shank to facilitate gripping and turning. A threaded nut fits onto the end of a bolt and is tightened by the application of torque. **2.** The protruding part of a lock that prevents a door from opening. **3.** Raw material used in the manufacture of shingles and shakes. A wedge-shaped split from a short length of log that is taken to a mill for manufacturing. **4.** Short logs to be sawn for lumber or peeled for veneer. **5.** Wood sections from which barrel staves are made. **6.** A large roll of cloth or textile. **7.** A single package containing two or more rolls of wallpaper.

bolt

bolted pressure switch A switch that uses movable blades and stationary contacts with arcing contacts to make and break switching action.

bolt sleeve A tube surrounding a bolt in a concrete wall to prevent the concrete from sticking to the bolt and acting as a spreader for the formwork.

bona fide bid A good faith bid that is essentially complete and complies with the bidding documents. Must be signed by a properly empowered party.

bond 1. The adhesion and grip of concrete or mortar to reinforcement or to other surfaces against which it is placed, including friction due to shrinkage and longitudinal shear in the concrete engaged by the bar deformations. **2.** The adhesion of cement paste to aggregate. **3.** Adherence between plaster coat or between plaster and a substratum produced by adhesive or cohesive properties of plaster or supplemental materials. **4.** In masonry, the connection between stones, bricks, or other materials formed by laying them in an overlapping arrangement, one on top of another, to form a single wall mass. **5.** The arrangement of, or pattern formed by, the exposed faces of laid masonry units. **6.** The layer of glue in a plywood joint. **7.** A written document, given by a surety in the name of a principal to an obligee to guarantee a specific obligation.

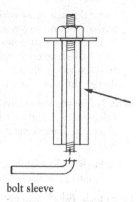

bolt sleeve

bond beam A horizontally reinforced concrete or concrete masonry beam built to strengthen and tie a masonry wall together. A bond beam is often placed at the top of a masonry wall with continuous reinforcing around the entire perimeter.

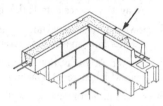

bond beam

bond blister A blister that sometimes forms between the base material and the coating, particularly on metal-clad products.

bond breaker A material used to prevent adhesion between freshly placed concrete and the substrate.

bond coat 1. That coat of plaster that is applied over masonry and is bonded to it. **2.** A primer coat of paint or sealer.

bond course In masonry, the course consisting of units that overlap more than one wythe of masonry.

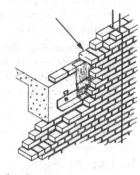

bond course

bonded member A prestressed concrete member in which the tendons are bonded to the concrete either directly (pretensioned) or by grouting (posttensioned).

bonded posttensioning A process in posttensioned construction whereby the annular spaces around the tendons are grouted

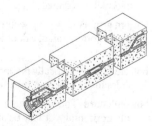

bonded posttensioning

after stressing in such a way that the tendons become bonded to the concrete section.

bonded roof A type of roofing guarantee offered by the manufacturer that may or may not be purchased by the owner, and that covers materials and/or workmanship for a stated length of time.

bonded tendon A prestressing tendon that is bonded to the concrete either directly or through grouting.

bonder (header) A masonry unit that ties two or more wythes of a wall together by overlapping.

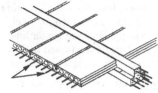

bonded tendon

bond face That face of a joint to which a field-molded sealant is bonded.

bonding In electrical circuitry, the connecting or joining of metal parts to form an electrical conductive path. Prevents the accumulation of static electricity and assures the safe conduction of electrical current.

bonding agent 1. A substance applied to a suitable substrate to create a bond between it and a succeeding layer, as between a subsurface and a terrazzo topping or subsequent application of plaster. **2.** A licensed professional who identifies the most appropriate bonding company from which a contractor should purchase construction bonds.

bonding capacity The maximum total contract value a bonding company will extend to a contractor in performance bonds. The total bonding capacity is the sum of all contracts being bonded.

bonding company A firm providing a surety bond for work to be performed by a contractor, payable to the owner in case of default of the contractor. The bond can be for work performance or for payment for materials and labor.

bonding jumper (bonding conductor) In a circuit, the connection between parts of a conductor to maintain the ampacity requirements of the circuit.

bond length (development length) The length of embedment of reinforcing steel in concrete required to develop the design strength.

bondstone In stone facing, a stone that extends into the backing to tie the facing wall with the backing wall. *See also* **bonder.**

bond strength 1. Resistance to separation of mortar and concrete reinforcing steel and other materials with which it comes in contact. **2.** Collective expression of all forces, such as adhesion, friction due to shrinkage, and longitudinal shear in the concrete engaged by the bar deformations that resist separation.

bond stress 1. The force of adhesion per unit area of contact between two bonded surfaces. **2.** Shear stress at the surface of a reinforcing bar, preventing relative movement between the bar and the surrounding concrete.

bonnet A wire mesh cover over the top of a chimney or vent.

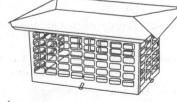

bonnet

bonus provisions Provisions in the contract between the owner and the contractor for granting monetary rewards to the contractor for achieving some savings that benefit the owner. For example, a stipulated bonus may be offered for early completion of the work, or the achievement of some savings in construction cost.

book matching (herringbone matching) Consecutive flitches of veneer from the same log, laid side by side so that the pattern formed is almost symmetrical from the common center line. Book matching is used in decorative paneling and cabinetry.

book value The net amount at which an asset is carried on the books and reported in the financial statements: the asset's cost at acquisition, reduced by the amount of accumulated depreciation on the asset.

boom A long, straight member, hinged at one end, and used for lifting heavy objects by means of cables and/or hydraulics. Booms can be of lattice construction or heavy tubular material.

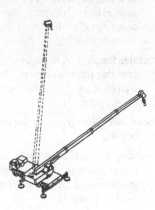

boom

boom crane A crane with a long slender boom, usually of lattice construction.

boom crane

boom hoist A lifting device with a vertical mast and an inclined boom commonly used to hoist materials in multistory building construction.

boom jack A short member on which sheaves are mounted to guide cables to a working boom on a crane or derrick.

boost-buck transformer A device that raises or lowers the voltage of a supply line.

booster Any device that increases the power, force, or pressure produced by a system.

booster fan An auxiliary fan that increases the air pressure in a system, such as an HVAC system, during certain peak requirement times.

boot 1. A term used to describe sleeves or coverings in many construction trades, such as a boot for pile driving, a boot for passing a pipe through a roof, and a boot for cold-air return to a furnace casing. 2. The money that compensates for differences in value in a 1031-exchange. Boot is taxable even though the exchange may be tax-free.

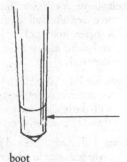

boot

bore 1. The inside diameter of a pipe, valve, fitting, or other hollow tubular object. 2. The circular hole left by boring.

bored latch A door latch manufactured to fit into a circular hole in a door.

bored lock A lock manufactured to fit into a circular hole in a door.

bored well A water well constructed by drilling a hole with an auger and then inserting a casing.

borescope A fiber-optic device that allows for nondestructive investigation of an area with restricted access, such as a housed mechanical assembly or a cavity inside a wall. Often used with a still or video camera.

boring (borehole) A hole drilled in the ground to obtain samples for subsoil investigation. Borings are important in determining the load-bearing capacity of the soil and the depth of the water table.

borrow (borrow fill material) In earth moving, fill acquired from an excavation source outside the required cut area.

borrowed light A glazed unit in an interior partition to allow light to enter from an adjoining room or passageway.

borrow pit An excavation site, other than a designated cut area, from which material is taken for use nearby.

boss 1. A projecting block, usually ornamental, placed at the exposed intersection or termination of ribs or beams of a structure. 2. The enlarged portion of a shaft. 3. In masonry, a stone that is left protruding from the surface for carving in place at a later time. 4. A projection left on a cast pipe fitting for alignment or for gripping with tools.

Boston hip (Boston ridge, shingle ridge finish) 1. A method of finishing a ridge or hip on a flexible shingled roof in which a final row of shingles is bent over the ridge or hip with a lateral overlap. 2. A method of finishing a ridge or hip on a rigid-shingled slate or a tile roof in which the last rows on either side overlap at the hip or ridge. Alternate courses overlap in opposite directions.

bottle brick A hollow brick shaped so that it can be mechanically connected to adjoining bricks. In each brick there is also a void available for inserting reinforcing steel, if required.

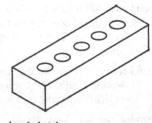

bottle brick

bottom arm A long strap secured to the bottom rail of a door to attach it to a floor closer or pivot hinge.

bottom bolt A vertically mounted bolt on the bottom of a door that slides into a socket in the floor.

bottom chord The lowest longitudinal member of a truss. It is usually horizontal, but may be at an incline depending on truss design.

bottom rail The lowest bottom member of a door. The bottom rail connects the stiles.

boundary survey A closed diagram (mathematically) depicting the complete outer boundary of a site. Shows dimensions, compass bearings, and angles. A licensed land surveyor's signed certification is required, and a *metes and bounds* or other written description may be included.

bottom rail

bow The longitudinal deflection of a piece of lumber, pipe, rod, or the like, usually measured at its center.

bow saw A hand tool with a thin saw blade held in place between the two ends of a bow-shaped piece of metal.

bowstring roof A roof supported by bowstring trusses and fabricated in the shape of a bow.

bowstring truss (bowstring beam, bowstring girder) A structural roof truss having a bow-shaped top cord and a straight or cambered bottom cord.

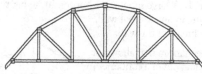

bowstring truss (bowstring beam, bowstring girder)

bow window 1. A window that projects from a wall in the shape of an arc. 2. A rounded bay window.

box casing In finished carpentry, the inner lining, as in a cabinet.

box column A built-up, hollow column, usually square in shape, used in porch construction.

box culvert A rectangular-shaped, reinforced concrete drainage structure either cast in place or precast in sections.

box drain A small rectangular-shaped drainage structure usually constructed of brick or concrete. It may be covered or have an open grate on top.

box culvert

boxed mullion A hollow mullion in a window frame built up from boards so as to appear solid; houses sash weights.

box frame A window frame containing hollow sections on either side in which the sash weights are suspended.

box gutter A rectangular-shaped wooden roof gutter recessed in the eaves to conceal them and to protect them from falling foliage.

boxing 1. Enclosing or casing, as in window frame construction. 2. In welding, continuing a principle fillet weld around a corner of the member. 3. Pouring paint back and forth from one pail to another to ensure a uniform consistency.

box out To make a form that will create a void in a concrete wall or slab when the concrete is placed.

box scarf A joint in a rectangular-shaped roof gutter formed by beveling the ends of the two pieces to be joined.

box sill A common method of frame construction using a header nailed across the ends of floor joists where they rest on the sill.

box stair An interior staircase in which the edges of the treads and risers are not revealed but closed in with a closed stringer. A box stair typically has a partition on both sides.

box stoop An elevated platform at the entrance to a building with stairs running parallel to the building. The underside of the platform and stairs is enclosed.

box strike A fastening on a door frame (a strike) with an enclosed recess to receive a lock bolt.

box wrench A mechanic's hand tool with a closed socket, usually at each end, that fits over the head of a bolt or a nut.

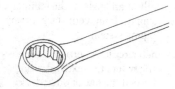

box wrench

brace (bracing) 1. A diagonal tie that interconnects scaffold members. 2. A temporary support for aligning vertical concrete formwork. 3. A horizontal or inclined member used to hold sheeting in place. 4. A hand tool with a handle, crank, and chuck used for turning a bit or auger.

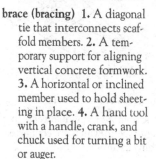

brace (bracing)

braced excavation Excavation in which the perimeter is supported by sheeting.

brace rod A round steel member used as a tension brace, especially to transfer wind or seismic loads.

brace table (brace scale) 1. The markings on a carpenter's square indicating the length of the hypotenuse for right triangles with various length legs. 2. A brace table used by carpenters to determine the length to cut braces.

bracket An attachment projecting from a wall or column used to support a weight or a structural member. Brackets are frequently used under a cornice, a bay window, or a canopy.

bracket pile A steel H pile driven next to an existing foundation. A steel bracket is welded to the pile and extends under and supports the foundation.

bracket saw A handsaw with a narrow blade attached to the open ends of a three-sided rectangular holder. Used for cutting curved shapes.

bracket scaffold (bracket staging) Scaffolding supported by brackets that are temporarily attached to the side of a building or column. Bolts or inserts are usually left in the previous construction to attach the brackets. This method saves the expense of shoring from the ground up.

brad A slender, smooth, wire nail with a small deep head. Used for finish carpentry work.

brad awl A carpenter's tool that looks like an ice pick and is used to make holes for brads and wood screws.

brad awl

brake 1. A device for slowing, stopping, and holding an object. **2.** A machine used to bend sheet metal.

brake drum A rotating cylinder with a machined surface (either internal or external) on which a brake band or shoe presses in order to slow, stop, or hold an object.

brake horsepower The horsepower output of an engine or other mechanical device, as measured at the flywheel or belt by applying a mechanical brake and measuring the work per unit time.

branch In plumbing, an inlet or outlet from the main pipeline, usually at an angle to the main pipeline. The pipe may be a water supply, drain, vent stack, or any other pipe used in a mechanical piping system.

branch circuit A portion of the electric system that extends wiring beyond the fuse or other device that is protecting that circuit.

branch duct In HVAC, a smaller duct that branches from the main duct. At each branch duct, the cross-sectional area of the main duct is reduced.

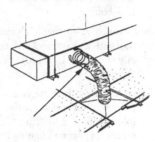
branch duct

branch sewer A sewer that receives sewage from a relatively small area and is connected to a main sewer or manhole.

branch vent 1. A vent connecting one or more individual vents to a vent stack or stack vent. **2.** A vent pipe to which are connected two or more pipes that vent plumbing fixtures.

brand name or equal A product description that identifies one or more commercial products by brand name and sets forth those characteristics of the named product that are essential to the owner's needs.

brander To nail furring strips to girders, joists, or walls in preparation for installing plaster or another finish material.

brass 1. A copper alloy with zinc as the principal alloying element. **2.** Slang for plumbing fittings and faucets, regardless of their actual material of composition.

braze To join two pieces of metal by soldering them together with a nonferrous metal such as brass.

breach of contract The failure, without legal cause, to perform some contractually described obligation in accordance with the terms of the contract.

breakdown A term used to define the separation of a project or an estimate into parts.

breakdown structure A hierarchical structure by which project elements are broken down, or decomposed.*

breakdown voltage Voltage that causes the failure of existing insulation, permitting the flow of electric current.

breaker panel Refers to the main electrical box and its circuit breakers, which distribute electricity to each branch circuit.

breakeven chart A graphic representation of the relation between total income and total costs for various levels of production and sales indicating areas of profit and loss.*

breakeven point 1. In business operations, the rate of operations output, or sales at which income is sufficient to equal operating costs or operating cost plus additional obligations that may be specified. **2.** The operating condition, such as output, at which two alternatives are equal in economy. **3.** The percentage of capacity operation of a manufacturing plant at which income will just cover expenses.*

breaking load The smallest load, determined by test, that would cause the failure of a structural system.

break joints (staggered joints) The arrangement of modular structural units, such as masonry or plywood sheathing, so that the vertical joints of adjacent units do not line up.

break joints (staggered joints)

break lines Lines used in drafting to omit part of an object so that the representation will fit on a drawing.

breast 1. The part of a wall that extends from the window stool to the floor level. **2.** A projecting part of a wall, such as at a chimney.

breast board A system of movable sheeting used to retain the face of an excavation, especially in tunnel work.

breastsummer 1. A large supporting horizontal timber that spans a wide opening in an external wall, usually the breast of the wall.

breather A mechanism that allows air to move in and out of a container to maintain even atmospheric pressure.

breech fitting A Y-shaped fitting in ductwork or pipework.

breeching The exhaust duct or pipe that leads from a furnace or boiler to the stack or chimney.

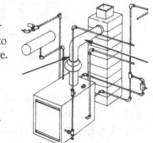
breeching

breeze brick A brick made from pan breeze and Portland cement, often laid into a brick wall because of its good nail-holding properties.

breezeway A covered passageway open at both ends, which connects two buildings or two portions of a building.

breezeway

brick A solid masonry unit of clay or shale, formed into a rectangular prism while plastic, and then burned or fired in a kiln.

brick anchor Corrugated fasteners designed to secure a brick veneer to a structural concrete wall.

brick and brick A method of laying brick so that units touch each other with only enough mortar to fill surface irregularities.

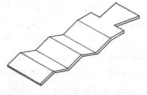
brick anchor

brick-and-half wall A brick wall that has the thickness of one header plus one stretcher.

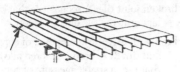

brick cement A waterproofed compound used in masonry and stucco work.

brick closure Short pieces of brick used in corners or jambs to maintain the pattern.

brick gauge A standard height for brick courses when laid, such as four courses in a height of 12″.

brick grade A durability rating given to brick. SW, MW, and NW ratings refer to a brick's ability to withstand severe weathering, moderate weathering, and negligible weathering, respectively.

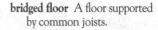

brick hammer A hand tool used by masons for breaking and dressing brick. One end of the head is square and flat and the other end is shaped like a chisel.

brick hammer

bricklayer's square scaffold A scaffold composed of framed wood squares that support a platform. A bricklayer's square scaffold is limited to light- and medium-duty.

brick molding The wood molding covering between the brick masonry and a door or window frame.

brick set A chisel-like tool used with a brick hammer to make exact cuts in brick.

brick trimmer A brick arch supporting a fireplace hearth or shielding a wood trimming joist from flames in front of a fireplace.

brick trowel A hand trowel with a pointed handle used to pick up mortar from a mortar tub and place and spread it in the joints.

brick type A rating given to facing brick based on its tolerance, chippage, and distortion. FBS, FBX, and FBA are designations for solid brick; HBS, HBX, and HBB are for hollow brick.

brick veneer A facing of brick laid against a structural wall but not bonded to the wall, and which bears no load other than its own weight.

brick veneer

brick whistle 1. A weep hole in a brick wall. **2.** A small hole in a mortar joint at the base of a wall for draining any moisture that penetrates the wall.

bridge 1. A structure built to span an obstruction or depression and capable of carrying pedestrians and/or vehicles. **2.** In an electric blasting cap, the wire that heats with current and ignites the charge. **3.** The temporary structure built over a sidewalk or roadway adjacent to a building to protect pedestrians and vehicles from falling objects.

bridge crane A crane often used in manufacturing or assembling heavy objects. A bridge crane requires a bridge spanning two overhead rails. A hoisting device moves laterally along the bridge while the bridge moves longitudinally along the rails.

bridge deck The load-carrying floor of a bridge that transmits the load to the beams.

bridge financing The short-term financing needed to bridge a short-fall between a loan presently committed or in place and the total financing required to complete a project.

bridged floor A floor supported by common joists.

bridge thrust The horizontal thrust from vertical loads on an arched bridge.

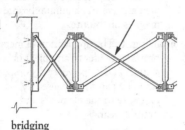
bridged floor

bridging 1. A method of lateral bracing between joists for stiffness, stability, and load distribution. **2.** A delivery method in which an owner selects a designer to develop partial design documents and then selects a design-builder to complete the design and construct the project.

bridging

brine In a refrigeration system, a liquid used as a heat transfer agent that remains a liquid and has a flashpoint above 150°. The liquid is usually a salt solution.

Brinell hardness A measure of the hardness of a metal determined in the laboratory by measuring the indentation made by a steel ball in the surface of the metal.

British thermal unit (Btu) A standard measurement of the heat energy required to raise the temperature of one pound of water one degree Fahrenheit.

brittle A characteristic of a material that makes it fracture easily without bending or deforming.

broach 1. To free stone block from a quarry ledge by cutting out the webbing between holes drilled close together in a row. **2.** To cut wide parallel grooves in a diagonal pattern across a stone surface using the point of a chisel, finishing it for architectural use. **3.** Any pointed structure, such as a steeple or spire, that is built for ornamental purposes. **4.** A spire that rises directly from a tower, often without an intervening parapet. **5.** A half pyramid constructed above the corners of a square tower, serving as an architectural transition from the slat of the tower to an octagonal spire.

broaching 1. A method of quarrying stone in which close holes are drilled around the breakline. A chisel, called a *broach*, is used to break the remaining material. The stone is then removed with wedges. **2.** A method of making shaped holes in metal by removing small pieces in succession with a reaming tool.

broadloom A seamless carpet woven on a wide loom, usually 6′ to 18′ (1.8 to 5.5 meters) wide.

broadscope A term describing the content of a section of the specifications, as established by the Construction Specifications Institute. A broadscope section covers a wide variety of related materials and workmanship requirements. (*Narrowscope* specifications denote a section describing a single material; *mediumscope* denotes a section dealing with a family of materials.)

broad tool A wide steel chisel used in the finish dressing of stone.

broken joints Vertical joints of masonry arranged in a staggered structure with no unit placed directly on top of another. Provides a more solid bond and increases structural strength.

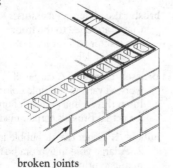
broken joints

B

broken joint tile A roofing tile that overlaps only the tile directly below it.

bronze 1. An alloy composed of copper and tin. **2.** A bronze-colored alloy that uses a sizable measure of copper to alter the properties of its chief elements, as in aluminum bronze or magnesium bronze.

bronzing 1. The application of metal bronze provider to an object or substance. **2.** The powdery decomposition of a paint film caused by exposure to the elements and natural wear.

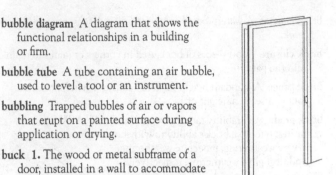

broken joint tile

broom-finish concrete Concrete that has been brushed with a broom when fresh in order to improve its traction or to create a distinctive texture.

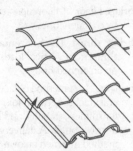

broom-finish concrete

brown coat (floating coat) 1. In two-coat wet-wall construction, the first rough coat of plaster applied as a base coat over lath or masonry. **2.** In three-coat work, the second coat of plaster applied over a scratch coat to serve as a base for the finish coat.

brownfield sites Abandoned or underused industrial sites that often require remediation of hazardous waste or other pollutant contamination prior to reuse.

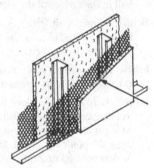

brown coat (floating coat)

brown millerite An oxide of calcium, aluminum, and iron commonly formed in Portland cement and high alumina cement mixtures.

brown-out 1. To apply a base coat of plaster. **2.** The setting process of base-coat plaster, which darkens to a brown hue as it dries. **3.** A partial loss of electrical power that dims lights. A brownout is less severe than a blackout.

brown rot A fungus that attacks the cellulose of wood, so called for the residual brown powder it leaves behind on the destroyed matter.

brush (brushed) finish A finish created by applying a rotating wire brush to a surface.

brush rake A large apparatus that is attached to the front of a tractor to clear land of brush and debris.

brush rake

Brussels carpet 1. A carpet of worsted yarn in multiple colors, attached to a backing of tough linen thread in a pattern of uncut loops. **2.** A less costly substitute for actual Brussels carpet, made in a single color of yarn.

bubble 1. Either the air bubble in a leveling tube or the tube itself. **2.** A large void in gypsum board caused by air entrapment during manufacturing.

bubble diagram A diagram that shows the functional relationships in a building or firm.

bubble tube A tube containing an air bubble, used to level a tool or an instrument.

bubbling Trapped bubbles of air or vapors that erupt on a painted surface during application or drying.

buck 1. The wood or metal subframe of a door, installed in a wall to accommodate the finished frame. **2.** One of a pair of four-legged supporting devices used to hold wood as it is being sawed.

buck

bucket A scoop-shaped attachment for an excavating machine that digs and transports loose earthen materials. Often outfitted with opening and closing mechanisms to facilitate unloading.

bucket trap A mechanical steam trap that operates on buoyancy and is designed with an inverted or upright cup that prevents the passage of steam through the system it protects.

buck frame (core frame, subframe) A wood frame built into the wall studs of a partition to accommodate a door lining.

buckle 1. The distortion of a structural member such as a beam or girder under load. This condition is brought on by lack of uniform texture or by irregular distribution of weight, moisture, or temperature. **2.** A flaw or distortion on the surface of a sheet of material, particularly asphalt roofing. **3.** A thin tree branch bent in the shape of a "U" to fasten thatch onto roofs.

buckling load In a compression member or compression portion of a member, the load at which bending progresses without an increase in the load.

buck opening A rough door opening.

budget A rough estimate of the value of a portion of, or total cost of, a construction project.

budgeted cost of work performed (BCWP) Measure of the amount of money budgeted to complete the actual work performed as of the data date. Represents the value of work performed, rather than the cost of the actual work performed. In current earned value management system usage, it is referred to as "earned value" or EV.*

budgeted cost of work scheduled (BCWS) Measure of the amount of money budgeted to complete the scheduled work as of the data date. In current earned value management system usage, it is referred to as "planned value" or PV.*

buff 1. To clean and polish a surface to a high luster. **2.** To grind and polish a floor of terrazzo or other exposed aggregate concrete.

buffer 1. Blasted rock left at a face to improve fragmentation and reduce scatter during a subsequent blast. **2.** A loose metal mat used to control scattering of blast rock. *See also* **spring buffer**

buff standard brick A buff-colored brick that may be either 2¼" × 3¾" × 7¾" or 2" × 4" × 8", nominal.

buggy (concrete cart) A two- or four-wheeled cart, with or without a motor, used to transport small amounts of concrete from hoppers or mixers to forms.

bug holes Small cavities, usually not exceeding ~5$\frac{1}{8}$" (15 mm) in diameter, at the surface of formed concrete. Bug holes are caused by air bubbles trapped during placing and compacting of wet concrete.

buggy (concrete cart)

builder's jack A temporary bracket, attached to a window-sill, that projects outward and supports scaffolding.

builder's risk insurance A special form of property insurance to cover work under construction.

builder's staging A heavy scaffold made from square timbers, usually used where heavy materials are handled.

building area The sum of the horizontal projected area of all buildings on a site. Terraces and uncovered porches are excluded, unless the stipulations of a mortgage lender or governmental program require their inclusion.

Building Automation System A network of integrated computer components that automatically control a wide range of building operations such as HVAC, security/access control, lighting, energy management, maintenance management, and fire safety control.

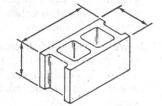

building block Any rectangular masonry unit (except brick) used in building construction. Typical materials are burnt clay, concrete, glass, gypsum, etc.

building block

building board Any board of building material, usually with a facing, which can be used as a finished surface on walls or ceilings.

building brick (common brick) Brick that has not been treated for color or texture. Used as an all-purpose building material.

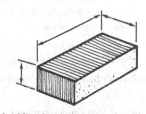

building brick (common brick)

building codes The minimum legal requirements established or adopted by a government such as a municipality. Building codes are established by ordinance, and govern the design and construction of buildings.

building drain That part of the lowest piping of a drainage system that receives the discharge from soil, waste, and other drainage pipes inside the walls of the building and conveys it to the building sewer.

building ecology The environment inside a building, such as air quality, lighting, and acoustics, and its effect on a building's occupants and the overall environment.

building envelope The elements of a building that enclose conditioned areas and through which thermal energy may be transferred to or from the outside environment.

Building Information Modeling (BIM) We use BIM as a verb or an adjective phrase to describe tools, processes and technologies that are facilitated by digital, machine—readable documentation about a building, its performance, its planning, its construction and later its operation. Therefore, BIM describes an activity, not an object. To describe the result of the modeling activity, we use the term *building information model*, or more simply *building model* in full.

Building Model (or Building Object Model) This consists of a digital database of a particular building that contains information about its objects. This may include its geometry (generally defined by parametric rules), its performance, its planning, its construction and later its operation. A Revit® model and a Digital Project® model of a building are examples of building models. *Building model* can be considered the next-generation replacement for *construction drawings* or *architectural drawings*. Downstream in the process, the term *fabrication model* is already in common use as a replacement for *shop drawings*.

Building model repository A building model repository is a database system whose schema is based on a building object based format. It is different from existing project data management (PDM) systems and web-based project management systems in that the PDM systems are file-based, and carry CAD and analysis package project files. Building model repositories are object-based, allowing query, transfer, updating and management of individual project objects from a potentially heterogeneous set of applications. *See also* **BIM Server**.

building inspector An official employed by a municipal building department to review plans and inspect construction to determine if they conform to the requirements of applicable codes and ordinances, and to inspect occupied buildings for violations of the same codes and ordinances.

building integrated photovoltaics (BIPV) Substituting a conventional part of building construction with photovoltaic material. Shingles, standing seam metal roofing, spandrel glass, and overhead skylight glass are some examples.

building line 1. The line, established by law, beyond which a building shall not extend, except as specifically provided by law. Exceptions may be made for terraces or uncovered porches. **2.** The outermost dimension of a building, generally used for building placement in relation to zoning or plot restrictions.

building main The water supply pipe (including fittings) that extends from the water main or other source of supply to the first distribution branch of a building.

building material Any material used in construction, such as sand, brick, lumber, or steel.

building occupancy A classification of the use or intended use of a structure, such as business or residential, for determining hazards, requirements, and restrictions.

building official (building inspector) An appointed government official who is responsible for enforcing building codes. The building official may approve the issuance of a building permit, review the contract documents, inspect the construction, and approve issuance of a certificate of occupancy.

building paper A heavy, asphalt-impregnated paper used as a lining and/or vapor barrier between sheathing and an outside wall covering, or as a lining between rough and finish flooring.

building paper

building permit A written authorization required by ordinance before construction on a specific project can begin. A building permit allows construction to proceed in accordance with construction documents approved by the building official.

building services The utilities, including electricity, gas, steam, telephone, and water, supplied to and used within a building.

building thermal envelope Parts of a building that enclose conditioned spaces, including exterior walls, roof, and floors.

building trades The skilled and semiskilled crafts employed in building construction, such as carpentry, masonry, or plumbing.

built-up 1. Fabricated of two or more pieces or sheets that are laminated. **2.** Assembled by fastening a number of pieces or parts to each other.

built-up air casing A field-fabricated enclosure for an air-handling system, with base, curbs, and drains.

built-up beam 1. A metal beam made of beam shapes, plates, and/or angles that are welded or bolted together. **2.** A concrete beam made of precast units connected through shear connectors. **3.** A timber beam made of smaller pieces that are fastened together.

built-up rib A rib made of laminations of various size timbers.

built-up roofing (composition roofing, felt-and-gravel roofing, gravel roofing) A continuous roof covering made up of various plies or sheets of saturated or coated felts cemented together with asphalt. The felt sheets are topped with a cap sheet or a flood coat of asphalt or pitch, which may have a surfacing of applied gravel or slag.

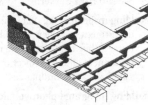

built-up roofing (composition roofing, felt-and-gravel roofing, gravel roofing)

built-up section In metal building systems, a structural member made from individual plates welded together.

bulb angle A hot-rolled angle with a formed bulb on the end of one leg.

bulb bar A rolled steel bar with a formed bulb on one edge.

bulb tee A rolled steel tee with a formed bulb on the edge of the web.

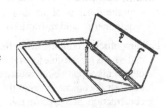

bulb tee

bulkhead 1. A horizontal or inclined door providing outside access to a cellar or shaft. **2.** A partition in concrete forms to separate placings. **3.** A structure on the roof of a building to provide headroom over a stairwell or other opening. **4.** A low structure on a roof covering a shaft or protruding service equipment. **5.** A retaining structure that protects a dredged area from earth movement.

bulkhead

bulking (moisture expansion) An increase in the volume of a quantity of granular material when moist, over the volume of the same quantity when dry.

bulk liquid A pumpable material that has a moisture content greater than 70%.

bulk material Material bought in lots. These items can be purchased from a standard catalog description and are bought in quantity for distribution as required. Examples are pipe (nonspooled), conduit, fittings, and wire.*

bulk solid A material with a moisture content less than 30%.

bulldog grip A U-bolt threaded at each end.

bulldozer 1. A tractor with a large, blunt blade attached to its front end by hydraulic-controlled arms. A bulldozer is used to push, shape, or move earth or rock short distances. **2.** A machine used to bend reinforcing bars into U shapes.

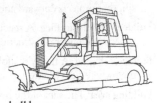

bulldozer

bullet catch A type of door latch consisting of a spring-loaded steel ball. The loaded ball holds the door closed but rolls free when the door is pulled.

bullet-resisting glass (bulletproof glass) A laminated assembly of glass, usually at least four sheets, alternated with transparent resin sheets, all bonded under heat and pressure. Bullet-resisting glass is also made of laminations of special plastics.

bull float A board of wood, aluminum, or magnesium mounted on a pole and used to spread and smooth freshly placed, horizontal concrete surfaces.

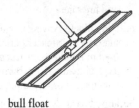

bull header (bull head) A brick with one rounded corner used to form a corner at a doorjamb or laid on edge to form a windowsill.

bull float

bullnose (bull's nose) 1. A rounded outside corner or edge. **2.** A metal bead used in forming a rounded corner on plaster walls.

bullnose plane A small, handheld carpenter's plane with the blade set forward.

bullnose step (bull stretcher) A step, usually the bottom one in a flight, with a semicircular end that projects beyond the railing.

bullnose stretcher 1. A masonry stretcher with one of the corners rounded. **2.** Any stretcher that is laid on edge, as at a windowsill.

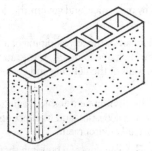

bullnose stretcher

bullnose trim A structural member or piece of trim that has a rounded edge, such as a stair tread, windowsill, or door sill.

bull pin A tapered steel pin used to align holes in steel members so bolts can be inserted.

B

bull-point A pointed steel hand drill that is struck with a hammer to chip off small pieces of rock or other masonry.

bulwark A low wall used as a defensive shield.

bundle Refers to a package of shingles. Typically, 3 bundles make up a square, and there are 27 shingles per bundle.

bundle of lath A quantity of lath for plastering, usually 50 pieces of wood lath, 5/16″ × 1½″ × 48″ (0.79 × 3.8 × 122 cm) or 6 sheets of gypsum lath, 16″ × 48″ (41 by 122 cm).

bundler bars An assembly of up to four parallel reinforcing bars connected to one another and enclosed in stirrups or ties. Used as a unit in reinforced concrete, particularly in columns.

bunker **1.** A compartmentalized storage container for aggregate or ore. **2.** Space in a refrigerator for ice storage and for the cooling element. **3.** A protective shelter usually constructed of reinforced concrete and located partially or fully below ground. **4.** A metal shield in a crushing or screening operation used to direct raw material to the feed belt.

buoyant foundation A reinforced concrete foundation designed and located such that its weight and superimposed permanent load is approximately equal to the weight of displaced soil and/or ground water.

burden **1.** The loose material that overlays bedrock. **2.** The depth of material to be moved or loosened in a blast. **3.** The distance from a blast face to a line of blast holes. **4.** The cost of maintaining an office with staff other than operating personnel. Includes also federal, state, and local taxes; fringe benefits; and other union contract obligations. In manufacturing, burden sometimes denotes overhead.*

burner That part of a boiler or furnace in which combustion takes place.

burner

burning **1.** Flame-cutting metal plates to a desired shape. **2.** A method of repairing lead roofs by replacing damaged sections.

burning-brand test A standard test to determine the resistance of a roof covering to exposure from flying brands.

burnt sienna Sienna that has been calcined to obtain a rich brown pigment. Used for coloring paint and stain.

burnt umber Umber that has been calcined to obtain a red to reddish-brown pigment. Used for coloring paint and stain.

burr **1.** An uneven or jagged edge left on metal by certain cutting tools. **2.** Partially fused brick. **3.** A batch of bricks that were accidentally fused together. **4.** A curly figure in lumber that was cut from an enlarged trunk of certain trees, such as walnut.

bus (bus bar) An electric conductor, often a metal bar, that serves as a common connection for two or more circuits. A bus usually carries a large current.

bus duct (busway) A prefabricated unit containing one or more protected busses.

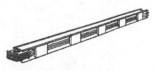
bus duct (busway)

bush hammer A hand or power hammer that has a serrated face with many pyramidal points. A bush hammer is used to dress concrete or stone.

bush hammer

bush-hammered concrete Concrete with an exposed aggregate finish that has been obtained by removing the surface cement using a percussive hammer with a serrated face.

bush-hammer finish A stone or concrete finish obtained through use of a percussive hammer with a serrated face. Used for decorative purposes or to provide a rough surface for better traction or adhesion.

bushing **1.** A threaded or smooth (for soldered tubing) pipe or tube fitting used to connect pipes or tubes of different diameters. **2.** A metal sleeve screwed or fitted into an opening to protect and/or support a shaft, rod, or cable passing through the opening. Usually the inside surface of the bushing is machined to close tolerance to reduce friction and abrasion. **3.** An insulating structure for a conductor, with provision for an insulated mounting.

business agent An official of a trade union who represents the union in negotiations and disputes and checks jobs for compliance with union regulations and union contracts.

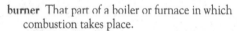

butt **1.** A short length of roofing material. **2.** The thick end of a shingle.

butt

butt casement hinge A type of toe-plate hinge intended for use on casement sashes.

butted frame A door frame with a depth equal to or less than the thickness of the wall in which the frame is mounted.

butt end The thicker end of a handle, tapered pole, pile, or other rod-shaped object.

butt-end treatment The process of protecting that part of a timber or post that will be exposed to soil and/or water by treating it with a preservative chemical.

butter (buttering) **1.** To apply mortar to a masonry unit with a trowel. **2.** To spread roofing cement smoothly with a trowel. **3.** To apply putty or compound sealant to a flat surface of a member before setting the member, such as buttering a stop before installing it.

butterfly hinge A decorative hinge.

butterfly spring A formed piece of spring metal set over the pin of a hinge to serve as a door closer.

butterfly valve A valve that contains a disk that rotates 90° within the valve body. Has excellent throttling characteristics.

butterfly valve

B

butterfly wall tie A masonry wall tie made from heavy wire in the shape of a figure eight.

buttering trowel A small trowel used to apply mortar to a brick before laying the brick.

butt fusion A method of joining thermoplastic resin pipe or sheets in which the ends to be joined are heated to the molten state, pressed together, and held until the material sets.

butt hinge The common form of hinge consisting of two plates, each with one meshing knuckle edge, connected by means of a removable or fixed pin through the knuckles.

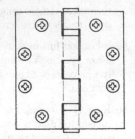

butt joint 1. A square joint between two members at right angles to each other. The contact surface of the outstanding member is cut square and fits flush to the surface of the other member. **2.** A joint in which the ends of two members butt each other so that only tensile or compressive loads are transferred.

butt hinge

buttonhead The head of a bolt, rivet, or screw that is shaped like a segment of a sphere and has a flat bearing surface.

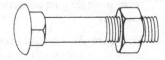

buttonhead

button punching Crimping the interlocking lap of metal ducting panels with a dull punching tool. Button punching has been largely replaced by spot welding overlapping edges.

buttress 1. An exterior pier of masonry construction, often sloped, which is used to strengthen or support a wall or absorb lateral thrusts from roof vaults. **2.** An A-shaped formwork of timber or steel used to strengthen or support a wall.

butt splice A butt joint secured by fastening a short piece of wood or steel to each side of the butted members.

butt weld joint 1. A joint made by welding two butted pieces or sheets together. **2.** A welded pipe joint made with the ends of the two pipes butting each other.

butt weld joint

butyl rubber (synthetic rubber) sealant A type of one-component sealant that may also contain fillers, such as calcium carbonate and talc powder, as well as polybutene and other additives that promote adhesion.

butyl stearate A colorless, oily liquid used for dampproofing concrete.

buy-in A bidder's attempt to win a contract by submitting a price that will result in a loss, with the hope of making the contract profitable through change orders or follow-on contracts.

BX Flexible metal conduit armored with spirals of galvanized steel and containing multiple insulated electric wires.

bypass 1. A pipe or duct used to divert flow around an element. **2.** A pipe used to divert flow around another pipe or piping system.

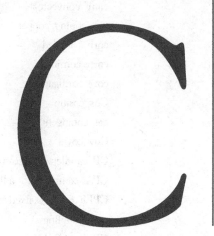

Abbreviations

c candle, cathode, channel, conductivity, cycle

C carbon, Celsius, centigrade, hundred

CAA Clean Air Act

Cab cabinet

CAB cabinet, cement-asbestos board

CAD computer-aided design

CADD computer-aided design and drafting

CAFM computer-aided facility management

Cair air tool laborer

cal calorie

calc calculated

cap capacity

carp carpenter

CAT catalog (catalogue)

CATV community antenna television

CATW catwalk

CB catch basin, center beaded, circuit breaker

CB1S center beam one side

CB2S center beam two sides

CBMA Certified Ballast Manufacturers Association

CBR California bearing ratio

C&Btr grade C and better (used in lumber industry)

cc cubic centimeter

C/C center to center

CCA chromate copper arsenate

CCF hundred cubic feet

CCTV closed-circuit television

CCW counterclockwise

cd candela

CD grade of plywood face and back

cd/sf candela per square foot

CDX plywood, grade C and D, exterior glue

Cefi cement finisher

ceil ceiling

Cem cement

cem fin cement finish

cem m cement mortar

cent central

cer ceramic

CERCLA Comprehensive Environmental Response, Compensation, and Liability Act

CF cubic feet, centrifugal force, cooling fan, cost and freight, hundred feet

CFCs chlorofluorocarbons

CFL compact fluorescent lamp

cfm, CFM cubic feet per minute

CFR Code of Federal Regulations

CFS cubic feet per second

cg center of gravity

CG ceiling grille, center of gravity, coarse gram, corner guard

CG2E center groove two edges

chfr chamfer

chkd checked

CHIM chimney

C

CHRIS Chemical Hazard Response Information System
CHU centigrade heat unit
CHW chilled water
CI cast iron, certificate of insurance
CIF cost, insurance, and freight
CIFE cost, insurance, freight, and exchange
CII Construction Industry Institute
cin bl cinder block
CIP cast-iron pipe, cast in place
CIR circle, circuit, circular
Circ circuit
CIRC circumference
cl closet
CL center line, carload lot
Clab common laborer
CLF hundred linear feet, current-limiting fuse
clg ceiling
clk caulk
CLP cross-linked polyethylene
clr clear
cm centimeter
CM center matched, construction management
CMMS computerized maintenance management software
CMP corrugated metal pipe
CMPA corrupted metal pipe arch
CMU concrete masonry unit
CND conduit
c-o cleanout
Co company
CO certificate of occupancy, change order, cleanout, cutout
CO₂ carbon dioxide
coef coefficient
col column
com common
COMB, Comb combination
comp compartment, compensate, component, composition
COMPF composition floor
COMPR, compr composition roof, compress, compressor
conc, Conc concrete
conc b concrete block
conc clg concrete ceiling
conc fl concrete floor
cond conductivity
const constant, construction
constr construction

Cont continuous, continued
CONTR contractor
conv convector
cop coping, copper
corb corbeled
corn cornice
corr corrugated
Cos cosine
cot cotangent
Cov cover
CP candlepower, cesspool
CPA control point adjustment
CPFF cost plus fixed fee
cplg coupling
CPM critical path method, cycles per minute
CPr hundred pair
CPU central processing unit
CPVC chlorinated polyvinyl chloride
CRC cold-rolled channel
Creos creosote
CRI color rendering index
crib cribbing
CRN cost of reproduction/replacement new
CRP controlled rate of penetration
cr pl chromium plate
Crpt carpet and linoleum layer
CRSI Concrete Reinforcing Steel Institute
CRT cathode ray tube
c/s cycles per second
CS cast stone, carbon steel, commercial standard, caulking seam
CSA Canadian Standards Association (CSA International)
CSC cosecant
CSF hundred square feet
csg casing
CSI Construction Specifications Institute
CSK countersink
ct coat (coats)
CT current transformer
CTB cement-treated base
c to c center to center
ctr center, counter
CTS copper tube size
cu cubic
cu ft cubic feet
cu in cubic inch

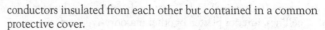

cur current

cu yd cubic yard (27 cubic feet)

CV center vee

CV1S center vee one side

CV2S center vee two sides

cw clockwise, continuous wave

CW cool white, cold water

CW pt cold water point

cwt hundred weight

CWX cool white deluxe

CY cycle, cubic yard (27 cubic feet)

CY/Hr cubic yards per hour

cyl, cyl cylinder

CYL L cylinder lock

cyp cypress

1/C single conductor

2/C two conductors

Definitions

cabana A small shelter of open, tent, or wood-frame construction placed near swimming pools or the shoreline.

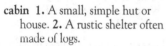

cabin 1. A small, simple hut or house. 2. A rustic shelter often made of logs.

cabin

cabinet 1. A small room or private apartment, especially for study or consultations. 2. A suite of rooms for exhibiting articles and curiosities. 3. A case, box, or piece of furniture with sets of drawers or shelves and doors, used primarily for storage. 4. An enclosure with doors for housing electrical devices and wiring connections.

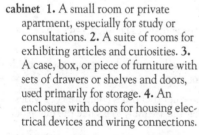

cabinet

cabinet filler The wood piece that closes the space between cabinets and adjacent walls or the ceiling.

cabinet finish A varnished, oiled, or polished wood surface finish, as distinguished from a painted surface finish.

cabinet head A decorative molding along the top of a door frame.

cabinet heater A metal housing enclosing a heating element, with openings to facilitate air flow. The heater frequently contains a fan for controlling the air flow.

cabinet lock A spring bolt or magnetic latch.

cabinet scraper A flat steel blade used for smoothing wood or removing a finish, such as paint or varnish, from a surface.

cabinet work Wood joinery used in the construction of built-in cabinets and shelves.

cable 1. A rope or wire comprised of many smaller fibers or strands wound or twisted together. 2. A group of electric

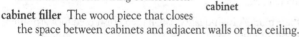

cable

conductors insulated from each other but contained in a common protective cover.

cable assembly A cable and its accompanying connectors to be used for a specific purpose.

cable ceiling heat A radiant heating system, installed in a ceiling, that sends current through encased conductors to produce heat.

cable duct (cable conduit) A rigid, metal, protective enclosure through which electrical conductors are run. For underground installations, concrete or plastic pipes are usually used.

cable length A standard surveying measurement equal to 720′.

cable pulling compound A lubricant used to facilitate the pulling of wires through a cable duct or conduit.

cable roof A system comprised of a roof deck and covering that are supported by cables.

cable sheath The protective cover that surrounds a cable.

cable support box An electrical box that is mounted on a wall and provides support for the weight of cables within a vertically installed conduit.

cable-supported construction A structure that is supported by a system of cables. This system is used for long-span roofs and for suspension bridges.

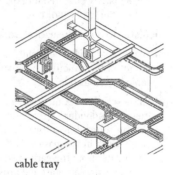

cable tray An open, metal framework used to support electrical conductors. Similar to cable duct except that the cable tray has a lattice-type construction and an open top.

cable tray

cable vault An underground structure utilized in the pulling and joining of underground electric cables.

cadmium plating A coating of cadmium metal applied over a base metal, usually for corrosion protection.

cage 1. The box or enclosed platform of an elevator or lift. 2. An enclosure for electrical lights or signals. 3. Any rigid open box or enclosure.

caisson 1. A drilled, cylindrical foundation shaft used to transfer a load through soft strata to firm strata or bedrock. The shaft is filled with either reinforced or unreinforced concrete. 2. A watertight box or chamber used for construction work below water level.

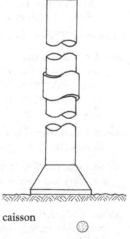

caisson

caisson drill A piece of boring equipment used to excavate a shaft (usually vertical) in the earth for construction of a building footing.

caisson pile A cast-in-place pile formed by driving a hollow tube into the ground and filling it with concrete.

caisson pile

C

calcimine (kalsomine) A low-cost wash, white or colored, used on ceilings, interior plaster, or other masonry-type surfaces.

calcine The process of heating a substance to a point just below its melting temperature in order to remove chemically combined water.

calcined gypsum Gypsum that has been partially dehydrated. In this form it is used as a base material to which mineral aggregate, fiber, or other material may be added to produce the desired plaster.

calcite The main raw material used in the manufacture of Portland cement. Calcite is a crystallized form of calcium carbonate and is the principal component in limestone, chalk, and marble.

calcium chloride A saltwater solution added to concrete and mortar during mixing. Acts as a drying agent and decreases setting time.

calcium silicate brick A concrete product made primarily from sand and lime that is hardened by autoclave curing.

calcium silicate insulation Hydrated calcium silicate that has been molded into rigid shapes and forms. The material is commonly used for pipe insulation where service temperatures reach 1,200°F. Particularly well suited for pipe insulation because it is not appreciably affected by moisture.

calculate To do mathematical work; to seek a numerical value.

calculated live load The useful load-supporting capability above the weight of the structural members, usually specified in the applicable building code.

calculate schedule A modeling process that defines all critical activities and individual activity scheduling data. The process applied in most scheduling software calculates the start and finish dates of activities in two passes. The first pass calculates early start and finish dates from the earliest start date forward. The second pass calculates the late start and finish activities from the latest finish date backwards. The difference between the pairs of start and finish dates for each task is the float or slack time for the task.*

calculation The product or result of mathematical work.

calendar Defined work periods and holidays that determine when project activities may be scheduled. Multiple calendars may be used for different activities, which allows for more accurate modeling of the project work plat (e.g., 5-day workweek calendar vs. 7-day workweek.*

calendar start date The date assigned to the first unit of the defined calendar; the first day of the schedule.*

calf's-tongue molding (calves-tongue molding) A molding with repeating semi-oval shapes, half pointing in the same direction or, when surrounding an arch, pointing to a common center.

caliber The internal diameter of a pipe. In construction, most pipe is specified by outside diameter, with a further designation for wall thickness.

caliche A term used to describe gravel or sand loosely cemented together by calcium carbonate or other salts.

California bearing ratio A ratio used to determine the bearing capacity of a foundation. It is a ratio of the force per unit area per minute required to penetrate a soil mass, to the force required for a similar penetration of a standard crushed rock material.

California brass ring A sampler used to collect soil samples during well installation. Samples are tested for volatile organic compounds.

caliper (calliper) An instrument with two hinged legs used to determine the thickness or diameter of objects.

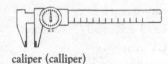
caliper (calliper)

calked rivet A rivet that has not been properly driven, but has been tightened to fit its hole by having the edge of its head driven under with a cold chisel.

calorie The amount of heat required, at a pressure of one atmosphere, to raise the temperature of one gram of water 1°C.

calorific value The amount of heat liberated by the oxidation or combustion of the unit weight of a solid material, or the unit volume of a gas or liquid.

cam 1. An eccentric wheel mounted on a rotating shaft and used to produce reciprocal or variable motion in an engaged or contacted part. **2.** In a lock, the rotating piece attached to the cylinder that actuates the locking mechanism.

camber A slightly convex curvature built into a beam or truss to compensate for deflection under a load. Camber is also built into structural components, roadways, or bridges to facilitate run off of water.

camber arch A flat arch with a very small upward curve of the undersurface. The upper surface may or may not be curved.

camber beam A beam constructed with a slight upward curve in the center.

camber diagram A construction drawing that shows the desired design camber along the beam, truss, or structural component.

camber piece The temporary wood support or template used to lay a slightly curved brick arch.

cambium The thin layer of tissue that lies between a tree's bark and its wood.

came A flexible, cast-lead rod used in stained glass windows to hold together panes or pieces of glass.

camelback truss A truss whose upper chord is comprised of a series of straight segments so that the assembly looks like the hump of a camel's back.

camelhair mop A high-quality, soft-haired brush used for varnishing, gilding, or painting.

cameo window A fixed oval window, often highly decorative.

cam handle (locking handle) On a hinged window, the handle that rotates against its keeper plate, thus pulling the window tight and locking it.

campaniform A term used to describe a structure in the shape of a bell.

Canadian Standards Association A nongovernmental body established in Canada for the purpose of developing uniform standards on products, processes, and procedures.

canal (canalis) 1. Any watercourse or channel. **2.** A channel or groove fluting in carved ornamentation.

candela A unit of luminous intensity; formerly called *candle*.

candlepower The luminous intensity of a light source expressed in *candelas*.

canopy **1.** An overhanging shelter or shade covering. **2.** An ornamental rooflike structure over a pulpit.

canopy

cant **1.** To slope, tilt, or angle from the vertical or horizontal. **2.** A log that has been slabbed on one or more sides.

cantilever **1.** A structural member supported at one end only. **2.** A beam, girder, or supporting member that projects beyond its support at one end.

cantilever footing **1.** A footing with a tie beam connected to another footing to counterbalance an asymmetrical load. **2.** In retaining wall construction, a wide, reinforced footing with reinforcing steel extending into the retaining wall to resist the overturning moment.

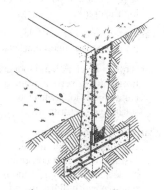

cantilever retaining wall A retaining wall that has a wide footing to resist its overturning moment.

cantilever retaining wall

cantilever truss A truss that is anchored at one end and overhangs a support at the other end.

cantilever wall A wall that resists its overturning moment with a cantilever footing.

canting strip A horizontal ledge near the bottom of an exterior wall sloped to conduct water away from the face of a building or its foundation.

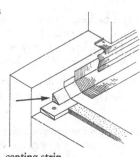
canting strip

cant strip (chamfer strip) A three-sided piece of wood, one angle of which is square, used under the roofing on a flat roof where the horizontal surface abuts a vertical wall or parapet. The sloped transition facilitates roofing and waterproofing. *See also* **chamfer**.

canvas A closely woven cloth, usually made of cotton, used for tarpaulins, awnings, sun covers, and temporary canopies.

canvassing A process used to repair uneven plaster. Performed by adhering canvas strips or other fabric-like material to the surface. Also provides a stable base for paint or wallpaper.

canvas wall A plaster wall with a canvas covering, that may be painted or wallpapered.

cap **1.** The top piece, often overhanging, of any vertical architectural feature or wall. A cap may be external, as on an outside wall or doorway; or it may be internal, as on the top of a column, pilaster, molding, or trim. **2.** A layered system of cover—whether natural soils, rock, synthetics, pavement, and/or polymeric liners—that controls hydrogeologic processes.

3. Decorative molded cornice or projection covering the lintel of a window.

capacitor A device used to introduce capacitance into an electric circuit. The device consists of conducting plates insulated from each other by a layer of dielectric material.

Cape Cod house (cape) A simple frame house with a shingled, steeply pitched roof, central chimney, and one or one-and-a-half floors. The style originated in colonial Cape Cod, Massachusetts, and has evolved over hundreds of years into several popular variations.

Cape Cod house (cape)

cap flashing A section of flashing attached to a vertical surface in order to prevent water from entering behind the base flashing.

capillary action In subsurface soil conditions, the rising of water above the horizontal plane of the water table.

capillary flow The flow of moisture through any form, shape, or porous solid by means of capillary action.

capillary tube **1.** A small-diameter tube used in refrigeration to restrict the flow of refrigerant from the condenser to the evaporator. **2.** The small-diameter tubing used to connect temperature- and pressure-sensing bulbs to a control mechanism.

capillary water Water that, by virtue of capillary action, is either above the surrounding water level or has penetrated where a body of water would not normally go.

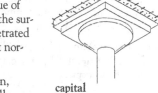

capital The top member of a column, pier, pillar, post, or pilaster; usually decorative.

capital

capital, fixed The total original value of physical facilities which are not carried as a current expenses on the books of account and for which depreciation is allowed by the federal government. It includes plant equipment, building, furniture and fixtures, and transportation equipment used directly in the production of a product or service. It includes all costs incident to getting the property in place and in operating condition, including legal costs, purchased patents, and paid-up licenses. Land, which is not depreciable, is often included. Characteristically, it cannot be converted readily into cash.*

capitalization The total value of a corporation's equity.

capitalize To increase the total value of a corporation's equity through expenditures that increase the evaluation of property, plant, and equipment.

capital project A project in which the cost of the end result or product is capitalized (i.e., cost will be depreciated). The product is usually a physical asset such as property, real estate or infrastructure, but may include other assets that are depreciable.*

capital, working The funds in addition to fixed capital and land investment that a company must contribute the project (excluding start-up expense) to get the project started and meet subsequent obligations as they come due. Working capital includes

C

inventories, cash, and accounts receivable minus accounts payable. Characteristically, these funds can be converted readily into cash. Working capital is normally assumed recovered at the end of the project.*

cap molding (cap trim) The molding or trim above a door or window casing.

capping brick Brick that is used to cap off the top of a wall.

capping piece (cap piece, cap plate) The timber used to cover the top of a series of uprights or other vertical members.

cap plate The plate on top of a column or post that supports the load. *See also* **capping piece**.

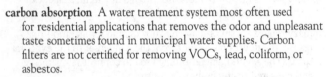
cap plate

cap sheet The top ply of mineral-coated felt sheet used on a built-up roof.

capstone A stone segment in a coping.

carbon absorption A water treatment system most often used for residential applications that removes the odor and unpleasant taste sometimes found in municipal water supplies. Carbon filters are not certified for removing VOCs, lead, coliform, or asbestos.

carbon-arc cutting A metal cutting process in which arc heat melts a path through metal. The arc is established between an electrode, which forms one terminal of an electric circuit, and the workpiece, which forms the other terminal.

carbon-arc welding An arc-welding process in which the heat for fusion is generated by an electric arc between a carbon electrode and the workpiece.

carbonation The process of burning or converting a substance through chemical reaction into a carbonate. For example, the reaction between carbon dioxide and the calcium compounds in cement paste or mortar to produce calcium carbonate.

carbon steel Steel that owes its distinctive properties (primarily its strength) to the carbon it contains, as opposed to the characteristics imparted by other alloying elements.

carcass roofing The roofing framework before the decking, membrane, shingles, etc., have been applied.

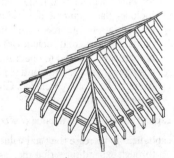
carcass roofing

carcinogen A cancer-causing substance.

cardinal change A series of changes in the work that are so extensive and significant as to change the entire character of the work required by the contract, thereby constituting a breach of contract.

care, custody, and control A feature of construction liability insurance policies that excludes damage to property in the care, custody, or control of the insured.

carillon A set of fixed bells sounded by striking with hammers operated either mechanically or from a keyboard. By extension, a bell tower or campanile.

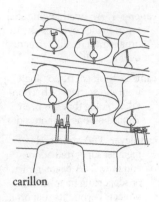
carillon

carpenter Gothic Gothic-style architecture and ornamentation constructed of wood.

carpenter's brace (bit brace) A crank-shaped tool with a chuck on one end to hold a drill bit or auger. A hole is bored by rotating the crank.

carpenter's bracket scaffold A scaffold that supports a platform using wood or metal brackets.

carpenter's level A hand tool used by carpenters to determine a horizontal or vertical plane or line. The level is a wood or metal bar about 2' long with four spirit levels set into it.

carpenter's square (framing square) A flat, metal, L-shaped tool that constitutes an accurate right angle and is engraved with divisions and markings useful to a carpenter laying out and erecting framing.

carpenter's square (framing square)

carpet backing The base material on the back of a carpet. The backing is usually made of jute, cotton, or carpet rayon, and may have a coating of latex.

carpet bedding Beds of low-growing plants, which may be used in ornamental design or as an erosion-preventing ground cover.

carpet cushion (carpet underlayment) A padding made of hair, felt, jute, foam, or sponge rubber that is placed on the floor before a carpet is laid. The cushion helps to provide resilience and extends the life of a carpet.

carpet density The number of pile rows per inch, for the length of the carpet.

carpet float A tool used by plasterers to give texture to a sand finish. The tool consists of a wood float covered with a piece of carpeting. The density of the carpet determines the type of finish.

carpet strip 1. A flat strip of molding used to fasten the edge of carpeting. **2.** A piece of wood or metal, approximately the same thickness as the carpet, installed at the edge of a carpet, as at a threshold.

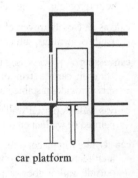
car platform

car platform The structural floor of an elevator car that supports the load.

C

carport A roofed shelter for automobiles, usually attached along the side of a dwelling, with one or more sides open.

carriage The sloping beam that is installed between the stringers to support the steps of a wooden staircase.

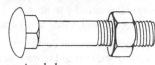

carport

carriage bolt A threaded bolt with a round, smooth head. The bolt is prevented from rotating in its hole by a square neck directly under the head.

carriage clamp A clamp used in carpentry and cabinetry.

carrier angle A metal angle attached to stair carriers or stringers to support the tread or riser.

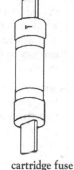

carriage bolt

carrier bar A flat metal bar used to support stair treads or risers in a manner similar to carrier angles.

carrying channel A three-sided metal molding used in the construction of a suspended ceiling.

cartridge fuse A low-voltage fuse consisting of a current-responsive element inside a cylindrical tube with a terminal on each end.

cartridge fuse

cartridge heater An electric heater in the shape of a cylindrical cartridge, with the heating coils at one end and a fan at the other.

carvel joint A flush, longitudinal joint between boards or planks with no tongues, grooves, or laps.

cascade refrigerating system Two or more refrigerating systems connected in series. The cooling coil of one system is used to cool the condenser of the next system. This type of system is an efficient way to produce and maintain ultra-low temperatures.

case 1. A box, sheath, or covering. 2. The process of covering one material with another; to encase. 3. A product or food display counter, such as a refrigerated case for displaying ice cream or frozen produce. 4. A lock housing.

case bay The area of a building floor or roof between two beams or girders.

cased beam An exposed interior beam encased in finished millwork.

cased opening (trimmed opening) An interior doorway or opening with all the trim and molding installed, but without a door or closure.

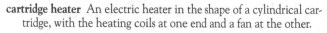

cased opening (trimmed opening)

cased post An exposed interior post or column encased in finished millwork.

case-hardened 1. Term used to describe a steel or iron alloy with a hard surface developed by a special heat-treatment process. 2. Timber whose outer fibers have dried too rapidly, thus causing checking and cracking.

case-hardened glass Glass that has been heated and quenched, thus giving it a hardness and strength several times greater than it originally possessed.

casein paint A paint using the white amorphous phosphoprotein emulsion from milk as a base adhesive and binder.

casement 1. A window sash that opens on hinges that are fixed on either side. 2. A ventilation panel that opens on vertical hinges like a door. 3. A casing.

casement door 1. A door consisting of a wooden frame around glass panels, which make up a major portion of the area of the door. 2. A French door.

casement stay The brace or bar used to hold a casement window open in any of several positions.

casement window

casement window A window assembly that has at least one casement or vertically hinged sash.

case mold A frame support, usually made of plaster, that holds smaller plaster pieces in position in a mold.

casework A term for assembled cabinetry or millwork.

cash allowance A sum included in the contract for construction that covers required items not specifically described. Any differences in cost between the total charges for these items and the cash allowance amount are handled through change orders.

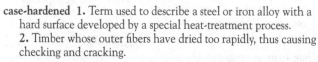

casework

cash flow The amount of cash a company or construction project generates, or fails to generate, over a specific period of time.

cash flow management The planning of project expenditures relative to income or authorized funding in such as way as to minimize the carrying cost of the financing of the project to keep within the constraints of a time-phased budget. This may be achieved by accelerating or delaying some activities, but at the risk of ineffective performance, late completion and consequent increased cost.*

casing 1. The exposed millwork enclosure of cased beams, posts, pipes, etc. 2. The exposed trim molding or lining around doors and windows. 3. The pipe liner of a hole in the ground, such as that used for a well, caisson, or pile.

casing bead A metal molding or wood strip that acts as a stop, an edge, or a separation between different materials.

casing-bead door frame A metal door frame used in plaster walls that includes a strip or molding along its outer edges as a ground or guide for the plaster.

casing knife A knife used to trim wallpaper around casing, molding, or other edges.

casing nail 1. A nail with a very small head. 2. A finishing nail. 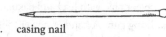 **casing nail**

casting Any object that has been cast in a mold. The object may be made of iron, steel, plaster, concrete, plastic, or any other castable material.

casting plaster A plaster mixture with additives to provide properties desirable for casting work.

cast-in-place concrete (in situ concrete) Concrete poured into forms at its final location.

cast iron An iron alloy cast in sand molds and machined to make many building products such as ornamentation, pipe and pipe fittings, and fencing.

cast molding A molding or molding component that has been cast, removed from the mold, and then fastened in place. The molding may be made of plaster, concrete, plastic, or some similar material.

catalyst 1. A substance that accelerates or retards chemical reactions. **2.** The hardener that accelerates the curing of adhesives, such as synthetic resins.

catch The fitting that mates with a latch or cam to lock a door, gate, or window.

catch basin (catch pit) A receptacle or reservoir that receives surface water runoff or drainage. Typically made of precast concrete, brick, or concrete masonry units, with a cast-iron frame and grate on top.

catch basin (catch pit)

catch drain A long drain installed across or along a slope to collect and convey surface water.

category One of the divisions of the breakdown in construction specifications, smaller than a trade.

catenary arch An arch constructed on the curve of an inverted catenary.

cathead The top piece of a hoist tower to which pulleys or sheaves are attached.

Catherine-wheel window 1. A round window with mullions like wheel spokes. **2.** A rose window.

Catherine-wheel window

cathodic protection A form of protection against electrolytic corrosion of fuel tanks and water pipes submerged in water or embedded in earth. Protection is obtained by providing a sacrificial anode that will corrode in lieu of the structural component, or by introducing a counteracting current into the water or soil. This type of protection is also used for aluminum swimming pools, cast-iron water mains, and metal storage tanks.

catslide 1. The term used in the southern United States for a *saltbox* home. **2.** The long pitched roof at the rear of a saltbox, which often sweeps down to the ground.

catwalk A small, permanent walkway, usually elevated, to provide access to a work area or service/maintenance access to lighting units, stage draperies, etc.

caulk (calk) 1. To fill a joint, crack, or opening with a sealer material. **2.** The filling of joints in bell-and-spigot pipe with lead and oakum.

caulking gun A hand or power tool that extrudes caulking material through a nozzle.

caulking gun

caustic An alkaline material that has a corrosive or irritating effect on living tissue.

caustic lime A material, white when pure, that is obtained by calcining limestone, shells, or other forms of calcium carbonate. Caustic lime is also called *quicklime* or *burnt lime* and is used in mortars and cements.

cavity The empty space between studs and joists.

cavity fill Material placed in the hollow of a wall, ceiling, or floor to provide insulation or sound deadening.

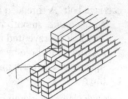

cavity fill

cavity flashing The continuous waterproof barrier that covers the longitudinal opening in a cavity wall.

cavity wall (hollow masonry wall, hollow wall) An exterior masonry wall in which the inner and outer wythes are separated by an air space, but tied together with wires or metal stays.

cavity wall (hollow masonry wall, hollow wall)

C/B ratio saturation coefficient The ratio of the weight of water absorbed by a masonry brick or block immersed in cold water to the weight absorbed when immersed in boiling water. This is a measure of the resistance of the masonry to spalling due to freezing and thawing.

C-clamp A C-shaped clamp frequently used in carpentry and joinery. Force is applied by rotating a threaded shaft through one jaw of the C to force the work against the other jaw.

C-clamp

cedar A softwood with reddish heartwood and white sapwood noted for decay resistance and used for shingles, shakes, posts, and gutters.

ceiling 1. The overhead inside lining or finish of a room or area. **2.** Any overhanging surface viewed from below.

ceiling cornice The molding installed at the intersection of the wall and ceiling planes.

ceiling diffuser Any air diffuser, located in a ceiling, through which warm or cold air is blown into an enclosure. The diffuser is designed to distribute the conditioned air over a given area.

ceiling height The distance between a room's finished ceiling and floor.

ceiling outlet In a ceiling, a metal or plastic junction box that supports a light fixture and encloses the connecting fixture wires.

ceiling plenum In air-conditioning systems, the air space between a hung ceiling and the underside of the floor or roof above. Acts as a return to the air-handling unit.

ceiling sound transmission The transmission of sound between adjacent rooms through the ceiling plenum above.

ceiling strap Wood strips nailed to the bottom of roof rafters or floor joists for the support or suspension of a ceiling.

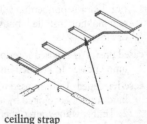

ceiling strap

ceiling strut A temporary brace suspended from an overhead structure or framing and used to hold door frames in place while the adjacent walls are being constructed.

ceiling suspension system A grid-work of metal rails and hangers erected for the support of a suspended ceiling and ceiling-mounted items, e.g., air diffusers, lights, fire detectors, etc.

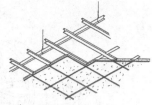

ceiling suspension system

cellar A room or set of rooms below or predominantly below grade, usually under a building.

cellular concrete (aerated concrete) Concrete that has had gas-forming chemicals mixed with the basic ingredients, so that the final set material is lighter than ordinary concrete, due to its porosity. This type of concrete is often used as insulation when it is placed over a roof slab.

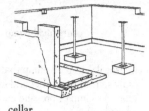

cellar

cellular-core door A hollow-core door filled with a lightweight, honeycomb-shaped, expanded material to provide rigidity and support.

cellular floor A flooring system that incorporates prefabricated raceways for telecommunications and electrical wiring.

cellular framing A method of framing in which the walls are composed of cells, with the cross walls transmitting the bearing loads to the foundation.

cellular plastic Plastic molding or extrusion with a pattern of cells that give it rigidity and reduce its weight.

cellular raceway A channel in a modular floor or wall that may be used as a raceway for electrical conductors.

celluloid A thermoplastic material often used in thin sheets because of its good molding properties.

cellulose A naturally occurring substance made up of glucose units. Constitutes the main ingredient in wood, hemp, and cotton, and is used in many construction products.

cellulose fiber tile A sound-absorbing acoustical tile made from cellulose fiber.

Celsius scale A thermometer in which the interval between the freezing point and boiling point of water is divided into one hundred parts or degrees. Named after Aners Celsius, a Swedish astronomer. Also called a *centigrade thermometer*.

Celtic cross A cross formed from a long vertical column, a shorter horizontal bar, and a circle about their intersection.

Celtic cross

cement Any chemical binder that makes bodies adhere to it or to each other, such as glue, paste, or Portland cement.

cement-aggregate ratio The ratio of cement to aggregate in a mixture, as determined by weight or volume.

cementation The firming and hardening of a cementitious material.

cement board (cement backer board) A water-resistant wallboard made of concrete faced with fiberglass on both sides. Commonly used as an underlayment in wet areas like bathtub surrounds and as a backer for ceramic tile. Brand names include Durock® and Wonder Board®.

cement block An inaccurate term, usually in reference to concrete block or to a hollow or solid-cast masonry unit.

cement-coated nail A nail coated with cement to increase its friction and holding power.

cement content (cement factor) The quantity of cement per unit volume of concrete or mortar, expressed in pounds or bags per cubic yard.

cement gravel Gravel bound together in nature by clay or some other natural binding agent.

cement grout 1. A thin, watery mortar or plaster that is pumped or forced into joints, cracks, and spaces as an adhesive sealer. 2. A mixture that is pumped into the soil around a foundation to firm it up and provide better load-bearing characteristics.

cement gun A tool used for spraying or injecting cement grout or mortar material.

cement gun

cementitious Capable of setting like a cement.

cement mixer 1. A concrete mixer. 2. A container used to mix concrete ingredients by means of paddles or a rotary motion. The container may be manually or power operated.

cement mortar A plastic building material made by mixing lime, cement, sand, and water. Cement mortar is used to bind masonry blocks together or to plaster over masonry.

cement mixer

cement paint (concrete paint) Paint containing white Portland cement and usually applied over masonry surfaces as a waterproofing.

cement paste A mixture of cement and water.

cement plaster Plaster containing Portland cement as the binder and commonly used on exterior surfaces or in damp areas.

cement, Portland A fine gray powder made from three main ingredients: silica, alumina and lime. May also include slag or flue dust. Used as a base material for mortar and concrete.

cement rock (cement stone) A natural, impure limestone that contains the ingredients for the manufacture of Portland cement.

cement stucco 1. A mixture of Portland cement, sand, and a small percentage of lime. Used to form a hard covering for textured exterior walls. 2. A fine plaster used for interior decorations and moldings.

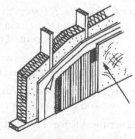

cement stucco

cement temper The addition of Portland cement to lime plaster to improve its strength and durability.

center 1. A point that is exactly halfway between two other points or surfaces. **2.** A point equidistant from all points on the circumference of a circle. **3.** The internal core of a built-up construction.

center bit A cutter used for boring holes in wood. The cutting end consists of a sharp, threaded screw for guidance and for pulling the tool deeper into the work piece. The bit also has a scorer for marking the outline of the hole and a lip for cutting away the wood inside the hole. The other end of the bit is usually a tapered rectangular block that is clamped in the chuck of a brace.

center-hung door (center-pivoted door) A door that is supported by, and swings about, pivot pins. The pins are inserted into or attached to the door on the center line of its thickness.

center-hung sash A window sash hung so that it rotates horizontally about hinge pins at its center. To open, the top part of the window swings into the room and the bottom part projects outside the building.

centering 1. The temporary support for a masonry arch while the arch is being built. **2.** The temporary forms for all supported concrete work.

center line A line of alternating long and short segments that indicates the center of a particular object. Frequently labeled with the letter C superimposed over the letter L.

center of gravity (center of mass) The location of a point in a body or shape about which all the parts of the body balance each other.

center pivot 1. The pins of a window sash that are centrally located in the frame so that the window rotates horizontally about its center line. **2.** The pivot pins of a door that support the door on the center line of its thickness.

center punch A tool for making starter indentations in metal where a hole is to be drilled. The punch has a sharp conical point on one end and is hit by a hammer on the other end.

center rail The horizontal portion or member that separates the upper and lower sections of a recessed panel door.

center stringer The structural member of a stair system that supports the treads and risers at the midpoint of the treads. **center-to-center (on center)** The distance from the center line of one beam, part, or component of a structure to the corresponding center line of the next similar unit (for example, center-to-center of rafters or joists).

centimeter A metric unit of length equal to one one-hundredth of a meter. One inch equals 2.54 cm.

central-mixed concrete Concrete that is mixed in a stationary mixer and then transported to the site, as opposed to *transit-mixed concrete*.

central system A system that utilizes a single source and distribution system for an area. Examples are central heating and/or air-conditioning systems, central power systems, and central water systems.

centrifugal chiller A gas compressor used for chilling in which the compression is obtained by centrifugal force.

centrifugal compressor An air or gas compressor in which the compression is obtained by centrifugal force.

centrifugal force The force away from the center of a rapidly spinning impeller.

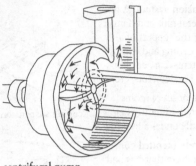

centrifugal pump

centrifugal pump A pump that draws fluid into the center of a rapidly spinning impeller and uses the force of the spin (centrifugal force) to impart pressure and velocity to the fluid as it leaves the periphery.

ceramic Items made of clay or similar materials that are baked in a kiln to a permanent hardness. Ceramic products include pottery, tile, and stoneware.

ceramic glaze A material that is applied over a clay product and then baked to produce a hard, vitreous, glassy surface. The glaze may be transparent or colored.

certificate A written document appropriately signed by responsible parties testifying to a matter of fact in accordance with a requirement of the contract documents.

certificate of insurance A document prepared by an insurance company or its agent that states the period of time for which the policy is in effect, along with the types and amounts of coverage for the insured.

certificate of occupancy A written document issued by the governing authority in accordance with the provisions of the building permit. The certificate of occupancy indicates that, in the opinion of the building official, the project has been completed in accordance with the building code. This document gives the owner permission from the authorities to occupy and use the premises for the intended purpose.

certificate of payment A statement issued by a design professional notifying the owner of the amount due to a contractor for work completed and/or the delivery or storage of building materials.

certified construction specifier A construction professional who, by experience and examination by the Construction Specifications Institute, has been certified as proficient in the knowledge and art of preparing technical specifications for the construction process.

certified forest product A product created from materials obtained from forests that have met specific environmental guidelines, and that are certified for use in green construction.

certified wood Wood from well-managed forests that replenish, rather than deplete, old growth timber.

chain 1. A unit of length used by land surveyors (*Gunter's Chain*): 7.92" = 1 link; 100 links = 66' = 4 rods = 1 chain; 80 chains = 1 mile.  chain **2.** An engineer's chain is equal to 100'. **3.** A series of rings or links of metal connected to each other to form useful lengths.

chain door fastener A chain that can be secured by a slide bolt between a door stile and a doorjamb to allow the door to be opened slightly while remaining securely fastened.

chain fall An assembly of chain with blocks or pulleys for hoisting or pulling a heavy load.

chaining pin A metal pin used for marking measurements on the ground.

chain intermittent fillet weld An interrupted fillet weld on both sides of a joint. The weld segments are opposite one another.

chain link fence

chain link fence A fence made of steel wire fabric supported by metal posts.

chain molding A molding carved to simulate the appearance of a chain.

chain-pipe wrench (chain tongs) A wrench for turning or restraining pipe. The wrench consists of a lever handle with a serrated jaw. The jaw is held in contact with the pipe by a roller chain that wraps around the pipe.

chain riveting A type of riveting in which the rivets are placed in parallel adjacent rows. Chain riveting is not as strong as the staggered riveting pattern, which has more tear material between rivets.

chain saw A gas- or electric-powered saw having cutting teeth running around a bar of flat steel. Used for cross-cutting logs, etc.

chain timber Large beams built horizontally into a masonry wall to provide support and a tie-in during construction.

chain trencher A machine that uses a chain belt to excavate a trench.

chair 1. A frame built into a wall to provide support for a sink, lavatory, urinal, or toilet. Also called a *carrier*. 2. In concrete construction, a small metal or plastic support for reinforcing steel. The support is used to maintain proper positioning during concrete placement.

chair rail A horizontal piece of wood attached to walls to prevent damage to plaster or paneling from the backs of chairs.

chalk A soft limestone, white, gray, or buff-colored, composed of the remains of small marine organisms. In construction, chalk is commonly dyed blue and used with a chalk line or snap line for marking lines.

chalk line 1. A light cord that has been surface-coated with chalk (usually blue) and is used for marking. 2. The line left by a chalked string.

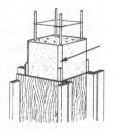
chamfer

chamfer The beveled edge formed at the right-angle corner of a construction member.

chamfer bit A tool bit for beveling the upper edge of a hole in wood so that a flat-head wood screw can be driven flush with the wood surface.

chamfer plane A carpenter's plane designed for cutting chamfers.

chancel arch An arch employed in some church designs to indicate the separation of the church body (*nave*) from the chancel or sanctuary.

chandelier A branched light fixture that is often ornate, and usually suspended from a ceiling.

change In construction, a deviation in the design or scope of the work as defined in the plans and specifications that are the basis for the original contract.

change, constructive An act or failure to act by the owner or the engineer that is not a directed change, but which has the effect of requiring the contractor to accomplish work different from that required by the existing contract documents.*

change documentation/log Records of changes proposed, accepted, and rejected.*

change in scope A change in the defined deliverables or resources used to provide them.*

change in sequence A change in the order of work initially specified or planned by the contractor. If this change is ordered by the owner and results in additional cost to the contractor, the contractor may be entitled to recovery under the changes clause.*

change management The formal process through which changes to the project plan are identified, assessed, reviewed, approved, and introduced.*

change order Written authorization provided to a contractor approving a change from the original plans, specifications, or other contract documents, as well as a change in the cost. With the proper signatures, a change order is considered a legal document.

changes in the work Changes to the original design, specifications, or scope of work as requested by the owner. All changes in the work should be documented on change orders.

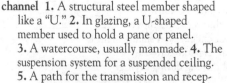
channel

channel 1. A structural steel member shaped like a "U." 2. In glazing, a U-shaped member used to hold a pane or panel. 3. A watercourse, usually manmade. 4. The suspension system for a suspended ceiling. 5. A path for the transmission and reception of electromagnetic signals between two or more points.

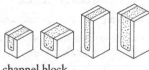
channel block

channel block A hollow concrete masonry block with a groove or furrow in it to provide a continuous recess for reinforcing steel and grout.

channel cutter A hand tool used to cut light-gauge metal framing.

channel glazing A window glazing system in which glass panels are set in a U-shaped channel using removable beads or glazing stops.

channeling Decorative fluting or grooving, as in the grooves in a vertical column.

channel iron (channel bar) A U-shaped, rolled, structural steel member, the web of which is always deeper than the width of the flanges. The web depth is the size of the channel.

channel runner A primary horizontal support in a suspended ceiling system.

charcoal filter **1.** A filter containing pulverized charcoal for removing odors, smoke, and vapors from the air. **2.** A filter for purifying and cleaning water.

charette **1.** The push or effort to complete an architectural problem within a specified time. **2.** A meeting held early in the design phase of a project during which the design team, contractors, end users, community stakeholders, and technical experts are brought together to develop goals, strategies, and ideas for maximizing the environmental performance of the project.

charge **1.** A quantity of explosives set in place. **2.** A load of refrigerant in a refrigeration or air-conditioning system.

charging door The door of a furnace or incinerator through which loads or charges are inserted.

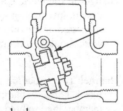

charging door

chase A continuous enclosure in a structure that acts as a housing for pipe, wiring conduits, ducts, etc. A chase is usually located in or adjacent to a column, which provides some physical protection.

chattel Any item of property, movable or immovable, except real estate or a freehold.

chattel mortgage A mortgage or security interest in personal property as collateral for a loan.

cheapener An additive in paint that is used to impart desirable characteristics to or extend coverage of the paint. Cheapener does not necessarily make the paint cheaper in cost.

check **1.** A flapper or valve that permits fluid to flow in one direction only. **2.** An attachment that regulates movement, such as a door check. **3.** A small split in wood that runs parallel to the grain. Caused by shrinking during drying.

check

check lock A small, supplementary lock designed to control the bolt or action of a larger lock.

check rail The horizontal meeting rail of a double-hung window.

check stop A molding strip or bead used to restrain a sliding unit such as a window sash.

check strip A bead or molding used for parting.

check throat The groove cut longitudinally along the underside of window and door sills to prevent rainwater from running back to the building wall.

check valve (back-pressure valve, clack valve, reflux valve) A valve designed to limit the flow of fluid to one direction only.

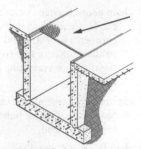

checkered plate A metal plate with a waffle-like pattern of squared projections cast or rolled into its surface.

checkered plate

cheek In general, the side of any feature, such as the side of a dormer.

cheek

cheek boards **1.** The vertical stops or bulkheads placed at the ends of a concrete wall form. **2.** Trim applied to the side of a dormer.

cheek cut (side cut) The oblique cut made in a rafter or jack rafter to permit a tight fit against a hip or valley rafter or to square the lower end of a jack rafter.

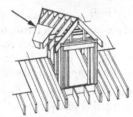

cheek cut (side cut)

chemical bond A bond between similar materials caused by the interlocking of their crystalline structures. Once formed, there is no defined parting line in the bond that would permit a clean separation.

chemical closet (chemical toilet) A toilet that contains or recirculates a chemical that breaks wastes down while serving as a disinfectant and deodorant. It does not use water to flush and carry away sanitary waste to a public septic system.

chemical flux cutting The use of a chemical compound or flux to expedite oxygen torch cutting.

Chemical Hazard Response Information System (CHRIS) A series of monographs prepared for the U.S. Coast Guard, each of which provides detailed information on the hazards of a particular chemical.

chemically foamed plastic Spongy or cellular plastic material formed by the gas-producing action of its component chemical compounds.

cherry picker A powered lift for raising workers or materials. Consists of a basket or a lift attached to the end of a telescoping boom, and may be either self-propelled or mounted on a truck.

chiding A method of gilding.

chilled-water refrigeration system A cooling system in which cold water is circulated to remote cooling coils/blower units.

chiller A piece of equipment that utilizes a refrigeration cycle to produce cold (chilled) water for circulation to the desired location or use.

chilling The process of applying refrigeration in moderation so as to achieve cooling without freezing.

chimney A vertical, noncombustible structure with one or more flues to carry smoke and other gases of combustion into the atmosphere.

chimney

chimney apron The metal flashing built into the chimney and roof at the point of penetration. Serves as a moisture barrier.

chimney arch The upper structure over a fireplace.

chimney back The inside back wall of a fireplace, which may be made of firebrick masonry or ornamental metal, and is intended to reflect heat out into the room.

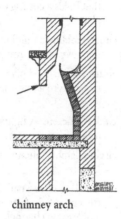

chimney arch

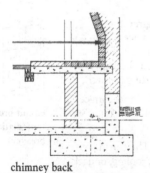

chimney back

chimney block Solid concrete blocks cast as segments of a circle for laying up or building round flues.

chimney bond A pattern of stretcher bond used in laying up the internal brickwork in masonry chimneys.

chimney breast (chimney piece) That part of the front of a fireplace that projects out into the room.

chimney can (chimney pot) A cylindrical extension placed on top of a chimney to improve the draft. It can be masonry or metal.

chimney cap (bonnet, chimney head) 1. The slab or masonry top piece on a chimney. 2. A metal top piece designed to minimize rainfall down the flue while improving the draft.

chimney cap (bonnet, chimney head)

chimney connector The smokepipe or metal breeching that connects the top of a stove, furnace, water heater, or boiler to a chimney flue.

chimney crane A swinging iron arm attached to the inside of a fireplace to support cooking pots over the fire.

chimney cricket (saddle) A small false roof built behind a chimney on the main roof to divert rainwater away from the chimney.

chimney effect (flue effect, stack effect) The process by which air, when heated, becomes less dense and rises. The rising gases in a chimney create a draft that draws in cooler gases or air from below. Stairwells, elevator shafts, and chases in a building often draw in cold air from lower floors or outside through this same process.

chimney flue The vertical passageway in a chimney through which the hot gases flow. A chimney may contain one or several flues. Flues are typically lined with fired clay pipes to resist corrosion and facilitate cleaning.

chimney gutter The flashing placed around a chimney for waterproofing the roof penetration.

chimney hood A covering top or cap that blocks rain from entering the flue.

chimney jamb The vertical sides of a fireplace opening.

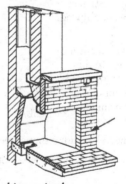

chimney jamb

chimney lining The noncombustible, heat-resisting material that lines the flue inside the chimney. Typically, the lining consists of round or rectangular fired clay material.

chimney throat (chimney waist) 1. The narrowest part of a flue. 2. The area above a fireplace and below the smoke shelf, where the damper is usually located.

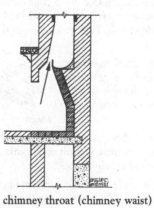

chimney throat (chimney waist)

chimney wing The sides of a fireplace chimney above the opening that close into the throat of the damper opening.

Chinese lacquer (Japanese lacquer, lacquer) A natural varnish, originating in China or Japan, and extracted from the sap of a sumac tree. This lacquer forms a high-quality surface.

chink **1.** A small, elongated cleft, rent, or fissure, such as the openings between logs in a log cabin wall. **2.** To install insulation around gaps in exterior walls and door and window frames.

chinking Any material used to fill in voids, such as those in stone walls or between the timbers of a log cabin.

chip **1.** A small piece of semiconducting substrate material, such as silicon, on which miniaturized electronic circuitry (LSI or VLSI circuits) is fabricated. **2.** A small fragment of metal, wood, stone, or marble that has been chopped, cut, or broken from its parent piece.

chipboard A flat panel manufactured to various thicknesses by bonding flakes of wood with a binder. Chipboard is an economical, strong material used for sheathing, subflooring, and cabinetry.

chip cracks (eggshelling) Similar to alligatoring or checking except that the edges of the cracks are raised up from the base surface.

chipper **1.** A small pneumatic or electric reciprocating tool for cutting or breaking concrete. **2.** Gasoline- or diesel-powered machine featuring a system of blades used to shred or chip logs or brush.

chisel A metal tool with a cutting edge at one end, used in dressing, shaping, or working wood, stone, or metal. A chisel is usually tapped with a hammer or mallet.

chisel

chisel bar A heavy metal bar with a sharp edge at one end.

chisel knife A sharp, narrow, square-edged knife used to remove paint or wallpaper.

chlorinated rubber A rubber-based product used in paints, plastics, and adhesives.

chlorinated solvent A material that dissolves greases and oils. Contains one or more chlorine atoms in its chemical structure.

chlorine A highly corrosive elemental gas.

chlorofluorocarbons (CFCs) Compounds consisting of chlorine, fluorine, and carbon atoms that are degraded only by the sun's radiation in the stratosphere, where released chlorine may contribute to ozone depletion. They can persist in the troposphere for 100 years or longer.

chord **1.** The top or bottom members of a truss (typically horizontal), as distinguished from the web members. **2.** A straight line between two points on a curve.

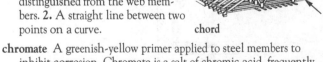

chord

chromate A greenish-yellow primer applied to steel members to inhibit corrosion. Chromate is a salt of chromic acid, frequently referred to as lead or zinc chromate.

chrome steel Carbon steel to which chromium has been added (in proportions of 0.5% to 2.0%) to increase the steel's hardness and yield strength.

chromium A grayish-white metallic element added to steel to enhance its strength and hardness. It is also used as a plating medium because of the rust-resisting surface that results. Chromium plating produces a highly polished finish.

chromium plating The application, by electroplating, of a thin coat of chromium to provide a hard, protective wearing surface or a surface that can take a decorative high polish.

chronic toxic chemical A substance that causes adverse health effects that may be delayed in their appearance, but usually cause permanent damage.

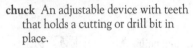

chuck An adjustable device with teeth that holds a cutting or drill bit in place.

chuck

churn molding A molding with a repeating wavy or chevron pattern.

chute An inclined plane or trough for sliding various items or bulk materials to a lower level, with the help of gravity.

cinder concrete A lightweight concrete that uses cinders for coarse aggregate.

circline lamp A fluorescent lamp with a circular tube.

circuit **1.** A closed path followed by an electric current. **2.** A piping loop for a liquid or gas. *See also* **channel.**

circuit breaker (automatic circuit breaker) An electrical device for discontinuing current flow during an abnormal condition. Unlike a fuse, a circuit breaker becomes reusable by resetting a switch.

circular arch An arch whose inner surface, and occasionally a portion of its outer surface, describe a circle.

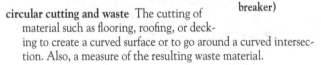

circuit breaker (automatic circuit breaker)

circular cutting and waste The cutting of material such as flooring, roofing, or decking to create a curved surface or to go around a curved intersection. Also, a measure of the resulting waste material.

circular mil An electrical term for the cross-sectional area of a wire. A conductor with a diameter of one mil (0.001″) has a cross-sectional area of one circular mil.

circular mil-foot A unit of measure for an electrical conductor that is one circular mil in diameter and 1′ long.

circular saw A circular steel blade fitted with cutting teeth and mounted on an arbor.

circular spike A metal timber connector with teeth set in a ring. The teeth grip the wood as a central axial bolt is tightened.

circular stair (spiral stair) A cylindrical stairway in which the staircase is built in the shape of a corkscrew.

circular stair (spiral stair)

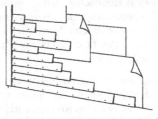

circulating head (circulating pressure) The pump-induced pressure in a piping system, such as one for hydronic heat, chilled water, or domestic hot water.

circulation **1.** The flow of a liquid or a gas within a closed series of pipes. **2.** The flow of air (fresh, heated, or cooled) in a building. **3.** The flow of people through a building. **4.** Any pipe forming a part of a liquid or gas circuit.

cistern A catch basin on a roof or an enlargement at the end of a gutter for catching and storing rainwater.

city planning (town planning, community planning) The conscious control of growth or change in a city, town, or community, taking into account aesthetics, industry, utilities, transportation, and many other factors that affect the quality of life.

civil drawings Site drawings, such as utility, grading and drainage, site improvements, and landscaping plans. Civil drawings encompass all work other than the structure itself.

civil engineer An engineer specializing in the design of public works, such as roads, buildings, dams, bridges, and other structures, as well as water distribution, drainage, and sanitary sewer systems.

C-label A door certified by Underwriters' Laboratories as meeting their class C level requirements.

cladding A covering or sheathing applied to provide desirable surface properties, such as durability, weathering, and corrosion or impact resistance.

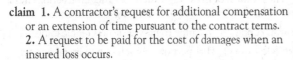
cladding

claim **1.** A contractor's request for additional compensation or an extension of time pursuant to the contract terms. **2.** A request to be paid for the cost of damages when an insured loss occurs.

claims examiner Relative to insurance claims, the supervisor who oversees the paperwork submitted by the field adjuster.

clamp A mechanical device used to hold items together or firmly in place while other operations are being performed. The clamping force may be applied by screws, wedges, cams, or a pneumatic/hydraulic piston.

clamping screw A screw used with wood or metal jaws to provide a clamping force for temporary or permanent fastening.

clamping screw

clamping time The amount of time a bonded joint must remain immobile in a clamp in order for the glue to set.

clamshell **1.** A bucket used on a derrick or crane for handling loose granular material. The bucket's two halves are hinged at the top, thus resembling a clamshell. **2.** A wood molding with a cross section that resembles a clamshell.

clapboard (bevel siding, lap siding) A wood siding used as exterior covering in frame construction. It is applied horizontally, overlapped with the grain running lengthwise. The thickest section of the board is on the bottom.

clapboard (bevel siding, lap siding)

clapper valve A swing check valve that permits fluid flow in one direction only.

clarification drawing An illustration provided by an architect or engineer to explain in more detail some area or item on the contract documents, or as a part of a job change order or modification.

clapper valve

Clarke beam A composite beam consisting of planks bolted with their flat sides together and then strengthened with a board nailed across the joint.

class **1.** A group of items ranked together on the basis of common characteristics or requirements. **2.** A division grouping or distinction based on grade or quality. **3.** The opening into which a door or window will be fitted.

Class Ia flammable liquid A liquid having a flash point below 73°F (22.8°C) and a boiling point below 100°F (37.8°C).

Class Ib flammable liquid A liquid having a flash point below 73°F (22.8°C) and a boiling point at or above 100°F (37.8°C).

Class Ic flammable liquid A liquid having a flash point at or above 73°F (22.8°C) and below 100°F (37.8°C).

Class II combustible liquid A liquid having a flash point at or above 100°F (37.8°C).

Class IIIa combustible liquid A liquid having a flash point at or above 140°F (60°C) and below 200°F (93.4°C).

Class IIIb combustible liquid A liquid having a flash point at or above 200°F (93.4°C).

Class A, B, C, D, E Fire-resistance ratings applied to building components such as doors or windows. Class A is an Underwriters' Laboratories classification for a component having a 3-hour fire endurance rating; Class B, a 1 or 1½-hour rating; Class C, a ¾-hour rating; Class D, a 1½-hour rating; and Class E, a ¾-hour rating.

Class A fire A fire involving ordinary combustibles, such as wood, paper, some plastics and textiles.

Class B fire A fire involving oil, gas, paint, or other flammable liquids.

Class C fire A fire involving live electrical equipment that requires the use of electrically nonconductive extinguishing agents.

Class D fire A fire involving combustible metals such as magnesium, titanium, and sodium that require an extinguishing medium that does not react with the burning metal.

classical architecture Characteristic of or pertaining to the styles, types, and modes of structural construction and decoration practiced by the Greeks and Romans.

classicism An architectural style derived from the basic principles and design of Greco-Roman or Italian Renaissance buildings.

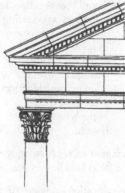

classical architecture

clause A subdivision of a paragraph or subparagraph within a legal document such as a contract. A clause is usually numbered or lettered for easy reference.

claw bar A steel bar that is straight at one end and bent at the other. The straight end has a chisel point, while the bent end has a notch for pulling out nails. A claw bar is used for general demolition work.

claw plate A round timber connector that has raised prongs to bite into the wood.

clay A fine-grained material, consisting mainly of hydrated silicates of aluminum, that is soft and cohesive when moist, but becomes hard when baked or fired. Clay is used to make bricks, tiles, pipe, earthenware, etc.

clay content In any mixture of soil or earth, the percentage of clay by weight.

clay-mortar mix A mixture consisting of pulverized clay that is added to masonry mortar to act as a plasticizer.

clay pipe Pipe made of earthenware and glazed to eliminate porosity. Clay pipe is used for drainage systems and sanitary sewers.

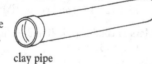

clay pipe

clay spade A wide, flat, chisel-like cutting tool used in a pneumatic hammer to cut into tightly compacted materials like clays.

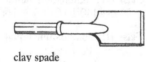

clay spade

clay tile A fired earthenware tile used on roofs. Called *quarry tile* when used for flooring.

Clayton Act Legislation enacted by Congress in 1914 to lessen the negative effects of the Sherman Anti-Trust Act by allowing labor to organize for purposes of negotiating with a single employer.

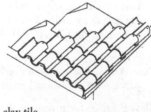

clay tile

clean aggregate Sand or gravel aggregate that is free of clay, dirt, or organic material.

clean energy Energy obtained from a renewable source that produces zero or low emissions and has minimal environmental impact.

cleanout 1. A pipe fitting with a removable threaded plug that permits access for inspection and cleaning of the run. **2.** A door in the base of a chimney that permits access for cleaning. **3.** A small door in a ventilation duct to permit access for removal of grease, dust, and dirt blockages.

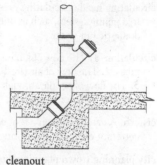

cleanout

cleanout door The door in a cleanout frame. *See also* **cleanout**.

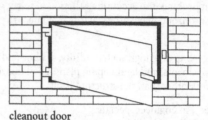

cleanout door

clean power Power that is relatively free of electrical noise and harmonic distortion.

clean room A special-purpose room that meets requirements for the absence of lint, dust, or other particulate matter. In a clean room, the filter systems are high efficiency and the air exchange is one-directional laminar flow.

clearance (clearage) 1. The distance by which one item is separated, or clear, from another; the empty space between them. **2.** An intentional gap left to allow for minor dimensional variations of a component or a part.

clear glaze A transparent glaze for ceramic applications.

clearing The removal of trees, vegetation, or other obstructions from an area of land. Also referred to as clearing and grubbing.

clearing arm A branch on a pipe to provide access for cleaning.

clearstory (clerestory) That part of a church or building that rises above the roofs of the other parts of the building, with windows for admitting light to the central interior area.

cleat A small block of wood nailed to the surface of a wood member to stop or support another member.

cleat wiring Electrical wiring that is exposed and supported on standoff insulators.

cleavage A natural parting or splitting plane found in slate, mica, and some types of wood.

cleavage plane In crystalline material, the plane along which splitting can easily be performed.

cleft timber A wooden beam that has been split to the desired size or shape.

clench bolt (clinch bolt) A bolt designed to have one end bent over to retain it in place.

clench nail (clinch nail) A nail made to have the protruding point bent over after being driven.

clerk of the works A representative of the architect or owner who oversees construction, handles administrative matters, and ensures that construction is in accordance with the contract documents.

clevis A U-shaped bar of metal with holes drilled through the ends to receive a pin or bolt. A type of shackle.

client 1. Party to a contract who commissions the work. On capital projects, may also be referred to as the "owner." 2. Customer, principal, owner, promoter, buyer, or end user of the product or service created by the project.*

clevis

climbing crane A hoisting device with a vertical portion that is attached to the structure of a building so that as the construction is built higher and higher, the crane rises along with it.

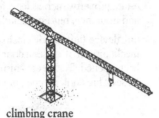
climbing crane

climbing form Vertical concrete formwork that is successively raised after each pour has hardened. The formwork is anchored to the concrete below. *See also* **slip form**.

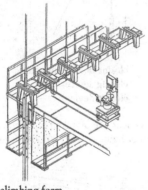
climbing form

clink 1. The sealed edge between sheets of metal roofing. 2. A pointed steel bar used to break up concrete.

clinker brick A hard brick that is frequently formed with a distorted shape and is used for paving.

clinometer An instrument for measuring angles of elevation or inclination.

clip 1. To cut off, as with shears. 2. A short piece of brick that has been cut. 3. A small metal fastening device.

clip joint A thicker-than-usual mortar joint used to raise the height of a masonry course.

clipped gable A gable roof with a shortened ridge pole creating a short hip roof at the gable.

clipped header (false header) In masonry, a half-brick placed in a wall to look like a header.

clipped lintel A lintel connected to another structural member to help carry the load.

cloister 1. A monastic establishment such as a monastery or convent. 2. A covered passageway on the side of an open courtyard, usually having one side walled and the other an open arcade or colonnade.

cloistered arch (cloistered vault) A structure whose sides have the appearance of four quarter rounds fitted together so that the corner joints appear as an "X" when viewed from above.

cloister garth An open courtyard surrounded by a cloister.

close 1. An enclosed place, such as the area around a cathedral. 2. A passage from a street to a court and the houses facing it. 3. The common stairway of an apartment building.

close-boarded (close-sheeted) 1. Sheathing with boards fitted tight against one another. 2. Privacy fencing in which adjacent boards are fitted tight against one another.

close-coupled tank and bowl A toilet in which the bowl and water tank are manufactured separately and then bolted tightly together with only a gasket at the joint.

close-coupled tank and bowl

close-cut The trimming of a roof covering (shingle, tile, or slate) where adjacent surfaces of a hip or valley meet.

closed-cell A type of foamed material in which each cell is totally enclosed and separate so that the bulk of the material will not soak up liquid like a sponge. Closed-cell material is characteristic of certain types or insulation board and flexible glazing gaskets.

closed-circuit grouting A system of injection that fills all voids to be grouted at each hole and returns excess grout to the grouting apparatus.

closed-circuit television A television circuit with no broadcasting facilities and a limited number of reception stations. Transmission is typically between two points. Can be integrated into a Building Automation System and used for building security purposes.

closed competitive selection (closed bidding) A process of competitive bidding whereby the private owner limits the list of bidders to persons he has selected and invited to bid. *See also* **closed list of bidders**.

closed list of bidders A list of those contractors who have been approved by the architect and owner as the only ones from whom bid prices will be accepted. *See also* **closed competitive selection**.

closed loop control An instrument that measures the changes in a controlled variable (temperature, pressure) and actuates the control device (valve, damper) to bring about a change.

closed loop system In Building Automation Systems, the arrangement of heating, ventilating, and air-conditioning components so that each component affects and can respond to the other, allowing system feedback.

closed shaft A vertical opening in a building that is covered over or roofed at the top.

closed sheathing (closed sheeting) A continuous framework with boards or planks placed side by side to support or retain the sides of an excavation.

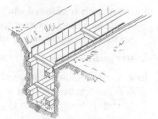

closed sheathing (closed sheeting)

closed shop A term applied to a trade or skill that requires membership in a particular union to the exclusion of nonunion members.

closed specifications Specifications that require certain trade name products or proprietary processes. Provisions for alternatives are not included.

closed stair (closed stair string, closed string) A stairway with covered strings on both sides so that the ends of the treads and risers are not visible.

closed system Any heating or cooling system in which the circulating medium is not expended, but is recirculated through the system.

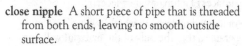
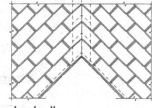

closed valley

closed valley A roof covering in a roof valley laid so that the flashing is not visible.

close-grained Wood in which the annual growth rings are close together, thus indicating slow growth.

close nipple A short piece of pipe that is threaded from both ends, leaving no smooth outside surface.

close nipple

closeout The completion of project work. The phase at the end of a project life cycle just before the operation begins.*

closer 1. A mechanism for automatically closing a door. 2. The final masonry unit (brick, stone, or block) that completes a horizontal course. The unit may be whole or trimmed to size.

closer

closer reinforcement A metal plate attached to a door and/or a frame to reinforce the area where a door closer is installed.

closer reinforcing sleeve A plate used to reinforce the corner joints of a metal door frame.

close studding A method of construction in which studs are spaced close together and the intervening spaces are filled with plaster.

closet 1. A small room or recess for storing utensils or clothing. 2. A water closet.

closet bend The 90° soil pipe fitting installed beneath a toilet.

closet bolt A special bolt for securing a floor-mounted toilet to the floor and soil pipe flange.

close tolerance A term describing smaller allowable dimension deviation than would normally be allowed.

closet pole (closet rod) A strong rod mounted horizontally in a closet to support coat hangers.

closet stop valve The valve underneath a tank-type water closet. The valve is used to shut the water off so that repairs can be made to the flushing mechanism inside the tank.

closing costs The costs associated with the sale/purchase of real estate property, such as legal fees, recording fees, and title search and insurance, but not including the cost of the property itself.

closing device (automatic closing device, self-closing device) 1. A mechanism that closes a door automatically and slowly after it has been opened. 2. A mechanism, usually fused, that releases and closes a fire door, damper, or shutter in the event of a fire.

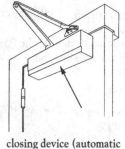

closing device (automatic closing device, self-closing device)

closing stile The vertical side of a door or casement sash that seats against the jamb or frame when the door or window is closed.

closing stile

closure bar A flat metal bar that is attached to a stair string next to a wall. The bar bridges or covers the opening between the string and the wall.

clout 1. A metal sheathing or protection plate on a moving wood member to protect it from wear. 2. A type of nail.

clustered column Two or more columns, joined together at the top and/or bottom, which share the structural load equally.

clustered pier A pier consisting of several shafts grouped together as a single unit, frequently surrounding a heavier central member.

cluster housing A planned residential development with dwelling units built close to one another and common open space for use by all residents.

clutch A friction or hydraulic coupling between a power source and a drive train.

coach bolt A bolt with a smooth, circular head. Rotation of the bolt in its hole during tightening is prevented by the square shoulder or neck under the head.

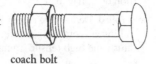
coach bolt

coal-tar pitch (tar) A black bituminous material made from the distillation of coal and used as a waterproofing material on built-up roofing and around elements that protrude through a roof.

coarse aggregate Aggregate that will not pass through a ~$1/4$″ sieve screen.

coarse filter A prefilter used in air-conditioning systems to remove large particles such as dust and lint. This type of filter is usually reusable after being vacuumed or washed.

coarse-grained 1. A term used to describe wood in which the annual growth rings are widely separated, thus indicating fast growth. *See also* **coarse-textured**. 2. The "pebbly" effect that sometimes occurs on a photographic enlargement.

coarse-textured (coarse-grained) A term referring to a porous, open-grained wood that usually requires priming and filling to produce a smooth surface.

coat A single covering layer of any type of material, such as paint, plaster, stain, or surface sealant.

coated base sheet (coated base felt) The underlying sheet of asphalt-impregnated felt used in built-up roofing.

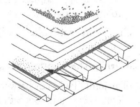

coated glass Insulating glass that has been treated to help retain building heat. Coated glass also blocks ultraviolet and infrared light.

coated base sheet (coated base felt)

coated nail A nail with a coating on it. The coating may be cement, copper, zinc, or enamel.

coat rack A piece of furniture that supports several coats (with or without hangers) and that may include a stand for umbrellas and a tray for galoshes.

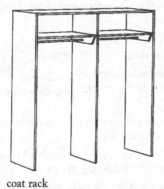
coat rack

coaxial cable A cable consisting of a tube of conducting material surrounding a central conductor. The tube is separated from

contact with the cable by insulation. Coaxial cable is used to transmit telephone, telegraph, television, and computer signals.

cock 1. A valve having a hole in a tapered plug. The plug is rotated to provide a passageway for fluid. 2. Any device that controls the flow of liquid or gas through a pipe. *See also* **sill cock** *and* **bibcock**.

cock bead A bead that projects higher than the adjacent surface.

cocking (cogging) The fitting together of notched timbers, beams, or pieces of wood to form a joint.

code The legal requirements of local and other governing bodies concerning construction and occupancy. The enactment and enforcement of codes is intended to safeguard the public's health, safety, and welfare.

Code of Federal Regulations A uniform system of listing all the regulations promulgated by any federal agency.

code review drawings Drawings that portray compliance to applicable codes and standards that can be developed by or through the contractor and are supplementary to a written scope of work (also called *incidental drawings*).

coefficient of conductivity The measure of the rate at which a material will conduct thermal energy, or heat.

coefficient of expansion For building materials, the increase in dimension per unit of length caused by a rise in temperature of one degree Fahrenheit, expressed in decimal form.

coefficient of friction A ratio of the force causing an object to slide to the total force perpendicular to the sliding plane. The ratio is a measure of the resistance to sliding.

coefficient of performance 1. In a heat pump, the ratio of heat produced to energy expended. 2. In a refrigeration system, the ratio of heat removed to energy expended.

coefficient of static friction The coefficient of friction required to make an object start sliding.

coefficient of subgrade reaction The ratio of load per unit area on soil to the corresponding deformation.

coffer (lacunar) 1. An ornamental recessed panel in the ceiling of a vault or dome. 2. A cofferdam; a caisson.

cofferdam A watertight enclosure used for foundation construction in waterfront areas. Water is pumped out of the cofferdam allowing free access to the work area.

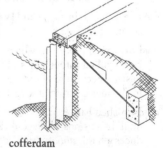
cofferdam

coffering A ceiling with ornamental recessed panels.

cog 1. In a cogged joint, the portion of wood left where a beam has been notched. 2. A nib on roofing tiles.

cogeneration The simultaneous production of power and another form of useful thermal energy in a single process. A common example is the use of combustion or co-combustion processes to produce both electricity and steam or hot water.

cohesion 1. In soils, the sticking together of particles. 2. In adhesives or sealants, the ability to stick together without cracking.

cohesionless soil A sod, such as clean sand, with no cohesive properties. cohesive failure The internal shearing of a joint sealant

that is weaker cohesively than the adhesion bond to the adjacent joint surfaces. The possible causes for failure include insufficient curing, excessive joint movement, improper joint dimensions, or improper selection of the sealant itself.

cohesive soil Soil that tends to hold together in a comparatively stable clump or mass, such as clay.

coil A term applied to a heat exchanger that uses connected pipes or tubing in rows, layers, or windings, as in steam heating, water heating, and refrigeration condensers and evaporators.

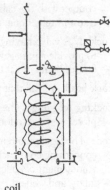
coil

coiled expansion loop A gradual bend, loop, or coil of pipe or tube intended to absorb expansion and contraction caused by temperature changes. By flexing like a spring, undesirable stresses on the piping system are relieved or minimized.

coinsurance penalty The penalty against a policyholder for not carrying enough insurance on his or her property. In these cases, the full cost of the claim will not be paid by the insurance company. Repair contractors should be aware that the policyholder will have to pay a substantial part of the repair bill.

cold-air return In a heating system, the return air duct that transports cool air back into the system to be heated.

cold-cathode lamp An electric-discharge lamp that operates with low current, high voltage, and low temperature. The color of the light depends on the vapor used in the lamp.

cold cellar An underground storage area where fruit and vegetables are stored during the winter. The constant ground temperature keeps the contents from freezing and retards rotting.

cold check Fine cracks that appear in some wood finishes, such as paint and varnish, after being subjected to alternating warm and cold temperatures.

cold chisel A steel chisel with a tempered edge for cutting cold steel.

cold deck (cold duct) In an HVAC system, the source of cold air for cooling interior zones.

cold-drawn A process in which metal is formed in shape and size by being pulled through dies while cold. This type of metal forming has an effect on final surface finish and hardness and is often used in the manufacture of wire, tubing, and rods.

cold-finished bar A metal bar that has been drawn or rolled to its final dimensions while cold. The process of cold finishing produces a hard, smooth, surface finish.

cold flow The continuing permanent change in slope of a structural system under constant load.

cold joint The joint between two consecutive pours of concrete if the time elapsed between the first and second pours is such that the first pour has started to set.

cold-laid mixture Any mixture that may spread at ambient temperatures without being preheated.

cold mix (cold patch) An asphaltic concrete made with slow-curing asphalts and used primarily as a temporary patching material when hot mix plants are closed. Cold mix is used to repair potholes, but is less durable than hot mix.

cold molding 1. Material that is shaped at ambient temperature and hardened by baking. 2. The material that is used in the cold-molding process.

cold-process roofing Built-up roofing consisting of layers of asphalt-impregnated felts that are bonded and sealed with a cold application of asphalt-based roof cement.

cold-rolled Descriptive of a metal shape that has been formed by rolling at room temperature, thus giving a dense, smooth, surface finish and high tensile strength.

cold room A walk-in refrigerator for storage of perishable materials.

cold saw A saw for cutting metal at room temperature. A cold saw may be either circular or reciprocating.

cold set A hardened hand tool used to flatten sheet metal seams.

cold-setting adhesive A glue that sets at ambient room temperature.

cold solder A soft-metal, air-hardening material used for bonding two metals at ambient temperatures. Cold solder, unlike a true solder, does not require a hot bond.

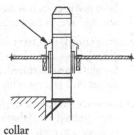

collar 1. A flashing for a metal vent or chimney where it passes through a roof. 2. A trim piece to cover the hole where a vent goes through a wall or ceiling. 3. A metal band that encircles a metal or wooden shaft.

collar

collar beam The horizontal board that joins the approximate midpoints of two opposite rafters in order to increase rigidity.

collar beam roof (collar roof) A roof with rafters connected by collar beams.

collaring The pointing of masonry or tile joints under overhangs.

collar joint The joint between the collar beam and roof rafters.

collar tie A horizontal tie connecting pairs of opposite rafters near the ridge of the roof.

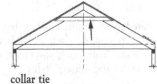
collar tie

collusion A secret agreement or action of a fraudulent, illegal, or deceitful nature.

colonial architecture 1. An architectural style from a mother country that has been incorporated into the buildings of settlements or colonies in distant locations. 2. Eighteenth-century English Georgian architecture as reproduced in the original colonies of the United States.

colonial casing A style of door and window trim molding.

color chart A standardized card or folder showing a display of colors or finishes.

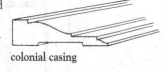
colonial casing

colored aggregate Aggregate material that has been selected for use because of its natural color.

colored concrete Concrete mixed with colored cement or to which pigment has been added during mixing.

color rendering index (CRI) A measure of the quality of a light, where the ability of the light to render an object's true color(s) is

compared to that of a reference light. Under a lamp with a low CRI rating, colors will appear unnatural.

color temperature The actual degree measurement, in Kelvin, of the color of a light source.

column A long, relatively slender, supporting pillar. A column is usually loaded axially in compression.

column

column baseplate A bearing plate beneath a column that distributes the load coming down through the column over a wider area.

column baseplate

column clamp A latching device for holding the sections of a concrete column form together while the concrete is being placed.

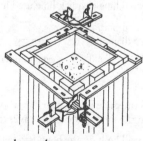

column clamp

column footing The foundation under a column that spreads the load out to an area large enough so that the bearing capacity of the soil is not exceeded and differential settling does not occur.

column schedule A production plan outlining the construction of columns in a structure. Provides column sizes, elevations (top and bottom), and type and size of reinforcement required.

column ties Steel reinforcement surrounding the vertical reinforcement in a concrete column to align the bars and provide additional strength. May be rectangular or ring-shaped.

comb 1. A tool used to produce a rough surface or pattern in plaster or concrete. 2. A comb board.

comb board A board with notches on its upper edge, used to cover the joint at the ridge of a pitched roof.

combed Descriptive of a surface that has been raked with a comb to produce a rough or patterned finish.

combination door An exterior door that has interchangeable panels of glazing and screening—one for summer and the other for winter use.

combination door

combination faucet A faucet that is connected to both hot- and cold-water supply pipes so that the water it dispenses can be of the desired temperature.

combination fixture 1. A combination kitchen and laundry sink in a single unit, with both deep and shallow bowls side by side. 2. An institutional or prison fixture that combines the functions of a sink and toilet into a single fixture.

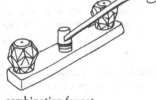

combination faucet

combination plane A carpenter's or joiner's hand tool that has interchangeable blades or guides so that it can be used for various shaping operations.

combination waste and vent system A plumbing system with an oversized waste pipe that also serves as a vent pipe.

combination window A window with replaceable screen and glass inserts for summer and winter use.

combined aggregate A mixture of both fine and coarse aggregate used in concrete.

combined footing A footing that receives loading from more than one column or load-supporting element.

combined frame A doorway with a fixed panel of glass on either or both sides of the door.

combined load The total of all the loads (live, dead, wind, snow, etc.) on a structure.

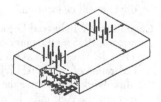

combined footing

combined sewer A sewer that receives both storm water and sewage.

C

combing **1.** The act of dressing a stone surface. **2.** The use of a serrated comb or brush to create a surface pattern or roughness in paint, plaster, or concrete. **3.** Those shingles on a roof that project above the ridge line. **4.** The top ridge on a roof.

combplate The fixed floor plate at either end of an escalator or moving walkway that has fingers that project into the serrations of the moving segments to prevent an object from wedging into the seam.

combustible Descriptive of a substance that will burn in the air, pressure, and temperature conditions of a burning building.

combustion A rapid chemical process, such as oxidation, that produces significant thermal energy (heat), and usually light as well.

combustion air Fresh, outside air brought in via ductwork to the furnace and/or hot water heater.

combustion chamber The section within a woodstove, boiler, or furnace where the burn takes place, typically lined with firebrick or insulation.

comfort chart A chart showing the limits of human comfort for various combinations of temperature, relative humidity, and circulating air velocity.

commercial projected window A type of steel window used in commercial and industrial applications where decorative trims and moldings are not used. A projected window has one or more panels that rotate either inward or outward on a horizontal axis.

commercial tolerances Allowable variations in the dimensions of items mass-produced for the commercial market.

comminution To reduce to minute particles, as in the grinding of coal or the screening of sewage.

commissioning The process of ensuring that the complex array of equipment that provides lighting, heating, cooling, ventilation and other amenities in facilities works together effectively and efficiently. Commissioning typically begins during the facility's conceptual design phase and ideally continues throughout the life of the facility.

commissioning agent A member of the project team who ensures the proper installation and operation of technical building systems. Normally, the commissioning agent is hired directly by the owner and works independently of both the designer and contractor.

committed cost A cost which has not yet been paid, but an agreement, such as a purchase order or contract, has been made that the cost will be incurred.*

Committee for Industrial Organizations (CIO) A labor union organized in 1935 for the purpose of representing industrial workers. The CIO was created as a result of a dispute with the American Federation of Labor (AFL). John L. Lewis, president of United Mine Workers and a member of the CIO, was instrumental in merging the AFL and CIO in 1955.

common area maintenance (CAM) charge A provision in a commercial lease that requires the tenant to pay a proportionate share of the common area operation costs.

common bond (American bond) A brick masonry wall pattern in which a header course of brick is laid after every five or six stretcher courses.

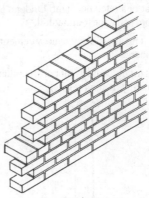

common bond (American bond)

common brick Brick not selected for color or texture, but instead used as filler or backing. Though usually not less durable or of lower quality than face brick, common brick typically costs less. Greater dimensional variations are also permitted.

common ground A strip of wood recessed in a masonry or concrete wall to provide a surface to which something else can be nailed.

common joist (bridging joist) One of a series of parallel beams laid on edge to support a floor or ceiling.

common joist (bridging joist)

common lime Hydrated lime or quick lime used in plaster or mortar.

common nail A general-purpose, headed nail with a diamond-shaped point used where appearance is not important, as in framing.

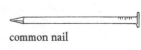

common nail

common rafter (intermediate rafter, spar) A wood framing member that reaches from the rafter plate at the eaves to the ridge board.

common wall **1.** A wall built on the boundary between separate lots of land so that it does not belong completely to either owner. **2.** A party wall between dwellings in a condominium or row house.

community antenna television (CATV) A single source of television signals transmitted to multiple receivers. Used in buildings that have television cables. Can be integrated in the building's communications system.

compacted volume The volume of any mass of material after it has been compressed, particularly soil in an embankment or fill area.

compact fluorescent lamp (CFL) Efficient lightbulbs that use far less energy than standard, incandescent bulbs. CFLs typically provide 8,000 hours of usage, compared to less than 2,000 for incandescent bulbs.

compact fluorescent lamp (CFL)

compaction 1. The compression of any material into a smaller volume; for example, waste compaction. In specifying compaction of embankment or fill, a percent compaction at optimum moisture content is often used. Another method is to specify the equipment, height of lift, and number of passes. **2.** The elimination of voids in construction materials, as in concrete, plaster, or soil, by vibration, tamping, rolling, or some other method or combination of methods.

compaction pile A pile driven into the ground to compact soil and increase its bearing capacity.

compactor A machine that compresses or compacts materials by using hydraulic force, weight, or vibration.

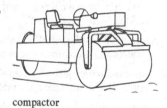

compactor

company Term used primarily to refer to a business first party, the purpose of which is to supply a product or service. In a capital project, typically refers to the contractor who is performing services for an owner or client.*

compartmentation The containment of smoke within a building by using a system of barriers and compartments.

compass 1. An instrument for locating the magnetic north pole of the earth. **2.** A mechanical drawing device used to draw circles.

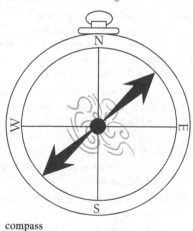

compass

compass brick A curved brick used in circular patterns or to form an arch.

compass plane A carpenter's plane with a curved baseplate and/or blade for finishing curved wood pieces.

compass rafter A rafter with one or both sides curved.

compass roof A roof built out of curved rafters.

compass saw (keyhole saw) A small tapering handsaw used to cut rounded shapes or circles.

compass survey A surveyor's traverse using bearings determined by a compass, as opposed to bearings determined by recording the angles turned by the instrument.

compass work (circular work) Carpentry or joinery based on a circular or curved motif.

compatibility agent A substance that, when added to another substance, improves the stability and even distribution of that substance. For example, compatibility agents allow the simultaneous application of a pesticide and fertilizer, or two or more pesticides.

compensation 1. The amount paid for goods, labor, or services provided. **2.** An amount paid to settle a claim for damages.

compensatory damages An award for damages intended to compensate the plaintiff by granting a monetary amount equal to the loss or injury suffered.

completed operations An insurance term referring to coverage for personal injuries or damage to property that occurs after construction has been completed. At least one of the following conditions must be met: all activities governed by the contract are finished; all activities at one project site are finished; or that part of the project that precipitated the claim is in use by the owner. This insurance does not cover the actual finished project itself.

complete fusion In welding, a total merging of the welded surfaces, including all of the weld filler material.

completion bond (construction bond, contract bond) The guarantee by a bonding company that a contract will be completed and will be clear of all liens and encumbrances.

completion date The date certified by the architect when the work in a building, or a designated portion thereof, is sufficiently complete, in accordance with the contract documents, so that the owner can occupy the work or a designated portion thereof for the use for which it is intended.

composite arch An arch formed from three or four centers of curvature.

composite beam A beam combining different materials to work as a single unit, such as structural steel and concrete or cast-in-place and precast concrete.

composite beam

composite column A column designed to combine two different materials or two different grades of material to form a structural member. A structural steel shape may be filled with concrete, or a structural steel member for reinforcing may be encased in concrete.

composite construction A building constructed of various building materials, or by using more than one construction method, such as a structural steel frame with a precast roof, or a masonry structure with a laminated wood beam roof.

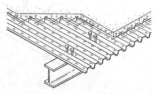

composite construction

composite door A door with a wooden or metal shell over a lightweight core. The core is usually made of foam or a material that is shaped like a honeycomb cell.

composite fire door A door made of metal or chemically treated wood surrounding a core material. The door is rated as resistant to fire for some specified period of time.

composite girder A girder assembled from component parts of similar or different materials; for example, a plate girder or a bar joist with wooden flanges.

composite joint **1.** A joint that is secured by using more than one method, such as riveting and welding. **2.** In plumbing, a general reference to joining bell-and-spigot pipe with packing materials such as rope and rosin, or cement and hemp.

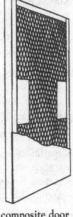

composite door

composite pile A pile consisting of more than one member or material, but designed to act as one unit.

composite truss A load-supporting framework of several members. Those members in compression are made of wood; the tension members are usually made of steel.

composition board (composite board) A manufactured board consisting of any of several materials, usually pressed together with a binder. Composition board is frequently used as sheathing, wall board, or as an insulation or acoustical barrier.

composition roofing (built-up roofing) A roof consisting of several layers, thicknesses, or pieces.

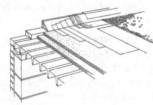

composition roofing (built-up roofing)

composition shingles (asphalt shingles) Shingles made from felt impregnated and covered with asphalt, and then coated on the exposed side with colored granules.

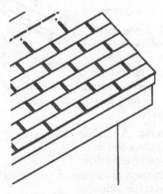

composition shingles (asphalt shingles)

compound beam (built-up beam) A beam consisting of smaller components that have been assembled and fastened together to function as a single unit.

compound pier (compound pillar) A column usually consisting of one central shaft surrounded by secondary shafts, or by two or more equal shafts working in conjunction with one another.

compound wall A wall constructed of more than one material.

Comprehensive Environmental Response, Compensation, and Liability Act (CERCLA) A federal law defining the government actions in the event of an uncontrolled release of hazardous materials into the environment.

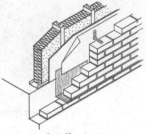

compound wall

comprehensive general liability insurance A type of insurance that provides blanket coverage for all types of liability with the exception of those specifically excluded. It covers contractual liability as well as unforeseen future hazards, and may also include automobile coverage. Completed operations liability, more extensive contractual coverage, and product liability insurance can be added to this type of policy.

compressed fiberboard Wood fiber that has been compressed into sheets and is commonly used as a paneling material.

compressed straw slab A board manufactured from straw compressed with a binder and cured.

compressibility The ease with which a fluid is reduced in volume by the application of pressure.

compression Structurally, it is the force that pushes together or crushes, as opposed to tension, which is the force that pulls apart.

compression bearing joint A joint consisting of a structural member in compression, pressing or bearing against another member also in compression.

compression chiller A system in which water is cooled by liquid refrigerant that vaporizes at a low pressure and proceeds into a compressor. The compressor increases the gas pressure so that it may be condensed in the condenser.

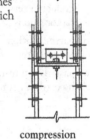

compression bearing joint

compression joint **1.** A bell-and-spigot joint that becomes watertight when the spigot is pressed into a compression gasket in the bell. **2.** Any joint that is engineered to employ pressure as a seal.

compression, schedule When an excessive amount of noncritical tasks are delayed until they become critical tasks and create chaos near the end of a schedule or phase.

compression tank In a hot water heating system, a tank that absorbs the water pressure exerted.

compression test **1.** A test of any material by compression to determine its useful strength characteristics. The test is most often performed in accordance with standards published by the American Society for Testing and Materials or some other professional or industrial organization. **2.** A determination of the compression developed in each cylinder of an internal combustion engine.

compressive stress The resistance of a material to an external pushing or shortening force.

compressor 1. A machine that compresses air or gases. **2.** In refrigeration/air-conditioning, a machine that compresses a refrigerant gas, which then goes to an evaporator. **3.** A truck- or trailer-mounted apparatus, usually diesel-driven, that generates compressed air for demolition hammers, sandblasting, and similar needs on a construction site.

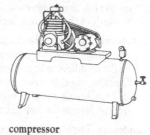

compressor

computer-aided design (computer-aided design and drafting) The use of computer technology to analyze, design, or model a building or system. Allows for greater efficiency and consistency, faster reproduction, and easier alteration than traditional methods.

computerized maintenance management software (CMMS) The use of computer technology to schedule plant and equipment maintenance, track assets, and efficiently carry out overall facility management. The purpose of CMMS is to simplify the planning and administrative functions of maintenance, purchasing, and inventory management.

concave joint A masonry joint formed by a curved pointing tool and used particularly on exterior masonry walls.

concave joint

concealed hinge A hinge mortised into both the jamb and edge of a door so that it is hidden when the door is closed.

concealed location As defined for the purpose of building codes, any area of a building that cannot be reached without doing permanent damage to other building components.

concept documents A series of drawings and other definitive documents that illustrate a design professional's concept for a project.

concourse An open area for the circulation of large crowds within a building, as in an airport terminal or shopping mall.

concrete A composite material consisting of sand, coarse aggregate (gravel, stone, or slag), cement, and water. When mixed and allowed to harden, it forms a stonelike material.

concrete

concrete admixture A special substance or chemical added to a concrete mix. Typically, an admixture is used to control setting, entrain air, impart color, control workability, or to waterproof.

concrete aggregate Granular mineral material that is mixed with cement and water to form concrete.

concrete block A masonry building unit of concrete that has been cast into a standard shape, size, and style.

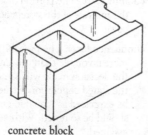

concrete block

concrete board A concrete and fiberglass panel typically used as a tile backing material.

concrete bond (bond plaster) A thin gypsum plaster coat applied to a concrete face as a base for subsequent coats of plaster.

concrete contraction The shrinkage of concrete that occurs as it cures and dries.

concrete curing compound A chemical applied to the surface of fresh concrete to minimize the loss of moisture during the first stages of setting and hardening.

concrete cylinder test A compression test for concrete strength. Wet samples of concrete are carefully placed in specially made containers 6″ in diameter and 12″ high. The cylinders are sent to a laboratory where a compression test is performed. This is done by putting the concrete in a hydraulic machine that measures the pressure(s) needed to crush it. Cylinder tests are usually performed for each pour on a project that requires concrete strength control.

concrete finish The smoothness, texture, or hardness of a concrete surface. Floors are trowelled with steel blades to compress the surface into a dense protective coat. Walls that are exposed to the weather are often ground with a carborundum stone or wheel, with cement then added to fill the small voids. A smooth surface is desired so that water cannot enter the small holes, freeze, and deteriorate the surface.

concrete finishing machine 1. A portable machine with large paddles like fan blades used to float and finish concrete floors and slabs. **2.** A large power-driven machine mounted on wheels that ride on steel pavement forms. Used to finish concrete pavements.

concrete finishing machine

concrete floor hardener An additive used to impart extra wear and chip resistance to concrete floors. The additive may be placed in the mix before the floor is cast, or it may be applied to the surface in liquid or granular form.

concrete, foamed Concrete to which a chemical foaming agent has been added. The result is a light, porous material, full of air holes, with low strength and good thermal properties.

concrete insert A device, such as a pipe sleeve, a threaded bolt, or a nailing block, that is attached to a concrete form before placing concrete. When the concrete forms are removed, the insert remains embedded in the concrete.

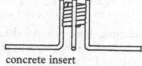

concrete insert

concrete masonry 1. Concrete blocks laid with mortar or grout

in a manner similar to bricks. **2.** Concrete that may be poured in place or as special tilt-up building walls.

concrete nail A steel nail that has been hardened and fitted with a diamond point so that it can be driven into concrete without bending.

concrete, normal weight Concrete made with the standard mix of component materials and weighing approximately 148 lbs. per cu. ft.

concrete paver 1. A machine used to pave roads with concrete. A paver is equipped with a loading skip, a rotating drum, and a discharge boom. **2.** A precast paving block used in landscaping for walks and patios.

concrete pile A slender, precast concrete-reinforced member that is embedded in the soil by driving, jetting, or insertion into a predrilled hole. May or may not be prestressed.

concrete pile

concrete planer A machine with cutters or grinders used to level and refinish old concrete pavement.

concrete plank A hollow-core or solid, flat beam used for floor or roof decking. Concrete planks are usually precast and prestressed.

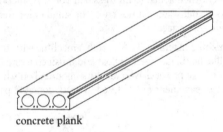

concrete plank

concrete, prestressed Concrete members with internal tendons that have been tensioned to put a compressive load on the members. When a load is applied to a prestressed member, compression is decreased where tension would normally occur.

concrete pump

concrete pump A pump that forces premixed concrete through a hose to a desired location.

concrete, ready-mixed (transit-mixed) Concrete delivered to a site, mixed and ready for placement.

concrete reinforcement Metal bars, rods, or wires placed within formwork before concrete is added. The concrete and the reinforcement are designed to act as a single unit in resisting forces.

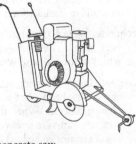

concrete saw A power saw used, for example, to cut concrete in order to remove damaged sections of pavement or to groove the surface to create a control joint.

concrete saw

concrete slump test A test to determine the plasticity of concrete. A sample of wet concrete is placed in a cone-shaped container 12″ high. The cone is removed by slowly pulling it upward. If the concrete flattens out into a pile only 4″ high, it is said to have an 8″ slump. This test is done on the job site. If more water is added to the concrete mix, the strength of the concrete decreases and the slump increases.

concurrent insurance Insurance under two or more policies that are exactly alike in their terms and conditions, even though they might be different in the dollar amounts of coverage or the dates the policies begin. (Nonconcurrent insurance differs in the terms and conditions as well.) This condition can complicate the collection process for contractors.

condenser The heat exchanger in a refrigeration system that removes heat from the high pressure refrigerant gas and transforms it into a cool liquid.

condensing unit In a refrigeration system, the mechanism that pumps vaporized refrigerant from the evaporator, compresses and liquefies it, and returns it to the refrigerant control. Includes the compressor, condenser, condenser fan motor, and controls.

conditioned air Building air whose temperature and relative humidity is controlled.

conditions of the contract A document detailing the rights, responsibilities, and relationships of the parties to the contract for construction.

condominium A legal system by which individual units of real property, such as apartments, stores, or offices, may be owned separately. Each unit owner obtains all the rights incidental to ownership of real property, and shares with the other owners rights to the common areas in the building, facilities, or land of the condominium.

conductive flooring Flooring designed to prevent electrostatic buildup and sparking. Typical uses include floors in computer areas and hospital operating rooms.

conductivity The rate at which heat is transmitted through a material.

conductor 1. Any substance that can serve as a medium for transmitting light, heat, or sound. **2.** In electricity, a wire or cable that can carry an electric current. **3.** A pipe that leads rainwater to a drain.

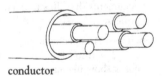

conductor

conduit (cable conduit) 1. A pipe, tube, or channel used to direct the flow of a fluid. **2.** A pipe or tube used to enclose electric wires to protect them from damage.

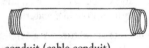

conduit (cable conduit)

conduit body The portion of a conduit or tubing system that makes the interior of the system accessible at a junction or terminal point.

confidence level The probability **1.** that results will be equal to or more favorable than the amount estimated or quoted; **2.** that the decision made will achieve the desired results; or **3.** that the stated conclusion is true. *Note:* Confidence level may also be expressed as "equal to or less favorable." If that is the case, it should be so noted. Without such a note, the definition shown is assumed.*

confidentiality agreement A legal document that allows the release of trade secret information to physicians and others who may need this information to protect workers from danger or to treat an exposed worker.

conglomerate A rock deposit consisting of stones of varying sizes naturally cemented together as a natural concretion.

connected load The total load, as measured in watts, that is connected to an electric supply system if all lights and equipment are turned on.

consequential damages Losses that do not flow directly from the actions of the defendant, but rather are the result of special or remote circumstances relating to those actions.

conservation In the context of historic buildings, efforts to bring old building finishes and features to a condition as close as possible to their original state when new. Conservation also involves planning and carrying out maintenance practices that will ensure the best appearance and condition of the building and its features for as long as possible.

conservation easement An easement restricting a land owner to land uses that are compatible with long-term conservation and environmental values.

consistency A reference to the relative mobility or plasticity of freshly mixed concrete or mortar, as measured by a slump test for concrete, and by a flow test for mortar and grout.

consistometer A device for measuring the firmness or consistency of concrete, mortar, grout, plaster, etc.

console 1. A control station with switches and gauges to govern the operation of mechanical, electrical, or electromechanical equipment. 2. An ornamental bracket-like member used to support a cornice.

consolidated formation Soil strata that may have been subjected to glacial or other consolidating loads in the geological past.

consolidation 1. Compaction of freshly poured concrete by tamping, rodding, or vibrating to eliminate voids and to ensure total envelopment of aggregate and reinforcement. 2. Compaction of soil in an embankment to achieve a higher bearing strength.

constant-voltage transformer A transformer designed to minimize or eliminate the variations in standard line voltage and to produce an unchanging voltage, as required by computers and some instrumentation.

constant-wattage ballast An efficient lamp ballast that provides high power factor by minimizing the effects of voltage variations.

constructability The input of construction knowledge and expertise throughout the planning, design, procurement, and field operations to improve the means and methods of improving the design intent.

constructability review The practice of peer reviewing the plans and specifications prior to issuing for bid with the intent of correcting errors, omissions, inconsistencies, or discrepancies.

construction administrator One who oversees the fulfillment of the responsibilities of all parties to the contract for construction, for the primary benefit of the owner. In the typical project, construction administration is usually provided by the design professional. However, the owner may employ a separate professional entity for this purpose.

construction change A change allowed to the contractor for the consequences of an owner failing to perform certain obligations in a contract, where such failure causes extra cost and schedule for the contractor. Examples include overinspection by the owner, the owner's failure to grant legitimate schedule extensions, or the owner's failure to coordinate other contractors and subcontractors on the job site.

construction change directive (CCD) An alternate mechanism for directing the contractor to perform additional work to the contract when time and/or cost of the work is not in agreement between the owner and contractor performing the work.

construction cost The sum of all costs, direct and indirect, inherent in converting a design plan for material and equipment into a project ready for start-up, but not necessarily in production operation; the sum of field labor, supervision, administration, tools, field office expense, materials, equipment, and subcontracts.*

construction documents The written specifications and drawings that provide the requirements of a construction project.

construction drawings The portion of the contract documents that gives a graphic representation of the work to be done in the construction of a project.

construction-grade lumber Good quality lumber that is generally free of defects.

construction joint The interface or meeting surface between two successive pours of concrete.

construction joint

construction management 1. Project management as applied to construction. 2. A professional service that applies to effective management techniques to the planning, design, and construction of a project from inception to completion for the purpose of controlling time, cost, and quality, as defined by the Construction Management Association of America (CMAA).*

construction manager One who directs the process of construction, either as the agent of the owner, as the agent of the contractor, or as one who, for a fee, directs and coordinates construction activity carried out under separate or multiple prime contracts.

construction manager-at-risk (CMAR) A project-specific delivery method that is suited for medium to large capital or renovation projects. CMAR provides technical assistance to the designer during the design phase, has a cost-capping feature, and allows construction to start before design documents are 100% complete. The CMAR contracts directly with subcontractors, fabricators, and material suppliers.

construction progress Construction progress is monitored and reported as percent complete. Actual work units completed are measured against the planned work units for each applicable account in the bill of materials or quantities. Usually reported against individual accounts by area and total project.*

construction wrench A wrench with a handle and jaw at one end for holding or turning fasteners, pipes and fittings. The other end is pointed and is used to align mating holes in steel construction.

construction wrench

constructive change A change to a contract resulting from conduct by the owner that has the effect of requiring the contractor to perform work different from that presented in the contract.

constructor One who is in the business of managing the construction process. A contractor is a constructor who is acting under the terms of a contract for construction.

consultant A person (or organization) with an area of expertise or professional training who contracts to perform a service.

consulting engineer A licensed engineer, employed to perform specific engineering tasks.

consumables Supplies and materials used up during construction. Includes utilities, fuels and lubricants, welding supplies, worker's supplies, medical supplies, etc.*

contact adhesive (dry-bond adhesive) A bonding agent that is applied to two surfaces and allowed to dry before being pressed together.

contact pressure The pressure or force that a footing and the structure it supports exerts on the soil below.

containment berm A physical barrier used to keep liquids from running outside of a protected area.

containment box A temporary container for hazardous liquid spills.

contemporaneous documentation Actual project invoices, plans, and specifications, used to document the outlay of capital by a particular taxpayer for a particular construction project. Generally considered the best supporting documentation for a quality cost segregation study.

contiguous Adjacent and touching properties.

contingency An amount included in the construction budget to cover the cost of unforeseen factors related to construction.

contingency allowance A specified sum included in the contract sum to be used at the owner's discretion, and with his approval, to pay for any element or service that was unforeseen or that is desirable but not specifically required of the contractor by the construction documents.

continuous beam A beam supported at three or more points, and thus having two or more spans.

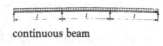

continuous beam

continuous footing A concrete footing supporting a wall or two or more columns. The footing may vary in depth and width.

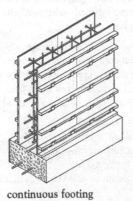

continuous footing

continuous girder A girder supported at three or more points.

continuous header Wood frame construction in which the top plate is replaced by a double member, such as a 2″ × 8″ on edge, that acts as a lintel over all wall openings.

continuous rating The constant load that a device can carry at rated primary voltage and frequency without damaging or adversely affecting it.

continuously reinforced pavement A longitudinally reinforced concrete pavement with no intermediate transverse expansion or contraction joints.

continuous span A rigid span extending over three or more supports so that bending movements are transmitted from one segment to the next.

continuous waste and vent A single pipe, the upper portion of which is a vent, while the lower portion functions as a waste drain.

contour line A line on a map or drawing indicating a horizontal plane of constant elevation. All points on a contour line are at the same elevation.

contract An agreement between two parties to perform work or provide goods, including an agreement or order for the procurement of supplies or services.

contract change An authorized modification to terms of a contract. May involve, but is not limited to **1.** a change in the volume or conditions of the work involved; **2.** the number of units to be produced; **3.** the quality of the work or units; **4.** the time for delivery; and/or **5.** the consequent cost involved.*

contract completion date The date established in the contract for completion of all or specified portions of the work. This date may be expressed as a calendar date or as a number of days after the date for commencement of the contract time is issued.*

contract dates The start, intermediate, or final dates specified in the contract that impact the project schedule.*

contract documents All the written and graphic documents concerning execution of a particular construction contract. These include the agreement between the owner and contractor, all conditions of the contract including general and supplementary conditions, the specifications and drawings, any changes to the specifications and drawings, any changes to the original contract, and any other items specifically itemized as being part of the contract documents.

contract for construction An agreement between owner and contractor whereby the contractor agrees to construct the owner's project in accordance with the contract documents, within a specified amount of time, and for consideration to be paid by the owner as mutually agreed.

contracting officer The representative of a government agency with authority to bind the government in contract matters.

contracting officer's decision The contracting officer's final ruling regarding a properly submitted claim.

contraction joint (control joint) A formed, sawed, or tooled groove in a concrete structure. The purpose of the joint is to create a weakened plane and to regulate the location of cracking resulting from the dimensional change of different parts of the structure.

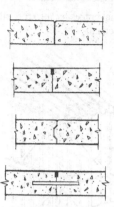

contraction joint (control joint)

contract modification Any unilateral or bilateral written alteration of the contract in accordance with the governing regulations and contract clauses.

contractor A constructor who is a party to the contract for construction, pledged to the owner to perform the work of construction in accordance with the contract documents.

contractor's option As laid out in the contract documents, the ability of the contractor to select predetermined materials, methods, or systems, at his or her option, with no change in the contract sum.

contractor's qualification statement A statement of the contractor's qualifications, experience, financial condition, business history, and staff composition. This statement, together with listed business and professional references, provides evidence of the contractor's competence to perform the work and assume the responsibilities required by the contract documents.

contract price The monies payable by the owner to the contractor under the contract documents as stated in the agreement.*

contract sum An amount representing the total consideration in money to be paid the contractor for services performed under the contract for construction.

contributory negligence A term used to describe legal responsibility for an error or fault by one or more parties who have allegedly contributed in whole or in part to a loss or damage suffered by another party as a result of a specific occurrence.

control joint Grooves manually made on concrete flooring to help control where the concrete should crack.

controlled fill A backfilling or embankment operation in which the moisture content, depth of lift, and compaction equipment are closely regulated by specification and inspection.

controller An electrical, electronic, pneumatic, or mechanical device designed to regulate an operation or function.

convection The movement of a gas or liquid upward as it is heated and downward as it is cooled. The movement is caused by a change in density.

conversion 1. The change in use of a building, as from a warehouse to residential units, that may require changes in the mechanical, electrical, and structural systems. 2. The sawing or milling of lumber into smaller units.

conveyance An instrument or deed by which a title of property is transferred.

conveyors Frame-mounted continuous belts that move aggregate, earth, or concrete.

coolant A fluid used to transfer heat from a heat source to a heat exchanger, where the heat is removed and the coolant is usually recycled.

cooling load The amount of air conditioning that would be required to keep a building at a specified temperature during summer months (usually 78°F), regardless of variations in outside temperature.

cooling tower An outdoor structure, frequently placed on a roof, over which warm water is circulated for cooling by evaporation and exposure to the air. A natural draft cooling tower is one in which the airflow through the tower is due to its natural chimney effect. A mechanical draft tower employs fans to force or induce a draft.

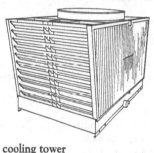

cooling tower

cool roof A roof made of materials (typically light in color) that have a high thermal emittance and solar reflectance. Cool roofs can substantially reduce a building's cooling load.

coordinator The device on a pair of double doors that permits the doors to close in the correct sequence. If the door with the overlapping astragal closed first, the other door would strike the astragal and not close properly.

cope 1. To cut the end of a molding to match the contour of the adjacent piece. 2. To cut structural steel beams so that they fit tightly together.

coping The protective top member of any vertical construction such as a wall or chimney. A coping may be masonry, metal, or wood, and is usually sloped or beveled to shed water in such a way that it does not run down the vertical face of the wall. Copings often project out from a wall with a drip groove on the underside.

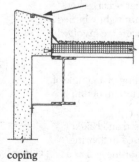

coping

copper A pure metal that is ductile, malleable, nonmagnetic, and high in electrical and thermal conductivity. Used for roofing in many historic buildings, copper is also a primary material in piping and electrical wiring.

copper fitting In piping, an elbow, tee, reducer, or other fitting made of wrought copper, cast brass, or bronze.

copper roofing A roof covering made of copper sheets joined by weatherproof seams.

copper roofing

corbel A course or unit of masonry that projects beyond the course below. A corbel may be used entirely for decoration or for a ledge to support a load from above.

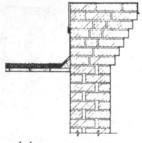
corbel

core 1. The interior structure of a hollow-core door. 2. A cylindrical sample of concrete or rock extracted by a core drill. 3. The void in a concrete masonry unit. 4. The center layer of a sheet of plywood. 5. The vertical stack of service areas in a multistory building. 6. The central part of an electrical winding. 7. Metal support for a handrail. 8. The rubble filling in a thick masonry wall. 9. The center material between facing papers in gypsum wallboard.

coreboard (battenboard) A manufactured board with a wood fiber or wood chip center and bonded veneer faces on both sides.

cored beam 1. A precast concrete beam with longitudinal holes through it. 2. A beam that has had core samples removed for testing.

core drill 1. A drill used to remove a core of rock or earth material for analysis. 2. The drilling of a hole through a concrete floor, wall, or ceiling for running pipe or conduit.

core gap An open joint extending through, or partly through, a plywood panel, that occurs when core veneers are not tightly butted. When center veneers are involved, the condition is referred to as a *center gap*.

coring The core drilling of a sample of rock, soil, or concrete to obtain a test sample.

corkboard Compressed and baked granulated cork used in flooring and sound conditioning and as a vibration absorber.

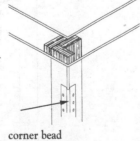
corner bead

cork tile Tile made from cork particles bound and pressed into sheets and covered with a protective wearing finish.

corner bead An L-shaped, light-gauge metal strip used to finish drywall and plaster wall angles and corners.

corner brace The diagonal braces let into the studs of wood frame structures for reinforcement.

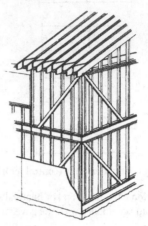
corner brace

corner reinforcement 1. The reinforcement in the upper corners of a metal door frame. 2. The metal angle strip used at the corners of plaster or gypsum board construction. 3. The bent reinforcing rods embedded at the corners of a cast-in-place concrete wall.

corner return block A concrete masonry block with solid faces on its sides and on one end. Used at the end of a run of blocks.

cornice 1. An ornamental molding of wood or plaster that encircles a room just below the ceiling. 2. An ornamental topping that crowns a structure. 3. An exterior ornamental trim at the meeting of the roof and wall. This type of cornice usually includes a bed molding, a soffit, a fascia, and a crown molding.

cornice

corporation An association of individuals established under certain legal requirements. A corporation exists independently of its members, and has powers and liabilities distinct and apart from its members.

corridor A long, interior passageway, usually with doors opening to rooms, apartments, or offices and leading to an exit.

corrosion The oxidation or eating away of a metal or other material by exposure to chemical or electrochemical action such as rust.

corrugated aluminum Sheet aluminum that has been rolled into a parallel wave pattern to impart stiffness.

corrugated fastener A small, wavy, steel fastener with one edge sharpened. The fastener is driven into two pieces of wood, bridging the joint in order to hold them together.

corrugated metal Sheet metal that has been rolled into a parallel wave pattern for stiffness and rigidity.

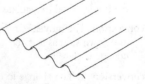
corrugated metal

corrugated roofing Corrugated metal or fiberglass mounted on rafters as sheet roofing.

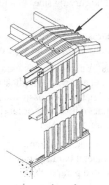

corrugated roofing

corrugated siding Siding made of sheet metal or asbestos-cement composition board and used for siding on factories and other nondecorative buildings. By corrugating the material, the structural strength is increased.

corrugated siding

cost The total expenditure in dollars approved after the completion of a project.

cost analysis A historical and/or predictive method of ascertaining for what purpose expenditures on a project were made and utilizing this information to project the cost of a project as well as costs of future projects. The analysis may also include application of escalation, cost differentials between various localities, types of buildings, types of projects, and time of year.*

cost engineer An engineer whose judgment and experience are utilized in the application of scientific principles and techniques to problems of estimation; cost control; business planning and management science; profitability analysis; project management; and planning and scheduling.*

cost estimate A prediction of quantities, cost, and/or price of resources required by the scope of an asset investment option, activity, or project. As a prediction, an estimate must address risks and uncertainties. Estimates are used primarily as inputs for budgeting, cost or value analysis decision making in business, asset and project planning, or for project cost and schedule control processes. Cost estimates are determined using experience and calculating and forecasting the future cost of resources, methods, and management within a scheduled timeframe.*

cost-plus-fee agreement An agreement between an owner and the contractor or a design professional that provides for payment of all costs associated with completion of their duties. This includes direct and indirect costs as well as a fee for services, which may be a fixed amount or a percentage of costs.

cost plus fee with guaranteed maximum price (CPF-GMP) A delivery method similar to CPF, except that it has a cost-capping feature that reduces the owner's risk. CPF-GMP is more popular than CPF for projects of higher value.

cost-reimbursement contract A type of contract in which the pricing arrangement involves the payment of allowable costs incurred by the contractor during performance.

cost segregation A process of identifying assets within a larger construction project and using cost estimating techniques that allocate costs to individual items of property, which can then be classified for depreciation purposes (e.g., land improvements, buildings, equipment, furniture and fixtures, etc.). The underlying incentive for building owners is the significant acceleration of tax benefits (depreciation deductions) resulting in improved current after-tax cash flow.

cotter pin A split metal pin that, when pushed through a hole in a shaft or a bolt (and having the ends bent), prevents parts or nuts from vibrating loose.

cotter pin

counter 1. A person or device that keeps a tally of the occurrences of some event, such as the number of loads of fill. **2.** A long, flat surface over which sales are transacted at a store or business. **3.** A flat work surface in a kitchen.

counterbrace A diagonal brace that balances the force from another brace as part of a web network in a truss.

counterflashing A thin strip of metal frequently inserted into masonry construction and bent down over other flashing to prevent water from running down the masonry and behind the upturned edge of the base flashing.

countersink 1. A conical depression cut to receive the head of a flat-head bolt or screw for driving it flush with the surrounding surface. **2.** A bit used to cut a conical depression.

countersunk rivet A rivet used in countersunk holes.

countertop The level surface material covering cabinetry, such as in a kitchen or bath.

counterweight A weight that balances another weight, for example, a sash weight that balances a window sash, or a large weight that balances a lift bridge.

couple Two equal, opposite, and parallel forces that tend to produce rotation. The movement equals the product of one force times the perpendicular distance between the two.

coupler A metal clamplike device used to join and extend tubular frames and braces in barricades or scaffolding.

coupling A fitting for joining two pieces of pipe.

course 1. A horizontal layer of bricks or blocks in a masonry wall. **2.** A row or layer of any type of building material, such as siding, shingles, etc.

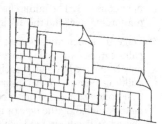

course

coursed masonry Any masonry units laid in regular courses, as opposed to rough or random rubble.

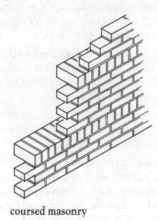

coursed masonry

coursed rubble Masonry construction consisting of roughly dressed stones of mixed sizes and shapes, with small stones used to fill irregular voids.

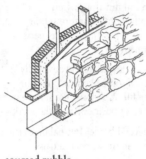

coursed rubble

coursing joint 1. The horizontal mortar joint between two courses of masonry. 2. In an arch, the arched joint between two curving courses.

court An open, unobstructed space surrounded on at least three sides by walls or buildings.

court deed A deed executed by a court official conveying title to a property purchaser following judicial foreclosure, often called a *sheriff's deed*.

cove A concave-shaped surface where a ceiling and wall meet.

cove lighting Indirect lighting in which the fixtures are behind a molding or valance, and thus out of sight.

covenant A term used to describe one or more specific points of agreement that may be set forth in a contract.

cover 1. That which envelops or hides, such as a protecting cover of paint. 2. That part of a tile or shingle that is overlapped by the next course. 3. The minimum thickness of concrete between the reinforcing steel and the outer surface of the concrete.

coverage 1. The nominal square feet of area that a can of stain or paint can be expected to cover. 2. The amount of surface that may be covered by a unit, such as a bundle, square, or ton of building material. 3. An insurance policy, or the dollar protection that policy provides. The amount of protection depends on many factors, including the type of insurance, amount of insurance purchased, policy limits, and exclusions. 4. Dependable estimates or firm bids for portions of a construction project.

coving 1. Concave molding such as that used at the intersection of a wall and the ceiling. 2. The outward curve of an exterior wall to meet the eaves. 3. The curving sides of a fireplace that narrows at the back.

cradle 1. A scaffold that is suspended outside the roof or top of a structure. 2. A U-shaped support for pipe or conduit.

cradle-to-grave analysis A methodology used to determine the energy, environmental, and waste implications of a product, material, or system. Analysis begins at the point of raw material extraction and includes refinement or fabrication, transportation, use, and eventual reuse or disposal.

cramp 1. A device for supporting a frame in place during construction. 2. A U-shaped metal bar used to lock adjacent blocks of masonry together, as in a parapet wall.

crane A machine for raising, shifting, or lowering heavy weights, commonly by means of a projecting swinging arm.

craneway Steel or concrete column and beam supports and rails on which a crane travels.

crank A twist or bend in a reinforcing bar.

crawler tractor A powered earth-moving vehicle that moves on continuous, segmented, cleated treads or tracks. The tracks provide low ground-bearing pressure and a high level of mobility and power.

crawler tractor

crawling A defect in the application of paint whereby the paint film raises, separates, or breaks apart. Usually caused by painting over a slippery or glassy surface, or from applying heavy or elastic coatings over a hard or brittle surface.

crawl space 1. In a building or portion of a building without a basement, the accessible space between the surface of the ground and the bottom of the first floor joists, with less-than-normal headroom. 2. Any interior space of limited height designed to permit access to components such as ductwork, wiring, and pipe fittings.

crawl space vents A system of openings in a crawl space to allow for an air exchange that will reduce moisture and harmful gas buildup.

creep 1. The slow but continual permanent deformation of a material under sustained stress. 2. The very slow movement of rock or soil under pressure. 3. Gradual deformation of a roofing membrane due to thermal stress.

crescent truss A truss in which the upper and lower chords are curved in the same direction, but with different radii of curvature, so that they meet at the ends, thus giving the assembly a crescent-shaped appearance.

crew tasks Tasks done by the crew that combine the performance of different individuals, versus a single individual. Each crew member performs a specific function to complete the overall task.

crib 1. A boxed-in area, the sides of which may be an open lattice, often filled with stone, and used as a retaining wall or support for construction above. 2. A retaining framework lining in a shaft or tunnel.

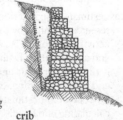

crib

C

cribbing A system of open boxes built with heavy timbers and filled with rocks or pervious fill. Used as a retaining wall.

cribwork An open construction of beams, at the face of an embankment, the alternate layers of which project to provide lateral stability, prevent erosion, and resist thrust or overturning.

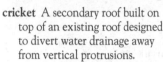

cricket A secondary roof built on top of an existing roof designed to divert water drainage away from vertical protrusions.

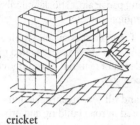

cricket

crimp A sharp bend in a metal sheet, as in the joint in metal roofing.

crimped wire Wire that has been deformed or bent to improve its bonding effectiveness when used to reinforce concrete.

cripple 1. In construction framing, members that are less than full length, for example, studs above a door or below a window. 2. In roofing, a bracket secured at the ridge of a pitched roof to carry the scaffold for roofers.

cripple jack rafter A short supporting beam that extends from a valley rafter to a hip rafter.

critical angle The maximum angle or rise of a stair or ramp before it becomes unsafe for public use. For stairs the angle is 50°; for ramps, it is 20°.

critical delay An event that results in the overall project completion occurring later than originally planned.

critical path A term used to describe the order of events (each of a particular duration) that results in the least amount of time required to complete a project.

critical path method (CPM) A system of construction management that involves the complete planning and scheduling of a project, and the development of an arrow diagram showing each activity, its appropriate place in the timetable, and its importance relative to other tasks, and the complete project.

critical task A task or activity that lies on the critical path of the schedule, which if delayed will delay the completion of the project.

cross-bar In a grating or grill, one of the bars perpendicular to the primary bar.

cross-beam (brow post) 1. A large beam that spans two walls or sides of a structure, frequently holding them together. 2. A brace between opposite whalings or sheeting in an excavation. 3. In bridge construction, a beam that spans the trusses and supports the stringers or deck.

cross-bracing (cross-bridging) Diagonal braces placed in pairs that cross each other.

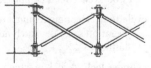

cross-bracing (cross-bridging)

cross-connection 1. A piping connection between two otherwise separate piping systems, one of which contains potable water, and the other water of unknown or questionable safety. 2. In a fire protection system, a connection from a Siamese fitting to a standpipe or sprinkler system.

crosscut Cutting wood at a right angle to the grain.

crosscut saw A saw with its teeth filed and set to cut across the grain of a piece of wood.

cross-furring (brandering, counterlathing) Strips or slats attached to the lower edge of joists to which plastering lathing is nailed.

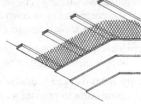

cross-furring (brandering, counterlathing)

cross-garnet hinge (T-hinge) A hinge that is shaped like the letter T, with the head attached to the door frame and the long part fastened to the door.

cross-laminated Laminated wood members in which the grain of some layers is at a right angle to the grain of other layers.

crosslap joint A wood joint in which two pieces are each cut to half their thickness at the overlap, so that the total thickness of the system does not change.

crossover 1. In an auditorium, a walkway that is parallel to the stage and connects with the aisles. 2. A pipe fitting used to bypass a section of pipe. 3. A connection between two piping systems.

cross peen hammer A hammer with one face flat and the other a wedge. Used to break or work hard surfaces.

cross-runner In a suspended or T-bar ceiling, one of a series of short pieces that span between the long runners.

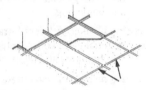

cross-runner

cross-section A diagram or illustration showing the internal construction of a part or assembly if the front portion were removed.

crown 1. An ornamental architectural topping. 2. The central top section of an arch or vault. 3. The high point in the center of a road that causes water to flow to the edges. 4. The top of a tree or flowering plant. 5. In plumbing, the section of a trap that changes direction from vertical to horizontal. 6. The convex curvature or camber in a beam.

crown

crown course 1. The top row of a roofing material. 2. The row that actually covers the ridge.

crown rafter The central common rafter in a hip roof.

crushed gravel Gravel that has been crushed and screened so that substantially one face of each particle is a fractured face.

crushed stone Stone crushed and screened so that substantially all faces result from fracturing.

crystallization As applied to absorption chillers, the precipitation of salt crystals from the absorbent. This causes a slushlike mixture that plugs fluid passages within the chiller and renders it inoperable.

cube strength An analytical strength test of Portland cement in which a standard-size concrete cube is loaded to failure.

cul-de-sac A dead-end street that ends in an enlarged turn-around area.

culvert A transverse drain under a roadway, canal, or embankment other than a bridge. Most culverts are fabricated with materials such as corrugated metal and precast concrete pipe.

culvert

cumulative batching The weighing of a batch of material by adding the components successively to the same container and balancing the scale at each new total weight.

curb cock (curb stop) A control valve that is placed in the water supply pipe that runs from a water main in a street to a building.

curb form 1. A form used with a curb machine for extruding concrete curbs to produce a desired shape and finish on the curb. **2.** A reusable metal form for cast-in-place concrete curbs.

curb level The elevation of the curb grade in front of a building, as measured at the center of its street frontage.

curb machine A self-propelled, hand-operated machine for extruding bituminous or concrete curbs.

curb plate 1. The wall plate at the top of a circular or elliptical structure. **2.** The plate on which the upper rafters of a curb roof rest.

curb roof (gambrel roof, mansard roof) A roof with the slope divided into two pitches on each side.

cure 1. A change in the physical and/or chemical properties of an adhesive or sealant when mixed with a catalyst or subjected to heat or pressure. **2.** To maintain the proper moisture and temperature after placing or finishing concrete to assure proper hydration and hardening.

curing blanket A layer of straw, burlap, sawdust, or other suitable material placed over fresh concrete and moistened to help maintain humidity and temperature for proper hydration.

curing membrane Any of several kinds of sheet material or spray-on coatings used to temporarily retard the evaporation of water from the exposed surface of fresh concrete, thus ensuring a proper cure.

curling A change in the shape of wood, such as straightness or flatness, due to drying or temperature differences.

current The movement of electrons through a conductor. Measured in amperes.

current assets Cash and other assets that will be consumed or converted into cash within one year.

current-carrying capacity (ampacity) The maximum rated current, measured in amperes, that an electrical device is allowed to carry. Exceeding this limit could lead to early failure of the device or create a fire hazard.

current density The current flowing to or from a unit area of an electrode surface.

current liabilities Liabilities to be paid within one year.

current-limiting fuse A fuse that acts as a protective device by interrupting currents in its current-limiting range and guarding against overcurrents in an electrical system.

current transformer A transformer generally used to convert a current supplied at one current rate to another current rate.

curtained doorway A doorway consisting of two overlapping sheets of plastic over a door frame, each attached to the frame at the top and down one side. Allows worker entry into and from an area while minimizing the air movement between rooms.

curtain wall The exterior closure or skin of a building. A curtain wall is nonbearing and is not supported by beams or girders.

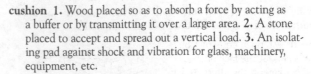

curtain wall

cushion 1. Wood placed so as to absorb a force by acting as a buffer or by transmitting it over a larger area. **2.** A stone placed to accept and spread out a vertical load. **3.** An isolating pad against shock and vibration for glass, machinery, equipment, etc.

cut 1. Material excavated from a construction site. **2.** A term for the area after the excavated material has been removed. **3.** The depth of material to be removed, as in a 5′ cut. **4.** To reduce a cost item.

cut and fill An operation commonly used in road building and other rock and earthmoving operations in which the material excavated and removed from one location is used as fill material at another location.

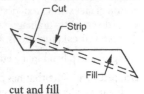

cut and fill

cutaway drawing A drawing of an area or object that shows what would be seen if a slice could be made into the area or object and a piece removed. *See also* **cross-section.**

cutback asphalt A bituminous roof coating or cement that has been thinned with a solvent so that it may be applied without heat to roofs or other areas that need sealing or cementing. Also used for dampproofing concrete and masonry.

cut nail (clasp nail) A hard, wedge-shaped nail used for nailing hardwoods such as oak flooring.

cut nail (clasp nail)

cutoff 1. A wall or barrier placed to minimize underground water percolation or flow. A cutoff is often used at the inlet and outlet of culverts. **2.** The design elevation at which the tops of driven piles are cut.

cutout 1. A mechanical or electrical device used to stop a machine when its safe limits have been exceeded. **2.** An opening in a wall or surface for access or other purposes. **3.** A piece stamped out of sheet metal or other sheet material.

cutout box A metal box used in electrical wiring to house circuit breakers, fuses, or a disconnect switch.

cutting in A painting technique used to paint around the edges of an object or area, such as trim or a light fixture.

cybertecture A term coined as the result of development of the "smart" house, which is wired to integrate power and telecommunications into one computer-directed system.

cycle The flow of alternating current as it travels in one direction, then reverses itself and travels in the opposite direction. A 60-cycle circuit completes 60 cycles per second.

cycle time The time required to complete a repetitive operation such as a batch mix or the refill of a toilet tank after flushing.

cylinder lock A lock in which the keyhole and tumbler mechanism are contained in a cylinder or escutcheon separate from the lock case.

cylinder unloaders Automatic devices used to hold open the reciprocating compressor valves of a number of cylinders in order to reduce compressor pumping capacity when not being used to full capacity.

cypress A wood of average strength and very good decay resistance and durability.

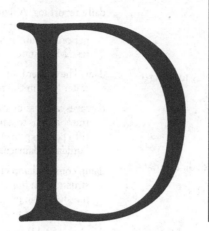

D

Abbreviations

d degree; density; penny (nail size)

D deep; depth; diameter; dimensional; discharge

DA discharge air

D/A digital-to-analog

DACS digital alarm communicator system

DBOM design, build, operate, maintain

DCC Digital Command Control

D&CM dressed and center matched

DCV demand-controlled ventilation

D&M dressed and matched

D&MB dressed and matched beaded

D&SM dressed and standard matched

D1S dressed one side

D2S dressed two sides

D2S&CM dressed two sides and center matched

D2S&SM dressed two sides and standard matched

D4S dressed four sides

DAD double acting door

db decibel

DBA a unit of sound level (as from the A-scale of a sound-level meter), doing business as

DB Clg double-headed ceiling

DBG distance between guides

Dbl double

DBT dry-bulb temperature

DC direct current, distribution center

DD design development, device description

DDC direct digital controller

DE digitally enhanced

DEC decimal

DEG degree, degrees

DEL delineation

Demo demolition

Demob demobilization

DEPT department

DER distributed energy resources

DET detached, detail, double end trimmed

DF damage free, direction finder, drainage free, drinking fountain, Douglas fir

dfl Douglas fir-larch

dflct deflection

dfu drainage fixture unit

DGI daylight, disability, or discomfort glare index

DH double-hung

DHW domestic hot water, double-hung window

DIA diameter

Diag diagonal

DIM dimension

DIN Dutch industry normal (German industry standard)

Dis, Disch discharge

Distrib distribution

DIV division

Dk deck

dkg decking

D

DL dead load, deadlight, diesel

DM demand management, document management

DN down

Do ditto

DOC delivery order contract

DOE U.S. Department of Energy

DOE-2 software used to predict the use and cost of energy for different types of buildings

DOT Department of Transportation

DOZ dozen

Dp depth

DP degree of polymerization, dew point, double-pitched, differential pressure

DPC dampproof course

DPST double pole, single throw

Dr, DR dining room, drain, dressing room, driver

DRG drawing

drink drinking

drn drain, drainage

drwl drywall

DS double strength, downspout, drop siding

DSA double strength A grade

DSB double strength B grade

DSGN design

DSIRE Database of State Incentives for Renewables and Efficiency

DT drum trap

DT&G double tongue and groove

DTS digital theater system

Dty duty

DU disposal unit

DUP duplicate

DV displacement ventilation

DVTL dovetail

dwg, DWG drawing

DWV drain waste vent

DX deluxe white, direct expansion

DXF AutoCAD Drawing Interchange Format or Drawing Exchange Format

dyn dyne

Definitions

dabber A soft, round-tipped brush for applying varnish or for polishing and finishing gilding.

dado 1. The flat-bottomed groove cut into one board (usually across the grain) to receive the end of another. If the groove is cut at the edge of a board, it is called a *rabbet*. **2.** An ornamental paneling above the baseboard of a room finish. **3.** Part of a column between the base and the cap or cornice.

daily allocable overhead The amount of main office overhead per day that is allocated to a delay as calculated using the Eichleay formula.

daily report log A daily report or historical record of the day-to-day events on a construction site, recording weather, temperature, personnel, subcontractors, accidents, visitors, and specific progress. Many states consider this a legal record.

dam The barrier built into the trapway of a toilet that controls the water level in the toilet bowl.

damages A measure of monetary compensation that a court or arbitrator awards to a plaintiff for loss or injury suffered by the plaintiff's person, property, or other legally recognizable rights. *See also* **liquidated damages**.

damp course (damp check, dampproof course) In masonry construction, an impervious course or any material that prevents moisture from passing through to the next course (usually a horizontal layer of material such as metal, tile, or dense limestone).

damper 1. A blade or louver within an air duct, inlet, or outlet that can be adjusted to regulate the flow of air. **2.** A pivotal cast-iron plate positioned just below the smoke chamber of a fireplace to regulate drafts.

damping 1. The force that acts to reduce vibrations in the same way that friction acts to reduce ordinary motion. **2.** The gradual dissipation of energy over a period of time.

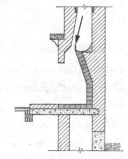

damper

dampproofing An application of a water-resisting treatment or material to the surface of a concrete or masonry wall to prevent passage or absorption of water or moisture. Can also be accomplished by using an admixture in the concrete mix.

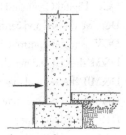

dampproofing

dap A notch made in one timber to fit the end of another.

darby (derby, derby float, derby slicker) A hand float or trowel, about 4′ long, used by concrete finishers and plasterers in preliminary floating and leveling operations. The float is usually made of wood or aluminum and has either one or two handles.

dash-bond coat A thick slurry of Portland cement, sand, and water applied to a concrete wall with a brush or whisk-broom to act as a bond for a subsequent plaster coat.

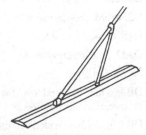

darby (derby, derby float, derby slicker)

database Part of a computer's software program that contains files and lists of related information organized for quick access. In Building Automation Systems, the database consists of digital and analog point names; descriptions such as temperature set points; and operating information such as programs and passwords.

data center A space that houses a facility's critical computer systems and has specialized protective features, such as backup power, fire suppression, HVAC, and communications.

date of acceptance Date on which the client agrees to final acceptance of the project. Commitments against the authorized funds usually cease at this time.*

date of agreement The date shown on the face of an agreement, or the date the agreement is signed (usually the date of the award).

date of commencement of the work The date established in the *notice to proceed* or, in the absence of such notice, the date of the agreement established by the parties for the work to begin.

date of substantial completion The date when work (or a portion of the work) is certified by the architect to be sufficiently complete as specified in the contract documents. The owner should be able to occupy the work fully or partially.

datum A base elevation to which other elevations are referred.

daub 1. To rough-coat with plaster. 2. To rough-coat a stone surface by striking it with a special hammer.

Davis-Bacon Act A federal labor law enacted in 1931 that requires laborers and mechanics (those who do physical work) on federally funded construction projects be paid no less than the local prevailing wages. The Davis-Bacon Act applies to all federally funded contracts over a specified value.

daylight compensation An energy-saving lighting system controlled by a photocell that decreases lamp intensity as daylight increases.

daylight factor A component in calculating the amount of natural outdoor light entering a building.

daylight harvesting Maximizing use of sunlight to illuminate interior building space. Photo sensors are used to measure natural light levels; artificial lighting can be adjusted accordingly.

daylighting Admitting natural light into a space, including distributing light at uniform levels, avoiding glare and reflections, and controlling artificial light to achieve energy and cost savings.

daylight lamp A lamp manufactured with daylight glass. The output is generally 35% less than that from an incandescent lamp.

daylight sensor A device that measures ambient light levels, used to help control lighting levels and conserve energy.

D-cracking The progressive formation of fine hairline cracks in concrete surfaces. In highway pavement these cracks run parallel to joints and edges and cut diagonally across corners. See also **D-line crack**.

deactivator A tank containing iron filings, through which hot water is passed to be purged of its active oxygen and other corrosive elements.

dead Refers to a conductor that is not connected to an electrical source.

dead air space Unventilated air space between structural elements. The space is used for thermal and sound insulation.

deadband In HVAC, a temperature range in which neither the heating nor cooling systems are activated.

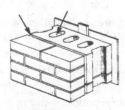
dead air space

dead band widening system A control system for boilers and furnaces that allows for a wide system dead band, or time delay, from set point. Burners can be shut off and stay on for longer periods of time, with fewer cycles. Avoiding short boiler cycling increases the net system efficiency.

dead bolt The bolt on a type of door lock that must be operated positively in both directions by turning a key or a thumb bolt.

dead end 1. In stressing a tendon, the end opposite the one to which stress is applied. 2. In plumbing, a drain line or vent that has been purposely terminated by a cap, plug, or other fitting.

deadening The use of insulating and dampening materials to restrict the passage of sound.

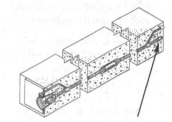
dead end

dead-front Refers to an electrical device in which the front part is insulated from voltage and can be touched without receiving an electric shock.

dead level 1. The state of being absolutely level, with no pitch or slope. 2. A grade of asphalt used on a level or nearly level roof.

deadlight A fixed window sash.

dead load A calculation of the weight of a building's structural components, fixtures, and permanently attached equipment (used in designing a building and its foundations).

deadlock A type of door lock employing a dead bolt.

deadman A heavy anchor, such as a concrete block or a log, usually buried underground and used for securing the end of a tie or guy. This device is frequently used with sheeting and retaining walls.

deal In the lumber industry, boards or planks usually more than 9" wide and 3" to 5" thick.

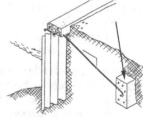
deadman

debarment The formal sanction by the government prohibiting a contractor from receiving contracts as a result of certain proscribed actions including crimes, fraud, etc.

debt service The periodic repayment of loans including interest and a portion of the principal.

decal A design that can be transferred from a special paper onto any of a number of different surfaces.

deceleration The opposite of acceleration. A direction, either expressed or implied, to slow down job progress.*

decenter To lower or remove shoring or centering.

decibel 1. The standard unit of measurement for the loudness of sound. 2. In closed-circuit television, a numerical unit used to express the difference in power levels, usually between acoustic or electric signals.

deciduous A term used to describe those trees that shed their leaves annually. Includes most hardwoods and some softwoods.

deck **1.** An uncovered wood platform usually attached to or on the roof of a structure. **2.** The flooring of a building. **3.** The structural system to which a roof covering is applied.

deck

deck curb A curb around the edge of a roof deck or around roof-mounted equipment.

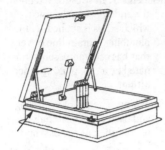

deck curb

decking **1.** Light-gauge, corrugated metal sheets used in constructing roofs or floors. **2.** Heavy planking used on roofs or floors. **3.** Another name for slab forms that are left in place to save stripping costs. **4.** The use of inert material to separate explosive charges.

decking

deck paint A special hard-surface paint that is resistant to abrasive wear and used particularly on decks and porches.

declining balance depreciation A method of computing depreciation in which the annual charge is a fixed percentage of the depreciated book value at the beginning of the year to which the depreciation applies.

decomposition Separation of the scope of work and requirements into smaller, component packages, so that work effort can be more effectively monitored and controlled.*

deconstruction A green building strategy that involves reuse of construction materials salvaged from buildings that are being demolished or remodeled. New building designs may include plans for deconstruction and later reuse of materials.

decontamination chamber A series of rooms separated from one another and from work areas by airlocks. Prevents contamination of sensitive work areas and/or the spread of dangerous substances used in the work area. Commonly used in asbestos abatement.

decorative block A concrete masonry unit manufactured or treated to have a desired architectural effect. The particular effect may be in color, texture, or both.

decoupling Separation of building elements to reduce or eliminate the transfer of sound, heat, or physical loads from one element to the other.

decorative block

deductible The dollar amount that the policyholder agrees to pay in the event of a loss. The insurance company pays the amount over the deductible, up to the limit of the policy.

deduction The amount of money deducted from the contract sum by a change order.

deductive alternate An alternative bid that is lower than that bidder's base bid.

deductive change A change resulting in a reduction in the contract price.

deed **1.** A legal document giving an individual rights or ownership to a property. **2.** A document that forms part of a contract and which, when signed by both parties, legally commits the contractor to perform the work according to the contract documents, and commits the client to pay for the work.

deed in lieu of foreclosure Refers to a deed given by a defaulting property owner to a creditor so that the lender does not have to proceed with foreclosure. Following a deed in lieu, the creditor may not be able to sue for a deficiency judgment, and the debtor may preserve credit reputation.

deed restriction A restriction on the use of a property as set forth in the deed.

deep-seal trap (antisiphon trap) A U-shaped plumbing trap having a seal of 4″ or more.

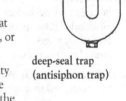

deep-seal trap (antisiphon trap)

default An omission of, or failure to perform, a contractual duty.

defect Any condition or characteristic that detracts from the appearance, strength, or durability of an object.

defective Work that is unsatisfactory, faulty or deficient, or does not conform to the contract documents, or does not meet the requirements of any inspection, reference standard, tests or approval referred to in the contract documents.*

defective specifications Specifications and/or drawings that contain errors, omissions, and/or conflicts that affect or prevent the contractor's performance of the work.*

defective work (deficiencies) Work that does not comply with the requirements of a contract.

defendant An individual against whom a claim or action is brought.

defense An explanation or reason why a person or entity should not be held legally responsible (liable) as a result of claims made against them by another party.

deflected tendons (draped tenon) Tendons in a concrete member that have a curved trajectory with respect to the gravity axis of the member.

deflection **1.** The bending of a structural member as a result of its own weight or an applied load. **2.** The amount of displacement resulting from this bending.

deflection angle In surveying, a horizontal angle as measured from the prolongation of the preceding transit line to the next line.

deflection limitation Maximum allowable deflection—dictated by the bending limit of the material under the required design load.

deflectometer An instrument for measuring the degree to which a transverse load causes a beam or other structure to bend.

deformation A change in the shape or form of a structural member that does not cause its failure or rupture.

deformed bar (deformed reinforcing bar) A reinforcing bar manufactured with surface deformations to provide bonding strength when embedded in concrete.

deformed bar (deformed reinforcing bar)

deformed reinforcement In reinforced concrete, reinforcement comprising bars, rods, deformed wire, welded wire fabric, and welded deformed wire fabric.

degradation Deterioration of a painted surface by heat, light, moisture, or other elements.

degree 1/360th of the circumference of a circle or a round angle.

degree-day A unit of measure for heating-fuel consumption. The unit is used to specify the nominal heating load of a building in winter. One degree-day is equal to the number of degrees, during a 24-hour day, that the mean temperature is below 65° F, which is the *base temperature* in the United States.

dehumidifier A device for removing water vapor from the air.

delamination Separation of plies of materials, usually due to failure of the adhesive. May occur, for example, in veneers, roofing, and laminated wood beams.

delay An event or condition that results in work activity starting, or the project being completed, later than originally planned.

deliquescence The absorption of moisture from the air by certain salts in plaster or brick, resulting in damp spots that appear darker than the surrounding material.

deliverable 1. A report or product of one or more tasks that satisfy one or more objectives and must be delivered to satisfy contractual requirements. 2. Another name for products, services, processes, or plans created as a result of doing a project. A project typically has interim as well as final deliverables.*

delivery Transfer or handover of a product from one party to another.*

delivery hose The hose through which concrete, mortar, grout, or shotcrete is pumped.

delivery hose

delivery order contract (DOC) A comprehensive procurement system for organizations to obtain construction services that is designed to lessen construction response time, reduce in-house workload and enhance quality control.

deluge sprinkler system A dry-pipe sprinkler system particularly well suited for areas that may experience temperatures below freezing. The system is actuated by a heat- or smoke-detection device, which then turns a valve to admit the water.

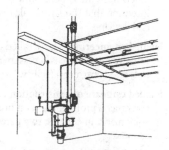
deluge sprinkler system

demand The electric load integrated over a specific interval of time, usually expressed in watts or kilowatts.

demand-controlled ventilation The adjustment of ventilation in response to actual occupancy and CO_2 readings.

demand factor In an electrical system, the ratio of the maximum demand to the connected load.

demand load The actual amount of electrical load on a circuit on any given time.

demand management Control of the load, or demand for electrical power, generally in keeping with utility rate fluctuations.

demand response Power management equipment that reduces a facility's power consumption during peak demand periods.

demolition The intentional destruction of all or part of a structure.

demountable partition (relocatable partition) A non-load-bearing wall made of prefabricated sections that can be readily disassembled and relocated. These partitions may be full height or partial height.

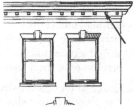
demountable partition (relocatable partition)

dense-graded aggregate Aggregate sized so as to contain a minimum of voids, and therefore to have the maximum weight when compacted.

densified impregnated wood (compregnated wood, compressed wood, resin-treated wood) Laminated wood that has been impregnated with resins and compressed to greatly increase its density and strength.

density 1. In urban planning, the number of people living within an acre (or sometimes a square mile) of land. 2. The ratio of the mass of a specimen of a substance to the volume of the specimen; the mass per unit volume of a substance. 3. The closeness of pile yarn; the amount of pile packed into a given area of carpet, usually measured in ounces per square yard. 4. The number of lots or building units per acre of land as determined by the zoning bylaw having jurisdiction.

density control The testing of concrete used in a structure to maintain the specified density.

dentil band A wood molding installed to create the effect of a series of dentils.

dentils Square, toothlike blocks used as ornaments under a cornice.

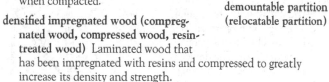

Department of Health and Human Services The federal agency responsible for establishing health and safety standards for the protection of persons.

Department of Labor The federal agency that oversees all laws associated with hiring, employing, and protecting workers.

dentils

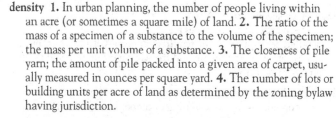

Department of Transportation (DOT) The federal agency responsible for regulating the safe transportation of goods and materials within the United States.

deposited metal The filler metal placed during welding.

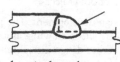
deposited metal

deposit for bidding documents A deposit required from the bidder for each set of the plans, specifications, and other bidding documents for a contract. Normally the deposit is returned to the bidder upon return of the documents in good condition and within a specified time.

deposition A formal method of obtaining information relevant to a lawsuit by verbally asking an individual questions under oath prior to trial.

depository A location where bids are received by an awarding authority.

depot **1.** A storehouse, warehouse, or transfer station. **2.** A railroad or bus station for ticketing, sheltering, transferring, and shipping passengers and freight.

depreciation The allocation of a part of the cost of a property, plant, or equipment item (that has a limited useful life) over its estimated useful life.

depreciation factor The ratio of initial illumination on an area to the present illumination of the same area, used in lighting calculations to account for depreciation of lamp intensity and reflective surfaces.

derrick A device consisting of a vertical mast and a horizontal or sloping boom operated by cables attached to a separate engine or motor. The device is used for hoisting and moving heavy loads or objects.

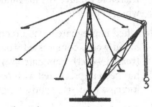

derrick

descriptive specification A type of specification that provides a detailed description of the required properties of a product, material, or piece of equipment, and the workmanship required for its proper installation.

desiccate To dry thoroughly or to make dry by removing the moisture content, as in the seasoning of timber by exposing it in an oven to a current of hot air.

desiccator An apparatus in which a substance is dried.

design **1.** To create a graphic representation of a structure. **2.** The graphic architectural concept of a structure. **3.** To make a preliminary sketch, drawing, or outline.

design-bid-build (DBB) A traditional method of construction project delivery involving the selection and award of professional design services followed by a separate process for construction services once the design documents are complete.*

design/build (design-construct) A method of construction in which the contractor provides both design and construction services to an owner.

design contingency An amount included in a construction budget to cover additional costs for possible design changes. The amount of contingency varies with the stages of design. As the design is finalized, the contingency should be reduced to near zero for most school projects.*

design development The process of reviewing and modifying construction drawings as a result of owner input, coordinating with other design disciplines, building code compliance and general fine-tuning.*

design development phase The second phase of a designer's basic services, which includes developing structural, mechanical, and electrical drawings, specifying materials, and estimating the probable cost of construction.

design load **1.** In structural analysis, the total load on a structural system under the worst possible loading conditions. **2.** In air-conditioning, the maximum heat load a system is designed to withstand.

design pressure The highest pressure expected during operation of a system, device, or piece of equipment.

design professional A term used generally to refer to architects; civil, structural, mechanical, electrical, plumbing, and heating, ventilating, and air-conditioning engineers; interior designers; landscape architects; and others whose services have either traditionally been considered "professional" activities, require state licensing or registration, or otherwise require the knowledge and application of design principles appropriate to the problem at hand.

design review A formal, documented, comprehensive and systematic examination of a design to evaluate design requirements and capability of the design to meet these requirements and to identify problems and propose solutions.*

design specification A type of specification that prescribes the materials and methods to be used for contract performance.

design strength **1.** The assumed value for the strength of concrete and yield stress of steel, used to develop the ultimate strength of a section. **2.** The load-bearing capacity of a structural member, determined by the allowable stresses assumed in the design.

design to budget A requirement in the contract between an owner and a design professional that requires the design professional to redesign the project at no additional cost to the owner if the contract price exceeds the owner's budget.

destination elevators A system that groups passengers in order to reduce individual elevators' travel time.

detached dwelling A structure intended for habitation that is surrounded by open space.

detached garage A parking structure whose exterior walls are surrounded by open space.

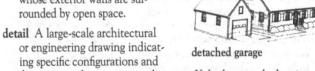

detached garage

detail A large-scale architectural or engineering drawing indicating specific configurations and dimensions of construction elements. If the large-scale drawing differs from the general drawing, it is the architect's or engineer's intention that the large-scale drawing be used to clarify the general drawing.

detailed engineering The detailed design, drafting, engineering, and other related services necessary to purchase equipment and materials and construct a facility.*

detailed estimate of construction cost A forecast of the cost to construct a project, based on unit prices of materials, labor, and equipment, in contrast to a parameter estimate or square-foot estimate.

detail(ed) schedule A schedule used to communicate the day-to-day activities to working levels on the project. The intent of this schedule is to finalize remaining requirements for the total project.*

detailer A draftsman who prepares shop drawings and lists of materials.

detention door A special steel door with fixed lights and steel bars, used to restrict passage in prisons and mental institutions.

detention window A narrow, metal awning window manufactured especially for the security of prisons and mental institutions.

detritus tank A settling tank in a sewage-treatment system for collecting sediment without interrupting the flow of sewage.

developed area An area or site on which improvements have been made.

developed length The length of a pipeline, including fittings, measured along its center line.

developed surface A curved or angular surface graphically represented as flattened out on a plane.

development A tract of land that has been subdivided for housing and/or commercial usage and includes streets and all necessary utilities.

deviation costs The sum of those costs, including consequential costs such as schedule impact associated with the rejection or rework of a product, process, or service due to a departure from established requirements. Also may include the cost associated with the provision of deliverables that are more than required.*

device In an electrical system, a component that carries but does not consume electricity, such as a switch or a receptacle.

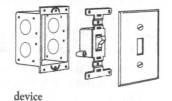

device

deviling The act of scratching a plaster coat prior to applying another coat.

dewater To remove water from a job site by pumps, wellpoints, or drainage systems.

dewpoint 1. The temperature at which air of a given moisture content becomes saturated with water vapor. **2.** The temperature at which the relative humidity of the air is 100%.

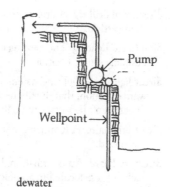

dewater

D-grade wood Used only for interior wood applications, such as for inner plies and backs, where specified.

diagonal A straight structural member forming the hypotenuse of a right triangle, as in the diagonal bracing of a stud wall.

diagonal bond A pattern of bricklaying in which every sixth course is a header course, with the bricks placed diagonally to the face of the wall to form a decorative herringbone pattern.

diagonal brace A member installed at an angle to make a rectangular frame more rigid.

diagonal bridging Crossing pairs of diagonal bracing that extend between the

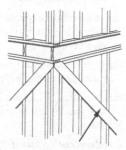

diagonal brace

top of one floor joist and the bottom of the adjacent joist. The bracing is used to distribute the load and to decrease deflection.

diagonal buttress A support structure constructed at a 45° angle to the right-angle corner extension of two exterior walls.

diagonal crack A crack forming at about a 45° angle to the longitudinal dimension of a concrete member, beginning at the tension surface of the member.

diagonal pitch The distance between one rivet or bolt in one row to the nearest rivet or bolt in the next row of a structural member with two or more rows of bolts.

diagonal rib A structural member that crosses the bay of a vault in a diagonal direction.

diagonal sheathing A covering of wooden boards placed diagonally over an exterior stud wall. Although slightly more expensive to install, this method provides a more rigid frame than horizontally installed boards, and may be more architecturally pleasing.

diagonal tension In concrete structural members, the tensile stress resulting from shearing stresses within the member.

diameter The line in a circle passing through its center. Also refers to the nominal diameter designated by an approved material standard.

diamond blade A high-strength circular saw blade used to cut brick, concrete, tile, and other hard surfaces.

diamond drill A high-strength rotary rock drill that cuts a core in rock to create blasting holes.

diamond lath (diamond mesh) A type of expanded metal strip used as a base for plaster. Manufactured by slitting and expanding metal sheets.

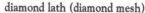

diamond lath (diamond mesh)

diamondwork Masonry laid up so as to incorporate diamond shapes, usually every sixth course.

diaphragm 1. A stiffening member between two structural steel members, used to increase rigidity. **2.** The web across a hollow masonry unit. **3.** An instrument to measure the flow of water in pipes.

diaphragm pump (mud sucker) A reciprocating water pump with a flexible diaphragm used for continuous dewatering of excavations containing mud and small stones.

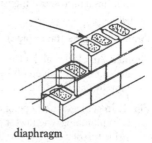

diaphragm

diaphragm valve A valve that is actuated by fluid pressure on a diaphragm.

die A tool for cutting threads on pipe and bolts.

die-cast A method of casting by forcing molten metal into a mold.

differential settlement Uneven, downward movement of the foundation of a structure, usually caused by varying soil or loading conditions and resulting in cracks and distortions in the foundation.

differing site conditions Unanticipated physical conditions at the site that differ materially from those set forth in the contract or ordinarily encountered in work of the same nature.

89

diffused light Light reflecting from a surface rather than radiating directly from a light source.

diffuser 1. Any device or surface that scatters light or sound from a source. **2.** A circular, square, or rectangular air distribution outlet, generally located in the ceiling, and composed of deflecting members to discharge supply air in various directions.

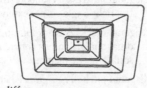

diffuser

digital Refers to communications equipment and procedures in which information is represented in binary form (1 or 0), as opposed to analog form (variable, continuous wave forms).

digital theater system A surround-sound system used in homes and commercial applications.

digital video recording (DVR) A video recording system used as a surveillance device that is capable of retaining voluminous records for long periods of time, limited only by the digital storage device.

dike (dyke) 1. An earth embankment for retaining water. **2.** A large ditch.

dilation The expansion of concrete during cooling or freezing.

dimension A distance between two points, lines, or planes.

dimension line A light, fine line with arrowheads or tick marks at each end, used to show the measurements of the main object lines.

dimension lumber (dimension stuff) Lumber that is cut from 2″ to less than 5″ thick, and 2″ or more wide.

dimension shingles Shingles cut to a uniform size, usually 5″ or 6″ wide, and used for special architectural effects.

diminished stile (diminishing stile, gunstock stile) A door stile that has a narrower dimension at a panel, particularly at a glazed panel.

diminishing pipe (taper pipe) A pipe reducer.

dimmer An electrical device that varies the amount of light given off by an electrical lamp by varying the current.

dip 1. In geology, the slope of a fault or vein. **2.** In plumbing, the lowest point of the inside top of a trap.

dipcoat A covering of paint or other coating applied by immersing an object in a liquid. Usually applied as an anticorrosive.

dipper The digging bucket attached to the arm of a power shovel or backhoe.

dimmer

dipper dredge A bucket used as part of a dredging operation for excavating under water.

dipper stick The straight arm that connects the dipper with the boom of a power shovel.

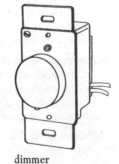

dipper stick

dipper trip The mechanism that releases the latch to the bucket door of a power shovel to discharge the load.

direct component The portion of light from a lamp or luminaire that arrives at the workplane without being reflected.

direct costs The labor, material, subcontractor, and heavy equipment costs directly incorporated into the construction of physical improvements.

direct current A term applied to an electrical circuit in which the current flows in one direction only.

direct damages Damages that can be expected to arise as a matter of course from a breach.

direct digital control (DDC) A system that receives electronic signals from sensors, converts those signals to numbers, and performs mathematical operations on the numbers inside a computer. The computer output (in the form of a number) is converted to a voltage or pneumatic signal to operate an actuator.

direct drive A system in which the driver and the driven element (motor and fan) are positively connected in line to operate at the same speed.

direct dumping The placing of concrete using a chute or bucket directly into open forms at or near its final position.

direct expansion (DX) Refrigeration systems that employ expansion valves or capillary tubes to meter liquid refrigerant into the evaporator.

direct expense All cost items that are directly incurred by or chargeable to a project, assignment, or task.

direct-indirect lighting A lighting system employing diffused light luminaires that emit little or no light horizontal to the lighting unit.

directional drill rig A drill rig that can drill holes in directions other than vertical.

directional lighting The lighting of an object or workplace from a desirable source location.

direct lighting A lighting system, usually emitting light in a downward direction, that distributes 90% to 100% of its light in the direction of the surface to be lit.

direct luminaire A luminaire that directs 90% to 100% of its light downward.

directory board An informational panel with changeable letters, such as those located in an office building lobby.

direct overhead costs Project overhead costs that are directly related to one individual project. These may include temporary facilities, trailer rental, superintendence, telephone usage, staging or lifts, and electrical power.*

direct personnel expense Salaries and wages, including fringe benefits, of all principals and employees attributable to a particular project or task.

direct solar gain Solar heating achieved when sunlight enters a room directly and heats the room or is stored in walls or floors.

direct system A heating, air conditioning, or refrigerating system that employs no intermediate heat exchanger to heat or cool a space.

dirt-depreciation factor A measurement of the reduction of light emitted by a luminaire as a result of accumulated dirt.

disappearing stair A hinged stair, usually folding, that is attached to a trap door in the ceiling and can be raised out of sight when not in use.

discharge head The pressure (in pounds per square inch or psi) measured at the center line of the discharge side of a pump and very close to the discharge flange converted into feet.

disconnecting means A device or group of devices that can disconnect the conductors of a circuit from their supply source.

discontinuous construction A method of construction whereby structural members are staggered to reduce sound transmission through a wall or floor.

disappearing stair

discounting Adjusting cash flow to a common point in time (often the present) when an analysis is performed. The conversion of all costs and savings to time-equivalent "present values" allows them to be added and compared in a meaningful way.

discount rate A percentage rate used to compute the present value of future cash flow.

discovery A pretrial procedure providing for the full disclosure of all facts and documents related to a contract dispute. The methods used include production of documents, depositions, interrogations, and requests for admission.

disintegration Deterioration or crumbling caused by oxidation, freezing, or exposure to other elements. The term is often applied to the deleterious effects suffered by exposed concrete.

disk sander A hand-held power tool that has a rotating, circular abrasive disk used for smoothing or polishing a surface.

disk sander

dismantling The process of taking apart a structure or building in sections in order to reassemble them elsewhere.

dispatching The selecting and sequence of jobs to be run at individual work stations and the assignment of these jobs to workers. In many companies, dispatching is done by the actual shop line supervisor, setup worker, or lead worker. A dispatcher is usually a representative of the production control department, which handles this job assignment task.*

dispersing agent An additive that increases the workability or fluidity of a paste, mortar, or concrete.

displacement 1. The volume displaced by a piston or ram moving to the top or bottom of its stroke. **2.** In hydraulics, the volume of water displaced by a floating vessel.

displacement pile A pile that displaces soil as it is driven by either shifting or compacting it.

displacement ventilation Ventilation that uses natural convection processes to move warm air up and out of a building. Displacement ventilation tends to use less energy than conventional forced air ventilation, as it works with natural convection processes.

disposal unit An electric motor-driven unit installed in a sink to grind food waste for disposal through the sanitary sewer system.

dispute A disagreement between the owner and the contractor as to a question of fact or contract interpretation which cannot be resolved to the mutual satisfaction of the parties.*

dispute procedure The administrative procedure for processing a contract dispute with the United States government. This procedure is provided for in the Contract Disputes Act.

disruption The result of actions or inactions that interfere with performance efficiency, thereby requiring additional effort in performance of the contract work.

distance separation The distance, as specified in fire-protection codes, from an exterior wall to an adjacent building, property line, or the centerline of an adjacent street.

distributed antenna system Antennas and active repeater amplifiers used to enhance wireless communications within a building.

distributed generation A facility's generation and supply of electricity or heat for its own use. Surplus electrical power can be delivered to the utility's grid.

distributed load A load distributed evenly over the entire length of a structural member or the surface of a floor or roof, expressed in weight per length or weight per area.

distribution 1. The movement of heated or conditioned air to desired locations. **2.** The delivery of electricity over a system from transmission points. **3.** The placement of concrete from where it is discharged to its final location.

distribution board (distribution panel, distribution switchboard) An electric switchboard or panel used to distribute electricity within a building. The switchboard is enclosed in a box and contains circuit breakers, fuses, and switches.

distribution box A container located at the outlet of a septic tank to distribute the effluent evenly to the drain tiles in the absorption field.

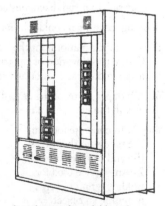
distribution board (distribution panel, distribution switchboard)

distribution line The main electrical feed line to which other circuits are connected.

distribution reinforcement The placement of small-diameter reinforcing steel at right angles to the main reinforcing steel in a slab. This method of reinforcement is used to distribute the load and prevent cracking.

distribution steel The small-diameter reinforcing steel used in distribution reinforcement.

distribution transformer A device that transforms primary voltage into a lower secondary voltage for use within a building.

district heating or cooling Distribution of heat or cooling from a large district plant, at greater efficiency than would be the case with numerous, smaller plants.

ditch A long, open trench used for drainage, irrigation, or for burying underground utilities.

ditch

diversion A pipe, trench, or channel, usually temporary, used to divert the flow of water from its usual course.

diversity factor In electrical systems, the sum of the individual demands of the subsystems divided by the maximum demand of the whole system.

dividers A compass with both legs terminating in points. Used for transferring measurements from a plan or map to a scale, or vice versa.

divider strips The nonferrous metal or plastic strips used as screeds in terrazzo work. Also used to divide the panels.

divider strips

division One of the standard major construction index classifications used in specifying, pricing, and filing construction data.

division wall (fire wall) A regulation fire-resistant wall extending from the lowest floor of a building through the roof to prevent the spread of fire.

D-line crack (D-crack) 1. One of several fine, closely spaced, randomly patterned cracks in concrete surfaces. 2. Similar cracks parallel to the edges or joints in a slab. 3. Larger cracks in a highway slab that run diagonally across its corners.

dobying Blasting a boulder by mud-capping rather than drilling.

dock A raised platform for the loading and unloading of trucks. Often, the elevation of a portion of the dock is adjustable to accommodate the differing heights of truck beds.

dock bumper A recoiling or resilient device attached to a dock to absorb the impact of a truck.

dog 1. In concrete forming, the hardware that holds the end of a snap tie. 2. Any device used for holding, gripping, or fastening an object.

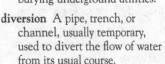

dog

dog anchor (dog iron) A short piece of iron with its two pointed ends bent at right angles, driven into two adjoining pieces of timber to hold them together.

dog bars Intermediate vertical members in the lower part of a gate, used to prevent animals from passing through.

dog-ear In roofing or flashing, an external corner formed by folding rather than cutting the metal.

dolly 1. One of a variety of small, wheeled

dolly

devices used to move heavy loads. 2. A block of wood placed on the upper end of a pile to cushion the blow of a hammer. 3. A steel bar with a shaped head to back up a rivet while the rivet is being driven.

Dolly Varden siding Beveled wood siding that is rabbeted along the bottom edge, such as on novelty siding.

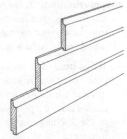

Dolly Varden siding

dome 1. A hemispherical roof such as that commonly seen on government structures. 2. A rectangular pan form used in two-way joist or waffle concrete floor construction.

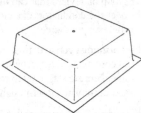

dome light A dome-shaped skylight made of glass or plastic.

domestic hot-water heater A packaged unit that heats water for household purposes.

dome

domestic sewage Household waste, including that originating from single- or multi-family housing.

door A movable member used to close the opening in a wall.

door area The total area of outside doors and facings. Used in computing heating and cooling requirements of a building.

dome light

door casing The finished visible frame into which a door fits.

door check An automatic door closer.

door class A fire-rating classification for doors.

door clearance The distance or space between the bottom of a door and a finished floor, or between two double doors.

door closer A device that controls the speed and force of closure of a door.

door contact (door switch) A switch in an electrical circuit that is opened or closed by opening or closing a door.

door frame The surrounding assembly into which a door fits. Consists of two uprights, jambs, and a head over the top.

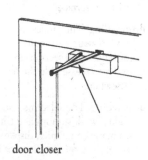

door closer

door hand (door swing) The direction of the swing of a door, expressed as either left-hand or right-hand.

door head 1. The horizontal upper member of a door frame. 2. A lintel.

door holder A device for holding a door open.

door jack A device for holding a wooden door in place while the hinges are being set or while the door is being planed to fit.

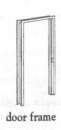

door frame

door latch A device for holding a door closed. Usually a beveled bolt self-activated by a spring or by gravity when hitting the strike.

door mullion The vertical member in the middle of a double door frame that contains the strike of the two leaves.

door opening The size of a door frame opening measured between jambs and from the finished floor (or threshold) to the head.

door operator A power-operated mechanism for opening and closing an elevator door.

door pivot The pin on which a swinging door rotates.

door pocket The opening in a wall or partition that receives a sliding door. The door is installed on tracks during the rough framing stage of construction.

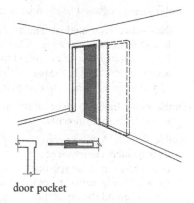

door rail Any of the horizontal members that connect the stiles of a door. In a flush door, the members are hidden. In a panel door, the members are exposed.

door pocket

door schedule A table in the contract documents listing all the doors by size, specifications, and location.

door sill See **threshold**.

door stile The outside, upright members of a door.

door strip A weatherproofing material used around a door.

door trim The casing or molding around a door frame that covers the joint between the frame and the wall.

doorjamb The vertical member on each side of a door frame.

doorstop The strip on the door frame against which the door closes.

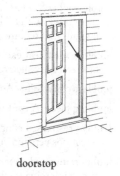

doorstop

dope 1. In plumbing, a material placed on pipe threads to make them waterproof. **2.** An additive in mortar or plaster to accelerate or retard the set.

dormant beam Any large beam that supports smaller ones.

dormer A projection through the slope of a roof for a vertical window.

dormer cheek The side wall of a dormer.

dormer window A window on the front vertical face of a dormer.

dormitory A residential building that includes sleeping facilities. Most commonly found at an educational institution.

dormer

dot A small amount of plaster placed on a surface between grounds to aid the plasterer in maintaining a constant thickness.

D.O.T. shipping name A material classification name assigned by the Department of Transportation to all transportable materials.

double-acting door (swinging door) A door with double-acting hinges so that it can swing both ways.

double-acting frame A door frame without stops, which allows a double-acting door to swing both ways.

double-acting hinge A door hinge that allows the door to swing both ways.

double back (double up) A method of applying two coats of plaster where the second coat is applied before the first coat has set.

double-break switch An electrical switch that opens and closes a conductor in two places.

double breasted A firm that operates or has ownership in both open shop and union companies.

double brick A masonry unit that measures 4″ × 5~THQ″ by 8″. The nominal measurement of standard bricks is 4″ × 2-2/4″ × 8″.

double bridging Bridging that breaks the span of joists into three sections.

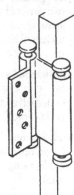

double-acting hinge

double cheek cut A two-sided cut that forms a "V" at the end of a rafter, especially in hip and gambrel roofs.

double corner block A concrete masonry unit that has flat, rectangular faces.

double course (doubling course) Two thicknesses of roofing or siding shingles laid one over the other so that there is complete coverage of at least two layers overall.

double-cut saw A saw with teeth that cut both on the push and the pull stroke.

double door Two single doors or leaves hung in the same door frame.

double egress frame A door frame designed for two doors, both of the same hand, swinging in opposite directions. The door that swings toward the viewer is called a *reverse-hand door*.

double-end-trimmed Refers to wood that has been passed through saws and trimmed smoothly at both ends, commonly in length increments of 2′.

double-extra-strong pipe A specification for steel pipe that indicates the wall thickness has been increased to give the pipe double strength.

double egress frame

double-faced hammer A hammer whose head has two striking faces.

double-faced stock AA and AB grades of sanded plywood, both interior and exterior, having a good appearance on both sides.

double Flemish bond Brickwork laid up so that it has a pattern of headers and stretchers laid alternately on both faces.

double floor A floor constructed with a subfloor under a finished wood floor.

double-framed floor A floor system constructed with both binding joists and common joists. The ceiling is attached to the binding joists and the floor is laid on the common joists. The binding joists are framed by girders.

D

double-framed roof A roof framing system in which both longitudinal and lateral members are used, such as one that employs purlins.

double-gable roof An M-shaped roof with a valley between two gables.

double glazing Window with two panes of glass with an air space between for increased thermal and sound insulation.

doublehanded saw A crosscut saw with a handle on each end, designed to be used by two people.

double header A header joist for an opening made stronger by using two pieces of lumber.

double-gable roof

double-hung window A window that has two vertical sliding sashes, one above the other, mounted on separate guides so that either or both can be opened at one time. This is the type of window most commonly used on wood-frame houses.

double layer Two layers of gypsum board installed over each other to improve fire resistance or sound-suppression characteristics.

double-lock seam In metal roofing and ductwork, a seam formed by double-folding the edges of adjacent sheets and laying them down.

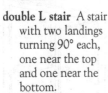

double-hung window

double L stair A stair with two landings turning 90° each, one near the top and one near the bottom.

double-margin door An extra-wide door constructed to look like two doors.

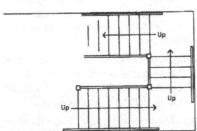

double L stair

double partition A partition constructed with two rows of studs to form a pocket door or a cavity for soundproofing.

double-pitched A roof with a second pitch or slope, such as a gambrel roof.

double-pole switch An electrical switch that has two blades and contacts that open or close both sides of a circuit simultaneously.

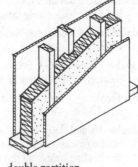

double partition

double pour In built-up roofing, a second, final application of bitumen without adding another layer of felt.

double rafter A pair of roof rafters installed alongside each other.

double return stair A stair with one flight to a landing, separating into two flights to the next floor.

double roof A roof-framing system in which the rafters rest on purlins.

double step In timber framing, a double notch cut into a tie beam supporting the rafters to reduce horizontal shear.

double-strength glass Plate or float glass that is $\frac{1}{8}''$ thick.

double-suction riser method The use of two pipes for the upward flow of refrigerant suction gas. During periods of low refrigeration load, the refrigerant flow of one of the risers is blocked by the accumulation of oil in the base of the riser, diverting refrigerant flow at a higher velocity to transport oil up the other riser.

double swing frame A frame for two matching doors, each of which swing in the same direction.

double T-beam A precast concrete member composed of two beams with a flat slab cast monolithically across and projecting beyond the top of the two beams.

double T-beam

double-throw switch An electrical switch that can make connections in either of two circuits by changing its position.

double-tier partition A partition that is continuous through two stories of a building.

double-wall cofferdam A cofferdam constructed from two rows of sheet piling with fill between them. Used principally for high cofferdams that require greater stability.

double-walled heat exchanger A heat exchanger with two walls between the collector fluid and the potable water.

double-welded joint A joint welded on both sides.

doubling piece A cant strip or fillet.

doubly prestressed concrete A concrete member prestressed in two directions perpendicular to each other.

doubly reinforced concrete A concrete member cast with compression reinforcing steel as well as tension reinforcing steel.

Douglas fir *Pseudotsuga taxifolia.* A coniferous softwood found throughout the western United States and Canada and grown abundantly on the western slopes of the Cascade Mountains in Oregon. This tree produces a strong, durable timber that is widely used in general construction as well as in finish applications.

dovetail (dovetail joint) In finish carpentry, an interlocking joint that is wider at its end than at its base.

dovetail cramp A dovetail-shaped metal fastener used for lifting masonry units.

dovetail cutter A rotary power tool used to shape dovetail joints and mortises.

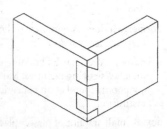

dovetail (dovetail joint)

dovetail half-lap joint A dovetail joint and mortise cut halfway through adjoining pieces to give the appearance of a butt or miter joint on the opposite side.

dovetail hinge A strap hinge with both leaves shaped like a dove's tail (wider at the end than at the base).

dovetail miter A joint having a concealed dovetail so that it appears to be a miter joint.

dovetail plane A wood-finishing tool used for preparing dovetail joints.

dovetail saw A small tenon saw or back saw with a stiffening strip along its back and fine teeth for preparing dovetails.

dowel A cylindrical piece of stock inserted into holes in adjacent pieces of material to align and/or attach the two pieces.

dowel-bar reinforcement Short sections of reinforcing steel that extend from one concrete pour into the next. Used to increase strength in the joint.

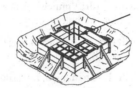

dowel-bar reinforcement

dowel lubricant A lubricant applied to dowels placed in adjoining concrete slabs to allow longitudinal movement in expansion joints.

dowel plate A steel plate with holes through which dowels are driven to trim them to size.

dowel screw A threaded dowel.

downlight A small direct-lighting unit, usually recessed in the ceiling and directing light downward only.

downpayment A partial payment of the construction project's contract price made by an owner to a contractor immediately at the beginning of the contract.*

downspout (drainpipe, conductor, leader) A vertical pipe used to carry rainwater from the roof to either the ground or a drainage system.

down-the-hole drill A percussion drill used for excavation powered from the surface that goes down the hole without the use of steels.*

downtime The amount of time a piece of equipment cannot be used due to failure, repair, or maintenance.

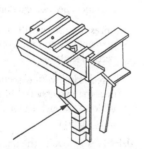

downspout (drainpipe, conductor, leader)

dozer shovel A tractor-mounted unit that can be used for pushing, digging, or loading earth.

draft 1. The drawing power of air and gases through a chimney flue. **2.** A note similar to a check except that banks will not cash it without certain preliminary processing. A draft takes 10 to 14 days to "clear," or be guaranteed by the bank. (During this period, the bank sends the draft back to the insurance company for validation.) **3.** A smooth strip worked on a stone surface to be used as a guide for finishing the surface.

draft curtain (curtain board) A noncombustible curtain surrounding a high-hazard area to contain and direct flame, smoke, and fumes.

draft gauge An instrument used to measure differences in air pressure.

draft hood A cap that fits over a chimney flue to prevent downdrafts.

drafting chisel A tool used most commonly to cut a narrow border or line at the end of a stone.

draft regulator A controlling device that maintains a desired draft level in an appliance.

drag 1. A long, serrated plate used to level and score plaster in preparation for the next coat. **2.** Any combination of logs, chains, or other materials dragged behind a tractor to fine-grade earth.

draft hood

dragging A textured finish similar to graining in which a wet, stippled paint surface is brushed with a clean, natural brush to produce subtle stripes.

dragline A bucket attachment for a crane that digs by drawing the bucket towards itself using a cable. The attachment is most commonly used in soft, wet materials that must be excavated at some distance from the crane. This device is used extensively in marsh or marine work.

dragon beam A short, horizontal membrane that bisects the angle of the wall plate at the corner of a wood-frame building, and bisects the angle brace at its other end. The member supports a hip rafter at one end and is supported by the angle brace at the other end.

drag saw A powered reciprocating saw used to cross-cut logs.

drag strut A structural member that transfers lateral force from one vertical member to another.

drain A pipe, ditch, or trench designed to carry away wastewater.

drainage 1. Wastewater. **2.** The process by which wastewater is carried away.

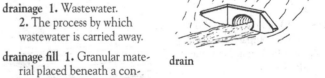

drain

drainage fill 1. Granular material placed beneath a concrete floor slab to inhibit the capillary rise of moisture. **2.** Lightweight concrete that is placed on a floor or roof to promote drainage.

drainage piping The piping comprising the plumbing of a drainage system.

drainage system All the components that convey the sewage and other wastewater to a point of disposal.

drain cock A cock or spigot at the lowest point of a water system where the system can be drained.

drain tile Short-length sections of tile laid with open joints, usually surrounded with aggregate and covered with asphaltic paper or straw. Used to drain the water from an area.

drapery track A system of tracks or bars that support draperies and allow them to be drawn.

drawbar The bar attached to the back of a tractor for pulling equipment.

drapery track

drawdown The distance by which the groundwater level is lowered by excavation, dewatering, or pumping.

drawer dovetail A dovetail joint in which the tenons of one member do not pass completely through the other member, such as the joint used on the front of a drawer.

drawer kicker A block that prevents a drawer from tilting downward when open.

drawer roller Prefabricated sliding drawer hardware employing rollers to facilitate the opening of a drawer.

drawer runner (drawer slip) The strips on which a drawer slides.

drawer slide The hardware on which a drawer slides.

drawer stop A device that prevents a drawer from sliding all the way out.

drawings 1. Graphic illustrations depicting the dimensions, design, and location of a project. Generally including plans, elevations, details, diagrams, schedules, and sections. **2.** The term, when capitalized, refers to the graphic portions of a project's contract documents.

drawknife (drawshave) A two-handled, curved, woodworking knife used by pulling it toward the user.

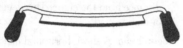
drawknife (drawshave)

drawn finish A smooth, bright finish on longitudinal stock obtained by drawing the stock through a die.

draw tongue The bar attached to a machine so that it can be pulled by another machine.

dredge 1. To excavate under water. **2.** The equipment used for excavating under water, such as a clamshell, dragline, or suction line.

drencher system A fire-protection sprinkler system that protects the outside of a building.

dressed Building material that has been prepared, shaped, or finished on at least one side or edge.

dressed and matched (tongued and grooved) Refers to boards or planks that have been machined so that each piece has a tongue on one edge and a groove on the other.

dressed and matched (tongued and grooved)

dressed lumber (dressed stuff) Lumber that has been processed through a planing machine in order to attain a smooth surface and uniformity of size on at least one side or edge.

dressed size The true dimensions of lumber after sawing and planing, as opposed to the nominal size.

dresser A plumber's tool for flattening sheet lead and straightening lead pipe.

dresser coupling A device for connecting unthreaded pipe.

dressing compound Liquid bituminous material used to cover and protect the exposed surfaces of roofing felt.

dress up To fasten and fabricate structural members on the ground before erection.

drier 1. A unit containing a desiccant that is installed in a refrigeration circuit to collect excess water in the system. **2.** An additive to paints and varnishes that speeds the drying process.

drift 1. In a water-spraying system, the entrained, unevaporated water picked up by the air movement through it. **2.** In aerial surveying, the angle at which a plane must crab, or turn its nose into the wind, in order to fly a predetermined line. **3.** A natural deposit of loose material such as rock or sand. **4.** The lateral movement or deflection of a structure.

drift control agent An additive used in spray mixtures to ensure a cohesive spray solution and reduce spray drift.

drifter A pneumatic drill used for drilling horizontal or sloped blastholes.

driftpin (drift punch) A tapered pin used in steel erection to align holes for bolting or riveting.

drift plug A hardwood plug that is driven into a lead pipe to straighten it or to flare the end.

drill 1. A handheld, manually operated or power-driven rotary tool for boring holes in construction materials. **2.** A large machine capable of drilling 4″ diameter blast holes 100′ deep in rock cuts or quarries. **3.** A machine capable of taking core samples in rock or earth.

drill

drill bit The part of a drill that does the cutting.

drilled-in caisson A caisson that is socketed into rock. *See also* **caisson**.

drill press A rotary drill, mounted on a permanent stand, that operates along a vertical shaft.

drill steel Hollow steel sections that connect the percussion drill with the bit. Air is blown through the steel to clean out the drill hole.

drilled-in caisson

drip 1. A groove in the underside of a projection, such as a windowsill, that prevents water from running back into the building wall. **2.** A condensation drain in a steam heating system. **3.** An interruption or offset in an exterior horizontal surface, such as a soffit, immediately adjacent to the fascia, designed to prevent the migration of water back along the surface.

drip cap (drip mold, drip molding) A horizontal molding placed over exterior door or window frames to divert rainwater.

drip edge 1. The edge of a roof from which rainwater drips into a gutter

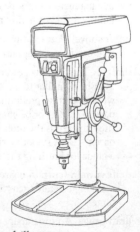
drill press

or away from the structure. **2.** The metal or wood strip that stiffens and protects this edge.

drip irrigation A landscape watering system using low water pressure and flexible tubing placed on the ground. These systems conserve water by targeting the roots of plants, rather than letting water evaporate as sprinklers would do.

drip pan A device for catching and containing drips and leaks from liquid storage drums.

dripstone A drip cap made of stone.

dripstone course A course of stone projecting out beyond the face of a wall to shed water away from the wall.

drive band (driving band) A steel band placed around the top of a wooden pile to protect it during driving.

driven pile A pile that is driven into place by striking it with a pile-driving hammer.

drivescrew (screw nail) A screw that is driven with a hammer and removed with a screwdriver.

drive shoe A cap placed over the bottom of a pile to protect the pile during driving.

driving cap A steel cap placed over the top of a pile to protect the pile during driving.

driving resistance The number of blows required to drive a pile a given distance.

drop In air conditioning, the vertical distance that a horizontally projected air stream has fallen when it reaches the end of its throw.

drop apron A metal strip that is bent downward at the edge of a metal roof to act as a drip.

drop-bottom bucket A concrete bucket whose bottom opens when it touches the surface where the concrete is to be deposited.

drop box An electrical outlet box hung from above, as from a ceiling.

drop ceiling (dropped ceiling, suspended ceiling) A non-structural ceiling suspended below the structural system, usually in a modular grid pattern. A drop ceiling usually contains a lighting system and can also conceal wiring, piping, and ductwork in the space above (the plenum).

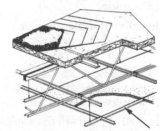

drop ceiling (dropped ceiling, suspended ceiling)

drop elbow A pipe fitting, at a 90° angle, with one or more flanges for attachment to a wall or stud.

dropout An event where voltage decreases to zero for at least five milliseconds.

dropout ceiling A suspended ceiling with heat-sensitive tiles that are designed to drop out in the event of a fire and sprinkler discharge. A dropout ceiling is aesthetically pleasing, as it does not require that sprinklers be extended below the ceiling.

drop panel The part of a cast-in-place flat slab that has been thickened on the underside at the location of the columns.

dropped girder (dropped girt)

dropped girder (dropped girt) A girder that runs below the floor joists and supports them.

drop siding A type of siding that is tongued and grooved or ship-lapped so that the edge of each board fits into the edge of the adjoining board.

drop siding

drop vent In plumbing, a vent that is connected at a point below the fixture.

drop wire An electrical conductor dropped from a pole to a building to supply electric power to the building.

drought tolerance The capacity of a landscape plant to function well in drought conditions.

drove A mason's broad, blunt chisel used for facing stone.

drum 1. One of the stone cylinders forming a column. **2.** The wall supporting a dome. **3.** The rotating cylinder used to wind up cable on a hoist. **4.** The cylinder in which transit-mixed concrete is transported, mixed, and agitated.

drum rasp A cylindrical abrasive device used with a power drill to smooth or contour wood, plastic, and soft metals.

drum sander A power tool with a cylindrical abrasive sleeve used to smooth or contour wood, plastic, and soft metals. Usually mounted on a stand.

drum trap A cylindrical plumbing trap usually set flush with a floor so that easy access can be gained by unscrewing the top.

drunken saw (wobble saw) A type of circular saw designed to operate with a built-in wobble so that the kerf it makes is greater than the thickness of the saw. This type of saw is used for grooving and for other special purposes in carpentry.

dry basin systems Condenser water systems in which the cooling tower basin is located indoors, and away from the cooling tower.

dry batch weight The weight of the materials, excluding water, used to make a batch of concrete.

dry-bond adhesive (contact adhesive) An adhesive that seems dry but that adheres on contact.

dry-bulb temperature The ambient temperature indicated by an ordinary thermometer with an unmoistened bulb.

dry clamping The practice of piecing together all parts of an assembly, such as a table, before fastening to confirm that all joints fit.

dry concrete Concrete that has a low water content, making it relatively stiff. The effects are a lower water-cement ratio, less pressure on forms, lower heat of hydration, and a consistency that allows for placement on a sloping surface.

dry course The first course of a built-up roof, which lies directly on the insulation without bitumen.

D

dry density The weight of a soil sample that has been heated to 105° centigrade (Celsius) to remove the moisture.

dry filter A device for removing pollutants from the air in a system by passing the air through various screens and dry, porous materials. Most dry filters in HVAC systems today are made of spun fiberglass.

dry glazing The installation of glass using dry, preformed gaskets in place of glazing compound.

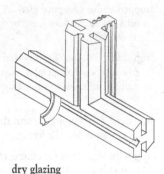

dry glazing

dry ice A substance made of solid carbon dioxide that is used in refrigeration applications. Dry ice sublimates, or changes directly from a solid to a gas.

drying shrinkage Contraction caused by the loss of moisture, particularly in concrete, mortar, and plaster.

dry joint A masonry joint without mortar.

dry masonry Masonry laid without mortar.

dry mix A concrete or mortar mixture that contains little water in proportion to its other ingredients.

dry mortar Mortar mixed to a very dry consistency, but still wet enough for hydration. This type of mortar is used to prevent compaction in the joints of high, narrow lifts.

dry-packing (dry-tamp process) The forceful injection of a damp concrete mixture into a confined space where high strength and little shrinkage is desired, such as under a bearing plate.

dry-pipe sprinkler system (dry sprinkler) A sprinkler system whose pipes remain empty of water until the system is activated. This type of system is particularly useful where there is danger of freezing.

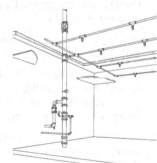

dry-pipe valve The valve that activates a dry-pipe sprinkler system.

dry-pipe sprinkler system (dry sprinkler)

dry-powder fire extinguisher A handheld fire extinguisher that discharges a dry powder by means of a compressed gas. This type of extinguisher is effective for Class B and Class C fires.

dry return In a steam heating system, a return pipe above the water level.

dry-rodded volume A standard method of compacting aggregate when designing a concrete mix.

dry-rodded weight The unit weight of aggregate compacted under standard conditions and used when designing a concrete mix.

dry rodding Compaction of aggregate in a container under standard conditions for determining the unit weight in a concrete mix design.

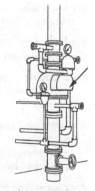

dry-pipe valve

dry rubble construction A method of construction using stone rubble but no mortar.

dry sampling In soil testing, the practice of obtaining an undisturbed sample of earth with a boring tool.

dry shake (dry topping) A concrete surface treatment, such as color, hardening, or antiskid, which is applied to a concrete slab by shaking on a dry, granular material before the concrete has set and then troweling it in.

dry rubble construction

dry strength The strength of an adhesive joint determined under standard laboratory testing conditions.

dry tape The application of joint tape to gypsum board using an adhesive other than the standard joint compound.

dry-type transformer A transformer whose core and coils are not immersed in an oil bath.

dry-volume measurement Measurement of the ingredients of a concrete or mortar mix by using their bulk volume. This method is used principally in mix design.

drywall

drywall Interior finish construction materials that are manufactured and installed in preformed sheets, such as gypsum wallboard. Drywall is an alternative to plaster.

drywall construction Interior construction using drywall rather than plaster.

drywall frame A door frame specifically designed to be used with drywall construction.

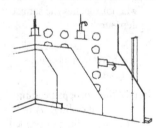

drywall construction

dry well A pit or well that is filled or lined with coarse aggregate or rocks and designed to contain drainage water until it can be absorbed into the surrounding ground.

dual bed system A gas-phase or liquid adsorption treatment system with two carbon absorbers, a pump, and associated piping, that can be configured in series or in parallel, depending on facility requirements.

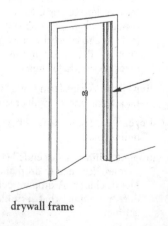

drywall frame

dual duct In electrical wiring, a duct that has two individual raceways.

dual-duct system An HVAC system using two ducts, one for hot air and one for cold air. The air from these ducts is blended in mixing boxes before distribution to each location.

dual-flush toilets Toilets that conserve water by allowing different settings for liquid versus solid waste.

dual-fuel system A heating unit that can use either of two types of fuel.

dual project gates Providing one gate for employees of union firms working on the project and a separate gate for employees of nonunion firms working on the project. In the event of a labor dispute, only the union gate can be picketed.

dual vent (common vent) A single plumbing vent that serves two fixtures.

dubbing (dubbing out) 1. Filling in irregularities in a plaster wall before applying the final coat. 2. Roughing in a plaster cornice before forming it with the final coat.

duckboard A wooden walkway across a wet area.

duct 1. In electrical systems, an enclosure for wires or cables, often embedded in concrete floors or encased in concrete underground. 2. In HVAC systems, the conduit used to distribute the air. 3. In post-tensioning, the hole through which the cable is pulled.

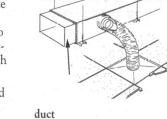

duct

duct liner Fiberglass sound and thermal insulation material used inside the sheet metal ducts of air conditioning systems.

duct static pressure 1. The pressure acting on the walls of a duct. 2. The total pressure less the velocity pressure. 3. The pressure existing by virtue of the air density and its degree of compression.

duct system The connected elements of an air-distribution system through ductwork.

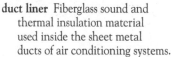

duct system

ductwork The ducts of an HVAC system.

due care A legal term defining the standards of performance that can be expected, either by contract or by implication, in the execution of a particular task.

dumbwaiter A small hoisting mechanism or elevator in a building used for hoisting materials only. The dumbwaiter was originally developed by Thomas Jefferson for use in his home at Monticello.

dummy In CPM (critical path method) scheduling, a restraint with no activity and no time.

dumped fill Soil that has been deposited, usually by truck, but has not yet been spread or compacted.

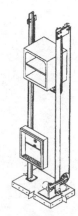

dumbwaiter

dumpster Portable container box normally placed on a construction site for the disposal of job-generated refuse.

dumpy level A surveying instrument consisting of a telescope rigidly attached to a vertical spindle. Used to determine relative elevation.

dunnage 1. Waste lumber. 2. Temporary timber decking. 3. Structural support for a system within a building that is independent of the building's structural frame. An example would be the supports for an air-conditioning cooling tower. 4. Strips of wood used in stowing cargo to provide air space between pieces or packages.

duplex (duplex house, two-family house) A house with two separate dwelling units often sharing the same parcel of land.

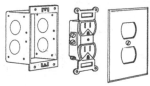

duplex (duplex house, two-family house)

duplex apartment An apartment with rooms on two floors connected by a private staircase.

duplex burner A gas heating system with two burners that can burn either simultaneously for rapid heating or separately during lower heating demand.

duplex cable Two conductors, separately insulated, encased in a common insulated cover.

duplex receptacle (duplex outlet) Two electrical receptacles housed in the same outlet box.

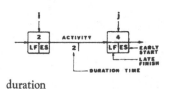

duplex receptacle (duplex outlet)

duramen The hard inner portion, or heartwood, of a tree.

duration The amount of time estimated to complete an activity in the time scale used in the schedule (hours, days, weeks, etc.). Planned production rates and available resources will define the duration used in a given schedule.*

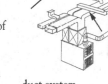

duration

Durham fitting A cast-iron, threaded, pitched fitting used on drainage pipes.

Durham system A wastewater system using all threaded pipes and pitched fittings.

durometer An instrument for determining the relative hardness of a material.

dust collector Any device used to collect the dust produced by a machine or tool, such as those required at all rock-drilling operations and those used with bench saws.

dust-free time The time required for a paint or varnish application to become dust-free under given environmental conditions.

dustproof strike A strike plate for a lock with a spring plunger that completely fills the bolt hole when the bolt is not in it.

Dutch door A door with two leaves, one above the other.

Dutch hip roof A style of roof framing similar to that in a hip roof, but with a small gable at the top.

Dutch door

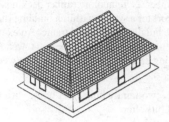

Dutch hip roof

Dutch lap A method of applying shingles with a head lap and a side lap.

dutchman In carpentry and joinery, a small piece inserted to cover a joint or a defect.

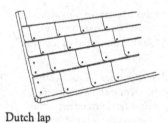

Dutch lap

dwarf partition A partition that does not extend to the full ceiling height.

dwarf wall Any wall less than one story in height, such as a subwall for a crawl space or a wall that supports sleepers for the floor above.

dwelling unit A room or rooms designed for permanent living accommodations for one or more persons.

dynamic Refers to information that will be updated and displayed automatically on a building automation system CRT or display panel when the status of the building equipment changes.

dynamic analysis The analysis of stresses in a framing system under dynamic loading conditions.

dynamic load A load on a structural system that is not constant, such as a moving live load or wind load.

dynamic loading Loading imposed by a piece of machinery through vibration, in addition to its dead load.

dynamic penetration test A test to determine the bearing capacity of soil where a testing device is delivered a series of blows to penetrate the soil.

dynamic resistance The resistance of a pile to blows from a pile hammer, expressed in blows per unit of penetration.

Abbreviations

e eccentricity, erg

E east, engineer, equipment only, modulus of elasticity, 90° elbow

ea each

EA exhaust air

EAM enterprise asset management

E and OE errors and omissions excepted

eastern S-P-F eastern spruce, pine, or fir

EB1S edge bead one side

E&CB2S edge and center bead two sides

EC electrical contractor

ECI energy conservation initiative

ecol ecology

Econ economy

ECM energy conservation measure

ECN engineering change notice

EDA emergency declaration area, equipment distribution area

EDP electronic data processing

EDR equivalent direct radiation

EE eased edges, electrical engineer, energy efficiency, errors excepted

EEM energy efficiency measure

EEO equal opportunity employer

EF entrance facilities

eff efficiency

EG edge (vertical) grain

ehf extremely high frequency

EHP effective horsepower, electric horsepower

EHS environmental health and safety

EIA Electronic Industries Alliance

EIS energy information system

EL elevation, elevator, electroluminescence

elec electric, electrical, electrician

elev elevation, elevator

Elev elevating, elevator

EM end matched

EMCS energy management and control system, energy monitoring and control system

EMF electromotive force

EMI electromagnetic interference

EMPP equal marginal performance principle

EMS energy (or emergency) management system

EMT electrical metallic conduit, thin wall conduit

emu electromagnetic unit

enam enameled

encl enclosure

eng engine

Eng engine

engr engineer

EOC energy operations center

EOL end of line termination

EPA Environmental Protection Agency

EPC Electronic product code; engineering, procurement, and construction

EPDM ethylene propylene diene monomer

EPO emergency power off

EPS expanded polystyrene

eq equal

Eq equation

Eqhv equipment operator, heavy

Eqlt equipment operator, light

Eqmd equipment operator, medium

Eqmm equipment operator, master mechanic

Eqol equipment operator, oilers

equip equipment

equiv equivalent

erec erection

ER equipment room

ERW electric resistance welding

ESCO energy service company

ESD emergency shutdown, electrostatic discharge

est estimate

EST estimated

esu electrostatic unit

ET ethyl

EV electron volt

evap evaporate

EV1S edge vee one side

EW each way

EWSD early warning smoke detector

EWT entering water temperature

ex example, extra

exc excavation, except

excav excavation

exh exhaust

exp expansion

Exp expansion

exp bt expansion bolt

ext exterior

extg extracting

extru extrusion

exx examples

E The rating of elasticity or stiffness of a material.

Definitions

ear (shoulder) 1. A small decorative or structural projecting member or part of a structure or piece. **2.** A small metal projection on a pipe by which it can be nailed to a wall.

earliest event occurrence time In CPM (critical path method) scheduling, the earliest completion of all the activities that precede the event.

earliest expected completion date The earliest calendar date on which the completion of an activity work package or summary item occurs.*

ear (shoulder)

early finish In CPM (critical path method) scheduling, the first day on which no work is to be done for an activity, assuming work began on its early start time.

early start In CPM (critical path method) scheduling, the first day of a project on which work on an activity can begin if all preceding activities are completed as early as possible.

early strength The strength that has developed in concrete or mortar, usually within 72 hours after placement.

early warning smoke detector A device designed to alert building occupants before they become endangered by a fire.

earnest money The deposit that is made by the purchaser when a purchase contract is made to show the purchasers serious intent.

earth auger (earth borer, earth drill) A screw-like drill (usually powered) for boring cylindrical holes in earth or rock.

earth berm A small, earthen, dike-like embankment, usually used for diverting runoff water.

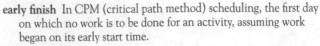

earth auger (earth borer, earth drill)

earth coupling The practice of building into the ground to take advantage of the vast thermal mass of the earth, which typically remains a constant temperature at a certain depth below grade (depending on the climate).

earth electrode A metal plate, water pipe, or other conductor of electricity partially buried in the earth so as to constitute and provide a reliable conductive path to the ground.

earthing lead The conductor that makes the final connection to an earth electrode.

earth pigment Pigment derived from the processing of materials or substances mined directly from the earth.

earth plate A buried metal plate that serves as an earth electrode.

earth pressure The horizontal pressure exerted by retained earth.

earthquake load The total force that an earthquake exerts on a given structure.

earth station (ground station) The equipment set up on earth to communicate with satellites.

earthwork A general term encompassing all operations relative to the movement, shaping, or compacting of earth.

eased edge Any slightly rounded edge.

easement 1. The legal right afforded a party to cross or to make limited use of land owned by another. **2.** A curve formed at the juncture of two members that otherwise would intersect on an angle.

easing Excavation that allows an allotted space to accommodate a foreign piece or part.

eastern red cedar A distinctively red, fine-textured, aromatic wood often used for shingles and closet linings.

eastern S-P-F Eastern spruce, pine, or fir—softwoods whose primary uses include structural lumber as in posts, beams, framing, and sheathing.

eaves 1. Those portions of a roof that project beyond the outside walls of a building. **2.** The bottom edges of a sloping roof.

eaves board A thick, feather board nailed across rafters at the eaves of a building to slightly raise the first course of shingles.

eaves course The first course of shingles or tiles at the eaves of a roof.

eaves fascia (fascia board) A board secured horizontally to, and covering the vertical ends of, roof rafters. The board may support a gutter.

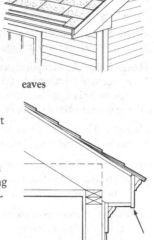

eaves

eaves fascia (fascia board)

eaves flashing A thin metal strip dressed into an eaves gutter from a roof in order to prevent overflowing water from running down the exterior siding.

eaves gutter (eaves trough) A long, shallow wood or metal trough fitting installed under and parallel to the eaves in order to catch and direct water dripping from a roof.

eaves plate A wood beam installed horizontally at the eaves and used in the absence of a rafter-supporting wall to support the feet of roof rafters between existing posts or piers.

eaves strut A structural piece spanning columns at the edge of a roof. The term is usually associated with preengineered steel buildings.

eaves vent Vent openings under the eaves of a house to allow the exchange of air in the attic.

eccentric Not having the same center or center line.

eccentric fitting A fitting whose opening or centerline is offset from the run of pipe.

eccentric load A load or force upon a portion of a column or pile not symmetric with its central axis, thus producing bending.

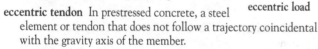

eccentric load

eccentric tendon In prestressed concrete, a steel element or tendon that does not follow a trajectory coincidental with the gravity axis of the member.

ecology The study of the relationship between living things and their environment.

economic life The term during which a structure is expected to be profitable. Generally shorter than the physical life of the structure.

economizer cycle A system of dampers, temperature and humidity sensors, and actuators, which maximizes the use of outdoor air for cooling.

economy brick A cored, modular brick with nominal dimensions of 4″ × 4″ × 8″ and intrinsic dimensions of about 3½″ × 3½″ × 7½″.

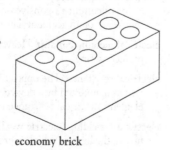

economy brick

economy wall A 4″-thick, back-mortared brick wall strengthened at intervals with vertical pilasters. Designed to support roof framing or walls.

edge beam A stiffening beam at the edge of a slab.

edge form Formwork used to limit the horizontal spread of fresh concrete on flat surfaces such as pavements or floors.

edge grain Lumber sawn so as to reveal the annual rings intersecting the wide face or surface at an angle of at least 45°.

edge guide An attachment to a router that ensures straight cuts by keeping the bit parallel to the work surface edge.

edge joint **1.** A joint between two veneers or laminations, formed in the direction of the grain. **2.** A joint created by the edges of two boards or surfaces united so as to form an angle or corner.

edge nailing Nailing through the edges of successive boards, such as floorboards, so that each board conceals the nails in the board adjacent to it.

edger (edging trowel) A tool used to fashion finishing edges or round corners on fresh concrete or plaster.

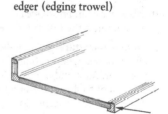

edger (edging trowel)

edge spacer A spacer that positions and holds a pane of glass in place within a window frame.

edgestone A stone used for curbing.

edge strip A plain or molded strip of wood or metal used to protect the edges of panels.

edge strip

edge treatment The finish used to conceal the plies of a plywood edge. Finish may involve molding, wood strips, or wood putty.

edge vent An opening at the perimeter of a roof for the relief of water vapor pressure within the roof system.

edging **1.** An edge molding. **2.** An edging strip. **3.** The process of rounding to reduce the possibility of chipping or spalling exposed edges of concrete slabs. **4.** The process of grinding the edge of flat glass to fit an application.

effective area **1.** The net area of an air inlet or outlet system through which air can pass. The effective area equals the free area of the device multiplied by the coefficient of discharge. **2.** The cross-sectional area of a structural member, calculated to resist applied stress.

effective area of reinforcement In reinforced concrete, the product derived by multiplying the cross-sectional area of the steel reinforcement by the cosine of the angle between its direction and the direction for which its effectiveness is considered.

effective date of the agreement The date indicated in the agreement on which it becomes effective, but if no such date is indicated, the date on which the agreement is signed and delivered by the last of the two parties to sign and deliver.*

effective depth The depth of a beam or slab section as measured from the top of the member to the centroid of the tensile reinforcement.

E

effective height of a column In calculating the slenderness ratio, multiply a valve by the actual column length. The valve varies depending on whether the column is fully or partly restrained at both ends.

effective length The distance between inflection points in a column when it bends.

effective modulus of elasticity The combination of elastic and plastic effects in an overall stress-strain relationship in a service structure.

effective opening The minimal cross-sectional area of the opening at the point of water supply discharge, expressed in terms of the diameter of a circle. The diameter of a circle of equivalent cross-sectional area is given in instances where the opening is not circular.

effective prestress The stress remaining in concrete resulting from prestressing after *loss of prestress*. Effective prestress includes the effect of the weight of the member, but not the effects from any superimposed loads.

effective reinforcement The reinforcement assumed to be active in resisting applied stress.

effective span The center-to-center distance between the supports of a beam, girder, or other structural member.

effective stress In prestressed concrete, the remaining stress in the tendons after the loss of prestress.

effective temperature A single-figure index reflecting the combined effects of temperature, humidity, and air movement on the sensation of warmth and/or cold felt by the human body. Numerically equivalent to the temperature of still, saturated air, which produces an identical sensation.

effective thickness of a wall The thickness of a wall, used in calculating the slenderness ratio.

efficacy The ratio of light output to energy consumption, measured in lumens per watt.

efficiency The ratio of actual performance to theoretical maximum performance.

effort The number of labor units necessary to complete work. Effort is usually expressed in staff hours, staff days, or staff weeks and should not be confused with duration.*

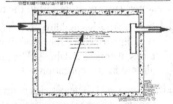

effluent

effluent 1. In sanitary engineering, the liquid sewage discharged as waste, as in the discharge from a septic tank. **2.** Generally, the discharged gas, liquid, or dust byproduct of a process.

egress An exit or means of exiting.

ejector 1. A pump for ejecting liquid. **2.** In plumbing, a device used to pump sewage from a lower to a higher elevation. **3.** In excavation, a dirt-pushing mechanism in a scraper bowl.

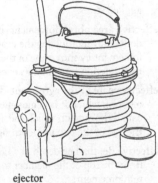

ejector

ejector grille A ventilating grille with slots shaped so as to force out air in divergent streams.

elastic Refers to any material that can return to its initial state after having been deformed by an outside force.

elastic curve In structural analysis, the curve showing the deflected shape of a beam with an applied load.

elastic design An analytical design method in which the design of a structural member is based on a linear stress-strain relationship that assumes that the active stresses are equal to only a fraction of the given material's limit of elasticity.

elasticity The ability of a material to return to its initial state after having been deformed by some outside force.

elastic limit The maximum amount of deforming stress a material can sustain and still return to its initial shape without having incurred any permanent deformation.

elastic loss In pretensioned concrete, the condition resulting in the reduction of prestressing load due to the elastic shortening of a member.

elastic shortening 1. A linearly proportional decrease in the length of a structural member under an imposed load. **2.** In prestressed concrete, the shortening of a member caused by, and occurring immediately upon, application of force induced by prestressing.

elastomer A term descriptive of various polymers that, after being temporarily but substantially deformed by stress, will return to their initial size and shape immediately upon the release of the stress.

elastomeric waterproofing A method of foundation moisture protection using impervious flexible sheet metal.

elbow A sharply bent or fabricated angle fitting, usually of pipe, conduit, or sheet metal. *See also* **knee.**

elbow

elbow catch A spring-loaded locking device having on one end a hook that engages a strike, and on the other end a right angle bend that provides a means for releasing the catch. Commonly used to lock the inactive leaf of a pair of cabinet doors.

electrical conductivity The ability to transmit an electrical current.

electrical drawings Graphic representations of the electrical requirements of a project, including power distribution, lighting, and low-voltage specialty wiring, such as for fire alarms, telephone/data, and technology wiring.

electrically supervised Refers to a closed-circuit wiring system that utilizes a current-responsive device to indicate a failure within the circuit or an accidental grounding.

electrical metallic conduit Heavy-walled conduit that encloses and protects wiring.

electrical resistance The opposition to the flow of an electric current as manifested by a device, conductor, substance, or circuit element. Electrical resistance is measured in *ohms*.

electric arc welding (electric welding) The joining of metal pieces by fusion, the heat for which is provided by an electric arc or the flow

of electric current between the electrode and the base metal. The electrode is either a consumable, melting and bead-depositing, or nonconsumable metal rod. The base is the parent metal.

electric baseboard heater A heating system with electric heating elements installed in longitudinal panels, usually along the baseboards of exterior walls.

electric baseboard heater

electric cable 1. An electric conductor comprising several smaller-diameter strands that are twisted together. **2.** A group of electric conductors that are insulated from one another.

electric-discharge lamp A lamp that emits light through a process where electric current flows through a vapor or gas.

electric drill A portable, handheld, motor-driven tool used for boring holes in a material. Powered either by direct or alternating current.

electric drill

electric eye A photoelectric cell whose resistance varies in response to the amount of light that falls on the cell. An electric eye is often incorporated into electric circuits where it is used in measuring or control devices that depend on illumination levels or on the interruption of a light beam.

electric feeder In power distribution, a set of electric conductors that originate at a primary distribution center and supply power to one or more secondary distribution centers, branch-circuit distribution centers, or a combination of these.

electric field A particular region or space characterized by the existence of a detectable electric intensity at every point within its perimeter.

electric fixture An electrical device that is fastened to a ceiling or wall and used to hold lamps.

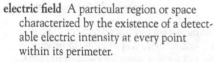
electric fixture

electric generator A mechanism that transforms mechanical power into electrical power.

electric lock A locking mechanism in which the motion of a latch or bolt is controlled by applying voltage to the terminals of the mechanism.

electric meter A device that measures the amount of electricity consumed over a certain period of time.

electric meter

electric outlet A point in an electric wiring system at which current is delivered through receptacles equipped with sockets for plugs, making it available to supply lights, appliances, power tools, and other electrically powered devices.

electric panelboard A panel or group of individual panel units capable of being assembled as a single panel and designed to include fuses, switches, and circuit breakers. The units are housed in a cabinet or cutout box positioned in or against a wall or partition and accessible only from the front.

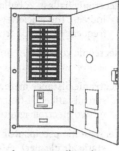
electric panelboard

electric receptacle A contact device, usually installed in an outlet box, which provides the socket for the attachment of a plug to supply electric current to portable power equipment, appliances, and other electrically operated devices.

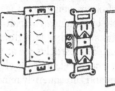
electric receptacle

electric space heater A portable, self-contained heating device in which electricity supplies the heat energy. The heat is blown into a given space or room by a powerful electric fan also contained within the unit.

electric strike An electrical mechanism that allows the release of a door at a remote location.

electric switch A device used for opening or closing electric circuits or for changing the connection of a circuit.

electric space heater

electrode 1. A solid electric conductor through which an electric current enters or leaves an electrolyte, vacuum, gas, or other medium. **2.** That component in an arc welding circuit through which electric current is conducted between the electrode holder and the arc itself. **3.** In resistance welding, that component through which the electric current in the welding machine passes directly to the work and that is usually accompanied by pressure, such as from a clamp.

electrolysis The process of producing chemical changes or the breakdown of chemical compounds into their constituent parts by passing a current between electrodes and through an electrolyte.

electrolyte 1. A substance that separates into ions when in a solution or when fused, thereby becoming electrically conducting. **2.** A conducting medium in which the flow of electric current occurs by the migration of ions.

electrolytic copper Copper refined through electrolytic deposition and used in the manufacturing of tough pitch copper and copper alloys.

electrolytic corrosion The deterioration of metal or concrete by the chemical or electrochemical reaction resulting from electrolysis.

electromagnetic interference (EMI) Radiation emission from a transmission medium caused by using high-frequency signal modulation. Can be reduced by shielding.

electromotive force (EMF) The force that moves electricity in a conductor. EMF, usually measured in volts, is the difference in electric potential, or voltage, between the terminals of a source of electricity.

electronic ballast A ballast that uses semiconductor components to increase the efficiency of a fluorescent system by increasing the frequency (in kilohertz) of its lamp operation.

electronic controls In an HVAC system, the electronically operated sensors and controls.

electronic distance meter (EDM) A surveying device used to measure distance. Distance is determined from the time it takes transmitted electromagnetic waves to reach a target.

electrostatic precipitator (electrostatic air cleaner) A type of filtering device that prevents smoke and dust from escaping into the atmosphere by charging the particles electrically as they pass through a screen. The particles are attracted to one of two electrically charged plates and subsequently removed.

elevated conduit An electrical conduit hung from structural floor members.

elevated floor Any floor system not supported by the subgrade.

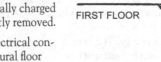

SECOND FLOOR

FIRST FLOOR

GROUND

elevated floor

elevated slab A floor or roof slab supported by structural members.

elevated water tank A domestic water tank supported on an elevated structural framework to obtain the required head of pressure.

elevation **1.** A vertical distance relative to a reference point. **2.** A view or drawing of the interior or exterior of a structure as if projected onto a vertical plane.

elevator A "car" or platform that moves within a shaft or guides and is used for the vertical hoisting and/or lowering of people or material between two or more floors of a structure. An elevator is usually electrically powered, although some short-distance elevators (serving fewer than six or seven floors) are powered hydraulically.

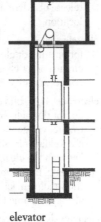

elevator

elevator car That part of an elevator that includes the platform, enclosure, car frame, and gate or door.

elevator interlock A device on the door of each elevator landing that prevents movement of the elevator until the door is closed and locked.

elevator machine beam A steel beam that is usually positioned directly over the elevator in the elevator machinery room and is used to support elevator equipment.

elevator pit That part of an elevator shaft that extends from the threshold level of the lowest landing door down to the floor at the very bottom of the shaft.

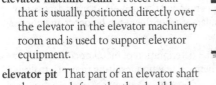

elevator pit

elevator recall The override of elevator operation by a building's fire safety system upon activation of a fire alarm. In a fire emergency, elevator cars automatically return to the ground floor and can be operated only by a fire department.

elevator shaft A hoistway through which one or more elevator cars may travel.

ell (el) **1.** An extension, addition, or secondary wing that joins the principal dimension of a building at a right angle. **2.** Same as *elbow*.

elliptical arch An arch in the shape of a semiellipse and whose intrados is in the form of an ellipse.

embankment A ridge constructed of earth, fill rocks, or gravel and used most commonly to retain water or to carry a roadway. The length of an embankment exceeds both its width and its height.

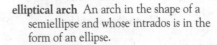

elevator shaft

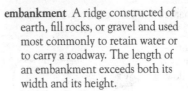

ell (el)

embedded length That length of embedded steel reinforcement that extends beyond a critical section.

embedded reinforcement In reinforced concrete, any slender steel members, such as rods, bars, or wires, that are embedded in concrete in such a way as to function together with the concrete in resisting forces.

embodied energy The energy needed to produce a building product, not accounting for transportation, durability, reuse, and recycling.

embossed Describes a material whose surface is raised or indented.

embedded reinforcement

emergency ballast A lighting system that uses a battery to function in the event of a power loss. When AC power fails, the emergency ballast senses the failure and immediately switches to the emergency mode to illuminate lamps at a reduced lumen output.

emergency declaration area (EDA) An area officially declared in need of hazardous waste or material cleanup.

emergency exit A door, hatch, or other device leading to the outside of the structure and usually kept closed and locked. The exit is used chiefly for the emergency evacuation of a building, airplane, or other occupied space when conventional exits fail, are insufficient, or are rendered inaccessible.

emergency lighting Temporary illumination provided by battery or generator and essential to safety during the failure or interruption of the conventional electric power supply.

emergency power Electricity temporarily produced and supplied by a standby power generator when the conventional electric power supply fails or is interrupted. Emergency power is essential in facilities such as hospitals, where even relatively short power outages could be life threatening to certain individuals.

emergency stop switch A safety switch in the car of an elevator that can be manually operated to cut off electric power from the driving machine motor and brake of an electric elevator, or from

the electrically operated valves and/or pump motor of a hydraulic elevation.

emery wheel An abrasive wheel similar to a grinding wheel but composed primarily of emery (or aluminum oxide) and rotated at high speeds for fine grinding or polishing.

eminent domain The legal right or power of a government to take for public use privately owned property, usually with some degree of compensation to its owner.

emissivity The relative ability of a surface to emit radiant heat.

emittance A percentage of the energy absorbed by a solar energy collector.

empty-cell process A procedure in which wood is impregnated with liquid preservatives under pressure.

emulsion 1. A mixture of two liquids that are insoluble in one another and in which globules of one are suspended in the other (such as oil globules in water). 2. The mixture of solid particles and the liquid in which they are suspended but insoluble, as in a mixture of uniformly dispersed bitumen particles in water, in which the cementing action required in roofing and waterproofing would occur as the water evaporates.

emulsion glue A usually cold-setting glue manufactured from emulsified synthetic polymers.

emulsion paint A paint comprising tiny beads of resin binder that, along with pigments, are dispersed in water. Evaporation of the water affects the coalescence of the resin particles, thus forming a film that adheres to the surface and binds the pigment particles. Vinyl, latex, or plastic paints are examples based on polyvinyl acetate emulsion.

emulsion sprayer An apparatus used in highway resurfacing to spray an emulsified asphalt tack coat on the existing surface.

enamel A type of paint composed of particles of finely ground pigments and a resin binder that dries to form a hard, glossy film with very little surface texture.

enameled brick In masonry, a brick or tile with a glazed ceramic finish.

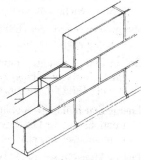
enameled brick

encapsulant (sealant) A liquid applied to asbestos-containing material that prevents the release of fibers into the air either by forming a membrane over the material or by penetrating the material and binding the fibers together.

encapsulate To encase a material or substance; to hide data (such as computer processing logic).

encapsulation A technique for trapping asbestos fibers in a dense chemical substance.

encase To cover with or enclose in a case or lining.

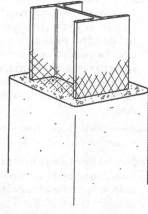
encase

encased beam A metal beam usually enclosed in concrete or some other similar material.

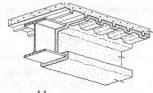
encased beam

encaustic 1. Descriptive of a process in which a material is covered with a mixture of paint solution and wax, then set by heat after application. 2. Relating to paint or pigment that has been applied to glass, tile, brick, or porcelain and then set or fixed by the application of heat.

encaustic tile A decorative or pavement tile whose pattern is created by inlaying clay of one color in a background of a different-colored clay.

enclosed knot A wood knot that is invisible from the surface of a wood member because it is completely covered by surrounding wood.

enclosed stair An interior staircase that has a closed string on each side, often encased in walls or partitions. There are also door openings at various floor levels, thus making it accessible to hallways of living units.

enclosure The containment behind airtight, impenetrable, permanent barriers of sprayed-on or troweled-on asbestos-based materials.

enclosure system A contamination-control system consisting of a work area, a holding area, a washroom, and an uncontaminated area.

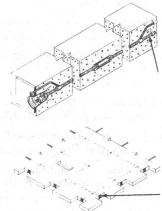

enclosure wall An interior or exterior non-load-bearing wall of skeleton construction, usually anchored to columns, piers, or floors.

encroachment The unauthorized intrusion or extension of a building or structure onto the property of another.

end anchorage A mechanism designed to transmit prestressing force to the reinforced concrete of a post-tensioned member.

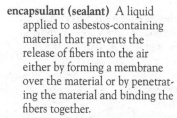
end anchorage

end bearing pile A pile supported mainly by point resistance. The pile's point (toe) rests on or is embedded in a posttensioned member.

end butt joint (end joint) 1. A joint formed when boards are connected square end to square end. 2. A joint between two veneers, formed perpendicular to the grain. 3. In masonry, a joint in which mortar connects the butt ends of two bricks.

end channel A metal stiffener welded horizontally into the tops and bottoms of hollow metal doors to supply strength and rigidity.

end checks Small cracks that develop in the end grain of drying lumber.

end distance The distance from an end of a member to the center of its nearest bolt hole, nail, or screw.

end grain The grain that is exposed when a piece of wood is cut perpendicular to the grain.

end-grain core Plywood or panel core made from wood blocks that are sawn and glued so that the grain is at right angles to the forces of the panels.

end grain

end lap The overlap in a lap joint, such as at the end of a ply of roofing felt.

end lap joint An angle joint involving two members, each having been cut to half its thickness and lapped over the other in such a way as to result in a change of direction.

endless saw A band saw.

end matched Descriptive of boards or strips that are tongued along one end and grooved along the other.

end post A post or other structural member in compression at the end of a truss.

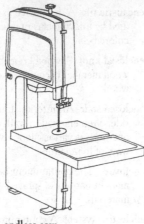

endless saw

end rafter A common rafter positioned in line with the ridge of the roof.

end scarf A scarf joint between two timbers made by notching and lapping the ends and then inserting the end of one into an end of the other.

end section In a drainage system, the prefabricated, flared metal end attached at the inlet and outlet to prevent erosion.

end span In a continuous-span floor design, the exterior span, which is often more heavily reinforced.

end stiffener One of the vertical angles connected to the web of a beam or girder at its ends, to stiffen the beam and transfer the end shear to the shoe, baseplate, or supporting member.

end-use/least cost A major green consideration when designing a building and selecting products. This factor involves an evaluation of the end users' needs and explores different ways of achieving them at the lowest cost with the greatest efficiency over time.

energized Electrically connected to a source of voltage.

Energy and Atmosphere (EA) A term used by LEED in reference to energy consumption and efficiency and greenhouse gas emissions.

energy audit A survey of heat loss through the components of a structure.

energy efficiency measure (EEM) A design, operation, or technology change for the purpose of reducing energy consumption.

energy efficiency ratio (EER) A measure of energy efficiency in the cooling mode that represents the ratio of total cooling capacity to electrical energy input.

energy-efficient design Design that emphasizes highly effective space conditioning equipment and controls, a tight building envelope, and an efficient ventilation system.*

energy-efficient lighting Light fixtures and lamps, such as CFL and LED, that consume substantially less energy than conventional incandescents.

energy harvesting Capture of renewable energy, such as solar, and conversion of that energy into usable electricity.

Energy Information Administration A division of the U.S. Department of Energy.

energy management system (EMS) A system that monitors and controls energy-consuming systems, in the most energy-efficient manner, through software programs.

energy modeling Using computer modeling software to analyze alternative energy systems to help determine the most efficient design. It typically involves not only mechanical and electrical systems, but building orientation, materials, lighting, and landscaping.

energy recovery A process whereby energy from one source is used to provide energy for another process.

energy recovery ventilator (ERV) Mechanical equipment that is added to the ventilation system, featuring a heat exchanger to provide controlled ventilation into a building, while increasing energy efficiency.

energy savings performance contract (ESPC) A contracting partnership between an agency/consumer and an energy services company (ESCO) in which the ESCO evaluates the facility's energy needs and consumption and identifies strategies for improvement. The ESCO often helps pay for the initial cost of implementation in exchange for a share of the energy savings.

energy service company (ESCO) A firm that provides energy management services in exchange for a set fee or an amount tied to a facility's energy cost savings.

energy service provider (ESP) A company that provides energy or energy management services.

Energy Star A volunteer rating and labeling program sponsored by the Department of Energy and the Environmental Protection Agency (EPA) to encourage the use of energy-efficient products.

Energy Star buildings A voluntary partnership between U.S. organizations and the U.S. Environmental Protection Agency (EPA) to promote energy efficiency in buildings.

Energy Usage Intensity (EUI) A measurement used for the purpose of rating a facility's energy consumption per square foot; used by the U.S. Department of Energy.

engineer A design professional who, by education, experience, and examination, is duly licensed by one or more state governments for practice in the profession of engineering. This practice may be limited to one or more specific disciplines of engineering. Construction-related engineering disciplines include civil, structural, mechanical, or electrical systems design.

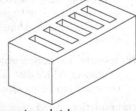

engineer brick

engineer brick Brick whose nominal dimensions are $3\frac{1}{5}'' \times 4'' \times 8''$.

engineered items Items that are purchased to be used for a particular purpose and are engineered to unique specifications, as opposed to commodity materials. This typically includes tagged items and materials that require detailed engineering data sheets.*

engineering survey A survey undertaken for the purpose of obtaining information essential to the planning of an engineering project.

engineer-in-training A person who is qualified to become a registered professional engineer in all respects but the necessary professional experience.

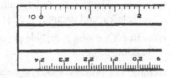
engineer's scale

engineer's scale A straightedge on which each inch is divided into uniform multiples of ten, thus enabling drawings to be made with distances, loads, forces, and other calculations expressed in decimal values.

engineer's transit 1. An instrument used in surveying to measure and to lay out vertical and horizontal angles, distance, directions, and differences in elevation. 2. A theodolite, the directions of whose alidade and telescope are reversible.

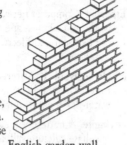
English garden wall bond (American bond, common bond)

English bond 1. Brickwork consisting of alternate courses of headers and stretchers. 2. A strong, easily laid bond.

English garden wall bond (American bond, common bond) Widely used brickwork that can be laid quickly because headers constitute only every fifth or sixth course, with all the other courses being stretchers.

English tile A smooth, flat, single-lap clay roofing tile with interlocking sides.

entablature In classical architecture, the total beam member spanning from column to column, including the architrave (bottom), frieze (middle), and cornice (top).

enterprise asset management Any software system designed to assist with facility management, including the maintenance of physical assets, proper and safe plant operation, and warranty and leasing information.

enthalpy The total amount of heat in a material or system, measured in Btus per pound.

enthalpy wheels Also referred to as heat wheels, a heat recovery system that removes moisture from the ambient air while also cooling the ventilated air by passing all incoming air over a desiccant-coated wheel. This process removes up to 85% of heat/moisture.

entitlement A claimant's right or grounds for recovering losses (as opposed to determining the amount of the losses).

entrained air Nearly spherical microscopic air bubbles that are purposely incorporated into mortar or concrete during mixing. Entrained air is added in order to lessen the effects of the freeze/thaw cycle on concrete that will be exposed to the weather.

entrance An exterior door, lobby, passage, or other designed point of entry into a building.

entrance facilities (EF) Equipment, such as hardware and cables, used to connect utilities to a facility.

entrapped air Voids in concrete that are of at least one millimeter in diameter, and which are the result of air other than entrained air.

envelope 1. A term used to denote the extreme outside surface and dimensions of a building. 2. An assembly of planes representing the limits of an area that may house a structure, i.e., a zoning envelope.

environmental audit An assessment of a facility's compliance with local and national environmental requirements.

environmental building consultant A specialist in sustainable building design who makes recommendations regarding the impact of building materials as they are produced, including waste generated in the construction process and over the product life cycle.

Environmental Protection Agency (EPA) An independent federal agency, formed in 1970, that establishes and controls rules for protecting the environment.

environmental remediation The process of treatment, containment, or removal of hazardous or toxic wastes to protect human health and the environment.

EOS Economy of scale.

EPAct The U.S. Energy Policy Act of 2005. Among its many components, it allows for tax incentives and loan guarantees for various energy-related projects and conservation.

EPA Identification Number A number assigned to each generator of hazardous waste that notifies the EPA of its activity.

EPDM roofing A single-ply roofing material manufactured of an elastomeric polymer synthesized from ethylene, propylene and a small amount of diene monomer. Because of its versatility in application methods, EPDM roofing may be used for any roof shape, slope, height, and climatic exposure.

epoxy Any of several synthetic, usually thermosetting, resins that provide superior adhesive power and hard, tough, chemical- and corrosion-resistant coatings.

epoxy joint In masonry, an unsealed joint in which epoxy resin is used in place of mortar.

epoxy resin A material containing an average of more than one epoxy group per molecule used in the preparation of special coatings or adhesives for concrete and as binders in epoxy resin mortars and concrete.

equalizing bed

equalizing bed A layer of sand, stone, or concrete laid in the bottom of a trench as a resting place for buried pipes.

equipment All the machinery, tools, and apparatus necessary for the proper construction and acceptable completion of a project.

equipment distribution area (EDA) A space housing equipment cabinets in a data center.

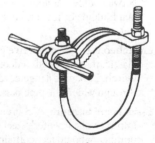
equipment ground

equipment ground 1. A connection that provides a path to ground a conductor within the equipment. 2. A ground connection to any metal part of a wiring system or equipment that does not carry current.

equity The residual value of a business or property, often calculated by subtracting the amount of outstanding liens or mortgages from the total value of the business or property.

equivalent direct radiation (EDR) A unit of heat delivery equal to 240 Btu per hour.

equivalent embedment length That longitudinal part of an embedded reinforcement that is capable of developing stress equivalent to that developed by a hook or mechanical anchorage.

equivalent length The resistance of a duct or pipe elbow, valve, damper, orifice, bend, fitting, or other obstruction to flow, expressed in the number of feet of straight duct or pipe of the same diameter that would have the same resistance.

equivalent round A term used to define the size of an oblong-shaped pipe. The equivalent round of the oblong pipe is equal to the diameter of a pipe with the same circumference.

equivalent thickness The solid thickness to which a hollow unit would be reduced if there were no voids and it had the same face dimensions.

erection The positioning and/or installation of structural components or preassembled structural members of a building, often with the assistance of powered equipment such as a hoist or crane.

erection bracing Temporary bracing used to hold framework in a safe condition during construction until enough permanent construction has been put in place to provide complete stability.

ergonomics The interaction between people and work, particularly the design of machines, chairs, tables, etc., to suit the body and to permit work with minimum fatigue.

erosion The gradual deterioration of metal or other material by the abrasive action of liquids, gases, and/or solids.

erratum An error in printing, writing, or editing, especially one included in a list of corrections appended to a book or publication in which the error appeared.

escalation An increase in the cost of performing construction work, resulting from performing the work in a later period of time and at a cost higher than originally anticipated in the bid.

escalator A continuously moving, power-driven, inclined stairway used to transport passengers between different floor levels of a building or into and out of underground stations.

escalator

escutcheon 1. The protective plate, usually metal or plastic, that surrounds a door keyhole or a light switch. 2. A pipe flange covering a floor hole through which the pipe passes. *See also* **faceplate**.

Essex board measure A chart appearing on a certain type of carpenter's steel square, listing the number of board feet contained in a board 1″ thick and of several standard sizes.

estimate The anticipated cost of materials, labor, equipment, or any combination of these for a proposed construction project.

escutcheon

estimated design load The sum of (a) the useful heat transfer, (b) the heat transfer to or from the connected piping, and (c) the heat transfer that occurs in any auxiliary apparatus connected to a heating or air-conditioning system.

estimated maximum load The calculated maximum heat transfer that a heating or air-conditioning system might have to provide.

estimating The process of determining the anticipated cost of materials, labor, and equipment of a proposed project.

estimator One who is capable of predicting the probable cost of a building project.

etching The process of using abrasion or corrosion (acid) to wear away the surface of glass or metal, often in a decorative pattern.

EtherCAT Ethernet for Control Automation Technology.

Ethernet Computer networking technology used for local area networks and wireless and metropolitan area networks. The Institute of Electrical and Electronics Engineers (IEEE) provides Ethernet standards.

ethylene glycol 1. Antifreeze. 2. A water-miscible alcohol used to transfer heat in heating and cooling provide stability in latex- and water-based paints during freezing conditions.

ettringite A naturally occurring mineral containing a large amount of sulfate calcium sulfoaluminate. Ettringite is also produced synthetically by sulfate attack on mortar and concrete.

eurythmy In architecture, orderliness or harmony of proportion.

evaporable water Water held by the surface forces or in the capillaries of set cement paste, measured as the water that can be removed by drying under specific conditions.

evaporative cooling Cooling achieved by the evaporation of water in air, thus increasing humidity and decreasing dry-bulb temperature.

evaporator The part of a refrigeration system in which vaporization of the refrigerant occurs, absorbing heat from the surrounding fluid and producing cooling.

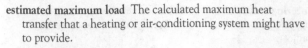

event

event In a CPM (critical path method) arrow diagram, the starting point for an activity that cannot occur until all work preceding it has been performed. Indicated on the arrow diagram by a number enclosed in a circle.

e-waste Electronic equipment that contains lead, mercury, or other hazardous material, and is subject to federal and state disposal requirements.

excavation

excavation 1. The removal of earth, usually to allow the construction of a foundation or basement. 2. The hole resulting from such removal.

excavator 1. A company or individual who contracts to perform excavation. 2. Any of several types of power-driven machines used in excavation.

exception report A report sent to the EPA by a generator of hazardous waste. The exception report is released when the generator does not receive copies of hazardous waste manifest forms from transporters or disposers of wastes that the generator has shipped off-site for disposal.

excess reprocurement costs Additional costs that the government incurs following a default termination to reprocure the defaulted quantity of supplies, services, or unfinished work.

exchange The centrally located arrangement of communications equipment governing the connection of incoming and outgoing lines, including signaling and supervisory tasks.

exchange, private branch (PBX) A telephone exchange system that provides manually operated private telephone service within an organization (a company, for example) as well as access to the public telephone network.

excusable delay A delay to contract performance that is beyond the control, fault, or negligence of the contractor. If a delay is determined to be excusable, the government cannot initiate a termination for default.

executed 1. Under contract law, a contract that has been signed by both parties. **2.** Work that has been completed.

exempt employees Employees exempt from overtime compensation by federal wage and hours guidelines.*

exfiltration The flow of air outward through walls, joints, or other apertures.

exfoliation 1. The flaking, scaling, or peeling of stone or other mineral surfaces usually caused by physical weathering, but sometimes by heating or chemical weathering. **2.** The heat treatment of minerals for the purpose of expanding their original volume many times over.

exhaust fan

exhaust fan A device used to draw unwanted, contaminated air away from a particular room or area of a building to the outside.

exhaust fume hood A prefabricated hood unit that serves to confine noxious or toxic fumes for subsequent exhausting or filtration.

exhaust grille A grate or louvers through which contaminated air exits to the atmosphere.

exhaust shaft A duct through which exhaust air is carried to the outside atmosphere.

existing building In codes and regulations, an already completed building or one that prior laws or regulations allowed to be built.

existing conditions survey Inspection and documentation of a building's composition, configurations, and as-built conditions. Includes visual inspection, laboratory analysis, and often more invasive, destructive investigational procedures.

exit That part of an exit system that, because of its separation from the rest of the building by devices such as walls, doors, and floors, provides a reasonably safe and protected emergency escape route from a building.

exit access The corridor or door leading to an exterior exit.

exit control alarm An electronic device that activates an alarm when a fire exit door is opened.

exit corridor An enclosed passageway or corridor that connects a required exit with direct or easy access to a street or alley.

exit device (panic exit device) An exit door locking device, consist-

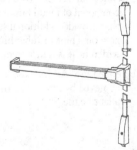

exit device (panic exit device)

ing of a bar across the inside of a door that, when pushed, releases the door latch.

exit discharge That part of an exit system beginning at the termination of the exit at the exterior of a building and ending at ground level.

exit light An illuminating sign above an exit that identifies it as an exit.

exothermic Descriptive of a reaction during which heat is released.

expanded aluminum grating An aluminum grating manufactured by cutting and mechanically stretching a single piece of sheet aluminum.

expanded blast-furnace slag A cellular material, used as lightweight aggregate, derived from treating molten blast-furnace slag with water, steam, and/or other agents.

expanded clay Clay expanded to several times its original volume by the formation of internal gas caused by heating it to a semiplastic state.

expanded glass (foam glass) A thermal insulation with a closed-cell structure, manufactured by foaming softened glass so as to produce a myriad of sealed bubbles, and then molding the glass into boards and blocks.

expanded metal A type of open-mesh metal lath made by slitting and stretching sheet metal. Comes in different patterns and thicknesses.

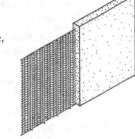

expanded metal

expanded metal partition A partition constructed from thin framing or support members covered with heavy expanded metal lath and plastered on both sides to produce a solid finished product 1½″ to 2½″ thick.

expanded polystyrene (EPS) A type of foamed styrene plastic that has a high resistance to heat flow and a high strength-to-weight ratio. EPS is used as insulation, as an inexpensive glazing, and in formwork for concrete, among many other applications.

expanded rubber Closed-cell rubber manufactured from a solid rubber compound.

expanding bit (expansion bit) A drilling bit whose blade can be adjusted to bore holes of different diameters.

expanding plug A drain stopper usually located at the lowest point of a piping system. When inflated, the plug seals the pipe.

expansion bead In plastering, a metal strip placed between materials to allow for movement.

expansion bearing An end support of a span, which allows for the expansion or contraction of a structure.

expansion bend (expansion loop) In a pipe run, a usually horseshoe-shaped bend inserted to allow thermal expansion of the pipe.

expansion bolt (expansion anchor, expansion fastener, expansion shield) An anchoring or fastening device used in masonry, which expands within a predrilled hole as a bolt is tightened.

expansion coil An evaporator fabricated from tubing or pipe.

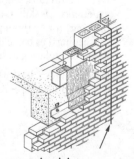

expansion joint

expansion joint In a building structure or concrete work, a joint or gap between adjacent parts, which allows for safe and inconsequential relative movement of the parts, caused by thermal variations or other conditions.

expansion joint cover The prefabricated protective cover of an expansion joint, which also remains unaffected by the relative movement of the two joined surfaces.

expansion joint filler Material used to fill an expansion joint to keep it clean and dry. Materials commonly used are felt, rubber, and neoprene.

expansion loop (expansion bend) A large radius loop in a pipeline that absorbs the longitudinal expansion and contraction in the line due to temperature changes.

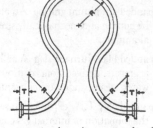

expansion loop (expansion bend)

expansion sleeve A short length of metal or plastic pipe built into a wall or floor to allow for the inconsequential expansion and/or contraction of the element (usually another, smaller pipe) that passes through it.

expansion strip 1. The material used in an expansion joint. **2.** Resilient insulating material used as a joint filler between a partition and a structural member.

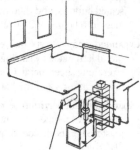

expansion tank

expansion tank 1. In a hot-water system, the tank that provides for the increased volume of the water heated in the heating tank below it. **2.** A device to control pressure in a fluid system by storing excess liquid.

expansion valve The valve that controls the flow of refrigerant to the cooling element in a refrigeration system.

expansive cement (expanded cement, expanding cement, sulfoaluminate cement) A type of cement whose set paste increases in volume substantially more than the set paste of Portland cement. Used in applications where the desired results include the compensation for volume decrease due to shrinkage, or the induction of tensile stress in reinforcement.

expediter One who monitors and facilitates the arrival of building materials or equipment to meet a progress schedule.

expert witness In a court case or other legal or arbitration proceeding, an individual who, because of his/her exceptional knowledge, experience, or skill in a particular field or subject is accepted by the court (or by those presiding over other legal or arbitration proceedings) as being qualified to render an authoritative opinion in matters relating to his/her field of expertise.

exploded view An illustration depicting the disassembled individual components in proper relationship to their assembled positions therein.

explosion-proof box Electrical equipment housing designed in compliance with hazardous location requirements to withstand an explosion that could occur within it and to prevent ignition by the explosion of flammable material, liquid, or gas that might externally surround it.

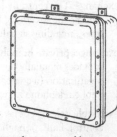

explosion-proof box

explosion-proof fixture Electric lighting fixture that will not explode or ignite when in contact with a flammable gas or liquid.

explosion welding The use of a controlled explosion to weld members together.

explosive rivet A rivet whose explosive-filled shank is exploded by striking it with a hammer after the rivet has been inserted.

exposed Refers to an unprotected or uninsulated "live" component of an electrical system whose location makes it susceptible to being inadvertently touched or approached by someone at an unsafe distance.

exposed aggregate The coarse aggregate in concrete work revealed when the surface layer of concrete paste is removed, usually before the full hardening of the concrete.

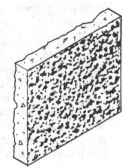

exposed aggregate

exposed area In roofing, the part of a shingle that is not covered by another shingle.

exposed masonry Any masonry construction whose only surface finish is a coat of paint applied to the face of the wall.

exposed nailing A method of nailing that leaves the nails exposed to the weather.

exposed suspension system A method of installing a suspended acoustical ceiling in which the panel-supporting grid is left exposed in a room.

exposure hazard The probability of a building's exposure to fire from an adjoining or nearby property.

extended overhead Overhead costs accumulated during compensable delay periods when full production was not achievable.

extended surface A fitting, usually composed of metal fins or ribs, that provides additional surface area on a pipe or tube through which heat is transferred.

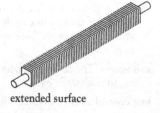

extended surface

extended use An increase in the function of a building, achieved by interventions such as rehabilitation or adaptive use.

extender 1. An opaque, white, inert mineral pigment, such as calcium carbonate, silica, diatomaceous earth, talc, or clay, that is added to paint to provide texture, lower gloss, or reduce paint cost by providing bulk. 2. A volume-increasing, cost-reducing additive in synthetic resin adhesives.

extensibility The capability of a sealant to be stretched in tension.

extension 1. A wing, ell, or other addition to an existing building. 2. The product of a quantity multiplied by a unit price. 3. The completion of a mathematical equation. 4. An agreement between parties extending to term of a contract.

extension casement hinge An exterior hinge on a casement window whose sash swings outward. The hinge is located so as to provide enough clearance to allow the cleaning of the hinged side from the inside when the window is open.

extension device Any device, other than an adjustment screw, used for accomplishing vertical adjustment.

extension flush bolt (extension bolt) A type of flush bolt whose head connects to the operating mechanism via a rod, which is inserted through a hole bored in the door.

extension ladder A ladder comprising more than one section, each of which slides within and locks on the other, thus allowing lengthening.

extension rule A rule containing an extendable, calibrated, sliding insert.

extension ladder

exterior-glazed Descriptive of glazing that has been set from the outside of a building.

exterior insulation and finish systems (EIFS) Exterior cladding assembly consisting of a polymer finish over a reinforcement adhered to foam plastic insulation that is fastened to masonry, concrete, building sheathing or directly to the structural framing. The sheathing may be cement board, gypsum sheathing or another acceptable substrate.

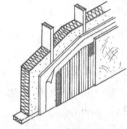
exterior insulation and finish systems (EIFS)

exterior paint Paint formulated with durable binder and pigment, making it suitable for exposure to weather.

exterior panel In a concrete slab, a panel with at least one of its edges not adjacent to another panel.

exterior plywood Plywood whose layers of veneer are bonded with a waterproof glue.

exterior stair An often legally required exit consisting of a series of flights of steps affixed to the structure but not enclosed.

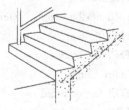
exterior stair

exterior trim Any visible or exposed finish material on the outside of a building.

exterior wall An outer wall or vertical enclosure of a structure.

external vibration Brisk agitation of freshly mixed concrete by a vibrating device strategically positioned on concrete forms.

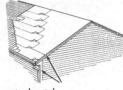
exterior trim

extra A work item performed in addition to the scope specified in the contract. Often involves an additional cost.

extraction The process of mechanically removing layers of old paint or other material and retaining a portion of the substrate for laboratory analysis.

extraction procedure toxicity The propensity of a substance to emit toxins under a specific test procedure, defined by the EPA, which uses an acid to extract the toxins from the substance in question.

extranet A shared communications network that uses Internet technology to allow collaborating parties to share information. Especially useful with large, fast-track projects that require extensive documents.

extra-strong pipe The common designation for steel or wrought iron pipe having wall thickness greater than that of standard-weight pipe.

extra work Any work, desired or performed, but not included in the original contract.

extruded section Structural sections formed by extrusion and used in light construction.

extruded tile A tile formed by pushing clay through a die and cutting it into specified lengths.

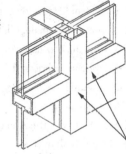
extruded section

extrusion coating A thin film of extruded molten resin, which is pressed onto a substrate to produce a coating with an adhesive.

eye 1. In architecture, the opening in the uppermost portion of a cupola. 2. The nearly circular center of the roll or volute of an Ionic capital. 3. The middle roundel of a pattern or ornament. 4. A hole through a member to provide access, such as for the passage of a pin. 5. In tools, the receiving orifice in the head of the implement.

eye bolt An anchoring device comprising a threaded shank with a lopped head, designed to accept a hook, cable, or rope.

eyebrow A window or ventilation opening through the surface of a roof. Unlike a dormer, it forms no sharp angles with the roof, but rather is incorporated into the general horizontal line of the continuous roof, which is carried over it in a wave line.

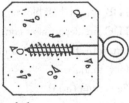
eye bolt

eye of a dome The opening at the very top of a dome.

Abbreviations

f fine, focal length, force, frequency, fiber stress

F Fahrenheit, female, fill, fluorine

FA fresh air, fire alarm

fab fabricate, fabricated

fac facsimile

FAI fresh air intake

FAIA Fellow of the American Institute of Architects

FAO finish all over

FAR floor-area ratio

FAS free alongside ship, firsts and seconds

FBGS fiberglass

FBM foot board measure

fc, FC footcandle, compressive stress in concrete, extreme compressive stress

FCC Federal Communications Commission

fcc face-centered cubic

FDB forced draft blower

FDC fire department connection

fdn, fdtn, fds, FDN foundation, foundations

fdry foundry

Fe ferrum (iron)

FE fire escape

F.E. front end

FEA Federal Energy Administration

FEMA Federal Emergency Management Agency

FEP fluorinated ethylene propylene (Teflon)

FFA full freight allowance

FG fine grain, flat grain, finished grain

FHA Federal Housing Administration

FHC fire hose cabinet

FHWA Federal Highway Administration

FID flame ionization detector

Fig figure

fill filling

Fin finish

Fixt fixture

fl floor, fluid

Fl floorline, floor, flashing

flash flashing

FLG flooring

fl oz fluid ounce

Flr floor

FLUOR fluorescent

fm fathom

FM frequency modulation, Factory Mutual

Fmg framing

FML flexible membrane liner

FMT flush metal threshold

FMV fair market value

fndtn foundation

FOB free on board

FOC free of charge

FOHC free of heart centers

FOK free of knots

fount fountain

fp fireplace, freezing point

fpfg fireproofing

fpm feet per minute

FPRF fireproof

fps feet per second

fr frame

FR, Fr fire rating

frmg framing, forming

FRP fiber-reinforced plastic

frt freight

frwy freeway

FS federal specifications

FSES fire safety evaluation system

FST flat-seam tin

ft foot, feet

ftc footcandle

ftg footing

fth fathom

ft lb foot pound

ft sm feet surface measure

Furn furnish, furnished

fus fusible

fv face velocity

FW flash welding

fwd forward, four wheel drive

Definitions

fabricate 1. To build, to construct, to manufacture. 2. To make by fitting together standardized parts. 3. To assemble in the shop, as to assemble a reinforced steel part.

Façade The exterior face of a building, sometimes decorated with elaborate detail.

face 1. The surface of a wall, masonry unit, or sheet of material that is exposed to view or designed to be exposed in finish work. 2. To cover the surface layer of one material with another, as to *face* a wall with brick or fieldstone. 3. The surface area to be excavated during a construction project. 4. The side of a hammerhead used to strike.

face block (faced block) A unit of concrete masonry with a plastic or ceramic face surface, often glazed or polished for special architectural uses.

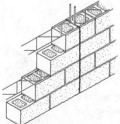

face block (faced block)

face brick (facing brick) 1. Brick manufactured to present an attractive appearance. 2. Any brick that is exposed, such as on a fireplace.

faced insulation Any batt insulation with a moisture barrier on one or both sides. The barrier may be paper, plastic, or foil.

faced plywood Plywood that is faced with plastic, metal, or any material other than wood.

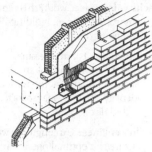

face brick (facing brick)

face frame cabinets Cabinets with a hardwood frame on the face of the cabinet that masks the raw edges of the sides, adds rigidity to the cabinet, and provides a strong base for attaching hinges.

face glazing The compound applied with a glazing knife after a light has been bedded, set, and fastened in a rabbeted sash.

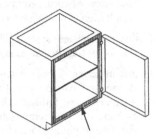

face frame cabinets

face grain The grain on the face of a plywood panel. The face grain should always be placed at right angles to the supports when applying it to a roof or subfloor.

face mark A mark made on the face of a piece of lumber, usually in pencil or crayon, near the edge or end of the piece that identifies the work face of a planed timber.

face measure The measurement of the area of a board or panel. Face measure is not the same as board measure, except when the piece being measured is 1″ thick.

face mix A concrete mix bonded to the exposed surface of a cast stone building unit.

face mold 1. A template used in ornamental woodworking to mark the boards from which pieces will be cut. 2. A device used to examine the shape of wood and stone faces and surfaces.

face nailing Nailing perpendicular to the face of wood and through the face of the material.

face oiling A light coating of oil on the face of concrete panels to aid in removing the forms after the concrete sets.

faceplate Any protective and/or decorative plate such as an *escutcheon.*

face side The better, wide side of a rectangular piece of lumber.

face string (finish string) An exterior string, usually of better material or finish, placed over the rough string in the construction of a staircase.

face veneer Those veneers of higher grade and quality that are used for the faces of plywood panels, especially in the sanded grades. Face veneer is selected for decorative qualities rather than for strength.

face width The width of the face of a piece of dressed lumber. In tongued or lapped lumber, the face width does not include the width of the tongue or lap.

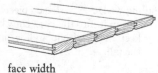

face width

facility In project work, this term usually refers to the constructed environment, e.g., buildings, structures, infrastructure, plant, and equipment.*

facility audit 1. An assessment of the physical condition and functional performance of an organization's facilities. **2.** A review of activities at a facility or location, the purpose of which is to identify the improper storage, handling, or disposal of hazardous materials or waste.

facility engineer An engineer who provides technical expertise to ensure the optimal operation of a facility's plants, grounds and offices. Responsibilities include maintaining and improving the efficiency of existing equipment and systems while overseeing the installation of new equipment and systems.

facility management system (FMS) An electronic control system that helps facility managers maximize a facility's performance.

facility manager A professional who coordinates the planning, design, and management of a facility, including its systems, equipment, and furniture. A facility manager helps an organization meet its long-term strategic and operational objectives while daily maintaining a safe, efficient working environment.

facing block A concrete masonry unit having a decorative exterior finish.

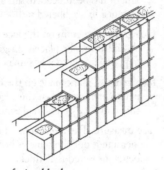
facing block

factor of safety 1. Stress factor of safety: the ratio of the ultimate strength, or yield point, of a material to the design working stress. **2.** Load factor of safety: the ratio of ultimate load, moment, or shear of a structural member to the working load, moment, or shear, respectively, assumed in design.

factory-built A reference to a construction, usually a dwelling, that is built, or at least partially preassembled, in a factory rather than *on site*. Most factory-built units are constructed in two or more modules, often complete with plumbing, wiring, etc. The modules are delivered to a building site and assembled there. Finish work is then performed. Factory-built houses are generally less expensive to build and quicker to erect because of savings gained through mass production and factory efficiencies.

factory finished A product that has been coated or stained as part of the manufacturing process.

factory primed A product to which an initial undercoat of paint has been applied.

factory select A grade of shop lumber containing 70% or more of No. 1 door stock, but having other defects, such as pitch or bark pockets, beyond the restrictions allowed in the No. 1 grade of lumber.

Fahrenheit (F) Temperature scale on which the freezing and boiling points of water are 32° and 212°, respectively.

failing wedge A V-shaped piece of steel, aluminum, or plastic that is driven it into the back cut when felling a tree. The purpose is to keep the saw from binding, and to help control the direction of the fall.

failure A deficiency of a structural element that makes it unable to continue the load-bearing function for which it was originally designed.

failure to perform When an owner or contractor, or both, does not perform each and all of his respective obligations in the construction contract.*

Fair Labor Standards Act (FLSA) An act of the Congress of the United States enacted in 1936, and the subject of numerous amendments to the present day. Commonly referred to as the Minimum Wage Law, it establishes a minimum wage for all workers with the exception of agricultural workers, and additionally provides a maximum of a 40-hour workweek for straight time pay for employees earning hourly wages.

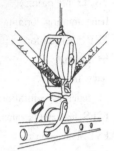

fairlead A device such as a ring or a block of wood with a hole in it, through which cable or rope is led for alignment.

fair market value A price that is fair and reasonable in light of current conditions in the marketplace.

fairlead

fall The slope in a channel, conduit, or pipe, stated either as a percentage or in inches per feet.

falldown Those lumber or plywood items of a lesser grade or quality that are produced as an adjunct to the processing of a higher quality stock.

false attic An architectural addition, built above the main cornice of a structure, that conceals the roof rafters but has no rooms or windows.

false door A nonoperable door placed in a wall for architectural appearance.

false front A front wall that extends beyond the sidewalls and/or above the roof of a building to create a more imposing façade.

false heartwood Dark innerwood that has been colored by disease or fungi so that it resembles heartwood in color.

false joint A groove in a solid masonry block or stone that simulates the appearance of a joint.

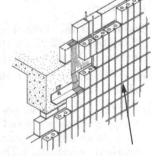
false joint

false rafter A short extension added to a main rafter over a cornice.

falsework The temporary structure erected to support work in the process of construction. Falsework consists of shoring or vertical posting formwork or beams and slabs, and lateral bracing. *See also* **centering.**

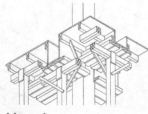
falsework

fan brake horsepower (Bhp) The horsepower output of a fan as measured at the pulley or belt. Bhp includes losses due to turbulence and other inefficiencies.

fan-coil unit An air-conditioning unit that houses an air filter, heating or cooling coil, and a centrifugal fan, and operates by moving air through an opening in the unit and across the coils.

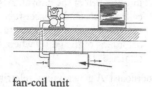

fan-coil unit

fancy butt A shingle with the butt end machined in a pattern. Fancy butt shingles are usually used on sidewalls to form geometric designs in the shingle pattern.

fanlight A semicircular or semielliptical window built above the opening of a door. A fanlight often has triangular panes of glass, in between radiating bars or leads, and resembles the form of an open fan.

fanlight

fan pressure curve The curve that represents the pressure developed by a fan operating at a fixed rpm at various air volumes.

fan suction box A specially designed 90° fitting installed at the fan inlet that reduces the inlet pressure loss created by system effect.

fan surge Turbulent flow caused when insufficient air enters the fan wheel to completely fill the space between the blades. Some of the air reverses its direction over a portion of the fan blade.

fan truss A truss with struts that radiate like the ribs of a fan, supported at their base by a common suspension member.

fascia (facia) 1. A board used on the outside vertical face of a cornice. **2.** The board connecting the top of the siding with the bottom of a soffit. **3.** A board nailed across the ends of the rafters at the eaves. **4.** The edge beam of a bridge. **5.** A flat member or band at the surface of a building.

fascia (facia)

fastener Any mechanical device used to hold together two or more pieces, parts, members, etc.

fast-track construction A building method in which construction is begun on a portion of the work for which the design is complete, while design on other portions of the work is under way.

fat edge A ridge of paint at the bottom edge of a surface when too much paint has been applied or because the paint runs too freely.

fatigue The weakening of a material caused by repeated or alternating loads. Fatigue may result in cracks or complete failure.

fatigue failure The phenomenon of rupture that occurs when a material is subjected to repeated loadings at a stress substantially less than the ultimate tensile strength.

fatigue life 1. The number of cycles of a specified loading that a given specimen can be subjected to before failure occurs. **2.** A measure of the useful life of similar specimens.

fatigue strength The maximum stress that can be sustained for a number of stress cycles without failure.

fat mix (rich mix) A mortar or concrete mix with a relatively high cement content. Fat mix is more easily spread and worked than a mix with the minimum amount of cement required for strength.

fat mortar (rich mortar) Mortar with a high percentage of cement. Fat mortar is sticky and adheres to a trowel.

fault 1. A defect in an electrical system caused by poor insulation, imperfect connections, grounding, or shorting. **2.** A shifting along a plane in a rock formation that causes differential displacement.

faulting Differential vertical displacement of a slab or other member adjacent to a joint or crack.

faux bois French term for "false wood," referring to simulated wood grain. Also called *graining*.

feasibility study A thorough study of a proposed construction project to evaluate its economic, financial, technical, functional, environmental, and cultural advisability.

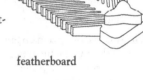

feather To blend the edge or finish of new material smoothly into an existing surface.

featherboard A safety device used to prevent table saw kickback.

featherboard

feather edge (featheredge) The beveling or tapering of the edge of one surface or surface coating where it meets another.

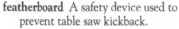

feather edge (featheredge)

feather joint A joint between two squared, closely butted boards made by plowing a groove along the length of each and fitting them with a common tongue.

Federal Communications Commission (FCC) A regulatory agency based in Washington, D.C., that was established by the Communications Act of 1934 to regulate broadcast communications (wire, radio, and television) in the United States.

Federal employer identification number A number assigned by the Internal Revenue Service to identify a business for tax purposes.

Federal Hazardous Substances Act The federal law that controls the use of hazardous constituents in consumer products.

Federal Home Loan Mortgage Corporation A government sponsored organization, nicknamed "Freddie Mac," that provides a secondary market for mortgages.

Federal Housing Administration (FHA) A division of the Department of Housing and Urban Development. The FHA works through lending agencies to provide mortgage insurance on private residences that meet the agency's minimum property standards. The FHA has also been charged with administering a number of special housing programs.

Federal Mediation and Conciliation Service An agency of the U.S. Department of Labor that acts as a mediator in the settlement of disputes as provided in the National Labor Relations Act of 1935 and the Labor Management Relations Act of 1947.

Federal National Mortgage Association A corporation that provides a secondary mortgage market for FHA-insured and VA-guaranteed home loans; commonly known as Fannie Mae.

fee Remuneration for professional services.

feeder **1.** An electrical cable or group of electrical conductors that runs power from a larger central source to one or more secondary or branch-circuit distribution centers. **2.** A tributary to a reservoir or canal. **3.** A bolt or device that pushes material onto a crusher or conveyor.

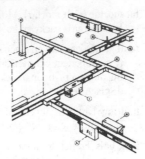

feeder

feeler gauge A gauge used to measure the thickness of a gap, consisting of a series of blades of graduated thicknesses.

fee simple An enduring, inheritable interest in land that may be legally honored until the death of all potential heirs of the original owner, and that the owner is free to convey at any time.

fell To cut down a tree. A faller fells a tree with a falling saw.

felt A fabric of matted compressed fibers, usually manufactured from cellulose fibers from wood, paper, or rags, or from glass fibers.

felt paper Paper sheathing used to insulate walls and to prevent infiltration of moisture.

felt

femerell A ventilator, often louvered, used to draw smoke through a roof when a chimney is not provided.

fence A straight-edge guide mounted parallel to a saw blade to guide a cant as it passes through the saw.

fence pliers Heavy-duty pliers used to cut thick wire. May also include a claw to remove pins and fasteners.

fender A protective bumper, sometimes of wood or old tires.

fenestration The layout and design of windows in a building.

feng shui An approach to interior design, based on Chinese ideals, that seeks to create harmonious surroundings, as in selecting light fixtures, arranging furniture, or determining the siting of a house.

ferrous metal Metal in which the principal element is iron.

fence pliers

ferrule **1.** A tube or metallic sleeve, fitted with a screwed cap to the side of a pipe to provide access for mainte-

nance. **2.** A metal sleeve or collar attached to the end of a short cable in making a choker. The ferrule fits into the bell of the choker. Also called a *nubbin*. **3.** Spacer tube used in hanging metal gutter to prevent deforming of the gutter when attached.

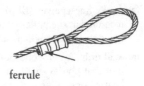

ferrule

fiberboard A general term referring to any of various panel products, such as particleboard, hardboard, chipboard, or other types formed by bonding wood fillers using heat and pressure.

fibered plaster Gypsum plaster reinforced with fibers of hair, glass, nylon, or sisal.

fiberglass (fibrous glass, glass fiber) Filaments of glass formed by pulling molten glass into random lengths that are either gathered in a wool-like mass or formed as continuous threads. The wool-like form is used as thermal and acoustical insulation. The threadlike form is used as reinforcing material and in textiles, glass fabrics, and electrical insulation.

fiberglass reinforced plastic (FRP) A coating of glass fibers and resins applied as a protective layer to plywood. The resulting composite is tough and scuff resistant. It is used to manufacture containers, cooling towers, concrete forms, and shower stalls.

fiber optics The transmission of light pulses through bundles of fine, transparent fibers.

fiber-optic waveguides Fine, transparent fibers, usually thin filaments of glass or plastic, through which a light beam can be transmitted.

fiber saturation point The point of wood moisture content when the cell walls are completely saturated with water, but there is no water in the cell lumens and intercellular spaces.

fiber stress The longitudinal compressive or tensile stress in a structural member.

fibrous Resembling or composed of fibers.

fibrous concrete Concrete with glass or other fibers added to increase tensile strength.

fibrous plaster Cast plaster reinforced canvas, excelsior, or other fibrous material.

FICA A federal law that imposes a Social Security tax on employees, employers, and the self-employed.

fiddleback grain A rippling or undulating grain common to certain hardwoods, such as maple and sycamore.

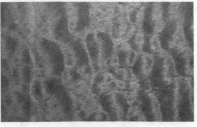

fiddleback grain

F

fiducial mark A standard reference point or line used as a base in surveying.

field 1. In masonry, an expanse of brickwork between two openings or corners. **2.** A term used to designate a construction project site. **3.** Occupation such as a trade, profession, or specialty. **4.** An area of floor, wall, or ceiling that is covered with tile or other preformed shapes that did not require cutting, as opposed to those areas along a border or edge.

field applied 1. The application of a material, such as paint, at a job site, as opposed to being applied at a factory. **2.** The construction or assembly of components in the field.

field cost Engineering and construction costs associated with the construction site rather than with the home office.*

field engineer A representative of certain government agencies who oversees projects at the site. Also called a *project representative* or *field representative*.

field gluing A method of gluing plywood floors in which specially developed glues are applied to the top edges of floor joists. Plywood is then laid on the joists and nailed in place. The combination of gluing and nailing results in a stiffer floor construction and tends to minimize squeaks and nail popping.

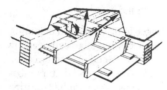
field gluing

field inspection The reinspection of lumber or plywood in the field, usually at the buyer's location. A field inspection is requested by a buyer when he believes a delivered product to be inferior to that specified.

field-molded sealant A joint sealant in liquid or semisolid form that can be molded to the desired shape as it is installed in the joint.

field order In construction, a written order passed to the contractor from the architect to effect a minor change in work, requiring no further adjustment to the contract sum or expected date of completion.

field painting The painting of structural steel or other metals after they have been erected and fastened.

field rivet A rivet driven in a field connection of structural steel.

field sound transmission class (FSTC) An onsite measurement of the ability of a specific construction element, such as a wall partition, to block sound.

fieldstone 1. Loose stone removed from the soil. **2.** Flat slabs of stone suitable for use in dry wall masonry.

field supervision The site supervisory work performed by a designated individual.

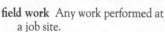
fieldstone

field work Any work performed at a job site.

figure The natural wood pattern that is created by growth rings, knots, and other deviations from regular grain.

figured glass Translucent sheet glass rolled with a bas-relief pattern on one face, providing high but obscure light transmission, with the degree of obscurity depending on the pattern.

filament A fine wire used in an incandescent lamp. The form is designated by a letter: S, straight wire; E, coil; CC, coiled coil.

file A handheld steel tool with teeth or raised oblique ridges, used for scraping, redressing, or smoothing metal or wood.

fill 1. The soil or other material used to raise the grade of a site area. **2.** A subfloor leveling material.

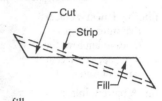
fill

filled particleboard Particleboard to which plastic material has been applied to fill gaps between particles on the edges or faces.

filler 1. Finely divided inert material, such as pulverized limestone, silica, or colloidal substances sometimes added to Portland cement, paint, or other materials to reduce shrinkage, improve workability, or act as an extender. **2.** Material used to fill an opening in a form.

filler block Concrete masonry units placed between joists or beams under a cast-in-place floor.

filler coat A coat of paint or varnish used as a primer.

filler metal Metal, added by a welding process, that has a melting point either approximately the same as or below the base metal.

fillet 1. A narrow band of wood between two flutes in a wood member. **2.** A flat, square molding separating other moldings. **3.** A concave junction formed where two surfaces meet.

fill insulation 1. Thermal insulation placed in prepared or natural cavities of a building. **2.** Any variety of loose insulation that is poured into place.

fill insulation

film-surfaced hardboard siding Hardboard to which a thin, dry sheet of paper, coated on both sides with a phenol-formaldehyde resin adhesive, has been applied. Such coatings are often printed with grain patterns in imitation of actual veneers, and are used as an inexpensive substitute for the higher priced veneer panels.

filter 1. A device to separate solids from air or liquids, such as a filter that removes dust from the air or impurities from water. **2.** Granular matter placed on an area to provide drainage while preventing the entry and flow of sediment and silt.

filter bed A bed of granular material used to filter water or sewage.

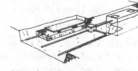
filter bed

filter block A hollow, vitrified clay masonry unit used for trickling filter floors in sewage treatment plants.

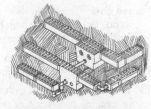

filter block

filter fabric Tough fabric used in excavation that screens out soil while allowing the passage of water. Also called *engineering fabric.*

filtering faucet A faucet that filters out particles to improve the taste, appearance, and safety of water.

filtering faucet

fin A narrow, linear projection on a formed concrete surface, resulting from mortar flowing out between spaces in the formwork.

final acceptance The formal acceptance of a contractor's completed construction project by the owner, upon notification from an architect that the job fulfills the contract requirements. Final acceptance is often accompanied by a final payment agreed upon in the contract.

final completion A term applied for a project that has been completed according to the terms and conditions set forth in the contract documents.

final inspection An architect's last review of a completed project before issuance of the final certificate for payment.

final payment The payment an owner awards to the contractor upon receipt of the final certificate for payment from the architect. Final payment usually covers the whole unpaid balance agreed to in the contract, plus or minus any amounts altered by change orders.

final stress 1. In prestressed concrete, the stress that remains in a member after substantially all losses in stress have occurred. **2.** The stress in a member after all loads are applied.

final waiver of liens A document prepared and executed by the contractor at the conclusion of the project stating that all subcontractors and material suppliers have been paid in full and that no mechanic's or materialmen's liens will be filed against the owner. Typically required to be filed with the owner as a condition of receiving the contractor's final payment or release of retention for the work he performed.*

fine aggregate 1. Aggregate passing the $^3/_8''$ (9.5-mm) sieve and almost entirely passing the No. 4 (4.75-mm) sieve and predominantly retained on the No. 200 (75-micrometer) sieve. **2.** That portion of an aggregate passing the No. 4 (4.75-mm) sieve and predominantly retained on the No. 200 (75-micrometer) sieve.

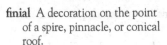

fine grading The final grading of ground to prepare for seeding, planting, or paving.

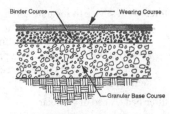

fine grading

fine-grained soil Soil in which the smaller grain sizes predominate, such as fine sand, silt, and clay.

fineness modulus A factor obtained by adding the total percentages by weight of an aggregate sample retained on each of a specified series of sieves, and dividing the sum by 100. In the United States the standard sieve sizes are No. 100 (150 micrometer), No. 50 (300 micrometer), No. 30 (600 micrometer), No. 16 (1. 18 mm), No. 8 (2.36 mm) and No. 4 (4.75 mm), and $^3/_8''$ (9.5 mm), $^3/_4''$ (19 mm), $1^1/_2''$ (38.1 mm), 3″ (75 mm), and 6″ (150 mm).

fines 1. Soil that passes through a No. 200 sieve. **2.** Fine-milled chips used in the production of particleboard. Fines are larger than sander dust or wood flour and are used on the faces of particleboard panels, with coarser chips used as the core of the board. **3.** An undesirable by product of cutting wafers and strands for waferboard and oriented strand board.

finger joint A method of joining two pieces of lumber end to end by sawing into the end of each piece a set of projecting "fingers" that interlock. When the pieces are pushed together, these form a strong glue joint.

finger joint

finial A decoration on the point of a spire, pinnacle, or conical roof.

finial

finish 1. The texture of a surface after compacting and finishing operations have been performed. **2.** A high-quality piece of lumber graded for appearance and often used for interior trim or cabinet work.

finish carpentry Carpentry that involves the installation of finish woods (and trim made of plastic or molded polyurethane materials) to provide a finished appearance to installed doors, windows, stairs, and other features of a building's interior. Elements include casing, baseboard, railings, mantels, louvers, paneling, and shelving.

finish carpentry

finished grade The top surface of an area after construction is completed, such as the top of a road, lawn, or walk.

finished size The final size of any completed object including trim.

finisher A tradesman who applies the final treatment to a concrete surface, including patching voids and smoothing.

finish floor The wearing surface of a floor.

finish flooring The material used to make the wearing surface of a floor, such as hardwood, tile, or terrazzo.

finish grinding 1. The final grinding of clinker into cement, with calcium sulfate in the form of gypsum or anhydrite generally being added. **2.** The final grinding operation required for a finished concrete surface, such as bump cutting of pavement, fin removal from structural concrete, or terrazzo floor grinding.

finishing Leveling, smoothing, compacting, and otherwise treating surfaces of fresh or recently placed concrete or mortar to produce the desired appearance and service.

finishing machine A power-operated machine used to give the desired surface texture to a concrete slab.

finishing machine

finishing nail A slender nail, with a narrow head, that may be driven entirely into a wood surface, leaving a small hole that can be filled with putty or a similar compound.

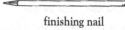

finishing nail

Fink truss (Belgian truss, French truss) A symmetrical truss, formed by three triangles, commonly used in supporting large, sloping roofs.

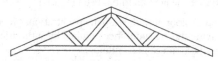

Fink truss (Belgian truss, French truss)

fir 1. A form of softwood indigenous to temperate zones, used principally for interior trim and framing. Varieties include Douglas fir, silver fir, balsam fir, and white fir. **2.** Although used most often to refer to Douglas fir, which is also a pseudo-fir, a general term for any of a number of species of conifers, including the true firs.

fire alarm system An electrical system, installed within a home, industrial plant, or office building that sounds a loud blast or bell when smoke and flames are detected. Certain alarms are engineered to trigger sprinkler systems for added protection.

fire area An area in a building enclosed by fire-resistant walls, fire-resistant floor/ceiling assemblies, and/or exterior walls with all penetrations properly protected.

fireback (chimney back) 1. The heat-resistant back wall of a fireplace, built of masonry, cast metal or wrought metal, that withstands flames and radiates heat into the area served. **2.** A fire-resistant attachment to the back wall of a fireplace that keeps the masonry from absorbing too much heat.

fireblocking Installed building materials designed to resist spread of fire in concealed spaces to other areas of a building.*

fire box In a chimney, the metal box that contains the fire.

firebreak 1. A strategic space between buildings, clusters of buildings, or sections of a city that helps to keep fires from leaping or spreading to surrounding areas before they can be contained. **2.** Any doors, walls, floors, or other interior structures engineered to go inside a building.

fire brick A flame-resistant, refractory ceramic brick used in fireplaces, chimneys, and incinerators.

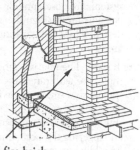

fire brick

fire clay An earthy or stony mineral aggregate having as the essential constituent hydrous silicates of aluminum with or without free silica. Fire clay is plastic when sufficiently pulverized and wetted, rigid when subsequently dried, and of suitable refractoriness for use in commercial refractory products.

fire control damper An automatic damper used to close a duct if a fire is detected.

fire cut A sloping cut on the end of a wood beam or joist supported on a masonry wall. If somewhere along its length the joist or beam burns through, injury to the wall is minimized.

fire detection system A series of sensors and interconnected monitoring equipment that detect the effects of a fire and activate an alarm system.

fire division wall A wall that subdivides one or more floors of a building to discourage the spread of fire.

fire door 1. A highly fire-resistant door system, usually equipped with an automatic closing mechanism, that provides a certain designated degree of fire protection when it is closed. **2.** The opening in a furnace or a boiler through which fuel is added.

fire-door rating A system of evaluating the endurance and fire resistance of door, window, or shutter assemblies, according to standards set by Underwriters Laboratories Inc., or other recognized safety authorities. Ratings of A through F are assigned in descending order of their effectiveness against fire.

fire endurance 1. The length of time a wall, floor/ceiling assembly, or roof/ceiling assembly will resist a standard fire without exceeding specified heat transmission, fire penetration, or limitations on structural components. **2.** The length of time a structural member, such as a column or beam, will resist a standard fire without exceeding specified temperature limits or collapse. **3.** The length of time a door, window, or shutter assembly will resist a standard fire without exceeding specified deformation limits.

fire escape A continuous, unobstructed route of escape from a building in case of fire, sometimes located on the outside of an exterior wall.

fire extinguisher A portable device for immediate use in suppressing a fire. There are four classes of fires: A, B, C, or D. A single device is designed for use on one or more classes.

fire-extinguishing system An installation of automatic sprinklers, foam distribution

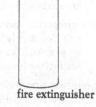

fire extinguisher

system, fire hoses, and/or portable fire extinguishers designed for extinguishing a fire in an area.

fire hazard The relative danger that a fire will start and spread, that smoke or toxic gases will be generated, or that an explosion will occur, endangering the occupants of a building or area.

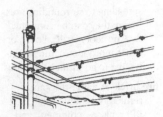

fire-extinguishing system

fire-hazard classification A designation of high, ordinary, or low assigned to a building, based on its potential susceptibility to fire, judged by its contents, functions, and the flame-spread rating of its inner furnishings and finishes.

fire hydrant A supply outlet from a water main, for use in firefighting.

fire load (fire loading) The amount of flammable contents or finishes within a building per unit of floor area, stated either in pounds per square foot or Btus per square foot.

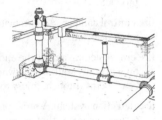

fire hydrant

fire management panel A fire alarm control unit consisting of multiple single-zone fire alarm control modules, a telephone module, an HVAC module, and an audio communications module.

fire partition An interior partition that does not fully qualify as a fire wall, but has a fire endurance rating of not less than two hours.

fireproof 1. Descriptive of materials, devices, or structures with such high resistance to flame that they are practically unburnable. **2.** To treat a material with chemicals in order to make it fire resistant.

fireproofing The use of fire-resistant materials for the protection of structural members to insure structural integrity in the event of fire.

fireproof wood Chemically treated wood that is fire resistive. No wood can be completely fireproof, but wood can be treated so that it is highly resistant to heat and flame and very difficult to ignite under ordinary circumstances.

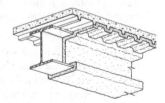

fireproofing

fire protection The practice of minimizing the probable loss of life or property resulting from fire, by fire-safe design and construction, the use of detection and suppression systems, establishment of adequate firefighting services, and training building occupants in fire safety and emergency evacuation.

fire-protection sprinkler system An automatic fire-suppression system, commonly heat activated, that sounds an alarm and deluges an area with water from overhead sprinklers when the heat of a fire melts a fusible link.

fire-protection sprinkler valve A valve, normally open, that is used to control the flow of water in a sprinkler system.

fire-rated system Wall, floor, or roof construction using specific materials and designs that have been tested and rated for conformance to fire safety criteria, such as the flame spread rate.

fire resistance 1. The property of a material or assembly to withstand fire, characterized by the ability to confine a fire and/or to continue to perform a given structural function. **2.** The property of a material or assembly that makes it so resistant to fire that, for a specified time and under conditions of a standard heat intensity, it will not fail structurally and will not permit the side away from the fire to become hotter than a specified temperature.

fire-resistant door A door designed to confine fire to one part of a structure, keeping it from spreading through an entire building. It may be a solid-core wooden door, or one sheathed in metal, depending on the intended location. Doors are rated for the projected time they could be expected to perform their function during a fire. Most building codes require that the door between living quarters and a garage be fire resistant.

fire-resistive construction Construction in which the floors, walls, roof, and other components are built exclusively of noncombustible materials, with fire-endurance ratings equal to or greater than those mandated by law.

fire-retardant finish Paint composed in part of noncombustible substances such as silicones, chlorinated waxes and resins, and antimony oxide, used to protect combustible surfaces, inhibiting the rapid spread of flames.

fire-retardant wood Wood chemically impregnated or otherwise treated so that the fire hazard has been reduced.

fire separation A floor or wall, with any penetrations properly protected, having the fire endurance required by authorities and acting to retard fire spread in a building.

fire shutter A complete metal shutter assembly with the fire endurance required by code. The rating depends on the location and nature of the opening.

firestat A fixed thermostat in an air-conditioning system, preset at a temperature required by code or other authorities, usually 125°F (52 °C).

firestop 1. A short piece of wood, usually a 2″ × 4″ or 2″ × 6″, placed horizontally between the studs of a wall. Firestops are equal in width to the studs. They are usually placed halfway up the height of the wall to slow the spread of fire by limiting drafts in space between the sheathing on the two sides of the wall. Building codes usually do not require firestops in walls that are less than 8′ high. **2.** Masonry units placed in wood framed walls.

firestop

fire tower A vertical enclosure, with a stairway, having the fire endurance rating required by code and used for egress and as a base for fire fighting.

fire wall An interior or exterior wall that runs from the foundation of a building to the roof or above, constructed to stop the spread of fire.

fir & larch A mixture of Douglas fir and western larch, sold as either individual species or as one species grouping. The two species are intermixed because they have similar characteristics and are used in similar ways.

firing The controlled-heat treatment of ceramic or brick ware in a kiln or furnace.

first coat The initial application of mortar, known as *base coat* in two-coat work, and *scratch coat* in three-coat work.

first in, first out (FIFO) A type of accounting for inventory in which items purchased first are assumed to be the first to be sold. An accountant will compute the cost and profit on the oldest or first in the inventory.

first mortgage The mortgage or security in a property that takes precedence over all other mortgages on the property.

fished joint A butted timber joint made using fishplates.

fish scale A fancy butt shingle pattern with the exposed portion of the shingle rounded. When used as siding, the overall effect is the pattern of fish scales.

fitch A thin, long-handled paintbrush used especially for touching up recessed areas.

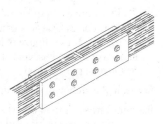

fished joint

fitting 1. A standardized part of a piping system used for attaching sections of pipe together, such as a coupling, elbow, bend, cross, or tee. **2.** The process of installing floor coverings around walks, doors, and other obstacles or projections. **3.** In electrical wiring, a component, such as a bushing or locknut, that serves a mechanical purpose rather than an electrical function.

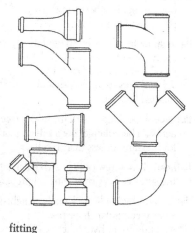

fitting

fit-up The construction necessary within the demising walls of a leased space, including partitions, finishes, fixtures, lighting, power, equipment, etc., allowing the space to be used for its leased purpose.

fixed assets Assets, such as real property, that are not readily converted to cash.

fixed beam A structural beam with fixed, as opposed to hinged, connections.

fixed costs These costs can be defined in two different ways. In total, fixed costs remain the same as volume changes. On a per-unit basis, fixed costs decline as volume increases. Examples include office rent, administrative salaries, insurance, and office supplies.

fixed fee A fixed amount specified in a contract to compensate a contractor for all materials and services required to complete the project.

fixed light A window or portion of a window that does not open. Also referred to as *fixed sash*.

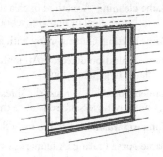

fixed light

fixed limit of construction cost A maximum amount to be paid for a construction job as specified in the agreement between owner and architect.

fixed plate heat exchanger A static device that transfers sensible heat through plates separating a warm air stream from a cold air stream.

fixed-price contract A type of contract in which the contractor agrees to construct a project for an established price, agreed in advance.

fixed-rate mortgage A mortgage on which the interest rate remains the same through the period of the loan.

fixed retaining wall A retaining wall supported both at its top and bottom.

fixing The installation of glass panels in a ceiling, wall, or partition. Fixing is distinct from glazing, which covers most other uses, such as the installation of glass in windows, doors, and showcases.

fixture 1. An electrical device, such as a luminaire, attached to a wall or ceiling. **2.** An article that becomes part of real property by virtue of being merged or affixed to it, although it was originally a distinct item of personal property.

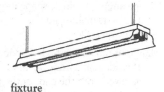

fixture

fixture drain A drain that leads from the trap of a fixture to another drainpipe.

fixtures Tangible goods that are permanently attached to, or embedded in, real property so that they are considered to be part of the real property.

fixture supply The pipe connecting a plumbing fixture to a supply line.

fixture unit A measure of the rate at which various plumbing fixtures discharge into a drainage system, stated in units of cubic volume per minute.

fixture supply

fixture whip A flexible wiring system used between the junction box and lighting fixture.

flagging 1. A walk or patio paved with flagstones. **2.** The process of setting flagstones. *See also* **flagstone**.

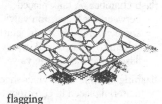

flagging

flagstone A flat, irregular-shaped stone, usually 1″ to 4″ thick, used as pedestrian paving or flooring. Flagstones are set in mortar or aggregate outdoors and in mortar indoors.

F

flame cleaning A method of cleaning mill scale, moisture, paint, and dirt from steel by applying a hot flame.

flame cutting Cutting metal with an oxyacetylene torch.

flame ionization detector An ionization detector that uses hydrogen flame to generate ions.

flameproof Able to resist the spread of flame and not easily ignited, but having less fire resistance than fire-retardant substances.

flame-retardant Treated to resist flame spread and ignition.

flame-spread rating A numerical value given to a material designating its resistance to flaming combustion. Materials can have a rating of 100, as in untreated lumber, to 0, as in asbestos sheets.

flame treating A process by which inert thermoplastic materials and objects are prepared to receive paints, adhesives, lacquers, and inks by immersion in open flame, causing surface oxidation.

flammability A material's capacity to burn.

flammable Capable of being ignited easily, burning intensely, or having a rapid rate of flame spread.

flange 1. A projecting ring, ridge, or collar placed on a pipe or shaft to strengthen, prevent sliding, or accommodate attachments. 2. The longitudinal part of a beam, or other structural member, that resists tension and compression.

flange angle An angle shape used as one of the parts making up the flange of a built-up girder.

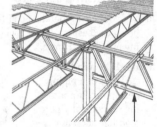

flange

flange union A method of connecting threaded pipe using a pair of flanges screwed onto the ends of the pipe and bolted together.

flapper The moving part of the flush valve in a toilet that allows water to leave the tank during the flush cycle.

flap trap (flap valve) A hinged flap within a plumbing system that restricts water flow to one direction, preventing backflow.

flaring tool A tool used to flare the end of a pipe for a fitting.

flash 1. To make a joint weathertight using flashing. 2. An intentional or accidental color variation on the surface of a brick. 3. A variation in paint color resulting from variable wall absorption. 4. The conversion of condensate into steam.

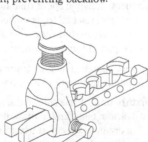

flaring tool

flash chamber A tank placed between the expansion valve and evaporator in a refrigeration system to separate and bypass any flash gas formed in the valve.

flashing A thin, impervious sheet of material placed in construction to prevent water penetration or direct the flow of water. Flashing is used especially at roof hips and valleys, roof penetrations, joints between a roof and a

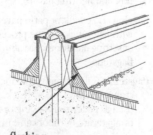

flashing

vertical wall, and in masonry walls to direct the flow of water and moisture.

flashing block A concrete masonry unit with a slot cast in one face to receive and hold one edge of a flashing.

flashing cement A mixture of solvent and bitumen, reinforced with inorganic glass or other fibers, and applied with a trowel.

flash point 1. The minimum temperature at which a combustible liquid will give off sufficient vapor to produce a combustible mixture when mixed with air and ignited. Defined flash points are for specific enclosure conditions and ignition energy. 2. The temperature at which a substance will spontaneously ignite.

flash welding A welding process that joins metals with the heat produced by the resistance to an electric current between the surfaces of the metals, followed by application of pressure.

flat 1. Descriptive of a structural element having no slope, such as a flat roof. 2. One floor of a multilevel building, or full-floor apartment. 3. Descriptive of low-gloss paint, used either as an undercoat or as a final coat. 4. A thin iron or steel bar with a rectangular cross section.

flat arch An arch with slight curvature.

flat-chord truss A truss with the top and bottom chords nearly flat and parallel.

flat coat An intermediate coat of paint applied between the primer and finish coat.

flat enamel brush A brush used to apply smooth films of enamel on woodwork, about 2″ to 3″ (5 to 7.5 cm) wide with flagged and tapered bristles.

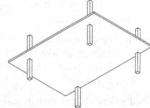

flat-chord truss

flat face Pipe flanges with the entire face of the flange faced straight across. These flanges use a full face gasket. Commonly employed for pressures of less than 125 lbs.

flathead 1. A screw or bolt with a flat top surface and a conical bearing surface. 2. A rivet with a flattened head.

flat jack A hydraulic jack made of light-gauge metal welded into a flat rectangular shape that expands under hydraulic pressure.

flat molding A thin, flat molding used only for interior and exterior finish work.

flat paint A paint that dries to a low gloss or flat finish.

flat plate A flat slab without column capitals or drop panels.

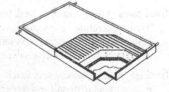

flat plate

flat plate collector A panel of metal or other suitable material used to convert sunlight into heat. Flat plate collectors are usually a flat black color and transfer the collected heat to circulating air or water.

flat plate collector

flat rolled Steel plates, sheets, or strips manufactured by rolling the steel through flat rollers.

flat roof A roof with only enough pitch to allow drainage.

flat sawing Sawing logs with parallel cuts, a conversion method that wastes less timber than others.

flat seam A seam between two joined sheets of metal, formed by turning up, folding, flattening, and finally soldering the edges.

flat skylight A skylight installed in a horizontal position, slanted only enough to permit rainwater runoff.

flat slab A concrete slab reinforced in two or more directions, generally with drop panels at supports, but without beams or girders.

flat spot A spot on a glossy painted surface that fails to absorb the paint properly, usually because of a porous place on the undercoat, resulting in a flat finish that flaws the appearance of the surface.

flatwork A general term that refers to any concrete slab that requires finishing, such as a walkway or floor.

Flemish bond A brick wall laid up with alternate headers and stretchers in each course. Headers in the next course are centered over the stretchers in the course below.

Flemish bond

flexible connector 1. In ductwork, an airtight connection of nonmetallic materials installed between ducts or between a duct and fan to isolate vibration and noise. 2. A connector in a piping system that reduces vibration along the pipes and compensates for misalignment. 3. An electrical connection that permits movement from expansion, contraction, vibration, and/or rotation.

flexible connector

flexible coupling A mechanical connection between rotating parts that adapts to misalignment, such as a universal joint.

flexible membrane liner (FML) A liner used to contain liquids or runoff in soil or on the soil surface.

flexible metal conduit A flexible raceway of circular cross-section for pulling electric cables through.

flexible-metal roofing Roof coverings of flat sheet metal, such as copper, galvanized iron, or aluminum.

flexible mounting A flexible support for machinery to reduce vibration between the machinery and its foundation or slab. Flexible mountings are usually made of rubber, neoprene, steel springs, or a combination of these.

flexible pavement A pavement structure that maintains intimate contact with and distributes loads to the subgrade and depends on aggregate interlock, particle friction, and cohesion for stability. Cementing agents, where used, are generally bituminous materials, as contrasted to Portland cement in the case of rigid pavement. *See also* **rigid pavement**.

flexible seamless tubing A seamless, welded, or soldered metal tube widely used for volatile gases and gases under pressure.

flexural bond In prestressed concrete, the stress between the concrete and the tendon that results from external loads.

flexural rigidity A measure of stiffness of a member, indicated by the product of the modulus of elasticity and moment of inertia divided by the length of the member.

flight A run of steps without intermediate landings.

flight header A horizontal structural member used to support stair strings at a floor or platform.

flight rise The vertical distance between two levels connected by a flight of stairs.

flight run The horizontal distance between top and bottom risers in a flight of stairs.

flitch 1. A log, sawn on two or more sides, from which veneer is sliced. 2. Thin layers of veneer sliced from a cross section of a log, as opposed to turning the log on a lathe and peeling from the outer edge in a continuous ribbon. Flitch veneers are often kept in order as they are sliced from a log. This provides a pattern to the veneer after it is laid up in panels. Panels that are laid up with matching flitches are said to have a flitch pattern. 3. A product cut from a log by sawing on two sides and leaving two rounded sides, often used for joinery.

flitch beam (flitch girder, sandwich beam) A beam built up by two or more pieces bolted together, sometimes with a steel plate in the middle. The pieces, or flitches, are cut from a squared log sawn up the middle. The outer faces of the log are placed together, with their ends reversed to equalize their strength.

flitch plate The steel plate pressed between components of a flitch beam.

float 1. A tool, usually of wood, aluminum, magnesium, rubber, or sponge, used in concrete or tile finishing operations to impart a relatively even but still open texture to an unformed fresh concrete surface. 2. To rest by a dozer blade's own weight or to be held from digging by upward pressure of a load of dirt against its mold board. 3. A body floating on water, which opens a valve in a water tank when the water level falls. 4. A tool used to polish marble. 5. In manufacturing, the amount of material in a system or process, at a given point in time, that is not being directly employed or worked upon. 6. In projects, the amount of time that an activity may slip in its start and completion before becoming critical.*

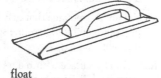

float

float coat A finish coat of cement paste applied with a float.

floated coat (topping coat) A layer of plaster applied to a surface with a float, usually the intermediate coat between the scratch coat and the finish coat.

float finish A rather rough concrete surface texture obtained by finishing with a float.

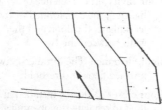

floated coat (topping coat)

float glass A high-quality, flat glass sheet with smooth surfaces, manufactured by floating molten glass on a bed of molten metal at a temperature high enough for the glass to flow smoothly.

floating The operation of finishing a fresh concrete or mortar surface by use of a float, preceding troweling when that is the final finish.

floating floor A floor used in sound-insulating construction. The finish floor assembly is isolated from the structural floor by a resilient underlayment or by resilient mounting devices for machinery. The construction isolates machinery from the building frame.

floating foundation A reinforced concrete slab designed to move with the surface of the ground without structural damage.

floating wood floor A floating wood floor constructed on a layer of resilient material that isolates it from the building structure.

float scaffold A large platform supported by ropes from overhead.

float trap A steam trap regulated by a float in the condensate chamber.

float valve (float-controlled valve) A valve controlled by a float in a chamber, such as the float valve in a tank for a water closet.

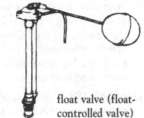

float valve (float-controlled valve)

flood coat The top layer of bitumen poured over a built-up roof, which may have gravel or slag added as a protective layer.

flooding The stratification of different colored pigments in a film of paint.

floodlight 1. A projector type of luminaire designed to light an area or object to a level of illumination higher than the surrounding illumination. **2.** A metal housing containing one or more lamps used to illuminate a stage uniformly.

floodplain A land area that may be subject to flooding. Used in plot plans and to determine suitability for construction and insurability.

floor 1. The surface within a room upon which one walks. **2.** The horizontal division between two stories of a building, formed by assembled structural components or a continuous mass, such as a flat concrete slab.

floor/area ratio The ratio of the total floor area of a building to the area of the building lot.

floor beam Any beam that supports the floor of a building or the deck of a bridge.

floor box A metal electrical outlet box providing outlets from conduits in or under the floor.

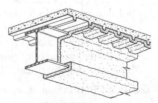

floor beam

floor cavity In the science of lighting, the area formed by the workplane, the floor, and the walls between the two.

floor clearance The distance between the bottom of a door and the finished floor or threshold.

floor closer A device installed in a recess in the floor below a door used to regulate the opening and closing swing of the door.

floor drain A fixture set into a floor, used to drain water into a plumbing drainage system.

floor framing Structural supports for floor systems, including joists, bridging, and subflooring.

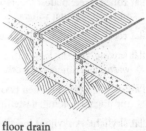

floor drain

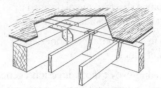

floor framing

floor furring Wood furring used to raise a finished floor above the subfloor and provide a space for piping or conduit.

floor hatch A unit with a hinged panel, providing access through a floor.

floor hole An opening in a floor, roof, or platform measuring less than 12″ in its least dimension, so that a person cannot fall through.

flooring 1. A tongued and grooved piece of lumber used in constructing a floor. Flooring is sold mostly in superior and prime grades, and is produced either as vertical grain or flat grain. **2.** Any finished floor material.

flooring saw A handsaw with teeth on both edges and a blade that tapers to a narrow point, used to cut holes in wood floors.

floor joist Any light beam that supports a floor.

floor load The live load for which a floor has been designed, selected from a building code, or developed from an estimate of expected storage, equipment weights, and/or activity.

floor panel A prefabricated section of floor, consisting of finish floor, subfloor, and joists.

floor pit Any deep recess in a floor used to provide access to machinery, such as an elevator pit.

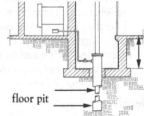

floor pit

floor plan A drawing showing the outline of a floor, or part of a floor, interior and exterior walls, doors, windows, and details such as floor openings and curbs. Each floor of a building has its own floor plan.

floor receptacle An electric outlet set flush with the floor or mounted in a short pedestal at floor level.

floor receptacle

floor sealer A liquid sealer applied with a brush, sprayer, or squeegee to seal floor surfaces, such as concrete or wood.

floor slab A structural slab, usually concrete, used as a floor or a subfloor.

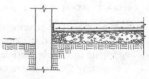

floor slab

floor span 1. The distance, in inches, between the centers of floor joists in a floor system. **2.** The span covered by floor joists between supports.

floor stop A door stop mounted on the floor.

floor system The entire system of components that makes up a floor, including joists, subfloor, finish floor, etc.

floor tile 1. Modular units used as finish flooring. Floor tile may be comprised of resilient (vinyl, rubber, or cork), ceramic, or masonry materials. **2.** Structural units used for floor or roof slab construction.

floor truss A manufactured truss used as a floor joist.

floor-type heater (floor furnace) An air-heating unit mounted below a floor grille or grate. Warm air rises from the center of the grille and cold air returns at the perimeter of the grille.

floor underlayment Particleboard, plywood, waferboard, or similar products used on a subfloor to provide a smooth surface on which to lay the finish floor.

floor underlayment

floor varnish A tough, durable varnish used on floors.

flow control valve A cylindrical pressure-compensating valve that regulates the flow of water. Rated in gpm or gpd.

flow-downs Clauses from a prime contractor's contract with the government that are incorporated into the prime's subcontracts.

flow pressure The pressure at or near an orifice during full fluid flow.

flow reducer A water-conserving mechanism attached downstream from a building's water shutoff valve that reduces the amount of water flow received from the street.

flow trough A sloping trough used to convey concrete by gravity flow from a transit mix truck, or a receiving hopper to the point of placement. *See also* **chute**.

flow trough

flow welding A welding technique that brings metal to welding temperature by pouring a molten filler metal over it.

flue 1. A noncombustible and heat-resistant passage in a chimney used to convey products of combustion from a furnace fireplace or boiler to the atmosphere. **2.** The chimney itself, if there is a single passage.

flue damper An automatic door in a furnace or boiler's flue that closes when the burner is off, reducing heat loss.

flue grouping The consolidation of multiple flues in one chimney or stack to limit the number of vertical shafts through a structure.

flue lining (chimney lining) The lining of a chimney flue composed of heat-resistant firebrick or other fireclay materials, which prevents fire, smoke, and gases from escaping the flue to contaminate surroundings.

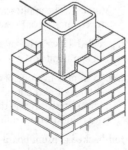

flue lining (chimney lining)

flue pipe A tight duct that conveys products of combustion to a chimney or flue.

fluidifier An admixture employed in grout to decrease the flow factor without changing water content.

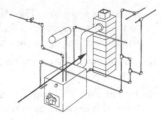

flue pipe

flume A channel made of wood, metal, or concrete and used for drainage purposes on a grade, trestle, or bridge.

fluorescence Visible light generated and emitted from a substance such as a phosphor by absorbing radiation of shorter wavelengths.

fluorescent lamp A low-pressure mercury electric-discharge lamp in which a phosphor coating on the inside of the tube transforms some of the ultraviolet energy generated by the discharge into visible light.

fluorescent lighting fixture A luminaire designed for fluorescent lamps.

fluorescent pigments Pigments that absorb ultraviolet radiant energy and convert it to brilliant, visible light.

fluorescent reflector lamp A fluorescent lamp with the coating on only a part of its cross-section so that the light produced will be directed.

fluorescent strip A luminaire in which fluorescent lamps are mounted on a wiring channel containing the ballast and lamp sockets. Usually reflectors or lenses are not used.

fluorescent strip

fluorescent U-lamp The tubular fluorescent lamp bent at 180° through its center, forming a U-shape.

fluorinated ethylene propylene (FEP) A soft plastic copolymer with relatively low tensile strength, high chemical resistance, and low coefficient of friction. Applications include coatings and protective linings, chemical and medical equipment, extruded insulation and glazing film for solar panels. A common brand name is Teflon®.

fluosilicate Magnesium or zinc silica-fluoride used to prepare aqueous solutions sometimes applied to concrete as surface-hardening agents.

flush 1. Having a surface or face even with the adjacent surface. **2.** To send a quantity of water down a pipe to clean, fill, or empty it.

flush bushing In plumbing, a bushing without a shoulder and engineered to fit flush into the fitting with which it connects.

flush-cup pull A door pull for a sliding door. The pull is mortised flush into the door and has a curved recess serving as a finger grip.

flush door A door with flush surfaces and concealed structural parts.

flush glazing Glazing in which glass is set in a channel, which may be formed by a rabbet

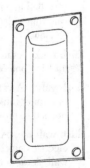

flush-cup pull

and stops, in a frame. The glazing, formed or a compound, is flush with the frame at the top of the channel.

flush-head rivet A rivet with a countersunk head.

flush joint Any joint with its surface flush with the adjacent surfaces.

flushometer (flushometer valve) A flushing valve, designed for use without a flush tank or cistern, that is activated by direct water pressure to deliver a certain quantity of water for flushing needs.

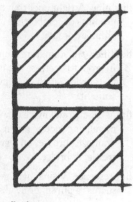

flush joint

flushometer (flushometer valve)

flush panel A panel in which the exposed surface is in the same plane as the exposed surfaces of the surrounding frame.

flush paneled door A door with one or both surfaces in the same planes as the surfaces of the rails and stiles.

flush plate In electricity, the metal or plastic cover that shields the flush wiring device in a wiring box and provides covering for an outlet or switch, with holes cut into its face to accommodate switch handles and plugs.

flush switch An electrical switch installed in a flush wall box in such a way that only its front face is exposed to view.

flush tank

flush tank A tank that holds water for flushing one or more plumbing fixtures.

flush valve A valve installed in the bottom of a toilet tank to discharge the water needed to flush the fixture.

flush wall box A wall box that houses an electric device and is embedded in a wall, floor, or ceiling with its exposed face in the same plane as the surrounding surface.

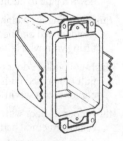

flush wall box

flute In architecture, one of multiple grooves or channels of semicircular to semielliptical sections, used to decorate and to embellish members, such as the shafts of columns.

flying bond A masonry bond formed by laying occasional headers at random intervals.

flying scaffold Staging suspended by ropes or cables from outrigger beams attached at the top of a structure.

flying shore A horizontal member that provides temporary support between two walls.

fly rafter A gable-end rafter on a roof overhang that runs parallel to the common rafters and is supported by the lookout rafter.

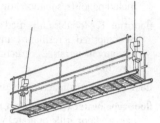

flying scaffold

flywheel effect The tendency of a building to retain heat due to its mass-thermal-inertia.

foam concrete A lightweight, cellular concrete made by infusing an unhardened concrete mixture with prepared foam or by generating gases within the mixture.

foam core The center of a plywood *sandwich panel*, consisting of plastic foam between wood veneers. The foam may be introduced in a liquid form that is forced under pressure into a space between the wood veneer skins, or the skins may be applied to a rigid plastic foam board.

foam glass (cellular glass, expanded glass) A thermal insulation made by foaming glass with hydrogen sulfide. Foam glass has a closed cell structure and is a low fire hazard material. It is manufactured in the form of block or board with a density of 9 to 10 lb. per cu. ft. (14 to 16 kg per cu m).

foaming agent A substance incorporated into plastic mixtures of concrete, rubber, gypsum, or other materials that helps to produce a light foamy consistency by releasing gases into the mixture.

foam-in-place insulation A plastic foam employed for thermal insulation, prepared by mixing an insulation substance with a foaming agent just before it is poured or sprayed with a gun into the enclosed receptacle cavities.

fog The visibility and path of the effluent air stream exiting a cooling tower and remaining close to the ground.

fog curing 1. The storage of concrete in a moist room in which the desired high humidity is achieved by the atomization of fresh water. 2. The application of atomized fresh water to concrete, stucco, mortar, or plaster.

fog sealed Descriptive of surfaces lightly treated with asphalt without a mineral cover.

foil 1. In masonry, one of multiple circular or nearly circular holes, set tangent to the inside of a larger arc, that meet each other in pointed cusps around the arc's inner perimeter. 2. A metal formed into thin sheets by rolling.

foil-backed gypsum board A gypsum board with aluminum foil on one face. The foil acts as a vapor barrier.

folded-plate construction A type of construction used with span roofs. Thin, flat elements of concrete, steel, or timber

are connected rigidly at angles to each other, similar to accordion folds, to form members with deep cross-sections.

folding door An assembly of two or more vertical panels hinged together so they can open or close in a confined space. A floor- or ceiling-mounted track is usually provided as a guide.

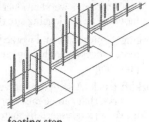

folding door

folding partition 1. Large panels hung from a ceiling track, sometimes supported also by a floor track, which form a solid partition when closed, but stack together when the partition is maneuvered into an open position. 2. A partition, faced with fabric and hung from a ceiling track, that folds up flexibly when opened like the pleated balloons of an accordion.

folding rule A rule made of lengths that are joined by pivots so it can be folded when not in use.

foliated joint A joint made between two boards by fitting their rabbeted edges together, forming a continuous surface on either face.

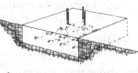

folding partition

foot 1. The bottom or base of an object. 2. A unit of measurement of length in the English system. 3. A projection on a cylindrical roller used to compact a layer of earth fill.

foot block A mat of concrete, steel, or timber used to distribute a vertical load from a post or column over an area of supporting soil.

footcandle The common U.S. unit of measurement of the illuminance on a surface. One footcandle is equal to one lumen per sq. ft. The international unit of measurement is the lux. One footcandle is equal to 10.76 lux.

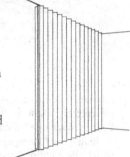

footing

footing That portion of the foundation of a structure that spreads and transmits the load directly to the soil.

footing beam 1. A reinforced concrete beam connecting pile caps or spread footings to distribute horizontal loads caused by eccentric loading. 2. The tie beam in a roof system.

footing course A broad course of masonry at the foot of a wall, which helps prevent the wall from settling.

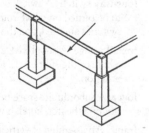

footing beam

footing step A change in elevation of a strip footing.

footing stop In concrete work, a temporary board used to close up concrete until work can resume.

footpiece An item of ductwork in a heating, air-conditioning, or ventilating system that serves to change the direction of the airflow.

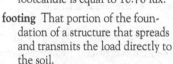

footing step

footplate A timber used in wood frame construction to disperse heavily concentrated structural loads among a number of supporting members, as a plate installed below a row of wall studs.

foot-pound A unit measure of work or energy in the English System equal to the force in pounds multiplied by the distance in feet through which it acts.

footprint The area that falls directly beneath and shares the same perimeter as a structure.

force, compression A force acting on a body, tending to compress the body (pushing action).

forced-air furnace A warm-air furnace, outfitted with a blower, that heats an area by transmitting air through the furnace and connecting ducts.

forced draft A draft of air that is mixed with fuel before the mixture is fed into the combustion chamber of a furnace or boiler.

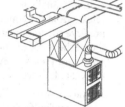

forced-air furnace

force majeure An unexpected or unanticipated event that, if it occurs, can provide the owner or contractor, or both, a legitimate reason to delay the project, cease work, or cancel the contract without penalty. A force majeure clause in a construction contract attempts to define those events. Such events may be, for example, natural disasters deemed "acts of God," unanticipated government mandates, civil disturbances, and so forth.

force, shear A force acting on a body which tends to slide one portion of the body against the other side of the body (sliding action).

force, tension A force acting on a body tending to elongate the body (pulling action).

force, torsion A force acting on a body that tends to twist the body (twisting action).

forecast 1. An estimate and prediction of future conditions and events based on information and knowledge available at the time of the forecast. 2. When in respect to resource requirements, considering future conditions and events, it is a synonym for an estimate.*

foreclosure The legal transfer of a property deed or title to a bank or other creditor because of the owner's failure to pay the mortgage, whereupon the owner loses the right to the property.

Forest Stewardship Council (FSC) An international organization formed in 1993 that is concerned with ecological, social, and economic aspects of the forest management practices used to produce wood products. In order to earn Forest Stewardship Council

certification, a wood product must meet certain criteria as it moves from the logging site through to the end user.

forging A piece of metal worked into a desired shape by one or more processes, including pressing, rolling, hammering, and upsetting.

forklift truck A self-powered vehicle equipped with strong prongs, or forks, that can be raised or lowered. A forklift truck is used to move objects, especially material on pallets, from one location and/or level to another.

form A temporary structure or mold for the support of concrete while it is setting and gaining sufficient strength to be self-supporting.

form anchor A device used to secure formwork to previously placed concrete of adequate strength. The device is normally embedded in the concrete during placement.

form board A board or sheet of wood used in formwork.

form coating A liquid applied to interior formwork surfaces for a specific purpose, usually to promote easy release from the concrete, to preserve the form material, or to retard the set of the near-surface matrix for preparation of exposed-aggregate finishes.

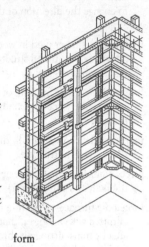
form

form hanger A device used to support formwork from a structural framework. The dead load of forms, weight of concrete, and construction and impact must be supported.

form insulation Thermal insulation, equipped with an airtight seal, that is applied to the exterior of concrete forms. Used to preserve the heat of hydration at required levels so that concrete can set properly in cold weather.

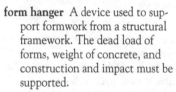
form hanger

form lining Selected materials used to line the face of formwork in order to impart a smooth or patterned finish to the concrete surface, to absorb moisture from the concrete, or to apply a set-retarding chemical to the formed surface.

form ply Plywood commonly used for constructing forms.

form pressure Lateral pressure acting on vertical or inclined formed surfaces, resulting from the fluid-like behavior of the unhardened concrete.

form scabbing The inadvertent removal of the surface of concrete as a result of adhesion to the form.

form tie A tensile unit adapted to prevent concrete forms from

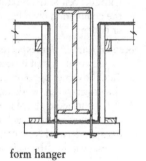
form tie

spreading due to the fluid pressure of freshly placed, unhardened concrete.

formwork The total system of support for freshly placed concrete, including the mold or sheathing that contacts the concrete, as well as all supporting members, hardware, and necessary bracing.

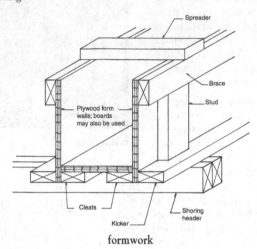
formwork

foundation The entire masonry substructure below the first floor or frame of a building, including the footing upon which the building rests.

foundation drainage tile Tile or piping, either porous or set with open joints, used to collect subsurface water or for the dispersion of effluent.

foundation engineering The category of engineering concerned with evaluating the ability of a locus to support a given structural load, and with designing the substructure or transition member needed to support the construction.

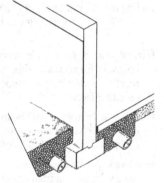
foundation drainage tile

foundation vent Vent openings in a foundation wall that allow for exchange of air and reduce moisture.

foundation wall That part of the foundation of a building forming a retaining wall for the portion of the building that is below grade.

four-way switch A switch used in a wiring layout to allow a circuit to be turned on or off from more than two places. Two three-way switches are used, and the remainder are four-way.

foxtail wedge A small wedge used to spread the split end of a bolt in a hole or the split end of a tenon in a mortise to secure the bolt or tenon.

foyer A subordinate space between an entrance and the main interior of a theater, hotel, house or apartment.

frame An assembly of vertical and horizontal structural members.

frame clearance The clearance between a door and the door frame.

frame construction A construction system in which the structural parts are wood or dependent on a wood framework for support. *See also* **framing**.

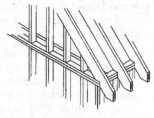
frame construction

framed building A type of building construction in which vertical loads are carried to the ground on a frame.

framed ground One of the vertical wood members fastened around an opening to which a door casing is attached, usually built with a tenon joint between the head and doorjambs.

framed joist A joist that is specially cut or notched in order to be joined securely with other joists or timbers.

frame gasket A strip of flexible material applied to the stop of a door frame to ensure tight closure.

frame house A house of frame construction, usually with exterior walls sheathed and covered with wood siding.

framer A carpenter who erects the wood frame of a home, including walls, trusses, rafters, decking, and floor systems.

frame wall

frame wall A wall of frame construction.

framework A network of structural members or components joined to form a structure, such as a truss or multilevel building.

framing 1. Structural timbers assembled into a given construction system. 2. Any construction work involving and incorporating a frame, as around a window or door opening. 3. The unfinished structure, or underlying rough timbers of a building, including walls, roofs, and floors.

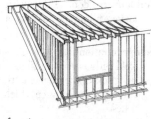
framing

framing anchor A sheet metal device used to join light wood members such as studs, joists, and rafters.

framing, balloon A system of framing a building in which all vertical structural elements of the bearing walls and partitions consist of single pieces extending from the top of the soleplate to the roofplate and to which all floor joists are fastened.

framing anchor

framing plan A drawing of each floor of a building showing exact locations of framing members and their connections. May include wall elevations and details.

framing, platform A system of framing a building in which floor joists of each story rest on the top plates of the story below or on the foundation sill for the first story, and the bear-ing walls and partitions rest on the subfloor of each story. *See also* **framing**.

f rating The measurement of stress (symbolized by the letter f) in a piece of lumber. Generally, the higher the f rating, the stronger the piece of lumber.

free fall 1. Descent of freshly mixed concrete into forms without drop chutes or other means of confinement. 2. The distance the concrete falls.

free float A term used in project management, planning, and scheduling methods such as PERT and CPM. The free float of an activity is the amount by which the completion of that activity can be deferred without delaying the start of the following activities or affecting any other activity in the network.

free moisture Moisture having essentially the properties of pure water not absorbed by aggregate in a test sample or a stockpile.

free on board (FOB) Refers to the point to which the seller will deliver goods without charge to the buyer. Additional freight or other charges connected with transporting or handling the product become the responsibility of the buyer.

freestanding

freestanding Said of a structural element that is fixed at its base and not braced at any upper level.

freezer A room or cabinet mechanically refrigerated to maintain a temperature of about 10°F (–12°C) used for food storage.

freezeback A roofing condition that occurs when water is forced back under the shingles by ice that has thawed and then refrozen.

freeze-thaw The cycle from freezing to thawing often detrimental to construction materials such as concrete. Air entrainment of concrete to be exposed to the elements is helpful in preventing damage from freeze-thaw cycles.

French door (casement door, door window)

French door (casement door, door window) A door, or pair of doors, with glass panes constituting all or nearly all of its surface area.

French drain A drainage ditch containing loose stone covered with earth.

French roof A mansard roof with nearly perpendicular sides.

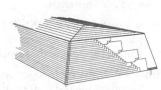
French roof

frequency A measure of oscillations per second, applied to the current or voltage of AC electrical circuits, sound waves or vibrating solid objects, and stated in hertz (HZ) or cycles per second (CPS).

frequency modulation A method of modifying a sound-wave signal to relay information.

F

F

fresh-air inlet 1. A connection to a building drain, located above the drain trap and leading to the atmosphere. **2.** An outside air vent for an HVAC system.

friction The resistance to relative motion, sliding or rolling, between two surfaces in contact. May produce heat.

friction catch Any catch held in position by friction when engaged with its strike.

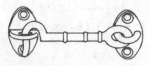

friction catch

friction head In hydraulics, the energy lost by friction in a pipe. This may include losses at elbows and bends.

friction loss The stress loss in a prestressing tendon resulting from friction between the tendon and duct or other device during stressing.

friction pile (floating pile foundation) A load-bearing pile that receives its principal vertical support from skin friction between the surface of the buried pile and the surrounding soil.

friction pile (floating pile foundation)

friction tape An insulating tape used by electricians, made of a fibrous base impregnated with a moisture-resisting compound that will stick to itself but not to most other materials.

friction welding A process by which thermoplastic materials are softened and welded together with heat produced by friction.

frieze A horizontal exterior trim member positioned between the siding of a structure and its soffit.

fringe benefits Compensation paid to a worker in addition to his or her wages. Usually consists of paid vacation and sick time, pension, health insurance, and so forth.

front-end loader A machine with a bucket fixed to its front end, having a lift-arm assembly that raises and lowers the bucket. A front-end loader is used in earth moving and loading operations and in rehandling stockpiled materials.

front-end loader

front hearth (outer hearth) The part of a hearth that occupies the room served by a fireplace, skirting the front of the fireplace opening.

frost boil 1. An imperfection on a concrete surface caused by the freezing of trapped moisture that swells and crumbles the affected concrete. **2.** The softening of soil caused by melting and release of subsurface water during a thaw.

frost line The depth to which frost penetrates the ground. This depth varies from one part of the country to another. Footings should be placed below the frost line to prevent shifting.

frostproof closet A water-closet bowl that retains no standing water. The trap and the valve for its water supply are located below the frost line.

fuel cell Cell that uses hydrogen as a fuel source through a catalytic process, producing electricity and giving off heat and small amounts of hot water as by products. The heat can be captured for space-heating needs, and the electricity generated can provide for power.

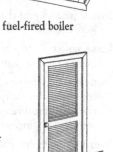

fuel-fired boiler An automatic, self-contained mechanical unit or system that produces heat by burning solid, liquid, or gaseous fuels.

fuel-fired boiler

full A reference to lumber that is slightly oversize.

full bond A masonry bond in which all bricks are laid as headers.

full-face respirator A respirator that covers the whole face with a protective shield.

full-flush door A door made of two sheets of steel assembled in hollow-metal construction with a top and bottom either flush or closed off with end panels, and seams that are visible only on the edge of the door.

full-louvered door

full glass door A door with glass in the area between rails and stiles except for dividing mullions. The glass is usually heat-treated or tempered.

full-louvered door A door with louvers filling the entire area between rails and stiles.

full-penetration butt weld A butt weld with a depth equal to the thickness of the smaller of the two members between which it lies.

full-penetration butt weld

full-surface hinge A hinge that may be installed on the surface of a door and jamb without need of mortising.

furlong A unit of measure equal to 220 yards or 1/8 of a mile.

furnace 1. That part of a boiler or warm-air heating plant in which combustion takes place. **2.** A complete heating unit that transfers heat from burning fuel to a heating system.

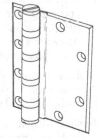

full-surface hinge

furnish The raw material used to make reconstituted wood-based nonveneer panel products.

furred Provided with furring to leave an air space, as between a structural wall and plaster or between a subfloor and wood flooring.

furring 1. Strips of wood or metal fastened to a wall or other surface to even it, to form an air space, to give the appearance of greater thickness, or for the application of an interior finish

such as plaster. **2.** Lumber 1″ in thickness (nominal) and less than 4″ in width, frequently the product of resawing a wider piece. The most common sizes of furring are 1″ × 2″ and 1″ × 3″.

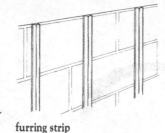

furring strip

furring strip Wood or metal channels used for attaching gypsum or metal lath to masonry walls as a finish or for leveling purposes.

furring tile Tile designed for lining the inside of exterior masonry walls. It is not intended to support superimposed loads and has a scored face for the application of plaster.

furrowing The process of making furrows in the mortar bed with the tip of a trowel to speed up bricklaying by better distributing mortar.

fuse

fuse A protective device, made of a metal strip, wire, or ribbon that guards against overcurrent in an electrical system. The device melts if too much current is generated and breaks the circuit.

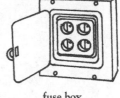

fuse box

fuse box A metal box with a hinged cover that houses fuses for electric circuits.

fusetron A special fuse that will carry an overload of current for a short time without "blowing" or opening the circuit. A fusetron is used where a heavy load, such as the starting of a motor, may overload a circuit momentarily.

fusible link A metal link made of two parts held together by a low-melting-point alloy. When exposed to fire-condition temperatures, the link separates, allowing a door, damper, or device to be closed.

fusible metal An alloy with a low melting point used in high temperature detecting devices.

fusion In welding, the melting and coalescence of a filler metal and base metal or two base metals.

FUTA The Federal Unemployment Tax Act, a federal law that imposes an unemployment tax on employers.

F

G

Abbreviations

g gram, gravity, gauge, girth

G gas

ga gauge

gal gallon

galv galvanized

GAO The U.S. General Accounting Office

GATT General Agreement on Tariffs and Trade

GB glass block

GC general contractor

GCF greatest common factor

gen general

GFCI ground-fault circuit-interrupter

GI galvanized iron

gl glass, glazing

GM grade marked

GMV gram molecular volume

GOB general obligation bonds

Goth Gothic

gov/govt government

gpd gallons per day

gpf gallons per flush

gph gallons per hour

gpm gallons per minute

GPS global positioning system

gr grade, gravity, gross, grains

G/R grooved roofing

gran granular

gr fl ground floor

gr fl ar ground floor area

grnd ground

gr wt gross weight

G/S grade-stamped

GT gross ton

gtd guaranteed

gup grading under pavement

GYP gypsum

Definitions

gabion A large compartmentalized container, usually cylindrical or rectangular, often fabricated from galvanized steel hexagonal wire mesh. When filled with stone it is used in the construction of foundations, dams, erosion breaks, etc.

gabion

gable The portion of the end of a building that extends from the eaves to the peak or ridge of the roof. The gable's shape is determined by the type of building on which it is used: triangular in a building with a simple ridged roof, or semioctagonal in a building with a gambrel roof.

gable

gableboard A board covering the timbers that extend over the gable end of a gable roof.

gable dormer A dormer that protrudes horizontally outward from a sloping roof and has its own gabled ends, whose base meets that of the sloping roof.

gable dormer

gable roof A ridged roof having one or two gabled ends.

gable vents Vent openings mounted in the top of a gable of a house to allow the exchange of air in the attic.

gable wall A wall whose upper portion is a gable.

gable window A window built into or shaped like a gable.

gain 1. The mortise or notch in a piece of wood into which a piece of wood, hinge, or other hardware fits. 2. An increase in transmission signal power from one point to another, expressed in decibels. 3. An increase in the volume of sound of a radio.

gallery 1. A long enclosure that functions as a corridor inside or outside a building or between different buildings. 2. The elevated (usually the highest) designated seating section in an auditorium, theater, church, etc. 3. A room or building in which artistic works are exhibited.

gallery apartment house An apartment house whose individual units can be entered from an exterior corridor on each floor.

galleting (gaffering) The use of chips of rock, stone, or masonry to fill the joints of rough masonry, either for appearance or to minimize the required amount of mortar.

gallon A standard liquid measure: The British imperial gallon contains 10 lbs. of water; the American gallon contains 8.33 lbs. of water.

gallon per minute (gpm) A unit measure of flow. Equals a flow rate of one gallon in one minute.

gallows bracket A triangular wall bracket often used for the support of shelving.

galvanic corrosion The electrochemical action that occurs as a result of the contact of dissimilar metals in the presence of an electrolyte.

galvanize The process of protectively coating iron or steel with zinc, either by immersion or electroplating (electrogalvanizing).

galvanized iron Zinc-coated iron sheet metal.

galvanized pipe Zinc-coated steel or wrought-iron pipe.

gambrel roof A roof whose slope on each side is interrupted by an obtuse angle that forms two pitches on each side, the lower slope being steeper than the upper.

gambrel roof

gang A working technique in which several machines and/or apparatuses are controlled by a single force and combined in such a way as to function as a single unit.

gang form Prefabricated form panels connected together to produce large reusable units. Gang forms are usually lifted by crane or rolled to the next location.

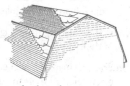
gang form

gang sawing A method of stone-cutting that made possible the use of stone as a veneer on buildings.

gantry crane A revolving crane situated atop a heavy framework that moves along the ground on tracks.

Gantt chart A time-scaled bar chart named after Henry L. Gantt.*

GAO The U.S. General Accounting Office that, under the direction of the Comptroller General of the United States, has jurisdiction over bid protests. The GAO is an arm of the Congress.

gap grading A particle-size distribution of aggregate that is practically or completely void of certain intermediate sizes.

garbage disposal unit An electrically motorized device that grinds waste food and mixes it with water before disposing of it through standard plumbing drainage pipes.

garden house In a garden, a small gazebo or other shelter.

garden tile Molded structural ceramic units used in a garden or on a patio as stepping stones.

garden wall bond Usually, ornamental brickwork that is only a single brick thick and is composed mostly of stretchers showing a fair face on each side.

gargoyle A waterspout, often with carvings of grotesque human or animal features, that projects from a roof gutter so as to drop rainwater clear of a wall, often through its open mouth.

garnet A silicate mineral with an isometric crystal structure. Garnet occurs in a variety of colors and is used as gemstones, abrasives, or gallers.

garnet paper A finishing and polishing paper covered with powdered garnet abrasive.

garrison house A style of house whose second story overhangs the face of the first story on one or more walls.

garrison house

gas A fluid (in the form of air) with neither independent shapes nor volume, that can expand indefinitely.

gas engine An internal-combustion engine that uses a gas rather than gasoline as fuel.

gas-filled lamp A type of incandescent lamp with a bulb atmosphere consisting of an inert gas within which the filament operates.

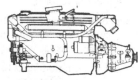
gas engine

gas furnace A furnace that burns gas to produce heat.

gasket Any of a variety of seals made from resilient materials and placed between two joining parts (as between a door and its frame, an oil filter and its seat, pipe threads and their fitting, etc.) to prevent the leakage of air, water, gas, or fluid.

gas main The public's gas supply as piped from the providing utility company into a community.

gas metal-arc welding A method of welding that achieves coalescence by arc-heating between the work and a consumable electrode.

gas meter A mechanical device that measures and records in cubic feet the volume of gas passing a given point.

gas pocket In a casting, a hole or void caused by the entrapment of air or gas that is produced during the solidification of the metal.

gas refrigeration A refrigeration system in which a gas flame is used to heat the refrigerant.

gas vent An exhaust pipe through which undesirable gaseous byproducts of combustion are vented to the outside.

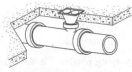
gas vent

gas welding Any of several welding methods in which a gas flame provides the heat necessary for coalescence and in which pressure and/or a filler metal may or may not be employed.

gate A usually hinged device of solid or open construction that is installed as part of a fence, wall, or similar barrier and that, when opened (either by swinging, sliding, or lifting), provides access through that barrier.

gate operator An electronically operated mechanism that, when activated, serves to open or close a gate.

gate valve A valve utilizing a gate, usually wedge-shaped, that allows fluid flow when the gate is lifted from the seat. Gate valves have less resistance to flow than globe valves, and should always be used fully open or fully closed.

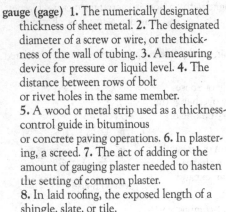
gate valve

gauge (gage) 1. The numerically designated thickness of sheet metal. 2. The designated diameter of a screw or wire, or the thickness of the wall of tubing. 3. A measuring device for pressure or liquid level. 4. The distance between rows of bolt or rivet holes in the same member. 5. A wood or metal strip used as a thickness-control guide in bituminous or concrete paving operations. 6. In plastering, a screed. 7. The act of adding or the amount of gauging plaster needed to hasten the setting of common plaster. 8. In laid roofing, the exposed length of a shingle, slate, or tile.

gauge (gage)

gauge box (gauging box) A container in which a batch of concrete, mortar, or plaster is measured and mixed.

gauged A term that refers to a material that has been ground so as to produce particles of uniform shape and/or thickness.

gauged brick Brick that has been sawn, ground, or otherwise shaped to accurate dimensions for special applications.

gauged mortar 1. Mortar made of cement, sand, and lime in specific proportions. 2. Any plastering mortar that is mixed with plaster of paris to hasten setting.

gauged skim coat A mixture of gauging plaster and lime putty applied very thinly as a final coat in plastering. The mixture is troweled to produce a smooth, hard finish.

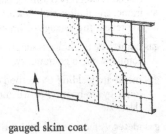
gauged skim coat

gauge glass A glass tubular device or vertical cylindrical device, often graduated, that indicates the level or amount of liquid in a tank or vessel.

gauge rod A stick used for measuring the gauge in brickwork. If used to mark floor and sill levels, the rod is termed a *story rod*, or *story pole*.

gauging plaster Plaster of paris (gypsum plaster) that is usually mixed with lime putty to produce a quick-drying finish coat.

gazebo A small, round, octagonal, or similarly shaped structure, usually roofed but open-sided, built in parks or large gardens to provide shelter or a place to view the surrounding area.

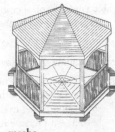
gazebo

gear A toothed wheel, cone, cylinder, or other machined element that is designed to mesh with another similarly toothed element for purposes that include the transmission of power and the change of speed or direction.

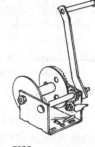
gear

General Agreement on Tariffs and Trade (GATT) A specialized agency of the United Nations that seeks to ease the barriers of international trade.

general & administrative costs (G&A) The fixed cost incurred in the operation of a business. G&A costs are also associated with office, plant, equipment, staffing, and expenses thereof, maintained by a contractor for general business operations. G&A costs are not specifically applicable to any given job or project.*

general conditions The portion of the contract document in which the rights, responsibilities, and relationships of the involved parties are itemized.

general contract In a single contract system, the documented agreement between the owner and the general contractor for all the construction for the entire job.

general contractor For an inclusive construction project, the primary contractor who oversees and is responsible for all the work performed on the site, and to whom any subcontractors on the same job are responsible.

general diffuse lighting Lighting from units that direct 40% to 60% of their emitted light upward and the remainder downward.

general foreman The general contractor's on-site representative, often referred to as the *superintendent* on large construction projects. It is the responsibility of the general foreman to coordinate the work of various trades and to oversee all labor performed at the site.

general industrial occupancy The designation of a conventionally designed building that can be used for all but high-hazard types of manufacturing or production operations.

general liability insurance Insurance that protects a project and any adjacent properties from damage or loss as a result of an accident during the construction process.*

general obligation bonds Instruments of obligation that, by permission of the public through referendum, are issued to investors by a subdivision of government. These bonds promise incremental payment of principal and interest from revenues collected annually by the government. In return, funds are supplied by investors and are used to pay for the construction of publicly owned buildings or other public works projects.

general partnership An association of two or more people to conduct business for profit as co-owners. The participants in a general

partnership, called *general partners*, share all aspects of the business including assets, profits/losses, liability, and management responsibility.

general-purpose branch circuit An electrical circuit that supplies a number of outlets.

general requirements The designation or title of Division 1 (the first of 16) in the Construction Specifications Institute's Uniform System. General requirements usually include overhead items and equipment rentals.

general terms and conditions 1. That part of a contract, purchase order, or specification that is not specific to the particular transaction but applies to all transactions. **2.** General definition of the legal relationships and responsibilities of the parties to the contract and how the contract is to be administered. They are usually standard for a corporation and/or project.*

generator 1. A mechanical or electromechanical device that converts mechanical energy into electrical power, as an alternator producing alternating current or a dynamo producing direct current. **2.** A person, firm, or entity whose activities create a hazardous waste.

geodesic dome A stable, dome-shaped structure fabricated from similar lightweight members connected to form a grid of interlocking polygons.

geodesic dome

Georgian architecture Neoclassical style of architecture that prevailed in 18th-century Great Britain. Characterized by orderly design, rectangular shapes, and symmetrical window placement.

geotechnical engineering A specialty in the field of civil engineering that focuses on geology and soil mechanics leading to the design of foundation structures.

geotextiles Synthetic fabrics that can be used to separate back filling materials to promote proper drainage. Commonly used in conjunction with high retaining walls and landscape design.

geothermal heat An energy-efficient heating system that uses subsurface ground temperatures to keep the system's water at a moderate temperature. Less energy is then required to heat the water enough warm the building.

gesso A plaster base coat, for gilding or decorative painting, consisting of gypsum plaster, calcium carbonate, and glue.

gilding 1. The application of gilt (as gold leaf or flakes) to a surface. **2.** A surface ornamented with gilt.

gimlet A small hand tool, of ancient origin, whose handle is perpendicular to a screw point, and that is used for boring small holes (less than 1/4") in wood.

gin A simple lifting device consisting of a vertical pole, tripod, or other frame.

gin block A simple tackle block, with a single pulley in a frame, having a hook on top for hanging.

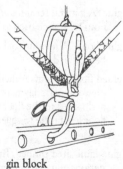
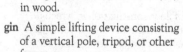
gin block

gingerbread An ornate style of architecture, usually employed in exterior house trim, and common in the United States during the 19th century.

gin pole A cable-supported vertical pole used in conjunction with blocks and tackle for hoisting.

girandole A branched fixture or bracket for holding lamps or candles, either freestanding or protruding from a wall, often having a mirror behind it.

girder A large principal beam of steel, reinforced concrete, wood, or a combination of these, used to support other structural members at isolated points along its length.

girder casing Material that encloses and protects (as from fire) the part of a girder extending below ceiling level.

girder post Any support member, such as a post or column, for a girder.

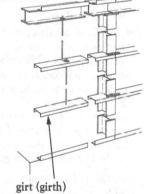
girder casing

girdle A horizontal band around the shaft of a column.

girt (girth) 1. A horizontal member used as a stiffener between studs, columns, or posts at intermediate level. **2.** A rail or intermediate beam that receives the ends of floor joists on an outside wall.

girth 1. The circumference of any circular object. **2.** The distance around a column.

give-and-take lines Straight equalizing lines used in calculating band areas and their boundaries.

glare Any unwanted brightness that brings discomfort or reduced visibility.

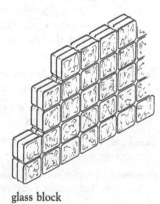
girt (girth)

glass A hard, brittle, inorganic product, ordinarily transparent or translucent, made by the fusion of silica, flux, and a stabilizer, and cooled without crystallizing. Glass can be rolled, blown, cast, or pressed for a variety of uses.

glass block A hollow, translucent block of glass, often with molded patterns on either or both faces, that affords pleasantly diffused light when used in non-load-bearing walls or partitions.

glass block

glass cement Any glue or other adhesive material that serves to bind glass to glass, or glass to another material.

glass cloth A wall covering made from woven vegetable fibers, especially arrowroot bark, that are laminated onto a paper backing.

glass cutter A small hand tool having a pointed diamond tip or a sharp, small, hardened steel wheel that scores glass.

glass door A door fabricated without stiles or rails and consisting entirely of thick, heat-strengthened, or tempered glass.

glass, obscure Sheet glass that has been made translucent instead of transparent. It is sandblasted or molded to make the surfaces irregular.

glasspaper A fine sandpaper or polishing paper whose abrasive is powdered glass.

glass size The size to which a piece of glass is cut so as to glaze a given opening. Its length and width should be ⅛″ less than the distance between the outside edges of the rebates.

glass stop **1.** A glazing bead. **2.** A fitting at the lower end of a patent glazing bar to prevent the pane from sliding down.

glass, structural Cast glass in squares or rectangles, 1″ to 2″ thick, sometimes laid up between concrete ribs, frequently as tile. Larger units are made in hollow or vacuum blocks. The wide use of the product is in colored and polished sheets for interior wall surfacing.

glass tile Transparent or translucent units installed in a roof surface that allow light to enter the room below.

glass wool Spun fiberglass used mostly for thermal and acoustical insulation and as a filtering medium in air and water filters.

glaze **1.** To install glass panes in a window, door, or another part of a structure by applying putty or other material to hold the glass in place. **2.** A hard, thin, glossy ceramic coating on the surface of pottery, earthenware, ceramics, and similar goods. Usually called *enamel*, a transparent solution that, when dry, provides tiles with a hard, shiny finish. Can be applied by either painting or dipping.

glaze

glaze coat **1.** The smooth top layer of asphalt in built-up roofing. **2.** A temporary, protective coat of bitumen applied to built-up roofing that is awaiting top-pouring and surfacing. **3.** In painting, the application of a nearly transparent coat that enhances and protects the coat below it.

glazed brick Brick or tile with a surface produced by fusing it with a glazing material.

glazed door Any door with glass panes or panels and top and bottom rails.

glazed tile Ceramic or masonry tile with an impervious, glossy finish.

glazement A waterproof surfacing used on masonry that creates a smooth, durable finish.

glazed door

glazier A person whose trade is to install glass in structures. The glazier removes old putty, cuts glass to fit the openings requiring it, and secures it there by whatever means are necessary or appropriate.

glazier's putty A glazing compound often made from a mixture of linseed oil and plaster of paris, and sometimes including white lead.

glazing **1.** Fixing glass in an opening. **2.** The glass surface of an opening that has been glazed.

glazing bead (glass stop) **1.** A narrow strip of wood, plastic, or metal fastened around a rebate and used to hold glass in a sash. **2.** At a glazed opening, removable trim that holds the glass in place.

glazing compound Any putty or caulking compound used in glazing to seal the joint at the edges of the glass.

glazing fillet A narrow wood or plastic strip fastened to the rebate of a glazing bar and used instead of face putty to hold glass in place.

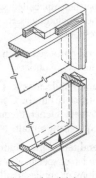

glazing bead (glass stop)

glazing gasket A narrow, sometimes grooved, prefabricated strip of material such as neoprene, that offers a dry alternative to glazing compound in glazing operations. The gasket is impervious to moisture and temperature and is often used with large panes or sheets of glass.

glazing sprig A small headless nail that holds a pane of glass in its wooden frame while putty hardens.

glazing tape A resilient tape used to seal glass into a frame, sash, or opening.

global positioning system (GPS) A global system of U.S. navigational satellites that locates precise points and coordinates on earth.

global warming An increase in the average temperature of the Earth's atmosphere, often used to refer to climatic changes due to increased emissions of greenhouse gases.

globe (light globe) **1.** A protective enclosure or covering over a light source. The globe is usually made of glass and can also serve to diffuse, redirect, or change the color of the light. **2.** An incandescent lamp.

globe valve A valve with a rounded body utilizing a manually raised or lowered disc that, when closed, seats so as to prevent fluid flow. Globe valves are ideal for throttling in a semi-closed position.

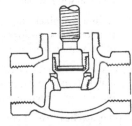

globe valve

gloss Surface luster, usually expressed in terms ranging from matte to high gloss.

glow lamp A type of low-consumption electric discharge lamp whose light is produced close to the electrodes within an ionized gas. A glow lamp is often used as an indicating light.

glue A general term for any natural or synthetic viscous or gelatinous substance used as an adhesive to bind or join materials.

glue bond A measure of how well articles are fastened together after being glued.

glued floor system A method of floor construction in which a plywood underlayment or other structural panel is both glued and nailed to the floor joists. This creates a stronger, stiffer floor less prone to squeaking than floors fastened only with nails.

glue-laminated timber Timber composed of wood layers, or laminations, glued together with the grains longitudinally parallel.

Commonly used for columns, headers, posts, beams, and truss components.

glue nailed A term applying to plywood joints and connections that have been both glued and nailed to produce the stiffest possible construction.

glue-up The process of spreading glue on the surfaces of veneers of similar sizes and pressing them together to form a sheet of plywood.

gob bucket (concrete bucket)

glycol Liquid with a very low freezing point that is miscible with water.

gob bucket (concrete bucket) On a crane, the bucket that is used to carry concrete.

going The horizontal distance between two consecutive risers or stairs. The horizontal distance between the first and last riser of an entire flight is called the *going of the flight* or the run.

going rod A rod used in planning the going of a flight of steps.

gold bronze Powdered copper or copper alloy used for bronzing, or in the manufacture of gold or bronze paint.

good one side (G1S) A grade of sanded plywood with a higher grade of veneer on the face than on the back. G1S is used in applications where the appearance of only one side is important.

good two sides (G2S) The highest grade of sanded plywood. G2S is allowed to have inlays or neat wood patches and is used in applications where the appearance of both sides is important.

gooseneck 1. In plumbing, a curved, sometimes flexible fitting connection or section of pipe. 2. In HVAC, a screened, U-shaped intake or exhaust duct. 3. The curved end of a handrail at the top of a stair. 4. The curved connector from a tractor to a trailer.

gothic arch A typically high, narrow arch with a pointed top and a jointed apex (as opposed to keystone).

gouge A cutting chisel having a long, curved blade and used for hollowing out wood, or for making holes, channels, or grooves in wood or stone.

government anchor A V-shaped anchoring device usually fabricated from ½″ round bar, and used in steelwork to secure a wall-bearing (or other) beam to masonry.

gouache A type of thick, water-based, decorative paint.

government anchor

Gow caisson (Boston caisson, caisson pile) A series of steel cylinders from 8′ to 16′ high used to protect workers and equipment during deep excavation in soft earth. Each successive cylinder is 2′ in diameter smaller than its predecessor, through which it is dropped or driven deeper into the surrounding soft clay or silt to prevent excessive loss of ground and to facilitate deep excavation. When excavation and construction are complete, the cylinders are withdrawn.

grab bar A short length of metal, glass, or plastic bar attached to a wall in a bathroom, near a toilet, in a shower, or above a bathtub.

grab bar

grab bucket A bucket-like device with hinged lower halves or "jaws" hydraulically or cable operated from a crane or used for rehandling granular materials.

grab crane A crane outfitted with a grab bucket.

Gradall A trade name for a wheel-mounted, articulated, hydraulic backhoe often used with a wide bucket for dressing earth slopes.

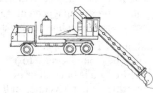

Gradall

gradation An assessment, as determined through sieve analysis, of the amounts of particles of different sizes in a given sample of soil or aggregate.

grade 1. The surface or level of the ground. 2. A classification of quality as, for instance, in lumber. 3. The existing or proposed ground level or elevation on a building site or around a building. 4. The slope or rate of incline or decline of a road, expressed as a percent. 5. A designation of a subfloor, either above grade, on grade, or below grade. 6. In plumbing, the slope of installed pipe, expressed in the fall in inches per foot length of pipe. 7. The classification of the durability of brick. 8. Any surface prepared to accept paving, conduit, or rails.

grade beam A horizontal end-supported (as opposed to ground-supported) load-bearing foundation member that supports an exterior wall of a superstructure.

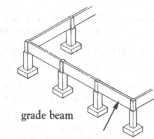

grade beam

grade course The first course of brick, block, or stone, at grade level, usually waterproofed.

graded aggregate 1. Aggregate having size ranging from coarse to graded sand. 2. A fine aggregate (under ¼″) having a uniformly graded particle size.

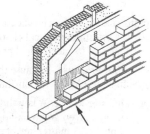

grade course

grade line 1. A line of stakes with markings, each at an elevation relative to a common datum and from whose elevations a grade between their terminal points can be established. 2. A strong string used to establish the top of a concrete pour or masonry course.

grader A multipurpose earth-working machine used mostly for leveling and crowning. A grader has a single blade, centrally located, that can be lifted from either end and angled so as to cast to either side.

grader

grade stamp (grade mark) A stamp on a manufactured piece of wood that designates it meets certain standards of quality as set forth by one of several standards organizations. Stamps for structural lumber, for instance, include information on species, moisture content, and lumber grade.

grade stake In earthwork, a stake that designates the specified level.

gradient 1. The change in elevation of a surface, road, or pipe, usually expressed in a percentage or in degrees. 2. The rate of change of a variable such as temperature, flow, or pressure.

grading 1. The act of altering the ground surface to a desired grade or contour by cutting, filling, leveling, and/or smoothing. 2. Sorting aggregate by particle size. 3. Classifying items by size, quality, or resistance.

grading

grading curve A line on a graph illustrating the percentages of a given sample of material that pass through each of a specific series of sieves.

grading instrument A surveyor's level having a telescope that can be adjusted upward or downward to lay out a required gradient.

grading plan A plan showing contours and grade elevations for existing and proposed ground surface elevations at a given site.

grading rules Quality criteria that determine the classification of lumber, plywood, or other wood products.

gradual load The gradual application of a load to the supporting members of a structure.

grain 1. The directional arrangement of fibers in a piece of wood or woven fabric, or of the particulate constituents in stone or slate. 2. The texture of a substance or pattern as determined by the size of the constituent particles. 3. Any small, hard particle (like sand). 4. A metric unit of weight; 7,000 grains equal one pound.

grain size 1. A size classification of mineral particles in soil or rock. 2. One of the physical characteristics of a particle of soil, which relates to its mechanical properties.

grandfather clause A clause that allows activities that were legal under an old law to continue under a superseding law.

grandmaster key 1. A key that not only operates all the locks within a given group having its own master key, but several such groups. 2. A master key of master keys.

granolithic concrete A type of concrete made from a hard aggregate whose particle shape and surface texture render it conducive for wearing-surface finishes on floors.

granolithic finish A concrete wearing surface, placed over a concrete slab, containing aggregate chips to improve its longevity.

granular 1. A technical term relating to the uniform size of grains or crystals in rock. 2. Composed of grains.

granular fill insulation An insulation material such as perlite or vermiculite that can be easily placed or poured because it comes in the form of chunks, pellets, or modules.

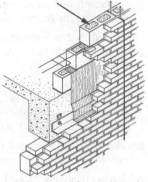
granular fill insulation

granular material A sandy type of soil whose particles are coarser than cohesive material and do not stick together.

graphite paint A type of paint made from boiled linseed oil, powdered graphite, and a drier. Graphite paint is used to inhibit corrosion on metal surfaces.

grass cloth A wall covering made from woven vegetable fibers, especially arrowroot bark. The cloth is laminated onto a paper backing.

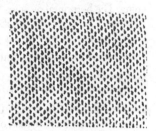
grass cloth

grate 1. A type of screen made from sets of parallel bars placed across each other at right angles in approximately the same plane. A grate allows water to flow to drainage, while covering the area for pedestrian or vehicular traffic. 2. A surface with openings to allow air to flow through while supporting a fuel bed, as in a coal furnace.

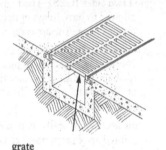
grate

gravel Coarse particles of rock that result from naturally occurring disintegration or that are produced by crushing weakly bound conglomerate. Gravel is retained on a No. 4 sieve.

gravel, bank run Naturally occurring gravel deposits that are often located along rivers or lakes.

gravel roofing Roofing composed of several (built-up) layers of saturated or coated roofing felt, sealed and bonded with asphalt or coal-tar pitch that, for solar protection and insulation purposes, is then covered with a layer of gravel or slag. Usually used on flat or nearly flat roofs.

gravel roofing

gravel stop A metal strip or flange around the edge of a built-up roof. The stop prevents loose gravel or other surfacing material from falling off or being blown off a roof.

gravel stop

gravity dam A pyramid-shaped dam whose own weight resists the force of the water behind it.

gravity separator A chamber in which oil rises and remains on the surface of wastewater until it is removed.

gravity wall A massive concrete retaining wall whose own weight prevents it from overturning.

gravity water supply (gravity water, gravity system) A water distribution system in which the supply source is located at an elevation higher than the use.

gray paper The rough textured, unsized paper that covers the back of regular gypsum board, or both face and back on backing board products.

gray water Wastewater recycled from appropriate sinks, showers, baths, and laundry, which can be used for watering plants, cooling HVAC equipment, and other purposes.

grease extractor A device installed in conjunction with a cooking exhaust system and employing grease-collecting baffles positioned so as to create a path of sharp turns through which the cooking exhaust is passed at high velocity. The grease, which is particulate and heavier than air, is collected on the baffles by centrifugal force, while the carrying air continues around the sharp turns on its way to being exhausted, thus becoming cleaner at each baffle.

grease interceptor (grease trap) A device installed between the kitchen drain and the building sewer to trap and retain fats and grease from kitchen waste lines.

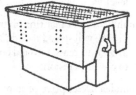

grease interceptor (grease trap)

green A term referring to unseasoned timber or to fresh, unhardened concrete, plaster, or paint.

green board (moisture-resistant) Gypsum board distinguishable by the green color of its face paper. Green board is designed to be used in areas that are often damp, such as in bathrooms for tile backing.

green building Design, construction and product selection that minimizes a structure's impact on the natural environment.

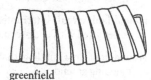

greenfield Flexible metal conduit for electrical wiring.

greenfield

greenhouse A glass-enclosed space with a controlled environment for growing plants, vegetables, and fruits out of season.

greenhouse effect **1.** The conversion of the sun's rays into heat that is retained by the glass roof of a greenhouse. **2.** The steady, gradual rise in temperature of the earth's atmosphere due to heat that is retained by layers of ozone, carbon dioxide, and water vapor.

greenhouse

Green Lights Program A government program sponsored by the Environmental Protection Agency that is aimed at promoting energy efficiency through investment in energy-saving lighting.

green power Refers to the production of electricity from environmentally friendly sources, including wind and solar power, photovoltaics, geothermal, biomass, hydrogen fuel cells, and so forth with the goal of reducing the use of fossil fuel and carbon dioxide generation.

green process Manufacturing products with consideration for the source of materials, energy-efficient manufacturing methods, use of recycled materials in packaging, and reclaiming manufacturing waste.

green products Sustainable products that minimize the impact on the natural environment.

green roof Also called a *living roof*, a roof with a layer of soil and plantings that dissipates solar heat, provides good insulation, absorbs rainwater runoff, generates oxygen, and protects and therefore extends the life of the roofing material below. It can also give the roof space a garden-like appearance. The roofing becomes its own ecosystem with soil and plantings.

greenwash Term referring to companies that overstate their environmental practices in order to conceal unsound procedures and appear environmentally responsible to the public.

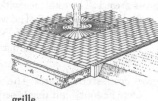

grid

grid **1.** In surveying, a system of evenly spaced perpendicular reference lines with its intersections used to measure elevations. **2.** The structural layout of a given building. **3.** A system of crossed reinforcing bars used in concrete footings.

grid ceiling **1.** A ceiling with apertures into which are built luminaries for lighting purposes. **2.** Any ceiling hung on a grid framework.

grid line Any line that is part of a reference pattern for surveying or a layout.

grillage Steel or wooden beams used horizontally under a structure to distribute its load over its footing or underpinning.

grille **1.** Any grating or openwork barrier used to cover an opening in a wall, floor, paving, etc., for decoration, protection, or concealment. **2.** A louvered or perforated panel used to cover an air duct opening in a wall, ceiling, or floor. **3.** Any screen or grating that allows air into a ventilating duct.

grille

grillwork In construction, any heavy framework of timbers or beams used to support a load on soil instead of on a concrete foundation.

grinder **1.** A device that sharpens or removes particles of material by abrasion. **2.** A machine or tool for finishing concrete surfaces by abrasion.

grinder pump A sewage pump that reduces solids and grinds and pumps material.

grit **1.** A granular abrasive used in making sandpaper or on grinding wheels to give a surface a nonslip finish. **2.** Particles of sand or gravel contained in sewage.

grizzly **1.** A screen or grate used to remove oversize particles during aggregate processing or loading. **2.** A gate or similar device on a chute.

groin **1.** In architecture, a ridge or curved line formed at the junction of two intersecting vaults. **2.** A structure built outward from a shore into water to direct erosion or to protect against it.

groin centering **1.** A method of supporting ribless groining during vaulting. **2.** In ribbed vaulting, the support of the stone ribs by timbers until construction is complete.

grommet A metal or plastic eyelet that provides a reinforced hole in a material, such as cloth or leather, that might otherwise tear from the stress on the hole when a fastener or other device is passed through it or attached to it.

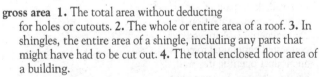

grommet

groove In carpentry, a narrow, longitudinal-channel cut in the edge or face of a wood member. The groove is called a *dado* when cut across the grain, and a *plow* when cut parallel to it.

groove joint A joint formed by the intentional creation of a groove in the surface of a wall, pavement, or floor slab for the purpose of controlling the direction of random cracking.

groove joint

groover A tool for creating grooved joints in unhardened concrete slabs.

groove weld A weld between the existing preformed grooves in members being joined.

groover

gross area **1.** The total area without deducting for holes or cutouts. **2.** The whole or entire area of a roof. **3.** In shingles, the entire area of a shingle, including any parts that might have had to be cut out. **4.** The total enclosed floor area of a building.

gross cross-sectional area The total area of that portion of a concrete masonry unit that is perpendicular to the load, inclusive of any areas within its cells and reentrant spaces.

gross floor area The total area of all the floors of a building, including intermediately floored tiers, mezzanine, basements, etc., as measured from the exterior surfaces of the outside walls of the building.

gross output The number of Btus available at the outlet nozzle of a heating unit for continuously satisfying the gross load requirements of a boiler operating within code limitations.

gross volume **1.** The total volume within the revolving part of a drum concrete mixer. **2.** The total volume of the trough on an open-top concrete mixer.

ground **1.** The conducting connection between electrical equipment or an electrical circuit and the earth. **2.** A strip of wood that is fixed in a wall of concrete or masonry to provide a place for attaching wood trim or burring strips. **3.** A screed, strip of wood, or bead of metal fastened around an opening in a wall and acting as a thickness guide for plastering or as a fastener for trim. **4.** Any surface that is or will be plastered or painted. **5.** Any electrical reference point.

ground area The area computed by the exterior dimensions of the structure.

ground beam **1.** A reinforced concrete beam or heavy timber positioned horizontally at ground level to support a superstructure. **2.** A ground sill.

ground bus An electrical bus to which individual equipment grounds are connected and which itself is grounded at one or more points.

ground coat The base or undercoat of paint or enamel, often designed to be seen through the topcoat or glaze coat.

ground course A first horizontal course of masonry at ground level.

ground cover A planting of low plants that in time will spread to form a dense, often decorative mass. Ground cover is also used to prevent erosion.

grounded Descriptive of an object that is electrically connected to the earth or to another conducting body that is connected to the earth.

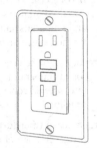

grounded system A system of electric conductors, at least one of which is intentionally grounded, either solidly or by means of a device that limits current.

ground-fault circuit interrupter

ground-fault circuit interrupter An electrical outlet fitted with code-required safety protection.

ground floor In a building, that floor closest to the level of the surrounding ground.

grounding electrode A conductor that is firmly embedded in the earth, and can thus function to maintain ground potential on the conductors connected to it.

grounding outlet An electrical outlet whose polarity-type receptacle includes both the current-carrying contacts and a grounded contact that accepts an equipment-grounding conductor.

ground plan The plan of a structure at ground level.

grounding plug A receptacle plug comprising a male member that, when plugged into a live grounding outlet, provides a ground connection for an electric device.

grounding outlet

grounding rod The rod that grounds an electrical panel.

ground iron Plumbing drain and waste lines installed under the floor of a basement.

ground joist Floor joists laid on sleepers, dwarf sills, or stones.

ground line The natural grade line or ground level from which excavation measurements are taken to determine excavation quantities.

ground pressure 1. The weight of a machine or a piece of equipment divided by the area, in square inches, of the ground that supports it. **2.** Pressure exerted on a structural member by the adjacent soil or fill.

ground-source heat pump A geothermal heating and cooling system that uses the moderating temperatures of the Earth to heat or cool a home.

groundwater 1. Naturally occurring water that moves through the Earth's crust, usually at a depth of several feet to several hundred feet below the Earth's surface. **2.** Water contained in the soil below the level of standing water.

groundwater recharge The reintroduction of water into the ground, as from trenches beyond the construction area or through injection.

groundwater table The top elevation of groundwater at a given location and at a given time.

ground wire 1. An electrical conductor leading directly or indirectly to the earth. **2.** Strong, small-gauge wire used in establishing line and grade, as in shotcrete work.

groundwork Batters used in roofing and siding as a base over which slate, tiles, and shingles are applied.

grout 1. A hydrous mortar whose consistency allows it to be placed or pumped into small joints or cavities, as between pieces of ceramic clay, slate, and floor tile. **2.** Various mortar mixes used in foundation work to fill voids in soils, usually through successive injections through drilled holes.

grouted-aggregate concrete Concrete that is produced by injecting grout into prepositioned coarse aggregate.

grouted masonry 1. Hollow masonry units with some or all of the cells filled with grout. **2.** Masonry comprising two or more wythes, the spaces between which are filled solidly with grout.

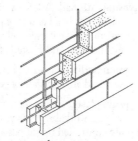
grouted masonry

grouting 1. The placing of grout so as to fill voids, as between tiles and under structural columns and machine bases. **2.** The injection of grout to stabilize dams or mass fills, or to reinforce and strengthen decaying walls and foundations. **3.** The injection of grout to fill faults and crevices in rock formations.

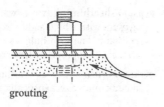
grouting

grout lift The height of grout placed in masonry wall voids during a single placement.

growth ring A ring that designates the amount of a tree's growth in a single year.

growth ring

grub In site work, the clearing of stumps, roots, trees, bushes, and undergrowth.

guarantee (guaranty, warranty) A legally enforceable assurance of quality or performance of a product or work, or of the duration of satisfactory performance.

guaranteed maximum cost The maximum amount above which an owner and contractor agree that cost for work performed (as calculated on the basis of labor, materials, overhead, and profit) will not escalate.

guaranteed maximum cost contract A contract for construction wherein the contractor's compensation is stated as a combination of accountable cost plus a fee, with guarantee by the contractor that the total compensation will be limited to a specific amount. This type of contract may also have possible optional provisions for additional financial reward to the contractor for performance that causes total compensation to be less than the guaranteed maximum amount.

guaranty bond A type of bond that is given to secure payment and performance. Each of the following four bonds are types of guaranty bonds: **1.** bid bond, **2.** labor and material payment bond, **3.** performance bond, and **4.** surety bond.

guard 1. Any bars, railing, fence, or enclosure that serves as protection around moving parts of machinery or around an excavation, equipment, or materials. **2.** A security guard hired to maintain safety and security at a construction site.

guard board

guard board A raised board around the edge of a scaffold or gantry crane to keep men and tools from falling off.

guard rail 1. A horizontal rail of metal, wood, or cable fastened to intermittent uprights of metal, wood, or concrete around the edges of platforms or along the lane of a highway. **2.** The rail that separates traffic entering or exiting through side-by-side automatic doors.

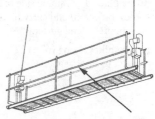
guard rail

G

143

guesstimate An educated guess or approximation of the cost of a project made by the cost estimator, without having performed a detailed quantity takeoff.

guide wire 1. A steel wire or cable used to guide the vertical movement of a stage curtain in a theater. **2.** A line or wire that guides the movement of a counterweight arbor. **3.** A wire placed along the edge of a roadway to be paved. A sensor on the pavement spreader adjusts the elevation of the pavement from the wire.

gumbo Soil composed of fine-grained clays. When wet, the soil is highly plastic, very sticky, and has a soapy appearance. When dried, it develops large shrinkage cracks.

gumwood Wood from a gum tree, especially eucalyptus, used mostly for interior trim.

gun 1. Equipment designed to deliver shotcrete. **2.** A pressure cylinder for pneumatic delivery of freshly mixed concrete. **3.** A spray gun. **4.** A slang expression for a transit, as it is used to *shoot* grades.

gun consistency The degree of viscosity of caulking or glazing compound that renders it suitable for application by a caulking gun.

gun finish The finish on a layer of shotcrete left undisturbed after application.

gunite Concrete mixed with water at the nozzle end of a hose through which it has been pumped under pressure. Gunite is applied or placed pneumatically, as shot, onto a backing surface.

gusset A metal or wood brace attached to structural members at their joints to add strength and stability.

gusset plate A plate fastened across a joint, as in wood or steel framework members. *See also* **chain**.

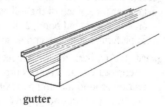

gusset plate

gutter 1. A shallow channel of wood, metal, or PVC positioned just below and following along the eaves of a building for the purpose of collecting and diverting water from a roof. **2.** In electrical wiring, the rectangular space allowed around the interior of an electrical panel for the installation of feeder and branch wiring conductors.

gutter

gutter bed A strip of flexible metal over the wall side of a gutter that prohibits any gutter overflow from penetrating the wall.

gutter hook A bent metal strip used for securing or supporting a metal gutter.

gutter plate 1. A single side of a box or valley gutter, lined with flexible metal and carrying the feet of the rafters. **2.** A beam that supports a lead gutter.

gutter spikes Long (6″ to 12″) aluminum or galvanized steel nails used to hang gutters.

gutter straps Metal bands used to support gutters.

guy A cable or rope anchored in the ground at one end and supporting or stabilizing an object at the other end.

gypsum A naturally occurring, soft, whitish mineral (hydrous calcium sulfate) that, after processing, is used as a retarding agent in Portland cement and as the primary ingredient in plaster, gypsum board, and related products.

gypsum backerboard A type of gypsum board, not as smooth as wall-board, surfaced with gray paper, and manufactured specifically as a base onto which tile or gypsum wallboard is adhered.

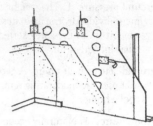

gypsum board

gypsum board A panel whose gypsum core is paperfaced on each side, and that is used to cover walls and ceilings while providing a smooth surface that is easy to finish. Used as a substitute for plaster.

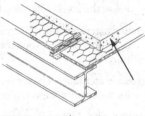

gypsum concrete A mixture of calcined gypsum binder, wood chips or other aggregate, and water. The mixture is poured to form gypsum roof decks.

gypsum concrete

gypsum fiber concrete A gypsum concrete whose aggregate is composed of wood shavings, fiber, or chips.

gypsum fiber panels Gypsum panels with fiber reinforcement concentrated on each face of the panel.

gypsum lath A gypsum board used as the base for application of gypsum plaster.

gypsum-lath nail A low-carbon steel nail with a large flat head and long point. The characteristics make it especially suitable for fixing gypsum lath and plasterboard.

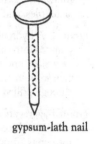

gypsum-lath nail

gypsum perlite plaster A base-coat plaster manufactured from gypsum and an aggregate of perlite.

gypsum plaster A plaster made from ground calcined gypsum. The set and workability of gypsum plaster are controlled by various additives. When mixed with aggregate and water, the resulting mixture is used for base-coat plaster.

gypsum sheathing A type of wallboard whose core is made from gypsum with which additives have been mixed to make it water-resistant. The sheathing is surfaced with a water-repellent paper to make it appropriate for exterior wall coverings.

gyratory crusher A rock-crushing mechanism whose central conical member moves eccentrically within a circular chamber.

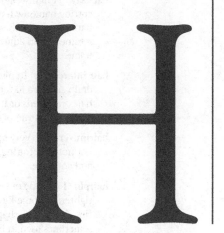

Abbreviations

h harbor, hard, height, hours, house, hundred

H "head" on drawings, high, high strength bar joist, Henry, hydrogen

HA hour angle

handhld handheld

H&M hit and miss

HASP health and safety plan

haz hazardous

HAZMAT hazardous materials

HB hollow bark

HC high capacity

HCFCs hydrochlorofluorocarbons

HCl hydrochloric acid

HD heavy duty, high density

hd head(s)

hdbrd hardboard

hdcp handicap

hdlg handling

HDO high-density overlaid

HDPE high-density polyethylene

Hdr header

hdw, hdwe, hdwr hardware

hdwd hardwood

He. helium

HE high explosive

Help helper average

hem hemlock

HEPA high-efficiency particulate air (filter)

HEW Department of Health, Education, and Welfare

hex hexagon

hf half, high-frequency

HF hot finished

HFCs hydrofluorocarbons

hg hectogram

Hg mercury

hgr hanger(s)

hgt height

hi high

HI height of instrument

HIC high interrupting capacity

HID high intensity discharge

hint high intensity

hip hipped (roof)

HIPS high-impact polystyrene

hl hectoliter

hm hectometer

HM hollow metal

hndrl handrail

HO high output

H-O-A hand-off-auto

hol hollow

hor, horiz horizontal

H or M hit or miss

hosp hospital

HOW homeowner's warranty

hp, HP horsepower

HP high pressure, steel pile section, handicapped person

HPF high power factor

HPLC high-performance liquid chromatography

hr hour

HRS hazard ranking system, hot rolled steel

Hrs/Day hours per day

HS high strength, hollow stem

HSC high short circuit

hsg housing

hst hoist

ht, Ht height, heat

HT high-tension

htd heated

htg, Htg heating

htr heater(s)

Htrs heaters

HTRW hazardous toxic radiological waste

HUD Department of Housing and Urban Development

hv, HV high voltage

HVAC heating, ventilating, and air conditioning

hvy, Hvy heavy

HW high-water, hot water, hazardous waste, heavy weight

HWM high-water mark

hwy highway

Hyd, Hydr hydraulic

hyd, hydraul hydraulic, hydrostatics

hyd exc hydraulic excavator

hydrocar hydrocarbons

hydrst hydrostatic

hyp, hypoth hypothesis; hypothetical

hz hertz (cycles)

Definitions

habitable space General living areas in a building, excluding bathrooms, storage, and utility spaces.

hachure One of the short parallel lines used on an architectural drawing for shading or for indicating a section of a drawn object, or on a topographic map for indicating the degree and direction of slopes and depressions.

hack 1. To cut or strike at something irregularly or carelessly, or to deal heavy blows. 2. A person who lacks, or does not apply, knowledge or skill in performing his job.

hacking 1. Striking a surface with a special tool so as to roughen it. 2. A style of bricklaying in which the bottom edge is set in from the plane surface of the wall. 3. In a stone wall, the breaking of a single course into two or more courses, sometimes for effect but

usually because of the scarcity of larger stones.

hacksaw A lightweight, metal-cutting handsaw having a narrow, fine-toothed blade retained in an adjustable metal frame.

hacksaw

hair interceptor In plumbing, a traplike device installed in the waste drain side of a fixture's plumbing system to capture and collect hair on screens or in perforated steel baskets that are removable from the bottom of the device.

hairline cracks Very fine, barely visible random cracks appearing on, but not penetrating, the finish surface of materials such as paint and concrete.

hairpin 1. A type of wedge used in tightening some kinds of form ties. 2. Hairpin-shaped rebar sometimes used in beams, columns, and prefabricated column shear heads.

half-bat (half-brick) A half-brick produced by cutting a brick in two, across its length.

half-brick wall A brick wall having the thickness of a brick laid as a stretcher.

half-brick wall

half header Half of a brick or concrete block made by cutting the unit longitudinally through its faces. Half headers are used to close the work at the end of a course.

half-landing (halfpace landing, halfspace landing) A platform in a stairway, where the stairs change direction halfway between the floors of a building.

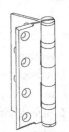
half-landing (halfpace landing, halfspace landing)

half-mortise hinge A door hinge with one plate surface-mounted on the jamb and the other plate mortised into the door stile.

half-pitch roof A roof with a pitch whose rise is equal to one-half the width of the span.

half-principal A roof rafter or similar member with the upper end not extending all the way to the ridgeboard, but instead supported by a purlin.

half-mortise hinge

half-rabbeted lock A type of mortise lock having a front turned into two perpendicular planes, used on a door with a rabbeted edge.

half-span roof (lean-to roof) A roof that slopes in only one plane and abuts a higher exterior wall.

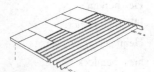
half-span roof (lean-to roof)

half-story An attic or story immediately below a sloping roof and usually having some partitions and a finished ceiling and floor.

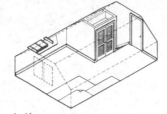
half-story

half-timbered Descriptive of a building style common in the 16th and 17th centuries, with foundations, supports, knees, and studs all made of timbers. The wall spaces between the timbers are filled with masonry, brick, or lathed plaster.

half-truss One side of a jack truss spanning from a main roof truss to a wall, usually at an angle to the main truss.

halide torch A device used to detect leaks of halocarbon refrigerant. The color of the sampling torch's normal, alcohol-produced, blue flame becomes a bright green when the refrigerant is detected.

hall 1. A large room in which people assemble for entertainment or meetings. **2.** A small entrance room or corridor. **3.** A term often used in the proper names of public or university buildings.

hammer A hand tool with a handle perpendicular to its head, for driving nails or other applications involving pounding or striking.

hammer beam (hammer-beam trusses) Either one of a pair of short horizontal members used in place of a tie beam in roof framing. A hammer beam is attached to the foot of a principal rafter and supported from below by a brace to the supporting column.

hammer brace The brace, often curved, between a hammer beam and pendant post.

hammer dressed Descriptive of stonemasonry having a finish created only by a hammer, sometimes at the quarry.

hammer drill A pneumatically powered mechanism using percussion to penetrate rock.

hammerhead crane A heavy-duty crane with a swinging boom and counterbalance, giving it a "T" shape.

hammer man The workman on a pile hammer who operates the hoist or controls the steam jet that, in turn, powers the hammer.

hammermill crusher An impact type crusher that breaks up and grinds materials to a finished size.

hance 1. A small arch or half arch connecting a larger arch or lintel to its jamb. **2.** Light-concealing trim member at the top of a window.

hand 1. Prefaced by "left" or "right" to designate how a door is hinged and the direction it opens. **2.** Preceded by "left" or "right" to designate the direction of turn one encounters when descending a spiral stair, with "right-hand" being clockwise.

hand brace A wood-boring hand tool made of a single frame of small diameter bar or rod bent to form a stationary bracing handle at one end and a bit-holding chuck at the other. A short distance from, but parallel to, the central axis, a handle repeatedly turns in wide circles, causing the bit to turn.

hand drill A hand-operated boring device made up of a central steel tube containing a shaft. At one end is a handle and at the other a bit-holding chuck.

hand float A wooden tool used to lay on and to smooth or texture a finish coat of plaster or concrete.

handicap door opening system A door equipped with a knob or latch and handle located approximately 36″ from the floor, and an auxiliary handle on the other side at the hinge edge, for convenience to persons in wheelchairs.

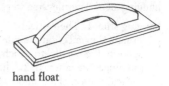
hand float

handicapped fixtures Refers to plumbing fixture connections complying with ADA requirements that are exposed and located in facilities for handicapped individuals.

handicap water cooler A water cooler set low and operated by push-bars or levers for convenience to persons in wheelchairs.

hand line 1. A line attached to a structural member or piece of building equipment being hoisted. Used to control the position of the item during erection or setting. **2.** A line manipulated to control stage rigging in a theater.

hand punch A struck tool designed for punching or marking metal, driving and removing pins, and aligning holes. Range in size from 1/4″ to 1″ in diameter, and 4½″ to 20″ long.

handrail A bar of wood, metal, or PVC, or a length of wire, rope, or cable, supported at intervals by upright posts, balusters, or similar members or, as on a stairway, by brackets from a wall or partition, so as to provide persons with a handhold.

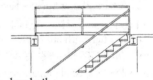
handrail

handsaw Any manual woodcutting saw having a handle at one end by which it is gripped and manipulated.

handsplit and resawn (HS and RS) A type of cedar shake. Handsplits are split from cedar bolts by a mallet and froe (a type of steel blade). The pieces are then ripped on a resaw to produce two shakes, each with a rough, split face and a smooth, sawn back.

hand tight Descriptive of couplings tightened by hand by the application of force roughly equal to the force an average man can exert.

hang To install a door or window within its respective frame and/or by its respective hardware.

hanger 1. A strip, strap, rod, or similar hardware for connecting pipe, metal gutter, or framework, such as for a hung ceiling, to its overhead support. **2.** Any of a class of hardware used in supporting or connecting members of similar or different material as, for instance, a stirrup strap or beam hanger for supporting the end of a beam or joist at a masonry wall. **3.** A person whose trade it is to install gypsum board products.

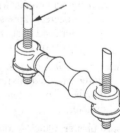
hanger

hanging gutter A metal gutter attached to the roof eaves with metal straps and sometimes further supported by the fascia.

hanging rail The horizontal section at the top and bottom of a door, to which the hinges are secured.

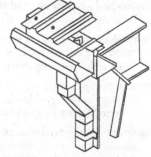
hanging gutter

hanging steps Cantilevered steps.

hanging stile 1. The vertical structural member on the side opposite the handle, to which the hinges are fastened. **2.** That vertical section of a window frame to which the casements are hinged.

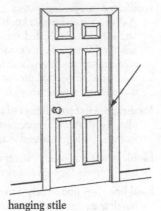

hanging stile

hardback Molding and BTR lumber which is D-select graded from the good face only, which must be clear. The back may contain knots that do not extend through the piece.

hardboard Dense sheets of building material made from heated and compressed wood fibers.

hard-burnt 1. Descriptive of clay products, such as bricks or tiles, having been burnt or fired at high temperatures, resulting in their durability, high compressive strength, and low absorption. **2.** A hard plaster, such as Keene's.

hard conversion The conversion from one system of measurement to another, with an inherent consequence being the necessity of changing the physical sizes of the products involved.

hard edge A special preparation used in the core of gypsum board under the papered edges to provide extra resistance.

hardener 1. Any of several chemicals serving to reduce wear and dusting when applied to concrete sustaining heavy traffic, such as a floor. **2.** The curing agent of a two-part synthetic resin, adhesive, or similar coating. **3.** A substance used to harden plaster casts or gelatin molds.

hard finish A mixture of gypsum, plaster, and lime applied as a finish coat, usually over rough plastering, then troweled to provide a dense, hard, smooth finish.

hardness 1. The resistance of a substance, material, or surface, to cutting, scratching, denting, pressure, wear, or other deformation. **2.** The degree, expressed as parts per million or grains per gallon of calcium carbonate in water, to which calcium and magnesium salts are dissolved in water.

hardpan Highly compacted soil, boulder clay, or other usually glacially deposited mixture, sometimes including sand, gravel, or boulders. The extreme density of hardpan makes its excavation difficult.

hard plaster Quick-setting calcined gypsum, usually used in finishing, often requiring a retarding agent to be incorporated in the mix to help control the set.

hardscape Nonplant material landscape work, such as paving and building structures, that occurs as part of a construction project.

hard solder Solder containing silver, copper, or aluminum, and thus requiring more intense heat for melting than does soft solder. Hard solder is usually applied with a brazing torch.

hard stopping A stiff paste having a calcined gypsum content, causing it to harden quickly. Hard stopping is used in painting operations to fill deep holes and wide cracks.

hardware 1. A general term encompassing a vast array of metal and plastic fasteners and connectors used in or on a building and its inherent or extraneous parts. The term includes rough

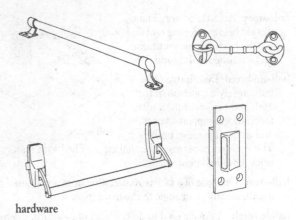

hardware

hardware, such as nuts, bolts, and nails, and finish hardware, such as latches and hinges. **2.** The mechanical equipment associated with data processing. In building automation systems, computer hardware includes the central processing unit (CPU), hard disk drive, monitor (CRT), keyboard, controllers, and analog or digital point modules. Digital equipment such as controls, sensors, and actuators are considered field hardware.

hardware cloth Usually galvanized, thin screen made from wire welded or woven to produce a mesh size of ⅛" to ¾".

hard water Water containing a concentration higher than 85.5 ppm of dissolved calcium carbonate and other mineral salts.

hard wired A communications link permanently joining two devices, nodes, or stations.

hardwood A general term referring to any of a variety of broad-leaved, deciduous trees, and the wood from those trees. The term does not designate the physical hardness of wood, as some hardwoods are actually softer than some softwood (coniferous) species.

harmonic The sinusoidal component of an arc voltage that is a multiple of the fundamental wave frequency.

harsh mortar A mortar that is difficult to spread due to an improper measure of materials.

Hartford loop The configuration of a steam boiler's return piping connections serving to equalize the pressure between the supply and return sides of the system, thus preventing water from backing out of the boiler and into the return line.

harvested rainwater Rainwater collected in a storage unit that can be treated or untreated and used for a variety of applications, such as flushing toilets, serving HVAC units, washing clothes, and irrigation.

hasp A metal fastening device made up of a staple secured to and protruding from one member, and a hinge with a slotted plate fastened to another member. The

hasp

slotted plate can be slipped over the staple and then locked with a tapered pin or a padlock.

hatch An opening in a floor or roof of a building, as in a deck of a vessel, having a hinged or completely removable cover. When open, a hatch permits ventilation or the passage of persons or products.

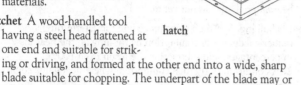

hatch

hat channel furring A light gauge metal furring strip used on vertical concrete surfaces to provide for fastening of finish materials.

hatchet A wood-handled tool having a steel head flattened at one end and suitable for striking or driving, and formed at the other end into a wide, sharp blade suitable for chopping. The underpart of the blade may or may not be notched for pulling.

hatchet iron The hatchet-shaped tip of a type of plumber's soldering iron.

haul distance The distance measured along the centerline or most direct practical route between the center of mass of excavation and the center of mass finally placed. It is the average distance material is moved by a vehicle. *

haunch **1.** A bracket built into a wall or column to support a load falling outside the wall or column, such as a hammer brace in a hammer-beam roof. **2.** Either side of an arch between the crown, or centerstone, and the springing, or impost. **3.** A thickening of a concrete slab to support an additional load, as under a wall.

haunch

haunched beam A beam or similar member broadened or thickened near the supports.

haunched floor A floor slab thickened around its perimeter.

hawk A flat, thin piece of wood or metal approximately 1 foot square and having a short, perpendicular handle centered on its underside. A hawk is used by plasterers for holding plaster from the time it is taken from the mixer to the time it is troweled.

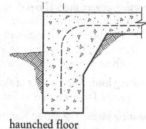

haunched floor

haydite Heated shale having an expanded cellular structure, making it a suitable lightweight aggregate for concrete.

hazard classes Nine classes established by the United Nations to categorize hazardous materials: flammable liquids, flammable solids, explosives, gases, oxidizers, radioactive materials, corrosives, poisonous infectious substances, and dangerous substances.

hazard insurance Insurance that protects against property damage from certain hazards such as storms and fire.

hazardous area **1.** The part of a building where highly toxic chemicals, poisons, explosives, or highly flammable substances are housed. **2.** Any area containing fine dust particles subject to explosion or spontaneous combustion.

hazardous ranking system (HRS) A scoring system used by the U.S. Environmental Protection Agency to assess various environmental aspects of sites, such as ranking disposal sites in need of cleanup.

hazardous substance Any substance that, by virtue of its composition or capabilities, is likely to be harmful, injurious, or lethal.

hazardous waste A material defined by any of several statutes and regulations, usually characterized by a propensity to cause an adverse health effect to humans.

H-bar A steel or aluminum bar used in structural systems, such as suspended ceilings, that has a cross-section in the shape of an "H."

H-beam A misused designation for an HP pile section.

H clip A small metal clip that resembles an "H" shape that is used to strengthen the joint of plywood roof sheeting or wafer board.

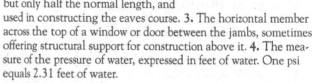

H-beam

head **1.** The top of almost anything, such as the head of a nail or a window head. **2.** In roofing, a tile of normal width but only half the normal length, and used in constructing the eaves course. **3.** The horizontal member across the top of a window or door between the jambs, sometimes offering structural support for construction above it. **4.** The measure of the pressure of water, expressed in feet of water. One psi equals 2.31 feet of water.

headache ball (breaker ball) The rounded, heavy, metal or concrete demolition device swung on a cable from the boom of a crane to break through concrete or masonry construction.

head casing The horizontally placed board at the head or top of a door or window opening between the two vertical casings.

header **1.** A rectangular masonry unit laid across the thickness of a wall, so as to expose its end(s). **2.** A lintel. **3.** A member extending horizontally between two joists to support tailpieces. **4.** In piping, a chamber, pipe, or conduit having several openings through which it collects or distributes material from other pipes or conduits.

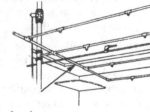

header

header block A concrete masonry unit from which part of one face shell has been removed to facilitate bonding with adjacent masonry, such as brick facing.

header course In masonry, a course comprising only headers.

header tile In a masonry-faced wall, a tile having recesses to accept headers.

head flashing In a masonry wall, the flashing over a projection, protrusion, or window opening.

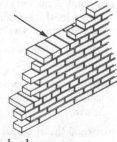

header course

heading 1. In mining, the digging face and its immediate work area in a tunnel, drift, or gallery. **2.** The increase of expansion of a localized cross-sectional area of metal bar due to hot-forging. **3.** A general classification of a category of data, under which follow more specific classifications. **4.** Pieces of lumber from which a keg, or barrel head, is cut. **5.** Stock after it has been cut and assembled to form a barrel head.

heading joint 1. The joint formed between two pieces of timber connected end-to-end, in a straight line. **2.** The joint between two adjacent masonry units in the same course.

head jamb The horizontal member that constitutes the doorhead or top of a door opening.

headlap That portion of a shingle not exposed to the weather, because it is covered by the shingle(s) in the course above it.

head mold The molding over an opening such as a door or window.

head pressure The operating pressure in the discharge line of a refrigeration system.

head room 1. The vertical distance, or space, allowable for passage, as in a room or under a doorway. **2.** The space between the top of one's head and the nearest obstacle above it, as inside a vehicle. **3.** The unobstructed vertical space between a stair tread and the ceiling or stairs above.

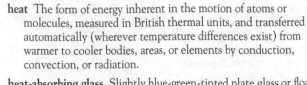

headlap

headstone Any principal stone in masonry construction, such as the keystone in an arch or the cornerstone of a building.

headstone

headwall The wall, usually of concrete or masonry, at the outlet side of a drain or culvert, serving as a retaining wall, as protection against the scouring or undermining of fill, or as a flow-diverting device.

heart The center portion of the cross-section of a log. The term usually refers to heartwood.

headwall

heart bond A masonry bond, used in walls too thick for through stones, in which a third header covers the joint of two headers meeting within a wall.

hearth The floor of a fireplace and the adjacent area of fireproof material.

hearthstone 1. A large stone used as the floor of a fireplace. **2.** Other naturally occurring or synthetic materials used to construct a hearth.

heartwood (heart wood) The core of a tree, which is no longer vital to the life growth of the tree and which is often darker and of a different consistency than the growing sapwood.

heat The form of energy inherent in the motion of atoms or molecules, measured in British thermal units, and transferred automatically (wherever temperature differences exist) from warmer to cooler bodies, areas, or elements by conduction, convection, or radiation.

heat-absorbing glass Slightly blue-green-tinted plate glass or float glass designed with the capacity to absorb 40% of the infrared solar rays and about 25% of the visible rays that pass through it. Cracking from uneven heating can occur if the glass is not exposed uniformly to sunlight.

heat-affected zone In welding or soldering, an area of metal that has been altered but has not yet melted.

heat balancing 1. An efficient procedure for determining a numerical degree of combustion by totaling all the heat losses, in percentages, and subtracting the result from 100%. **2.** A condition of thermal equilibrium where heat gains equal heat losses.

heat capacity The amount of heat required to increase the temperature of a given mass by one degree. The capacity is arrived at numerically by multiplying the mass by the specific heat.

heated space An area of a building that is directly supplied with heat.

heater 1. A general term including stoves, appliances, and other heat-producing units. **2.** A person who heats something, such as a steelworker who heats rivets on a small forge before passing them to the sticker.

heat exchanger A device designed to transfer heat between two physically separated fluids. The fluids are usually separated by the thin walls of tubing.

heat gain The net increase in Btus, caused by heat transmission, within a given space.

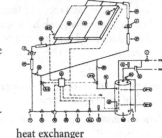

heat exchanger

heating degree days (HDD) A quantitative comparison of the average daily outdoor air temperature to the standard of 65°F used to analyze energy needed to heat or cool a space. One degree day equals one day with an average temperature one degree above 65°F.

heating load The number of Btus per hour required to maintain a specified temperature within a given enclosed space.

heating plant The entire heating system of a building or complex, including either a boiler, piping, and radiators, or a furnace, ducts, and air outlets.

heat island effect A pattern of elevated temperatures in urban areas caused by structural and pavement heat fluxes and pollutant emissions. In some cases, urban temperatures can be as much as 10°F warmer than nearby rural areas, which can result in increased energy demands and heat-related illnesses.

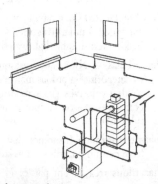

heating plant

heat loss **1.** The net decrease in Btus within a given space, by heat transmission through spaces around windows, doors, etc. **2.** The loss by conduction, convection, or radiation from a solar collector after its initial absorption.

heat of fusion The amount of heat needed to melt a unit mass of a solid at a specified temperature.

heat of hydration **1.** Heat resulting from chemical reactions with water, as in the curing of Portland cement. **2.** The thermal difference between dry cement and partially hydrated cement.

heat pump A refrigeration system designed to utilize alternately or simultaneously the heat extracted at a low temperature and the heat rejected at a higher temperature.

heat recovery The extraction of heat from any source not primarily designed to produce heat, such as a chimney or light bulb.

heat-reflective glass Window glass in which the exterior surface has been treated with a transparent metallic coating to reflect substantial portions of the light and radiant heat striking it.

heat sealing The use of heat and pressure to bond plastic sheets or films.

heat seaming In roofing, joining thermoplastic films or sheets together by heating contact areas so that they fuse together.

heat sink **1.** The substance or environment into which heat is discharged after its removal from a heat source, as by a heat pump. **2.** Any medium capable of accepting discharged heat.

heat source **1.** Any area, environment, or device that supplies heat. **2.** The area from which a refrigeration system removes heat.

heat tracing system A heating system with an externally applied heat source, usually a heating cable, that traces the object to be heated.

heat transfer fluid The liquid substance used to carry heat away from its source to be cooled, usually by another fluid, as in a heat exchanger.

heat transmission The rate at which heat passes through a material by the combination of conduction, convection, and radiation.

heat transmission coefficient Any of several coefficients used to calculate heat transmission by conduction, convection, and radiation through a variety of materials and structures.

heave The localized upward bulging of the ground due to expansion or displacement caused by phenomena such as frost or moisture absorption.

heavy concrete Concrete having a high unit weight up to 300 pounds per cubic foot, primarily due to the types of aggregate employed and the density of their ultimate incorporation. Such diverse materials as trap rock, barite, magnetite, steel nuts, and bolts can be used as aggregate. The density makes heavy concrete especially suitable for protection from radiation.

heavy construction Construction requiring the use of large machinery, such as cranes or excavators.

heavy-duty scaffold A scaffold constructed to carry a working load not to exceed 75 pounds per square foot.

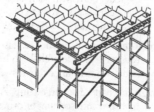
heavy-duty scaffold

heavy-edge reinforcement In highway pavement slabs, reinforcement made of wire fabric with up to four edge wires that are heavier than any of the other longitudinal wires.

heavy joist A timber at least 4″ thick and 8″ wide.

heavy metal A naturally occurring elemental metal with a high molecular weight.

heavy timber **1.** A type of construction requiring noncombustible exterior walls with a minimal fire-resistance rating of two hours, solid or laminated interior members, and heavy plank or laminated wood floors and roofs. Also called *mill construction*. **2.** Rough or surface pieces with a least dimension of 5″.

heavyweight aggregate Aggregate possessing a high specific gravity, such as barite, magnetite, limonite, ilmenite, iron, and steel, making it suitable for use in heavy concrete.

hectare A measurement of land area equal to 2.471 acres or about 107,637 square feet.

hedge In master production scheduling, a quantity of stock used to protect against uncertainty in demand. The hedge is similar to safety stock, except that a hedge has the dimension of timing as well as amount.*

heel **1.** The lower end of a door's hanging stile or of a vertically placed timber, especially if it rests on a support. **2.** A socket, floor brace, or similar device for wall-bracing timbers. **3.** The bottom inside edge of a footing or a retaining wall. **4.** The back end of a carpenter's plane.

heel strap A steel fastening device for connecting a rafter to its tie beam.

height **1.** The distance between two points in vertical alignment or from the top to the bottom of any object, space, or enclosure. **2.** The vertical distance between the average grade around a building, or the average street curb elevation, and the average level of its roof. **3.** The rise of an arch.

height board A measuring device for setting the heights of stair risers.

height of instrument The height of a leveling instrument above the datum being used in the survey.

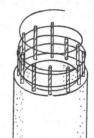
helical reinforcement

helical reinforcement Column reinforcement bent in the form of a helix. More commonly called *spiral reinforcement*.

helical rotary compressor (screw-type compressor) A device that compresses gas by trapping it in the space formed by the flutes of meshing screws, thus reducing the gas volume.

helical staircase (spiral staircase) A staircase built in the form of a helix.

heliostat An instrument having an automatically adjusting mirror that follows the apparent movement of the sun and continuously reflects its rays onto a collector.

helical staircase (spiral staircase)

helix 1. A reinforcing rod bent to form a spiral used for reinforcing the circumference of concrete columns. **2.** A volute found on a Corinthian or Ionic capital. **3.** Any spiral structure, ornament, or form.

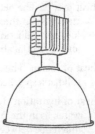

helix

helve 1. A tool handle, such as that of an ax, hatchet, or hammer. **2.** The handle of a wagon.

hem-fir A species combination used by grading agencies to designate any of various species, such as white fir and western hemlock, having common characteristics. The designation is used for identification and standardization of recommended design values and because some species, in lumber form, cannot be visually distinguished.

hemihydrate Any hydrate having only half a molecule of water for every molecule of compound. The most common hemihydrate is partially hydrated gypsum, or plaster of paris.

hemlock A coniferous North American tree, the wood of which is used in general construction and for pulp.

hemp A natural fiber once widely used in cordage, but now almost totally replaced by synthetic fibers, such as nylon and dacron. Hemp is still laminated to a paper backing to produce a type of wallcovering.

HEPA-filtered vacuum A vacuum device fitted with a high efficiency particulate removal system.

hepatoxin A substance that causes liver damage.

herbicide A substance that kills plants on contact.

hermaphrodite caliper A tool with two hinged legs used to lay out lines that are parallel with the edges of the workpiece. It can also be used to locate the center of cylindrical shaped workplaces.

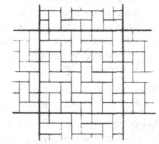

herringbone bond In masonry, a type of raking bond in which a zigzag effect is created by laying rows of headers perpendicular to each other.

herringbone bond

herringbone drain (chevron drain) A V-shaped drain.

herringbone work In masonry, the zigzag pattern created by laying consecutive courses of masonry units at alternating 45° angles to the general run of the course.

hertz A unit of measurement of frequency equal to one cycle per second.

hewn Roughly cut, fabricated, shaped, or dressed.

hex roofing Hexagonally shaped asphalt roofing shingles.

hickey (hicky) 1. A threaded electrical fitting for connecting a light fixture to an outlet box. **2.** An apparatus used to bend small pipe, conduit, or reinforcing bar.

hick joint In masonry, a mortar joint cut in any direction to be flush with the face of the wall, resulting in a hairline crack that render the joint no longer watertight.

high-bay lighting Usually an industrial lighting system having direct or semidirect luminaires located high above floor or work level.

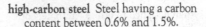

high-build coating A coating composed of a series of films that are thicker than those normally associated with paints (minimum 5 mils) and thinner than troweled material.

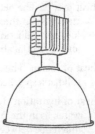

high-bay lighting

high-carbon steel Steel having a carbon content between 0.6% and 1.5%.

high chair Slang for a heavy, wire, vaguely chair-shaped device used to hold steel reinforcement off the bottom of the slab during the placement of concrete.

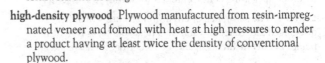

high chair

high-density overlay A cellulose fiber sheet impregnated with a thermosetting resin and bonded to plywood, rendering a hard, smooth, waterproof, wear-resistant surface for use in concrete formwork and decking.

high-density plywood Plywood manufactured from resin-impregnated veneer and formed with heat at high pressures to render a product having at least twice the density of conventional plywood.

high-density polyethylene (HDPE) An extensively polymerized plastic used for pipe. Polyethylene material is highly abrasion and corrosion resistant, and thereby able to handle a wide variety of slurries and abrasive materials as well as nearly all acids, caustics, salt solutions and other corrosive liquids and gases.

high-early-strength concrete Concrete containing high-early-strength cement or admixtures causing it to attain a specified strength earlier than regular concrete.

high-efficiency particulate air filter (HEPA filter) A high-efficiency (99.9%) dry filter made in an extended surface configuration of deep space folds on submicron glass fiber paper.

high gloss Descriptive of a substantial degree of luster or of a paint that dries with a lustrous, enamel-like finish.

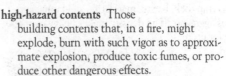

high hat A recessed lighting fixture that sheds its light vertically downward.

high hat

high-hazard contents Those building contents that, in a fire, might explode, burn with such vigor as to approximate explosion, produce toxic fumes, or produce other dangerous effects.

high-impact polystyrene A plastic with good toughness, moderately high heat resistance, and good resistance to stress cracking.

high-intensity discharge lamp A mercury, high-pressure sodium, or other electric discharge lamp requiring a ballast for starting and for controlling the arc, and in which

high-intensity discharge lamp

light is produced by passing an electric current through a contained gas or vapor.

high-mass construction A building construction approach using masonry, adobe, or other building materials that can lessen the extremes of diurnal flux, especially in arid climates.

high-output fluorescent lamp A rapid-start fluorescent lamp with greater flux as a result of its operation on higher current.

high-performance building A building that is energy-efficient, healthy, and comfortable for its occupants.

high-pressure laminate Laminate manufactured at pressures between 1,200 and 2,000 pounds per square inch during its molding and curing processes. Used in furniture, paneling, and cabinet manufacturing, high-pressure laminate is also an ideal material for access flooring around sensitive equipment because of its excellent ability to dissipate static electricity.

high-pressure mercury lamp A mercury vapor lamp designed to function at a partial mercury vapor pressure of about 1 atmosphere or more (usually 2 to 4).

high-pressure overlay A plastic laminate consisting of layers of melamine sheet or phenolic-impregnated kraft paper onto which a melamine-impregnated printed pattern sheet and/or a translucent melamine overlay may have been impressed. The laminate is produced at a temperature above 300°F and a pressure of about 400 psi, resulting in a hard, smooth, wear-resistant surface that is often bonded to wood and used in doors and on tabletops.

high-pressure sodium lamp A sodium vapor lamp, operating at a partial vapor pressure of 0.1 atmosphere, that produces a wide-spectrum yellow light.

high-pressure steam curing (autoclave curing) The steam curing of products made from cement, sand-lime, concrete, or hydrous calcium silicate in an autoclave at temperatures of 340° to 420°F.

high-pressure steam heating system A steam heating system in which heat is transported from a boiler to a radiator by steam at pressures above 100 psi.

high rib lath An expanded metal lath used as a backup for wet wall plaster and as formwork for thin concrete slabs.

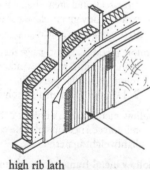

high rib lath

high-rise 1. A building having many stories and serviced by elevators. 2. A building with upper floors higher than fire department aerial ladders, usually ten or more stories.

high side (high pressure side) Those parts in a refrigeration system that are exposed to pressure at least as great as condenser pressure.

high-strength bolts Bolts made from high-strength carbon steel or from alloy steel that has been quenched and tempered.

high-rise

high-strength steel Steel with an inherent high yield point.

high-tensile bolt (high-tension bolt) A bolt made from high-strength steel and tightened to a specified high tension. High-tensile bolts have replaced the use of steel rivets in steel-frame construction.

high transmission glass Glass that transmits an exceptionally high percentage of visible light.

high-velocity duct system A duct system carrying air at more than 2,400' per minute.

hinge A flexible piece or a pair of plates or leaves joined by a pin so as to allow swinging motion in a single plane of one of the members to which it is attached, such as a door or gate.

hinge backset The horizontal distance from the edge of a hinge to the face of the door that closes against a rabbet or stop.

hinge, electric A hinge designed to pass electric wires from the frame to the door for use in an electric-controlled lock.

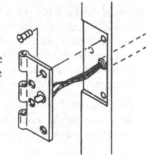

hinge, electric

hinge jamb The doorjamb to which the hinges are fastened.

hinge joint Any joint allowing action similar to that permitted by a hinge and only a very slight separation between the adjacent members.

hinge, security 1. A hinge with a pin that cannot be removed. 2. A hinge with a stud in one leaf projecting into the other leaf when the door is closed so the door cannot be moved with the pin removed. 3. See **hinge, electric**.

hip 1. The exterior inclining angle created by the junction of the sides of adjacent sloping roofs, excluding the ridge angle. 2. The rafter at this angle. 3. In a truss, the joint at which the upper chord meets an inclined end post.

hip-and-valley roof A roof incorporating both hips and valleys.

hip bevel 1. The angle between two adjacent sloping roofs separated only by a hip. 2. The angle at the end of a rafter that allows its conformation to the oblique construction at a hip.

hip capping The top layer of a hip's protective covering.

hip hook See **hip iron**.

hip iron (hip hook) A galvanized steel or wrought iron bar or strip secured to the foot of a hip rafter to hold the hip tiles in place.

hip jack In a hip roof, a rafter shorter in length than most of the other rafters used in the same construction, whose upper end is secured to a hip rafter.

hip molding The molding on a hip rafter.

hipped end Either of the triangular ends of a hipped roof.

hipped gable (jerkin head) A modified gabled end that is gabled only about halfway to the ridge and then inclines backward and forms hips where it meets the two principal slopes.

hip rafter

hip rafter The rafter that, in essence, is the hip of a roof, by virtue of its location at the junction of adjacent inclined planes of a roof.

H

hip roof A roof formed by several adjacent inclining planes, each rising from a different wall of building, and forming hips at their adjacent sloping sides.

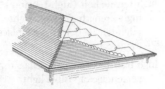

hip roof

hip vertical The upright tension member in a truss, the lower end of which carries a floor beam, and the upper end of which joins an inclined end post and an upper cord at a hip.

historical database Records accumulating past project experience stored as data for use in planning, estimating, forecasting and predicting future events. Often includes data that has been processed so as to facilitate planning and other purposes such as validation and benchmarking (e.g., metrics, etc).*

historic records Documentation from past projects that can be used to predict trends, analyze feasibility and highlight problem areas/pitfalls on future similar projects.*

hit-and-miss window A two-part window with the lower sash containing movable ventilation panels. One panel slides in front of the other to permit air flow.

H-molding An H-shaped trim piece used in a butt joint assembly. The two channels of the "H" hold the members in place.

hod In masonry construction, a V-shaped, trough-like container with a pole handle projecting vertically downward from the bottom to allow steadying with one hand while being carried on the shoulder of a laborer (hod carrier) at a construction site. A hod is used to transport bricks or mortar.

hoe 1. An implement similar to a garden hoe but having a larger blade to facilitate the mixing of cement, lime, and sand in a mortar tub at a construction site. 2. A backhoe.

hoe

hog 1. In masonry, a course that is not level, usually because the mason's line was incorrectly set and/or pulled. 2. A closer in the middle of a course. 3. A machine to grind waste wood into chips for fuel or other purposes.

hog-backed Cambered. The term is often used in reference to a sagging roof.

hogging The sagging of the end extremities of a beam or timber supported only in the middle.

hoist 1. Any mechanical device for lifting loads. 2. An elevator. 3. The apparatus providing the power drive to a drum, around which cable or rope is wound in lifting or pulling a load. Also called a *winch*.

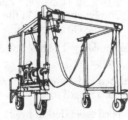

hoist

holdback 1. A safety device on a conveyor to prevent reverse motion of the belt automatically. 2. Funds retained until specific events occur or work is completed.

hold down A connector device used to resist overturning caused by uplift on the chords of shear walls.

hold-down clip A fastener used in an exposed suspension acoustical ceiling system or in roofing to join and anchor adjacent sections of capping.

holding tank A tank used for temporary storage of chemicals or materials being processed.

holiday A small area inadvertently missed during painting or other surfacing applications.*

holistic design An approach that emphasizes the functional relationship between the various building parts and the facility as a whole. May include protection of Earth's resources, as well as an element of spirituality, aiming to create spaces that enrich the quality of the environment *and* the lives of those who use the building.

hollow-backed Descriptive of the unexposed surface of a piece of wood, stone, or other material, intentionally hollowed to render a snug fit against an irregular surface.

hollow bed In masonry, a bed joint in which mortar is placed so as to provide contact only along the edges.

hollow brick A hollow clay masonry unit in which the net cross-sectional area is at least 60%.

hollow chamfer Any concave chamfer.

hollow-core door A flush door with plywood or hardwood faces secured over a skeletal framework, the interior remaining void or honeycombed.

hollow masonry unit (hollow block) A masonry unit in which the net cross-sectional area is less than 75% of the gross cross-sectional area when compared in any given plane parallel to the bearing surface.

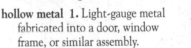

hollow masonry unit (hollow block)

hollow metal 1. Light-gauge metal fabricated into a door, window frame, or similar assembly. 2. Descriptive of an assembly thus produced.

hollow metal door A hollow-core door constructed of channel-reinforced sheet metal. The core may be filled with some type of lightweight material.

hollow metal frame A door frame constructed of sheet metal with reinforcing at hinges and strikes.

hollow plane A woodworking plane with a convex blade for fashioning hollow or concave molding.

hollow roll A process of joining two flexible metal roofing sheets in the direction of the roof's maximum slope by lifting them at the joint and bending them there to create a cylindrical roll. The fastening of the roll sometimes requires a fastener or metal clip.

Home Energy Rating System (HERS) A scoring system established by the Residential Energy Services Network (RESNET) in which a home built to the specifications of the HERS Reference Home (based on the 2006 International Energy Conservation Code) scores a HERS Index of 100, while a net zero energy home scores

a HERS Index of 0. The lower a home's HERS Index, the more energy-efficient it is.

home office Office of a company in the country of origin or centralized location. Usually synonymous with head office.*

home office cost Those necessary costs, typically not incurred at the project site, involved in the conduct of everyday business, which can be directly assigned to specific projects, processes, or end products, such as engineering, procurement, expediting, legal fees, auditor fees inspection, estimating, cost control, taxes, travel, reproduction, communications, etc.*

homogeneous Similar in structure, composition, appearance, or texture.

hone A smooth, fine-grained stone against which a tool's cutting edge is worked to achieve a finish edge much sharper than that yielded by the coarser stone used in preliminary sharpening procedures. Usually, an oil is used in the process to carry off minute particles of loose stone and metal to prevent them from clogging the pores on the stone's surface.

honed finish The very smooth surface of stone effected by manual or mechanical rubbing.

honeycomb 1. In concrete, a rough, pitted surface resulting from incomplete filling of the concrete against the formwork, often caused by using concrete that is too stiff or by not vibrating it sufficiently after it has been poured. 2. Voids in concrete resulting from the incomplete filling of the voids among the particles of coarse aggregate, often caused by using concrete that is too stiff. 3. In sandwich panel construction or in some hollow-cored doors, resin-impregnated paper is fabricated into a network of small, interconnected, open-ended, tubular hexagons laminated between two face panels to provide internal support.

hood 1. A protective cover over an object or opening. 2. A cover, sometimes including a fan, a light fixture, fire extinguishing system, and/or grease filtration/extraction system, and supported, hung, or secured to a wall such as above a cooking stove chimney, or to draw smoke, fumes, and odors away from the area and into a flue.

hood

hook 1. Any bent or curved device for holding, pulling, catching, or attaching. 2. A terminal bend in a reinforcing bar.

hook-and-butt joint A scarf joint between timbers with their ends fashioned to lock together positively and resist tension.

hook bolt A bolt with an unthreaded end bent into an "L" shape.

hook bolt

hooked bar In reinforced concrete, a reinforcing bar which has a hooked end to facilitate its anchorage.

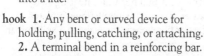
hooked bar

hook knife (hook bill knife, linoleum knife) A knife with a blade in which the cutting edge is bent back toward itself in the same plane, forming a hook shape.

hoop reinforcement Closely spaced steel rings providing circumferential or lateral reinforcement to prevent buckling of vertical reinforcing bars in concrete columns.

hopper 1. A top-loading, bottom-discharging funnel or storage bin, as for crushed stone or sand. 2. One of a pair of draft barriers at the sides of a hopper light. 3. A toilet bowl, usually funnel-shaped.

hopper lite (hopper light) 1. A bottom-hinged, inward-opening window sash which allows air to pass above it when open. 2. A side-hinged, inward opening window sash which, when open, allows most of the passing air over its top, but also allows the passage of some air through a narrow opening along its bottom.

horizon 1. The apparent intersection of the earth and sky, as perceived from any given position. 2. The same illusion as it might be portrayed in a perspective drawing.

horizontal Parallel to the plane of the horizon and perpendicular to the direction of gravity.

horizontal application A method of installing gypsum board with its length perpendicular to the framing members.

horizontal auger A drilling machine with a horizontally mounted auger, used to drill blast holes in strip mining.

horizontal boring Soil-boring on the horizontal as opposed to the vertical.

horizontal bracing Any bracing lying in a horizontal plane.

horizontal branch A branch drain accepting waste products delivered to it vertically, by gravity, from one or more similar but usually smaller fixtures, and conducting them horizontally to the primary disposal drain.

horizontal bridging Perpendicular braces between joists or beams placed horizontally to stiffen the system and distribute the load.

horizontal bridging

horizontal control In surveying, a control system in which the relative positions of points has been established by traverse, triangulation, or another system.

horizontal distance The distance between points anywhere on a horizontal plane.

horizontal panel A wall panel in which the major dimension is horizontal.

horizontal pipe Any pipe placed or laid horizontally or at an angle to the horizontal of less than 45°.

horizontal sheeting In excavation, any type of earth-restraining sheeting placed horizontally between and supported by soldier piles.

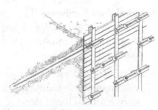
horizontal sheeting

horizontal shoring 1. Extendible beams or trusses capable of providing concrete form support over fairly long spans, thus reducing the number of vertical supports required. **2.** The collective support provided by several horizontal shores in an application.

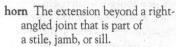

horizontal shoring

horn The extension beyond a right-angled joint that is part of a stile, jamb, or sill.

horse 1. Framework functioning as a temporary support, such as a sawhorse. **2.** In a stair, one of the slanting supports or strings carrying the treads and risers.

horsehead 1. A frame-like device for supporting a pulley back over a pit so that people and materials may be lowered into it and raised out from it. **2.** Forepole support when tunneling through soft material.

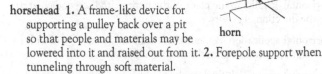

horn

horsepower A unit measurement of power or energy in the United States Customary System. Mechanically, a single horsepower represents 550 foot-pounds per second. Electrically, a single horsepower represents 746 watts.

horsepower hour A unit representing the amount of work performed by one horsepower in one hour.

hose bib (hose cock) An outdoor water faucet protruding from a building at about sill height, which is usually threaded to accept a hose connection.

hose cabinet Identifiable cabinet to house folded hose and valve, partially recessed and wall-mounted.

hose bib (hose cock)

hose coupling A connection between hoses or between a hose and a pipe.

hose station In a fire safety system, a storage rack that includes a valve, hose, and nozzle.

hose stream test A test that measures an assembly's ability to withstand lateral impact from falling debris during a fire endurance period and before active fire suppression efforts begin.

hospital arm pull A door handle having an extension to allow opening by hooking one's arm around it, thus leaving the hands free.

hospital door A flush door through an opening large enough to allow passage of beds and/or other large equipment.

hospital door hardware The special hardware with which hospital doors are often equipped, such as arm pulls, hinges, terminated stops, latches, and strategically placed protective metal strips or plates.

hospital hinge A fast pin hinge furnished with a special tip to eliminate the possibility of injuries caused by the projection of conventional hinge tips.

hot Slang for a *live* or electrically charged wire or other electrical component.

hot-air furnace A heating unit in which air is warmed and from which the warmed air is drawn into ducts to be carried throughout a building or selected portion thereof.

hot-cathode lamp A type of fluorescent, electric discharge lamp in which the electrodes operate at incandescent temperatures and in which the arc and/or circuit elements provide the energy required to maintain the cathodes at incandescence.

hot chisel A type of chisel used to cut red-hot steel. Has a sharp edge usually 2″ to 2¼″ wide.

hot-dip galvanized Descriptive of iron or steel immersed in molten zinc to provide it with a protective coating.

hot driven rivet Any rivet heated just prior to placement.

hot glue A glue requiring heating before being used.

hotmelt A thermoplastic substance almost always heated before being applied as a coating, sealer, or adhesive.

hot mix Paving made of a combination of aggregate uniformly mixed and coated with asphalt cement. To dry the aggregate and obtain sufficient fluidity of asphalt cement for proper mixing and workability, both the aggregate and asphalt must be heated prior to mixing.

hot press The method of producing plywood, laminates, particleboard, or fiberboard, in which adhesion of layers in the panel is accomplished by the use of thermosetting resins and a heat process, under pressure, to cure the guidelines.

hot rolled Descriptive of structural steel members or sections shaped from steel fillets or plates, heated to a plastic state, by passing them through successive pairs of massive steel rollers, each of which serves to bring the product closer to its final, intended shape, such as an angle, channel, or plate.

hot-setting adhesive An adhesive whose proper setting necessitates a minimum temperature of 212°F.

hot spraying The spraying of paints or lacquers in which the viscosity has been reduced by heat rather than by thinners, allowing formation of a thicker coat, requiring less spraying pressure, hence less overspray.

hot surface 1. A highly alkaline or highly absorbent surface. **2.** A surface having a high temperature.

hot water As defined for the purpose of building codes, water temperature that is 110°F or greater.

hot water boiler Any heating unit in a hot water heating system in which or by which water is heated before being circulated through pipes to radiators or baseboards throughout a building or portion thereof.

hot water heating system One in which hot water is the heating medium. Flow is either gravity or forced circulation.

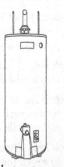

hot-water supply The combination of equipment and its related plumbing supplying domestic hot water.

hot wires An electrically charged or power-carrying wire.

hot work Any construction operation involving flames or hot air, or use of heat-producing equipment, such as arc welders and cutting equipment, brazing and soldering equipment, blow lamps, or bitumen boilers.

hot-water supply

housed joint A usually perpendicular joint formed where the full thickness of one member's edge or end is accepted into a corresponding housing, groove, or dado in another member.

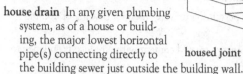
housed joint

house drain In any given plumbing system, as of a house or building, the major lowest horizontal pipe(s) connecting directly to the building sewer just outside the building wall.

house pump In a gravity supply system, the pump that is used to fill the gravity tank that supplies water to the building.

house sewer The exterior horizontal extension of a house drain outside the building wall leading to the main sewer, either public or private, and connecting directly to the sewer pipe.

housing Casings that enclose heating, ventilation, and air conditioning equipment, such as coils, filers, or fans. Made most often of sheet metal. May also be used to enclose other types of equipment.

Howe truss A truss with vertical and diagonal webs. The members of the vertical web absorb tension while those of the diagonal web absorb compression.

H-pile A type of steel beam driven into the earth by a pile driver.

H-runner A lightweight, H-shaped, metal member used on its side in a suspended ceiling system, so that its flat top fastens to a channel and the flat bottom fits into the kerfs in the ceiling tiles.

hub 1. The central core of a building, usually the area into which stairs and/or elevators are incorporated, and from which hallways or corridors emanate. **2.** The usually strengthened central part of a wheel, gear, propeller, etc. **3.** The end of a pipe enlarged into a bill or socket. **4.** A rotating piece within a lock, through whose central aperture the knob spindle passes to actuate the mechanism. **5.** In surveying, a stake designating a theodolite position. **6.** Caulking or cement connections between pipe joints.

hue 1. The designation of color. **2.** Characteristic by which one color differs from another.

humidifier A mechanical apparatus to add moisture to the air or other material.

humidistat (hygrostat) The automatic regulating device of a humidifier or dehumidifier that is sensitive to and actuated by changes in humidity.

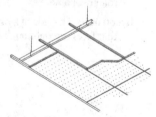
hung ceiling

humidity The water vapor contained in a given space, area, or environment.

hung ceiling A nonstructural ceiling having no bearing on walls, being entirely supported from above by the overhead structural element(s) from which it is suspended.

hung sash A sash hung from its sides by cords or chains, whose other ends are secured to counterweights to allow movement in the vertical plane.

hung window

hung window A window containing one or more hung sashes.

hurricane clips Metal anchor used in pole construction to fasten floor joists to a supporting beam or other structural member.

hurricane-prone regions As defined for the purpose of building codes, coastal regions of the United States susceptible to hurricanes with wind speed greater than 90 mph, including the coasts of the Gulf of Mexico and Atlantic Ocean, as well as Hawaii, Puerto Rico, American Samoa, Guam, and the Virgin Islands.

hurricane protection shield A durable guard that folds out accordion-style to protect windows and sliding glass doors from the high winds of hurricanes.

hybrid beam A fabricated beam having flanges made from steel with a specified minimum yield strength different from that of the steel used in the web plate.

hybrid photovoltaic generator system A power system that combines solar photovoltaics with a conventional generator system to minimize life cycle costs. It takes advantage of the low operating cost of a photovoltaic array and the on-demand capability of a generator. To optimize cost, a PV system can incorporate a generator to run infrequently during cloudy periods. The PV array typically provides 70% to 90% of the annual energy, and the generator provides the remainder.

hydrant A discharge connection to a water main, usually consisting of an upright pipe having one or more nozzles and controlled by a gate valve.

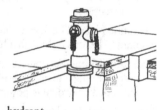
hydrant

hydrated lime A dry, relatively stable product derived from slaking quicklime.

hydration 1. Any chemical action occurring as the result of combining a material with water. **2.** The chemical reaction that occurs when cement is mixed with water.

hydraulic Characterized or operated by fluid, especially under pressure.

hydraulic cement Cement whose constituents react with water in ways that allow it to set and harden under water.

hydraulic dredge A floating dredge or pump by which water and soil, sediment, or seabed are pumped, either on board for sifting, as for clams or oysters before they are discharged overboard, or through a series of floating pipes for discharge on shore.

hydraulic ejector A pipe through which the working chamber of a pneumatic caisson is cleansed of sand, mud, or small gravel.

hydraulic excavator A powered piece of excavating equipment having a hydraulically operated bucket.

hydraulic fill Fill composed of solids and liquid, usually water, and usually delivered by a dredge. After placement, the water eventually drains to leave only the solid fill.

hydraulic friction Resistance to flow, effected by roughness or obstructions in the pipe, channel, or similar conveying device.

hydraulic glue Glue unaffected by water.

hydraulic hydrated lime The dry, hydrated, cementitious product resulting from the process of calcining a limestone containing silica and alumina to a temperature just below incipient fusion. The resultant lime will harden under water.

hydraulic jack A mechanical lifting device incorporating an external lever to which force is applied to cause a small internal piston to pressurize the fluid, usually oil, in a chamber. The pressure exerts force on a larger piston, causing it to move vertically upward and raise the bearing plate above it.

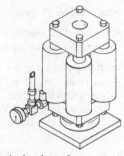

hydraulic jack

hydraulic lift An elevator car or platform moved by a piston, or plunger, powered by a pressurized fluid, usually oil, in a cylinder.

hydraulic lime Lime composed of at least ten silicates and which will set and harden under water.

hydraulic mortar A mortar capable of hardening under water, hence used for foundations or underwater masonry construction.

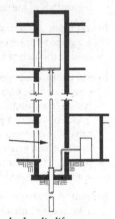

hydraulic pump The device causing the fluid to be forced through a hydraulic system.

hydraulic lift

hydraulic splitter A concrete- or rock-cracking mechanism incorporating a wedge inserted into a predrilled hole and then expanded by hydraulic power to cause the cracking.

hydraulic test Employing pressurized water to test a plumbing line for pressure integrity.

hydrocarbons Class of chemical compounds consisting of hydrogen and carbon.

hydrochlorofluorocarbons (HCFCs) Compounds composed of hydrogen, chlorine, fluorine, and carbon atoms. They do not persist to the same extent as chloroflourocarbons, and so do not pose as great a threat to the ozone layer.

hydrofluorocarbons (HFCs) Compounds composed of hydrogen, fluorine, and carbon atoms. Because they do not contain chlorine, they are not involved in ozone depletion.

hydrogeologic testing A means of determining the structure and characteristics of subsurface soils and rocks, and the way water flows through them.

hydronic A term pertaining to water used for heating or cooling systems.

hydrophobic cement A treated cement with a reduced tendency to absorb moisture.

hydro-seeding The liquid application of a combined mixture of grass seed, fertilizer, pesticide, and a moisture-retaining binder sprayed under pressure over an area requiring lawn or grass cover.

hydrostatic head The pressure in a fluid, expressed as the height of a column of fluid, which will provide an equal pressure at the base of the column.

hydrostatic pressure Pressure exerted by water, or equivalent to that exerted on a surface by water in a column of specific height.

hydrostatic strength The capability of a pipe to resist internal pressure buildup, measured under specific conditions.

hygrometer An instrument used for measuring the moisture content of air.

hygrometric expansion The expansion of a material as it takes on moisture.

hygroscopic Having the tendency to absorb and retain moisture from the air.

Hypalon® roofing An elastomeric roof covering available commercially in liquid, sheet, or putty-like (caulking) consistency in several different colors. Hypalon® roofing is more resistant to thermal movement and weathering than neoprene.

hypotenuse The side of a right triangle that faces the right angle.

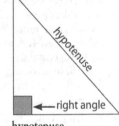

hypotenuse

H

158

Abbreviations

I moment of inertia

IARC International Agency for Research on Cancer

IBI Intelligent Buildings Institute

IC interrupting capacity, ironclad, incense cedar

ICBO International Conference of Building Officials

ICC Interstate Commerce Commission, International Code Council

ID inside dimension, inside diameter, identification

IDHA International District Heating Association

IDLH immediately dangerous to life and health

IEEE Institute of Electrical and Electronics Engineers, Inc.

IESNA Illuminating Engineering Society of North America

IF inside frosted

IFB invitation for bids

Ihp indicated horsepower

IMC intermediate metal conduit

imp imperfect

in inch

inc included, including, incorporated, increase, incoming

incan incandescent

incl included, including

Ins insulate, insurance

inst installation

insul insulation, insulate

int intake, interior, internal

IP iron pipe

IPS iron pipe size

IPT iron pipe threaded

IR inside radius at the start, or initiation, of an activity

IRS Internal Revenue Service

ISO Insurance Services Office, International Organization for Standardization

IWP Idaho white pine

Definitions

i In the critical path method (CPM) of scheduling, the symbol that represents the event at the start, or initiation, of an activity.

I-beam A structural member of rolled steel whose cross section resembles the capital letter "I."

I-beam

ice dam An accumulation of ice and snow at the eaves of a sloping roof.

idle equipment cost The cost of equipment that remains on site ready for use but is placed in a standby basis. Ownership or rental costs are still incurred while the equipment is idle.*

idler A gear or wheel used to impart a reversal of direction or rotation of a shaft.

idle time A time interval during which either the worker, the equipment or both do not perform useful work.*

igneous rock A rock formed by the solidification of molten materials.

illuminated sign A sign that is illuminated by an internal or external source, often used to mark a path of emergency exit.

illuminated sign

illumination The intensity of light on a surface exposed to incident light.

I-joist An engineered wood product created with two flanges joined by a web that develops certain structural capabilities. I-joists are also used for rafters.

impact cost Added expenses due to the indirect results of a changed condition, delay, or changes that are a consequence of the initial event. Examples of these costs are premium time, lost efficiency, and extended field and home office overhead.*

impact crusher A crushing machine that utilizes a series of hammers to break up materials.

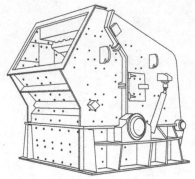

impact crusher

impact damages Losses that affect the overall performance and cost of contract work, such as delay to the project, lost labor productivity, and acceleration. Distinguished from direct damages.

impact factor A number by which a static load is multiplied to approximate that load applied dynamically.

impact isolation 1. The use of insulating material and structures that reduce the transmission of impact noise. 2. The degree of effective reduction of impact noise transmission accomplished by the structures and materials designed and used specifically for that purpose.

impact load The dynamic effect on a stationary or mobile body as imparted by the short, forcible contact of another moving body.

impact noise The sound created when a building surface is struck by an object.

impact noise rating The single-number rating used to evaluate and compare the effectiveness of assemblies and floor/ceiling constructions in isolating impact noise. The higher the number, the greater the effectiveness in suppressing noise.

impact resistance The resistance of a member or assembly to dynamic loading. Also refers specifically to an insulations ability to withstand damage or abuse.

impact sound transmission Sound that originates by contact with the structure and travels through the structure.

impact test Any of a number of dynamic tests (usually a load striking a specimen in a specified manner) used to estimate the resistance of a material to shock.

impact wrench An electric or pneumatic wrench with adjustable torque that is supplied to a nut or bolt in short, rapid impulses.

impedance Measured in ohms, the total opposition or resistance to the flow of current when voltage is applied to an alternating-current electric circuit.

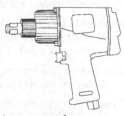

impact wrench

impeller 1. The vaned member of a rotary pump that employs centrifugal force to convey fluids from intake to discharge. 2. A related device used to force pressurized

gas in a given direction. 3. In ventilation, a device that rotates to move air.

impermeable Descriptive of a material that does not allow the passage of liquid.

impervious Highly resistant to penetration by water.

impervious soil A very fine-grained soil, such as clay or compacted loam, that is so resistant to water penetration that slow capillary creep is the only means by which water can enter.

imposed load Any load that a structure must bear, exclusive of dead load.

impost The often distinctively decorated uppermost member of a column, pillar, etc., that supports an end of an arch.

impoundment A water storage area or tank.

impost

impracticability Inability to perform work because of extreme and unreasonable difficulty, expense, injury, or loss involved. This is sometimes considered practical impossibility.*

impregnation 1. The penetration of a (timber) product under pressure with an oil, mineral, or chemical solution, usually for preservation. 2. Treating soil with a liquid waterproofing agent to reduce leakage.

improvement A physical change or addition made on a property so as to increase its value or enhance its appearance.

impulse radar The process of using a generator to produce an electromagnetic pulse that travels through a structure to record and measure the changes in wave velocity without causing damage to the structure. Used to investigate the stability of structures such as bridges, castles, roads, and modern concrete structures.

inactive leaf (inactive door) In a pair of doors, the stationary leaf to which the strike plate is secured. It is usually bolted at the head and sill.

inbark (bark pocket) Ingrown bark enclosed in the wood of a tree by growth and subsequently exposed by manufacture.

inbond A masonry bond across the entire thickness of a wall, and usually consisting of headers or bondstones.

incandescence The emission of visible light as a consequence of being heated.

incandescent lamp A lamp in which electricity heats a (tungsten) filament to incandescence, producing light.

inch A measure of length equal to $\frac{1}{12}$ of a foot (2.54 cm).

inch of water A unit of pressure that is equal to the pressure exerted by a 1″-high column of liquid water at a temperature of 39.2°F (4°C).

inch-pounds 1. A unit of work derived by multiplying the force in pounds by the distance in inches through which it acts. 2. A unit of energy that will perform an equivalent amount of work.

incident radiation Solar energy, both direct and diffuse, upon its arrival at the surface of a solar collector or other surface.

incident sound Noise that is directly received from the source, as distinguished from sound that is reflected from a surface.

incinerator A type of furnace in which combustible solid, semisolid, or gaseous wastes are burned.

incipient decay The early stage of decay in timber in which the disintegration has not proceeded far enough to affect the strength or hardness.

incising 1. Cutting in, carving, or engraving, usually for decorative purposes. **2.** Cutting slits into the surface of a piece of wood prior to preservative treatment to improve absorption.

inclination 1. The deviation of a surface or line from the vertical or horizontal. **2.** The angle produced by the deviation of a surface or line from the vertical or horizontal.

inclination

incline A slope, slant, or gradient.

inclined-axis mixer A truck-mounted concrete mixer. A revolving drum rotates around an axis that is inclined from the horizontal axis of the truck's chassis.

increaser In plumbing, a coupling with one end larger than the other. Generally, the small end has outside threads and the large end has inside threads.

increaser

incremental cost (benefit) The additional cost (benefit) resulting from an increase in the investment in a project.*

incrustation Mineral, chemical, or other deposits left in a pipe, vessel, or other equipment by the liquids that they convey.

indent In masonry, a gap left in a course by the omission of a masonry unit. An indent is used for bonding future masonry.

indented bolt A type of anchor bolt comprising a plain bar into which indentations have been forged to increase its grip in concrete or grout.

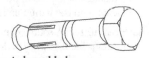
indented bolt

indented wire Wire whose surface has been provided with indentations to increase its bond when used as concrete reinforcement or for pretensioning tendons.

indenture 1. An official agreement between a bond issuer and his bondholders. **2.** Any deed or contract between two or more parties.

independent contractor A contractor who, under written contract, provides services to an owner but is not considered an employee of the owner for tax or other legal reasons. The independent contractor controls the means, method, and manner of producing the result to be accomplished.*

independent cost estimate (ICE) A cost estimate made in order to validate other prior estimates for a project.

index mark In surveying, a mark used as a sighting reference to take measurements.

index of plasticity The numerically expressed difference between the liquid and plastic limits (of a cohesive material).

Indiana limestone A durable, easily sawn, planed, carved, and lathed limestone quarried in, and exported from, the state of Indiana.

indicated horsepower The horsepower, determined by an indicator gauge, that is developed in the cylinders exclusive of losses sustained due to engine friction.

indicator pile A pile driven to evaluate future pile.

indicator valve A valve that includes some device indicating its open or closed condition.

indigenous Descriptive of any product, substance, growth, outcropping, characteristic, etc., that is geographically native to the area where it occurs, as opposed to having been introduced there.

indirect costs In construction, all costs that do not become a final part of the installation, but are required for the orderly completion of the installation and may include, but are not limited to, field administration, direct supervision, capital tools, startup costs, contractor's fees, insurance, taxes, etc.*

indirect expense Overhead or other indirect costs incurred in achieving project completion, but not applicable to any specific task.

indirect gain/loss In passive solar design, heat gain or loss that occurs at the surface of a thermal storage wall. Typical materials include brick, concrete, and water.

indirect lighting Lighting achieved by directing the light emitted from a luminaire toward a ceiling, wall, or other reflecting surface, rather than directly at the area to be illuminated.

indirect luminaire A luminaire that distributes 90% to 100% of its emitted light upward.

indirect waste pipe A waste pipe that discharges through an air break or air gap into a trapped receptacle or fixture, rather than directly into the building drainage system.

indirect water heater A water heater system that increases the temperature of the water via a remote heat exchanger.

individual vent A pipe that vents a fixture drain and that is connected to the main vent system at some point higher than the fixture.

indoor air quality (IAQ) The quality and general healthfulness of air within a building, as affected by temperature, humidity, and airborne contaminants.

indoor environmental quality (IEQ) An important criterion for green, or sustainable, building design, this refers to general overall building occupant comfort. Includes humidity, ventilation, and air circulation, acoustics, and lighting.

induced draft A process in which air is drawn through the cooling tower into the fan.

induced draft boiler A boiler that uses a fan at its discharge end to pull air through the burner and oiler, and transfer the exhaust products into the atmosphere through a chimney.

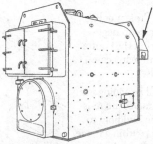
induced draft boiler

induced draft water-cooling tower A water-cooling tower incorporating one or more fans in the path of the saturated air stream leaving the tower.

inductance The process that occurs when one conductor is placed next to another carrying AC current. The ever-changing magnetic field that results will induce a current in the first conductor.

induction The entrainment of air in a room by the strong flow of primary air from an air outlet.

induction air terminal units A factory assembly consisting of a cooling coil and/or heating coil that receives preconditioned air under pressure that is mixed with recirculated air by the induction process.

induction motor A motor that operates on alternating current. Its primary winding is usually on the starter, which is connected to the electric power source. Its secondary winding, usually on the rotor, conveys the induced current.

induction welding A type of welding in which coalescence is achieved by heat derived from the work's resistance to an induced electric current, either with or without applied pressure.

inductive loads Loads whose voltage and current are out-of-phase. Most loads in modern electrical distribution systems are inductive.

industrial hygienist In asbestos abatement, a professional hired by the building owner to sample and monitor the air, and for other safety-related tasks.

industrial waste Liquid waste from manufacturing, processing, or other industrial operations, which might include chemicals but not rainwater or human waste.

industry standard Readily available information in the form of published specifications, technical reports and disclosures, test procedures and results, codes and other technical information and data. Such data should be verifiable and professionally endorsed, with general acceptance and proven use by the construction industry.

inelastic behavior Deformation of a material that remains even after the force that caused it has been relieved or removed.

inert 1. Chemically inactive. **2.** Resistant to motion or action.

inertia The tendency of a body at rest to remain at rest, or of a body in motion to remain in motion in a straight line unless directly influenced by an external force.

inertia block Usually a concrete block supported on some sort of resilient material and used as a base for heavy, vibrating mechanical equipment, such as pumps and forms, to reduce the transmission of vibration to the building structure.

inert pigment A pigment or extender that does not react chemically with the materials with which it is being mixed.

inexcusable delays Project delays that are attributable to negligence on the part of the contractor, which lead, in many cases, to penalty payments.*

infilling Fill material providing insulation, stiffness, and/or fire resistance. It is used in buildings to fill the void areas, within a frame, between structural members.

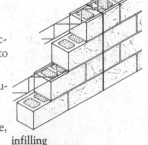
infilling

infiltration The leakage of air into a building through the small spaces around windows, doors, etc., caused by pressure differences between indoor and outdoor air.

inflammable Easily ignitable and highly combustible. The preferred term is *flammable*. *See also* **combustible**.

inflatable gasket A type of gasket whose effective seal results from inflation by compressed air.

inflatable structure An airtight structure of impervious fabric, supported from within by slightly greater than atmospheric pressure generated by fans.

inflation A persistent increase in the level of consumer prices, or a persistent decline in the purchasing power of money, caused by an increase in available currency and credit beyond the proportion of available goods and services.*

inflection The bending of a straight line from concave to convex, or the converse.

inflecture point That point in a flexural structural member where reversal of curvature occurs and the bending moment is zero.

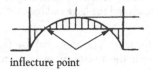
inflecture point

information stake A stake used on-site to convey surveying measurements to an excavation crew.

infrared camera A device used to detect uninsulated or underinsulated areas and gaps in the building envelope where heat is escaping in the winter and entering in the summer.

infrared heater A source of heat-producing wavelengths, longer than visible light, which do not heat the air through which they pass, but only those objects in the line of sight.

infrared lamp A type of incandescent lamp that often has a red glass bulb to reduce its radiated visible light. Such a lamp emits more radiant power in the infrared region than does a standard incandescent lamp, and it has a lower filament temperature, which contributes to its longer life.

infrastructure Improvements supporting an area, including transportation, roadways, communication, and utility systems.

ingle 1. A fireplace. **2.** A hearth.

ingle

ingot A large mass of molten metal cast in a vertical mold but requiring further processing before becoming a finished product.

inhibitor Oxidants added to coatings to retard drying, skinning, and other undesirable effects or conditions.

initial drying shrinkage The difference between the length of a concrete specimen when first poured and the final, permanent length of the same specimen after it has dried, usually expressed as a percentage of the initial moist length.

initial prestress The stress or force applied to prestressed concrete at the time of its stressing.

initial set That point in the setting of a concrete and water mixture when it has attained a certain degree of stiffness, but is not yet finally set. Initial set is usually expressed in terms of the time required for a cement paste to stiffen enough to resist a pre-established degree of penetration by a weighted test needle.

initial stress The stress existing in a prestressed concrete member prior to the occurrence of any loss of stress.

initiating circuit A circuit in a fire safety system that contains sensors to detect a fire condition.

injection burner A gas burner in which gas and air for combustion are forced into the burner by a gas jet.

inlay 1. To decorate with inlaid work. 2. An ornamental design cut into the surface of linoleum, wood, or metal, and filled with a material of different color, often by gluing.

inlet 1. The surface connection to a closed drain or pipe. 2. The upstream end of any structure through which there is a flow.

inorganic material Substances of mineral composition, not carbon compounds of animals or vegetables.

input system A ventilating system utilizing a fan to draw air from a roof into rooms through ducts. An air cleaner or filter and an automatic air heater are usually included.

insert 1. A patch, plug, or shim used to replace a defect in a plywood veneer. 2. A unit of hardware embedded in concrete or masonry to provide a means for attaching something. 3. A nonstructural patch in laminated timber, made for the sake of appearance.

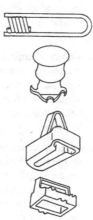

insert

inside-angle tool In masonry and plastering, a float designed especially for shaping inside (internal) angles.

inside caliper A caliper with pointed legs turned outward that is used for measuring inside diameters.

inside casing The interior trim around the frame of a door or window, which might consist of dressed boards, molding, and trim.

inside corner molding Concave or canted molding used to cover the joint at the internal angle of two intersecting surfaces.

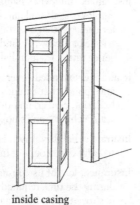

inside casing

inside lining Any part of a cased frame that has its face toward the building or structure.

inside stop Usually, a beaded or molded strip of wood secured to the casing along the inside edge of the inner sash to hold it in place and restrict its movement to the vertical plane.

inside thread The threaded inner surface of a pipe or fitting that accepts the outside threads of another pipe or fitting.

inside thread

inside trim Any trim used inside a building.

in situ 1. In place, as natural, undisturbed soil. 2. Descriptive of work accomplished on the site rather than in prefabrication elsewhere, as in cast-in-place concrete.

in situ concrete Concrete placed where it will harden to become an integral part of the structure, as opposed to precast concrete.

in situ soil test Soil testing performed in a borehole, tunnel, or trial pit, as opposed to being performed elsewhere on a sample that has been removed.

inspection 1. A visual survey of construction work—either completed or in progress—to ensure that it complies with the contract documents. 2. Examination of the work by a public official, owner's representative, or others.

inspection eye A pipe fitting equipped with a plug that can be removed to allow examination or cleaning of the pipe run.

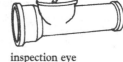

inspection eye

inspection manhole A covered shaft leading from the outside down to a sewer or duct, so constructed as to allow a person to enter it from the surface.

inspector A person authorized and/or assigned to perform a detailed examination of any or all portions of the work and/or materials.

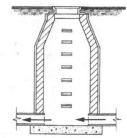

inspection manhole

instant lock (instant locker) 1. An automatic lock that is actuated by the closing of the door. 2. A time lock or chromatic lock working on the same principle.

instructions to bidders A document included as part of the bidding requirements that sets forth specific instructions to candidate constructors on procedures, expectations, and disclaimers of the owner, and other necessary information for the preparation of proposals for consideration by the owner for a competitive bid.

insulate To provide with special features and/or materials that afford protection against sound, moisture, heat, or heat loss.

insulated cavity wall A hollow masonry wall with a cavity containing some type of insulation.

insulated concrete forms (ICF) Highly insulated, pre-engineered forms for reinforced concrete walls with R-values as high as R-40. Most often made of polyvinyl chloride (PVC) or expanded polystyrene foam (EPS).

insulated conductor An electrical conductor that is contained within an NEC-approved, nonconductive material.

insulated flange A coupling that interrupts the electrical transmission between two metal pipes.

insulated metal roofing A type of roofing panel made from mineral fiber, insular glass, foamed plastic, etc., and faced with light-gauge flexible metal.

insulated roof membrane assembly (IRMA) A roofing system where the roof membrane is laid directly on the roof deck and then covered with extruded foam insulation and ballasted with stone.

insulating board A thin, lightweight, rigid or semirigid board, usually of processed plant fibers, that offers little structural strength but does provide thermal insulation. It is usually applied under a finish material.

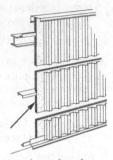

insulating concrete Concrete possessing low thermal conductivity and used as thermal insulation.

insulating fiberboard Building board manufactured from processed plant fiber in a variety of sizes, thicknesses, densities, and strengths.

insulating board

insulating formboard Insulation board that serves as a permanent form for poured-in-place gypsum or lightweight concrete roof decks.

insulating glass Glazing comprising two or more panes of glass, between which there exist(s) a hermetically sealed airspace(s), joined around the edges.

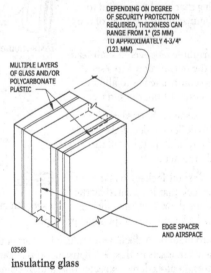

DEPENDING ON DEGREE OF SECURITY PROTECTION REQUIRED, THICKNESS CAN RANGE FROM 1" (25 MM) TO APPROXIMATELY 4-3/4" (121 MM)

MULTIPLE LAYERS OF GLASS AND/OR POLYCARBONATE PLASTIC

EDGE SPACER AND AIRSPACE

03568
insulating glass

insulating resistance The measure of a material's ability to inhibit electric current flow.

insulating sheathing Insulating board with a minimum thermal resistance of R-2 of the core material.

insulation 1. Material used to reduce the effects of heat, cold, or sound. 2. Any material, device, or technique that provides protection against fire or the transfer of electricity, heat, cold, moisture, or sound. 3. Thermal insulation is a material used for covering pipes, ducts, vessels, etc., to effect a reduction of heat loss or gain.

insulation batt Flexible insulation of loosely matted plant or glass fibers faced on one or both sides with kraft paper or aluminum foil and usually available in specifically sized sections.

insulation batt

insulation blanket Usually composed of the same materials and the same widths and thicknesses as batts, but is available in rolls.

insulation lath Gypsum lath with an aluminum foil backing that provides resistance to heat flow and moisture.

insulation resistance The resistance of an insulator to a directly applied current.

insulator An insulating device designed and used to physically support a conductor and electrically separate it from other conductors or objects.

insurance A contract, typically referred to as an insurance policy, in which the insurer, in return for the premium stated in the policy, agrees to pay the insured up to the limits specified in the policy for losses or damages incurred by the insured.

insurance, builder's risk A specialized form of property insurance that provides coverage for loss or damage to the work during the course of construction.

insurance, certificate of A document, issued by an authorized representative of an insurance company, stating the types, amounts, and effective dates of insurance in force for a designated insured.

insurance, completed operations Liability insurance coverage for injuries to persons or damage to property occurring 1. when all operations under the contract have been completed or abandoned; 2. when all operations at one project site are completed; or 3. when the portion of the work out of which the injury or damage arises has been put to its intended use by the person or organization for whom that portion of the work was done. Completed operations insurance does not apply to damage to the completed work itself.

insurance, comprehensive general liability A broad form of liability insurance covering claims for bodily injury and property damage that combines, under one policy, coverage for all liability exposures (except as specifically excluded) on a blanket basis and automatically covers new and unknown hazards that may develop.

insurance, contractor's liability Insurance purchased and maintained by the contractor to protect from specified claims that may arise out of, or result from, the contractor's operations under the contract, whether such operations are by the contractor or by any subcontractor, or by anyone directly or indirectly employed by any of them, or by anyone for whose acts any of them may be liable.

insurance, employer's liability Insurance protection for the employer against claims by employees or employees' dependents for damages that arise out of injuries or diseases sustained in the course of their work, and that are based on common law negligence rather than on liability under workers' compensation acts.

insurance, extended coverage An endorsement to a property insurance policy that extends the perils covered to include windstorm, hail, riot, civil commotion, explosion (except steam boiler), aircraft, vehicles, and smoke.

insurance, liability Insurance that protects the insured against liability on account of injury to the person or property of another.

insurance, loss of use Insurance protecting against financial loss during the time required to repair or replace property damaged or destroyed by an insured peril.

insurance, owner's liability Insurance to protect the owner against claims arising out of the operations performed for the owner by the contractor and arising out of the owner's general supervision.

insurance, personal injury Bodily injury, and also injury or damage to the character or reputation of a person. Personal injury insurance includes coverage for injuries or damage to others caused by specified actions of the insured, such as false arrest, malicious prosecution, willful detention or imprisonment, libel, slander, defamation of character, wrongful eviction, invasion of privacy, or wrongful entry.

insurance policy A contract that provides insurance against specific loss. *See also* **insurance**.

insurance, professional liability Insurance coverage for the insured professional's legal liability for claims for damages sustained by others allegedly as a result of negligent acts, errors, or omissions in the performance of professional services.

insurance, property Coverage for loss or damage to the work at the site caused by the perils of fire, lightning, extended coverage perils, vandalism and malicious mischief, and additional perils (as otherwise provided or requested).

insurance, property damage Insurance covering liability of the insured for claims for injury to or destruction of tangible property, including loss of use resulting therefrom, but usually not including coverage for injury to, or destruction of, property that is in the care, custody, and control of the insured.

insurance, public liability Insurance covering liability of the insured for negligent acts resulting in bodily injury, disease, or death of persons other than employees of the insured, and/or property damage.

insurance, special hazards Insurance coverage for damage caused by additional perils or risks to be included in the property insurance (at the request of the contractor, or at the option of the owner). Examples often included are sprinkler leakage, collapse, water damage, and coverage for materials in transit to the site or stored off the site.

insurance, workers' compensation (workmen's compensation insurance) Insurance covering the liability of an employer to employees for compensation and other benefits required by workers' compensation laws with respect to injury, sickness, disease, or death arising from their employment.

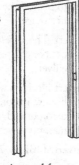

integral frame

intake The opening or device through which a gas or fluid enters a system.

integral frame A metal door frame whose trim, backhands, rabbets, and stops are all fabricated from one piece of metal for each jamb and for each head.

integral lock A type of mortise lock in which the cylinder is contained in the knob.

integral waterproofing The waterproofing of concrete achieved by the addition of a suitable admixture.

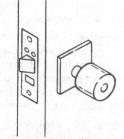

integral lock

integrated building system A building designed to efficiently use climatic resources for heating, cooling, lighting, and electric power generation.

integrated ceiling A suspended ceiling system in which the grid and individual acoustical, illumination, and air-handling elements are combined to form a single, integrated system.

integrated design Also referred to as holistic or whole building design. A design method that integrates, early in the process, the whole building team, including all disciplines. For a sustainable building, resource efficiencies, indoor air quality and other goals can be achieved most effectively with this approach.

integrated project delivery A project delivery method in which the three parties, owner, designer, and contractor, execute a single relational contract to work together collaboratively and share in the risks associated with the project.

integration An essential concept in sustainable building that views the building as a system and allows the discovery of synergies and potential tradeoffs or pitfalls with design choices. An integrated design approach helps maximize synergies and minimize unintended consequences.

intelligent building (smart building) A building that contains some degree of automation, such as centralized control over HVAC systems, fire safety and security access systems, and telecommunication systems.

intercept In surveying, the length of the staff visible between the two stadia hairs of a transit's telescope.

interceptor An apparatus that functions to trap, remove, and/or separate harmful, hazardous, or otherwise undesirable material from the normal waste that passes through it, allowing acceptable waste and sewage to discharge by gravity to the disposal terminal.

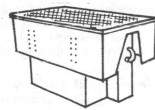

interceptor

intercom A general term for an internal communication system, such as one employing telephones, or microphones and loudspeakers that function without a central switchboard.

interconnection Any arrangement of pipes whereby water may flow from one system to another as determined by the pressure differential between the two systems.

interest expense A contractor's cost of borrowing funds or use of equity capital.

interface A common boundary shared by two adjacent parts of a system.

interference 1. Any obstruction that prevents planned or normal usage or operations. **2.** Cooling tower effluent that enters the intake of an adjacent tower.

interior door A door installed inside a building, as in a partition or wall, having two interior sides.

interior finish The interior exposed surfaces of a building, such as wood, plaster, and brick, or applied materials such as paint and wallpaper.

interior plywood Plywood in which the laminating glue is adversely affected by moisture; hence, it should be restricted to indoor or interior applications.

interior span An uninterrupted beam or slab that is continuous with neighboring spans.

interior stop In glazing, a removable bead or molding strip that serves to hold a light or panel in position when the stop is on the interior of the building, as opposed to an exterior stop.

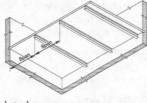

interior span

interior wall A wall having two interior faces and existing entirely within the exterior walls.

interlocked 1. Firmly joined. **2.** Closely united. **3.** Placed in close relative proximity, or in a specific relationship with another or others.

interlocking joint 1. In ashlar or other stonework, a joint accomplished by joggles in the joining units. **2.** In sheet metal, a joint between two parts whose preformed edges engage to form a continuous locked splice.

intermediate floor beam In floor framing, any other floor beams that are positioned between the end floor beams.

intermediate rail In a door, any rail, one of which might be a lock rail, between the top and bottom rails.

intermediate stiffener On a beam or girder, any stiffener between the end stiffeners.

intermittent-flame-exposure test An ASTM test of roof covering that involves exposing a sample of the roof material to a gas flame for a specified interval.

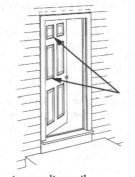

intermediate rail

intermittent weld A weld whose continuity is interrupted by recurrent unwelded spaces.

internal glazing Glazing placed in interior partitions or.

internal partition trap A plumbing trap that relies on an internal partition of some form, rather than the water used in a standard trap, for a seal.

internal pressure The pressure within a building. Dependent on outside conditions, such as wind velocity, and the number and location of openings in the building.

internal treatment The treatment of water by feeding chemicals into the boiler rather than into the preheated water itself.

internal vibration Rapid agitation of freshly mixed or placed concrete performed by mechanical vibrators inserted at strategic locations.

intern architect An apprentice architect who works under the direction of registered architects.

International Building Code (IBC) The building code established in 2000 by the International Code Council (ICC) that, when adopted by a jurisdiction, covers all buildings other than one- and two-family dwellings and multiple single-family dwellings not more than three stories in height.

International Residential Code (IRC) The International Code Council's (ICC) building code that, when adopted by a jurisdiction, covers all one- and two-family dwellings not more than three stories in height.

interoperability The ability of BIM tools from multiple vendors to exchange building model data and operate on that data. Interoperability is a significant requirement for team collaboration and data movement between different BIM platforms.

interpolate To estimate untested values that fall between tested values.

interrogatories A formal method of obtaining information relevant to a lawsuit from a party by submitting written questions that must be answered under oath within a certain time period.

interruption A stopping or hindering of the normal process or flow of an activity.*

interstate commerce The buying, selling, or exchange of goods or property across state lines.

interstitial condensation Condensation that forms within an element of a building, such as within a wall.

intrados The undersurface or interior curve of an arch or vault.

intrusion alarms Sensors that detect break-ins or forced entries into a facility.

intumescent Descriptive of a material, like a paint or caulk, that forms a passive fire protection system when applied to another material, like steel or pipe. The intumescent material remains inactive until subjected to heat, when it swells and chars and creates a fire barrier.

intrados

inventory control 1. A management plan designed to minimize the number and quantity of hazardous substances on a construction project. **2.** A managerial process to control the quantity of items in storage.

invert The lowest inside surface or floor of a pipe, drain, sewer, culvert, or manhole.

invert block A wedge-shaped, hollow, masonry tile incorporated into the invert of a masonry sewer.

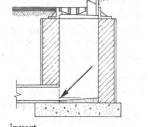

invert

inverted ballast A lamp ballast that operates on direct current.

inverted crown The fall or pitch from the sides to the center of a road, driveway, etc.

invert elevation The elevation of an invert (lowest inside point) of pipe or sewer at a given location in reference to a benchmark.

inverter Power conditioning equipment for photovoltaic systems used to convert DC power from photovoltaic arrays, wind turbines, water turbines, fuel cells, or batteries to AC power. A rotary inverter is a DC motor driving an AC generator. More common are static inverters, which use power transistors to achieve the conversion electronically.

invitation to bid A written notice of an owner's intention to receive competitive bids for a construction project wherein a select group of candidate constructors are invited to submit proposals.

invited bidders (preferred bid list) A group of bidders chosen by the owner to submit bids for a project.

invoice An itemized bill listing the items and charges of merchandise and/or work.

ion An atom with one or more electrons missing.

ion exchange The process of either gaining or losing electrons from compounds in a chemical exchange.

Ionic order An order of architecture characterized by a column that has spiral volutes at its capital.

ionization detector A device that indicates the presence of specific gaseous compounds in air by subjecting the compound to ultraviolet light or hydrogen flame, and recording the number of ions created.

iron A lustrous, malleable, magnetic, magnetizable, metallic element mined from the earth's crust as ore in hematite, magnetite, and lemonite. These minerals are heated together to 3,000°F in a blast furnace to produce pig iron, which emerges from the furnace as 95% iron, 4% carbon, and 1% other elements.

iron, cast An iron alloy usually containing 2½% to 4% carbon and silicon, possessing high compressive but low tensile strength, and that, in its molten state, is poured into sand molds to produce castings.

iron, malleable Cast iron that has undergone an annealing process to reduce its brittleness.

iron oxide A primary ingredient in a whole range of inorganic pigments.

ironwork A comprehensive term for iron fashioned to be used decoratively or ornamentally, as opposed to structurally.

iron, wrought The purest form of iron metal, which is fibrous, corrosion-resistant, easily forged or welded, and used in a wide variety of applications, including water pipes, rivets, stay bolts, and water tank plates.

ironwork

irrigation 1. The process or system, and its related equipment, by which water is transported and supplied to otherwise dry land. **2.** The use of water thus supplied for its intended purpose.

irritant A substance that causes discomfort, such as tearing, choking, vomiting, rashes, reddening of the skin, itching, or other topical responses.

isobar A line drawn on a map to indicate the limits of equal contaminant or pressure concentrations.

Isocyanate A glue used in making some building materials, such as bamboo flooring. Once dry, it does not produce toxic pollutants.

isolated heat gain Solar heating achieved using an attached sunspace, such as a greenhouse.

isolated solar gain Passive solar heating in which heat to be used on one area is collected in another area.

isolated spread footing A footing that transmits a load from columns to the supporting soil. If the soil is weak or the column loads heavy, isolated spread footings must be larger.

isolating switch A switch that isolates a circuit from its power source.

isolation joint A joint positioned so as to separate concrete from adjacent surfaces or into individual structural elements that are not in direct physical contact, such as an expansion joint.

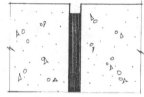

isolation joint

isolator The device on a circuit that can be removed to break the circuit in the absence of flowing current.

isometric drawing A type of projection drawing showing three dimensions. The horizontal planes generally appear at 30° from the standard horizontal axis, while the vertical lines are drawn parallel to the actual vertical axis.

isometric drawing

isotherm A line on a graph or map joining points of equal temperature.

isotropic Exhibiting the same properties in all directions.

item A subdivision of the breakdown, smaller than a category, but larger than an element.

itemize To list or state each item separately, as in an itemized bid.

I

J

Abbreviations

J&P joists and planks

jct junction

JIC Joint Industrial Council

jour journeyman

JP jet propulsion

jt, jnt joint

jtd jointed

jsts joists

junc junction

Definitions

j A symbol used to describe the event at the head of an activity arrow in a critical path method (CPM) schedule.

jack 1. A portable mechanism for moving loads short distances by means of force applied with a lever, screw, hydraulic press, or air pressure, as in applying the prestressing force to the tendons, or making small adjustments in the elevations of forms or form support, as in lift slab or slipform operations. **2.** In electricity, a female connecting device or socket to which circuit wires are attached and into which a plug may be inserted (as on a telephone switchboard).

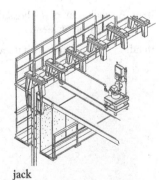

jack

jacked pile A pile that is forced into the ground by jacking against the building above it.

jacket 1. A covering, either of cloth or metal, applied to exposed heating pipes or ducts, to exposed or unexposed casing pipes, or over the insulation of such pipes. **2.** A watertight outer housing around a pipe or vessel, the space between being occupied by a fluid for heating, cooling, or maintaining a specific temperature. **3.** In wire and cable, the outer sheath or casing that protects the individual wires within from the elements and provides additional insulation properties.

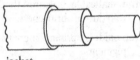

jacket

jackhammer A handheld, pneumatically powered device that hammers and rotates a bit or chisel. A jackhammer is used for drilling rock or breaking up concrete, asphalt paving, etc.

jacking device A mechanism used to stress the tendons for prestressed concrete, or to raise a vertical slipform. *See also* **jack.**

jacking force The temporary force applied to tendons by a jacking device to produce the tension in prestressing tendons.

jacking plate A steel-bearing plate used during jacking operations to transmit the load of the jack to the pile.

jacking stress The maximum stress that occurs during the stressing of a precast concrete tendon.

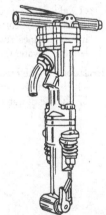

jackhammer

jack lagging The timber-bearing members employed in building the forms for arches or for other unusual shapes.

jackleg An outrigger post.

jack plane A medium-sized carpenter's plane used to do the coarse work on a piece of timber, such as truing up the edges in preparation for finish planing.

jack post An adjustable metal structural support most often used to replace inadequate supporting members.

jack rafter Shortened rafter generally found in hip roofs. A hip jack rafter spans from the plate to the hip rafter. A hip-valley cripple jack spans between a hip rafter and a valley rafter. A valley jack spans from a valley rafter to the ridgeboard.

jack screw A screw-operated mechanical device equipped with a load-bearing plate and used for lifting or leveling heavy loads.

jack shaft A short drive shaft, as between a clutch and a transmission.

jack shore An adjustable, usually telescopic, single-post metal shore.

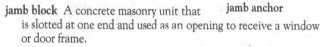

jack shore

jack timber In framework, a timber that is shorter than the rest because it has been intercepted by another member.

jack truss A roof truss that is smaller than the main trusses, such as a truss in the end slopes of a hip roof.

jamb An exposed upright member on each side of a window frame, door frame, or door lining. In a window, these jambs outside the frame are called *reveals*.

jamb anchor A metal anchoring device that secures the frame to the wall when inserted in the rear of the jamb of a door frame or window frame.

jamb anchor

jamb block A concrete masonry unit that is slotted at one end and used as an opening to receive a window or door frame.

jamb depth The face-to-face depth of a door frame.

jamb horn In a window frame, that portion of the jamb that extends beyond the sill or head jamb.

jamb post The vertical timber that serves as a jamb at the side of an opening.

jamb stud A wood or metal stud adjacent to the doorjamb.

japan 1. A dark-colored, short-oil varnish used to provide a hard, glossy surface. 2. A type of resin varnish often used in paints as a drying agent.

japanned Painted with black japan and then baked.

jaw One of a pair of opposing members of a device used for holding, crushing, or squeezing an object, as the jaws of a vise or a pair of pliers.

jaw crusher A rock-crushing device comprising one fixed inclined jaw and one movable inclined jaw. It is used to reduce rock to specific sizes.

J channel A plastic or metal channel, shaped like a J, that is used to support building trim material.

jedding ax A stonemason's hand tool with two faces, one flat and the other pointed.

jemmy (jimmy) A short (less than 18″) pinch bar or crowbar, both ends of which are curved.

jerrybuilt Constructed in a shoddy or flimsy manner.

jet 1. A high-velocity, pressurized stream of fluid or mixture of fluid and air, as emitted from a nozzle or other small orifice. 2. The nozzle or orifice that shapes the stream. 3. An orifice or other feature of a toilet that starts the siphon action by directing water into the trapway.

jetting The process by which piles or well points are sunk with the aid of high-pressure water or air. Jetting is the option usually employed in locations where nearby buildings might be adversely affected during pile-driving operations.

jetty 1. Any portion of a building that protrudes beyond the part immediately below it, such as a bay window or the second story of a garrison house. 2. A dike-like structure, usually of rock, that extends from a shore into water, usually to provide some kind of protection, but sometimes to induce scouring or bank-building. 3. A deck on pilings constructed for landing at the edge of a body of water.

jib The hoisting arm of a crane or derrick, whose outer end is equipped with a pulley, over which the hoisting cable passes.

jib boom The hinged extension attached to the upper end of a cranes boom. Its purpose is to extend the reach of the crane or the height of the boom.

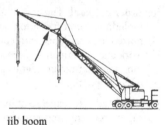

jib boom

jib crane 1. A crane having a swinging jib, as opposed to an overhead traveling crane, which does not. 2. A cantilevered boom or horizontal beam with hoist and trolley capable of picking up loads in all or part of a radius around the column to which it is attached.

jib door A door constructed and installed so as to be flush with its surrounding wall and whose unobtrusiveness is furthered by the lack of hardware on its interior face.

jig A device that facilitates the fabrication or final assembly of parts by holding or guiding them in such a way as to ensure their proper mechanical and relative alignment. A jig is especially useful in ensuring that duplicate pieces are identical.

jigsaw An electrically powered, table-mounted saw with a small, narrow, vertically reciprocating blade that is capable of cutting a tighter radius than a band saw.

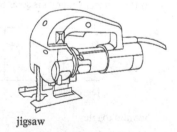

jigsaw

jimmer A hinge comprising two leaves that are permanently joined with a fixed pin.

jitterbug A grate tamper used to cause sand and cement grout to rise to the surface of wet concrete during placement of slabs. May be motorized or hand operated.

job (job site) 1. Term commonly used to indicate the location of a construction project. 2. An entire construction project or any component of a construction project.

jobber 1. A person reasonably knowledgeable and somewhat skilled in most of the more common construction operations,

J

such as carpentry, masonry, or plumbing. **2.** In construction, a jack-of-all-trades.

job condition Those portions of the contract documents that define the rights and responsibilities of the contracting parties and of others involved in the work. The conditions of the contract include general conditions, supplementary conditions, and other conditions.

job-made Made or constructed on the construction site, as a job-made ladder.

job order authorization Written authorization by the owner to accomplish the described work.

job order contracting (JOC) An indefinite delivery/indefinite quantity project delivery method used for construction, remodeling, repair, and landscaping projects. It can also be used for maintenance services. Pricing structures are based on competitively bid coefficients applied to preestablished unit prices. JOC contracts usually have options for annual renewal.

job order proposal The design, along with a detailed scope of work including a project's performance times and price proposal, submitted in accordance with contract requirements. The scope of work and performance times are mutually agreed on before the job order contractor submits a lump-sum, fixed-price, detailed price proposal.

job overhead The expense of such items as trailer, toilets, telephone, superintendent, transportation, temporary heat, testing, power, water, cleanup, and similar items, possibly including bond and insurance associated with the particular project.*

job site The area within the defined boundaries of a project.

joggle 1. A notch or protrusion in one piece or member that is fitted to a protrusion or notch in another piece to prevent slipping between the members. **2.** A protrusion or shoulder that receives the thrust of a brace. **3.** A horn or stub tenon at the end of a mortised piece to strengthen it and prevent its lateral movement. **4.** The enlarged portion of a post by which a strut is supported. *See also* **key.**

joggle joint In masonry or stonework, a joint that employs joggles in the adjacent members so as to prevent their lateral movement.

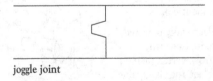

joggle joint

See also **joggle.**

joggle truss A roof truss comprising a single, centrally positioned post whose upper end is joggled to connect with the overhead chord, and whose lower end is supported by two braces that angle upward to join the ends of the chord.

joiner's chisel (paring chisel) A long-handled woodworking chisel whose cutting is accomplished without the aid of a striking tool, but by hand force only.

joiner's chisel (paring chisel)

joiner's hammer A hammer whose head has a flat end for striking and a clawed end for pulling nails.

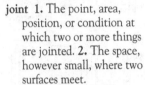

joiner's hammer

joinery 1. Woodworking that deals more with joining and finishing, such as that required by doors, cabinets, or trim. **2.** A European designation for quality grades of lumber suitable for cabinetry, millwork, or interior trim.

joint 1. The point, area, position, or condition at which two or more things are jointed. **2.** The space, however small, where two surfaces meet. **3.** The mortar-filled space between adjacent masonry units. **4.** The place where separate but adjacent timbers are connected, as by nails or screws, or by mortises and tenons, glue, etc.

joint

joint banding The visible striping of each panel joint—usually a result of oversanding or uneven absorption of the primer due to differences in surface texture. More noticeable under critical lighting.

joint bolt (handrail bolt) A threaded metal rod or bolt having a nut at each end, used to bring two ends of a handrail together, as well as in joinery to bolt two mating surfaces together, as in a butt joint.

joint box A cast-iron box constructed around a joint between the ends of two electric cables. The cables' protective lead (or other) sheathing is secured by bolted clamps on the exterior of the box, which may be filled with insulation.

joint compound (joint cement) A premixed finishing material for embedding joint tape and filling and finishing gypsum panel joints, corner bead, trim, and fasteners. Sanded to a smooth finish.

jointer (jointer plane) 1. A power-driven woodworking tool or long, hand-operated, bench plane used to square the edges of lumber or panels. **2.** An offset metal tool used to smooth or indent mortar joints in masonry. **3.** A metal tool about 6″ long and 2″ to 4″ wide, having interchangeable depth-regulating bits and used for cutting joints in fresh concrete. **4.** In masonry, a bent strip of iron used in a wall to strengthen a joint.

joint filler 1. A powder that is mixed with water and used to treat joints, as in plasterboard construction. **2.** Any putty-like material similarly used. **3.** A compressible strip of resilient material used between precast concrete units to provide for expansion and/or contraction.

jointing 1. Finishing the surface of mortar joints, as between units or courses, by tooling before the mortar has hardened. **2.** The finishing, as by machining, of a squared, flat surface on one face or edge of a piece of wood. **3.** The initial operation in sharpening a cutting tool, consisting of filing or grinding the teeth or knives to the desired cutting circle.

jointing compound In plumbing, any material, such as paste, paint, or iron cement, used to ensure a tight seal at the joints of iron or steel pipes.

joint knife A hand tool with a wide flexible blade used to apply and smooth joint compound to drywall.

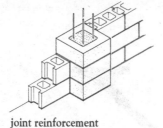
joint reinforcement

joint reinforcement Any steel reinforcement used in or on mortar joints, such as reinforcing bars or steel wire.

joint rule A metal rule having one end formed at a 45° angle. It is used in plastering to form and shape miters at the joints of cornice moldings.

joint sealant An impervious substance used to fill joints or cracks in concrete or mortar, or to exclude water and solid matter from any joints.

joint tape Paper, paper-faced cotton tape, or plastic mesh fabric, used with mastic or plaster to cover the joints between adjacent sheets of wallboard.

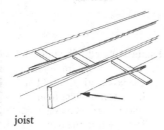
joist

joist 1. A piece of lumber 2″ or 4″ thick and 6″ or more wide, used horizontally as a support for a ceiling or floor. Also, such a support made from steel, aluminum, or other material. **2.** Parallel beams of lumber, concrete, or steel used to support floor and ceiling systems.

joist anchor A beam, wall anchor, or metal tie used to anchor beams or joists to a wall. An example is a metal strip with one end embedded in a concrete or masonry wall and the other end secured to a joist or rafter, so as to provide a lateral tie between the wall and a floor or roof. This type of anchor acts in shear and in tension.

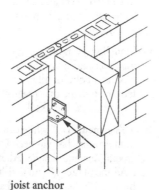
joist anchor

joist bridging Lateral braces inserted between joists to evenly distribute loads.

joist hanger A metal angle or strap used to support and fix the ends of wood joists or rafters to beams or girders. This type of anchor acts in shear and in tension.

joist hanger

joule The metric unit used to measure heat, work, and energy. One joule is the amount of work done or energy expended by a force of 1 newton acting through a distance of 1 meter.

journal 1. That section of a shaft or axle that rotates within a load-supporting bearing. **2.** The project superintendent's daily report of activities on a project.

journeyman The second or intermediate level of development of proficiency in a particular trade or skill. As related to building construction, a journeyman's license, earned by a combination of education, supervised experience, and examination, is required in many areas for those employed as intermediate level mechanics in certain trades (e.g., plumbing, mechanical, and electrical work).

jumbo brick 1. A brick manufactured larger than standard size, measuring 8″ × 4″ × 4″, including mortar joints.

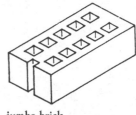
jumbo brick

jumper 1. The short length of wire or cable used to make a usually temporary electrical connection within, between, or around circuits and/or their related equipment. **2.** A steel bar used manually as a drilling or boring tool. **3.** A stretcher covering two or more vertical joints in square rubble. **4.** The inverted mushroom-shaped component of a domestic water tap, on which the washer fits.

junction box A metal box in which splices in conductors or joints in runs of raceways or cable are protectively enclosed, and which is equipped with an easy access cover.

junction chamber That section in a sewer system where the flow from one or more sewers joins or converges into a main sewer.

jurisdictional dispute An argument between or among labor unions over which entity should perform certain work.

just-in-time A "pull" logistical system driven by actual demand. The goal is to produce, provide, or deliver parts or supplies just in time for the next operation. The approach reduces stock inventories or storage costs, but leaves no room for error.*

junction box

jut Any protruding part of a building or structure, such as a jut window.

jute A plant fiber, from which a strong, durable yarn is made and used mostly for carpet backing, burlap, and rope.

K

Abbreviations

k kilo, knot

K Kalium

Ka cathode

kc kilocycle

kcal kilocalorie

kc/s kilocycles per second

KD kiln-dried

KDN knocked down

kg keg, kilogram

kHz kilohertz

KIT. kitchen

kl kiloliter

KLF kips per lineal foot

km kilometer

kmps kilometers per second

kn knot

Kr krypton

kv kilovolt

KVA kilo volt ampere

kvar kilovar

kw kilowatt

kwhr/kwh kilowatt-hour

Definitions

kalamein door A fire door whose solid wood core is usually covered with galvanized sheet metal.

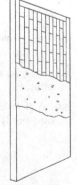

kalamein door

kaolin A usually white mineral found in rock formations, composed primarily of low-iron hydrous aluminum silicate, and used as a basic ingredient in the manufacture of white cement and as a filler or coating for paper and textiles.

keel molding A molding having two ogee curves that meet at a point or fillet, forming a shape that resembles that of a ship's keel.

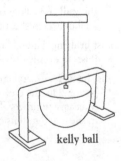

keel molding

Keene's cement 1. A white cementitious material manufactured from gypsum that has been burned at a high temperature and ground to a fine powder. Alum is added to accelerate the set. The resulting plaster is hard and strong and accepts and maintains a high polish, hence it is used as finishing plaster. 2. Anhydrous calcined gypsum.

keeping the gauge In masonry, maintaining the proper spacing of courses of brick.

keeping the perpends In masonry, the accurate laying of the units so that the perpends (end joints) in alternating courses line up vertically.

kelly ball A round-bottomed metal plunger that is dropped into fresh concrete, the degree of penetration indicating the consistency of the concrete.

kelly ball

Kelvin scale A temperature scale on which absolute zero is measured (0°K, equivalent to 273°C). It uses Celsius degrees.

kerf 1. A saw-cut in wood, stone, etc., that is usually performed crosswise and usually not completely through the member. Cuts are usually made to allow for bending. **2.** A groove cut into the edges of acoustical tiles to accommodate the splines or supporting elements in a suspended acoustical ceiling system.

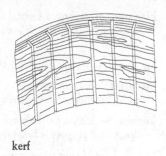

kerf

kerfed beam A beam in which several kerfs have been cut so as to permit bending.

kettle 1. The storage container for asphalt to be used for "hot mopped" roof construction. **2.** Any open vessel used to contain paint or in which glue is melted.

key 1. The removable actuating device of a lock. **2.** A wedge of wood or metal inserted in a joint to limit movement. **3.** A wedge or pin through the protruding part of a projecting tenon to secure its hold. **4.** The tapered last board in a sequence of floorboards, which, when driven into place, serves to hold the others in place. **5.** The roughened underside of veneer or other similar material intended to aid in bonding. **6.** In plastering, that portion of cementitious material that is forced into the openings of the backing lath. **7.** A small, usually squared piece that simultaneously fits into the keyways or grooves of a rotating shaft and the pulley.

key brick A brick whose proper fit in an arch is attained by tapering it toward one end.

keyed Fastened or fixed in position in a notch or other recess, as forms become keyed into the concrete they support.

keyed beam 1. A lap-jointed beam with joggles or slots cut into both components. Keys are driven into the joggles or slots to increase the bending strength of the joint. **2.** A compound beam whose adjacent layers possess mating grooves to help resist horizontal shearing stress at the interfaces.

keyed brick A brick, one of whose faces has been supplied with a usually dovetail-shaped recess, that serves as a mechanical key for plastering or rendering.

keyed joint 1. A joint between two timbers that employs a key to ensure its security. **2.** The concave pointing of a mortar joint.

keyhole saw (hole saw) A thin, narrow-bladed saw used to cut holes in panels or other surfaces.

key plan A small plan that depicts the units in a layout.

key plate An escutcheon.

keystone The usually wedge-shaped uppermost, hence last, set stone or similar member of an arch, whose placement not only completes the arch but also binds or locks its other members together.

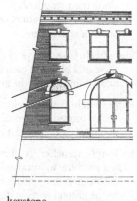

keystone

key switch An on-off switch in an electric circuit, which is actuated by a removable key rather than a toggle or button.

key valve A valve operated by a key rather than a handle or lever.

keyway 1. A recess or groove in one lift or placement of concrete that is filled with concrete of the next lift, giving shear strength to the joint. Also called a *key*. **2.** In a cylindrical lock, the aperture that receives and closely engages the key for its entire length, unlike a keyhole of a common lock. **3.** A key-accepting groove in a shaft, pulley, sprocket, wheel, etc.

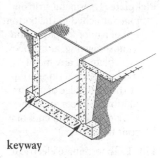

keyway

keyway forms Metal forms for pouring connected concrete slabs. Holes allow a keyway to be bolted on, then removed.

K factor The measure of a transformer's ability to withstand the heating effects of nonsinusoidal harmonic currents.

kibble A bucket-like device in which material, water, tools, and/or men are raised from a shaft.

kick 1. In brick, a shallow depression, fray, or panel. **2.** The raised fillet of a brick mold that forms the frog. **3.** The pitch variation between patent glazing and the surrounding roof.

kickback 1. When a rotating sawblade is pinched by the material it is cutting, momentarily stopping the blade and causing it to pull away from the material. **2.** The illegal return of part of the purchase price by the seller to induce purchase or to improperly influence future purchase of goods or services.

kicker 1. A wood block or board attached to a formwork member in a building frame or formwork to make the structure more stable. In formwork, a kicker acts as a haunch to take the thrust of another member. Sometimes called a *cleat*. **2.** A catalyst. **3.** An activator, as the hardener for a polyester resin. **4.** A luminaire used to accent or highlight a subject.

kicker

kick lift A jacking wedge that raises or adjusts a piece of gypsum board into position for nailing.

kick-off meeting A meeting that takes place at the beginning of a project. Its purpose is to introduce the project team members, review the overall project, and to discuss items such as construction site logistics, the phasing plan, and the schedule.*

kickpipe A short length of pipe that provides protection for an electric cable where it protrudes from a floor or deck.

K

kickplate 1. A metal strip or plate attached to the bottom rail of a door for protection against marring, as by shoes. **2.** A plate, of any metal, used to create a ridge or lip at the open edge of a stair platform or floor, or at the back edge or open ends of a stair tread.

kickplate

kill 1. To terminate electrical current from a circuit. **2.** To shut off an engine. **3.** To prevent resin from bleeding through paint on wood by the preliminary application to knots of a shellac or other resin-resistant coating.

kiln A furnace, oven, or heated enclosure for drying (wood), charring, hardening, baking, calcining, sintering, or burning various materials.

kiln-dried 1. Control-dried or seasoned artificially in a kiln. **2.** Lumber that has been seasoned in a kiln to a predetermined moisture content.

kiloampere A unit of electric current equal to 1,000 amperes.

kilocalorie The amount of heat needed to raise the temperature of one kilogram of water 1° Celsius.

kilovolt A unit of electric potential difference equal to 1,000 volts.

kilovolt-ampere A unit of apparent power equal to 1,000 volts or amperes.

kilowatt A measurement or unit of power equal to 1,000 watts or approximately 1.34 horsepower.

kilowatt consumption The amount of electrical power used over a specified time.

kilowatt demand The maximum electrical power usage required to operate a facility over a specified time.

kilowatt-hour A unit of measurement equal to the amount of energy expended in one hour by one kilowatt of power.

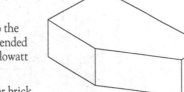
king closer

king closer A rectangular brick, one corner of which has been removed diagonally to leave a 2″ end, and that functions as a closer in brickwork.

king pile In strutted sheet pile excavation, a long guide pile driven at the strut spacing in the center of the trench before it is excavated.

kingpin A vertically mounted swivel, pivot, or hinge pin usually supported both above and below.

king post In a roof truss, a member placed vertically between the center of the horizontal tie beam at the lower end of the rafters and the ridge, or apex of the inclined rafters.

king stud In framing, a vertical support member that extends from the bottom to top plate alongside an opening for a door or window.

kip A unit of weight equal to 1,000 pounds.

kiss mark Marks on the faces of bricks where they were in contact with one another during their firing in a kiln.

kitchen As defined for the purpose of building codes, a space whose designated purpose is food preparation.

kite A sheet of kraft paper applied to a sheet of coated roofing during manufacturing to measure the weight of the granules applied to the surface of the roofing.

knapping hammer A steel hammer whose head design may vary, but which is used for breaking and shaping stone, splitting cobbles, etc.

knee 1. A naturally or artificially bent piece of wood, as used for a brace or haunch. **2.** A sharp, right-angled bend in a pipe. Also called an *elbow*. **3.** A convex handrail.

knee brace A brace between vertical and horizontal members in a building frame or formwork to make the structure more stable. In formwork, it acts as a haunch.

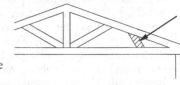
knee brace

knee bracket A brace used to provide extra support under bow and angle bay projecting windows.

kneeler 1. The pattern-breaking stone or brick at the point where a normal masonry wall changes and begins to form the curve or angle of an arch or vault. **2.** In a gable, the sloped-top, level-bedded stone that supports the inclined coping. **3.** Ecclesiastical furnishing that may be provided for kneeling in a pew. May be attached to the frontal or to the back of the pew.

knee rafter A rafter whose lower end is bent downward to rest more firmly against a wall. Sometimes called a *knee piece*.

knee wall 1. A wall that shortens the span of the roof rafters by acting as a knee brace, in that it supports the rafters at some intermediate point along their length. **2.** A short wall constructed to extend the height of an existing foundation or other wall system.

knife-blade fuse A type of cartridge fuse in which the metal blades at each end of the cylinder make contact with the fuse within.

knife-blade fuse

knife switch A type of electric switch designed with a hinged or removable blade that enters or embraces the contact clips.

knife-type trencher A vibratory plow attachment used on a trenching machine to install telephone and power cable, television cable, irrigation systems or other light weight cable-type products in the ground without digging a trench.

knob lock A door lock whose spring bolt is operated by one or more knobs, but whose dead bolt is actuated by a key.

knob rose The usually raised round plate that is attached to a door face so as to surround a hole in the door and form a knob socket.

knob shank The stem of a doorknob, into whose hole or socket the spindle is received and fastened.

knocked down Descriptive of precut, prefitted, and premeasured, but unassembled construction components, such as might be delivered to a job for on-site assembly.

knocked-down frame A door frame that comes from the manufacturer in three or more parts.

knockout A prestamped, usually circular section in an electrical junction box, panel box, etc., which can be easily removed to provide access for a fitting or raceway cable.

knot The hard, cross-grained portion of a tree where a branch meets the trunk.

knot cluster A compact grouping of two or more knots surrounded by deflected wood fibers or contorted grain.

knot sealer Any sealer, such as shellac, used to cover knots in new wood to prevent sap or resin bleed-through.

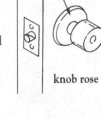

knob rose

knockout

knotty pine Pine wood sawn so as to expose firm knots as an appearance feature. Knotty pine is used for interior paneling and cabinets.

knuckle One of the enlarged, protruding, cylindrical parts of a hinge through which the pin is inserted.

knuckle joint A hinged joint by which two rods are connected.

knurling Very small ridges or beads as machined on a surface to facilitate gripping.

kraft paper A strong brown wrapping paper made from sulfate wood pulp that is sometimes impregnated with asphalt or resin for better moisture resistance when used in construction.

K series A standard Steel Joist Institute designation for long-span steel joists.

K truss A truss in which the arrangement of panels, chords, and web members resembles the letter K.

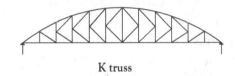

K truss

k value The thermal conductivity of a substance or material.

K

Abbreviations

L lambert, large, medium wall copper tubing
l labor only, left, length, liter, long, lumen
L & E labor and equipment
L & H light and heat
Lab. labor, laboratory
Lad ladder
LAG lagging
LAM laminated, laminate
LAN local area network
LAS laboratory analytical service
LAT latitude, lattice
Lath lather
LAV lavatory
lb, lbs pound, pounds
lbf/sq in pound-force per square inch
lb/hr pounds per hour
lb/LF pounds per linear foot
lbl label
LBL conduit body, load bearing
Lbr lumber
lc load center
lcd liquid crystal display
LCL less-than-carload lot
LCM least common multiple, loose cubic meter
L&CM lime and cement mortar
LCY loose cubic yard
ld load

lbbrk loadbreak
ldg loading
LDG landing
LDPE low-density polyethylene
Ldr loader
LE leading edge, lead equivalent
LEED Leadership in Energy amd Environmental Design
LECA light expanded clay aggregate
LED light-emitting diode
LEMA Lighting Equipment Manufacturers' Association
Len lens
lf lightface, light framing, lineal foot, linear foot, low frequency
lg large, length, long
Lg. large, length, long
LG liquid gas
Lge large
LGLCS landfill gas and leachate control system
lgp low ground pressure
lgr longer
lgs lengths
lgt lighting
lgth length
L&H light and heat
LH left hand, long-span, high strength bar joist, labor hours
LIC license
LIFO last in, first out
lin lineal, linear
lin ft linear foot, lineal feet

lino linoleum

liq liquid

LJ obsolete designation for long- span, standard-strength bar joist

lknt locknut

LL live load

L&L latch and lock

LL&B latch, lock, and bolt

LLD lamp lumen depreciation

LLRW low-level radioactive waste

lm lumen

LM lime mortar

lm/sf lumens per square foot

lm/W lumens per watt

Lm³ loose cubic meter

lnd lined

lng, Lng lining

LNG liquefied natural gas

lnr liner

L&O lead and oil (paint)

LOA length overall

log logarithm

L-O-L lateralolet

log logarithm

lox liquid oxygen

LP liquid petroleum, low pressure

L&P lath and plaster

LPF low power factor

LPG liquid petroleum gas

l-pull line pull

LR living room, law reports, long radius

LS left side, loudspeaker, lump sum

LSA low specific activity

L/s liters per second

LT long ton, light

ltd limited

ltg lighting

Lt Ga light gauge

LTL less than truckload lot

Lt Wt lightweight

LUST leaking underground storage tank

LV low voltage

lvl level

LVL laminated veneer lumber

lvr louver

LW low water

LWC lightweight concrete

LWM low water mark

Definitions

label 1. A projecting molding along the sides or top of a window. 2. Manufacturer identification located on a product that contains the manufacturer's name, the product's performance characteristics, and any testing information by inspection agencies.

labeled door A door that carries a certified fire-rating issued by Underwriters' Laboratories, Inc. 3-hour fire doors (A) are used in walls separating buildings or dividing a single building into fire areas. 1½-hour (B and D) fire doors are used in openings in 2-hour rated vertical enclosures such as stairs, elevators, etc., or in exterior walls subject to severe fire exposure from outside the building. 1-hour fire doors are for use in openings in 1-hour rated vertical enclosures. ¾-hour fire doors (C and E) are for use in openings in corridor and room partitions or in exterior walls which are subject to moderate fire exposure from outside the building. ½-hour fire doors and ⅓-hour fire doors are used where smoke control is a primary consideration, and for the protection of openings between a habitable room and a corridor when the wall has a fire-resistance rating of not more than one hour.

labeled frame A door frame that conforms to standards and tests required by Underwriters' Laboratories, Inc., and has received its label of certification.

labeled window A fire-resistant window that conforms to the testing standards of Underwriters' Laboratories, Inc., and bears a label designating its fire rating.

labor Effort expended by people for wages or salary. Generally classified as either direct or indirect. Direct labor is applied to meeting project objectives and is a principal element used in costing, pricing, and profit determination; indirect labor is a component of indirect cost, such as overhead or general and administrative costs.*

labor and material payment bond (payment bond) A bond procured by a contractor from a surety as a guarantee to the owner that the labor and materials applied to the project will be paid for by the contractor. Those who have direct contacts with the contractor may be considered claimants.

labor burden Taxes and insurances the employer is required to pay by law based on labor payroll on behalf of or for the benefit of labor. (In the United States, these are federal old-age benefits, federal unemployment insurance tax, state unemployment tax, and workers' compensation.)*

labor cost Gross direct wages paid to the worker (bare labor).*

laborer Ordinarily denotes a construction worker who has no specific trade and whose function is to support the activity of the licensed trades.

labor hour A worker hour of effort, synonymous with work hour.*

labor productivity A measure of production output relative to labor input. In cost estimating, inverse measures such as work hours/quantity or unit hours are common (where lower values reflect higher productivity or efficiency). Labor productivity (or efficiency) is improved by increasing production for a given work hour or decreasing work hours for a given production.*

labor rate Labor cost expressed on a per unit of labor effort basis (e.g., labor costs/labor hour).*

labor union An organization or confederation of workers with the same or similar skills who are joined in a common cause (such as collective bargaining) with management or other employers for work place conditions, wage rates, and/or employee benefits.

laced corner A method of laying shingles at interior and exterior corners on sidewalls. The corner shingles of each course are laid alternately on the faces of the two walls in order to overlap each other and eliminate the need for corner boards.

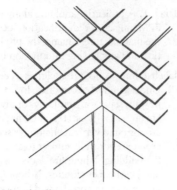

laced valley On a roof, the interweaving of shingles or tiles where two planes meet at a downward angle.

laced valley

lacing A system of members used to connect the different elements of a composite column or girder in such a way that they structurally act in unison. Also refers to securing insulation materials via hooks or wire, etc.

ladder core A hollow structure of wood or insulation board used as the core of interior doors and built with strips running vertically or horizontally through the core area.

ladder jack scaffold A simple scaffolding system that uses ladders for support. The ladder jack is attached to the ladders to provide support for the staging plank.

ladder ties Long parallel reinforcing rods studded with cross rods so as to resemble a ladder. Used in masonry installations.

lag Time that an activity follows, or is delayed from the start or finish of its predecessor(s). Sometimes called an offset.*

lag bolt (coach screw, lag screw) A threaded screw or bolt with a square head.

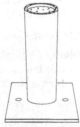
lag bolt (coach screw, lag screw)

lagging 1. Heavy wood boards used to line the sides of excavations and prevent cave-ins. **2.** Preformed insulation for pipes and tanks.

laitance In concrete, a weak, crumbly, and dusty surface layer caused by excessive water that has bled to the surface and subsequently weakened it. Overworking the surface during finishing can aggravate the problem. If laitance forms between pours, it must be brushed and washed away.

lally column A trade name for a pipe column from 3″ to 6″ in diameter, sometimes filled with concrete.

laminate 1. To form a product or material by bonding together several layers or sheets with adhesive under pressure and sometimes with nails or bolts. **2.** Any material formed by such a method.

lally column

laminated beam (laminated veneer lumber) A straight or arched beam formed by built-up layers of wood. The method of lamination may be gluing under pressure, mechanical nailing or bolting, or a combination.

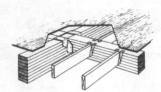
laminated beam
(laminated veneer lumber)

laminated glass (safety glass, shatterproof glass) A shatter-resistant safety glass made up of two or more layers of sheet glass, plate glass, or float glass bonded to a transparent plastic sheet.

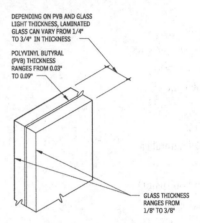

DEPENDING ON PVB AND GLASS LIGHT THICKNESS, LAMINATED GLASS CAN VARY FROM 1/4″ TO 3/4″ IN THICKNESS

POLYVINYL BUTYRAL (PVB) THICKNESS RANGES FROM 0.03″ TO 0.09″

GLASS THICKNESS RANGES FROM 1/8″ TO 3/8″

laminated glass (safety glass, shatterproof glass)

laminated wood (laminated veneer lumber) Any of several products formed by built-up layers (plies) of wood. Thin wood veneers may be laminated to a wood subsurface, several plies may be laminated together to form plywood, or thicker pieces may be used to form structural members such as beams or arches.

laminate floor A composite flooring product notable for its durability that is made of plastic resin and cellulose paper with a decorative finish layer. Its core material is a high-density fiberboard.

laminboard A compound board consisting of a core of small strips of wood glued together and covered by veneer faces.

lamp Any device that converts electric energy into light.

lamp depreciation The loss of luminous output in a lamp over time due to the accumulation of dirt on lamps, reflectors, and room walls and ceilings.

lamp life A rating that provides the life expectancy of a lamp. Obtained by testing a sample of lights and noting when 50% are no longer operating.

land-clearing rake A device outfitted with blades and attached to the front of a tractor to cut, collect, and remove brush from the site of proposed construction.

landfill An engineered disposal system characterized by the burial of waste material in alternating layers with an approved fill material.

landing An intermediate platform between flights of stairs, or the platform at the top or bottom of a staircase.

landing tread That board or portion of a stair landing that is closest to the next step down. The tread has an appearance and

dimensions similar to the other treads, but it is actually part of the landing, and not a true tread.

land reclamation Gaining land from a submerged or partially submerged area by draining, filling, or a combination of these procedures.

landing tread

Landrum-Griffin Act Enacted by Congress in 1959, this act requires labor union management to be subject to audit for the funds of union members for which they are responsible.

landscape architect A person whose professional specialty is designing and developing gardens and landscapes, especially one who is duly licensed and qualified to perform in the landscape architectural trade.

landscaping The combined grounds work tasks that improve the appearance of a plot of land, including adding plantings and lawn, constructing walkways and patios, and regrading as necessary.

land survey *See* **boundary survey** *and* **survey**.

land surveyor A person (usually registered in the state where survey is being done) whose occupation is to establish the lengths and directions of existing boundary lines on landed property, or to establish any new boundaries resulting from division of a land parcel.

land treatment area A defined parcel of land on which wastes are deposited for the purpose of allowing natural cleansing actions to occur.

land-use analysis A systematic study of an area or region that documents existing conditions and patterns of use, identifies problem areas, and discusses future options and choices. A part of the general planning process, such an analysis might cover topics such as traffic flow, residential and commercial zoning, sewer services, water supply, solid-waste management, air and water pollution, or conservation areas. In short, any factors that could affect how particular areas of land should or should not be used.

lane striping The application of lines on pavement to define parking spaces, traffic lanes, etc.

lanyard A safety line tying a worker to a stable element of a structure to prevent a long fall in the event of an accident above ground level.

lap cement Asphalt used in roll roofing as an adhesive between the laps.

lap joint In construction, a type of joint in which two building elements are not butted up against each other, but are overlapped, with part of one covering part of the other. Typical examples include roof and wall shingles, clapboard siding, welded metal sheets or plates, and concrete reinforcing bars lapped together at their ends.

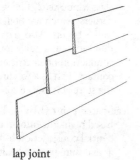

lapped joint A lapped joint, like a van stone joint, is a type of pipe joint

lap joint

made using loose flanges on lengths of pipe. The ends of this pipe are lapped over to give a bearing surface for a gasket or metal-to-metal joint.

lapping 1. The overlapping of reinforcing bars or welded wire fabric for continuity of stress in the reinforcing when a load is applied. **2.** The smoothing of a metal surface using a fine abrasive.

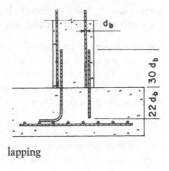
lapping

lap-riveted The riveting together of two metal members or plates where they have been deliberately overlapped, thus forming a lap joint.

lap seam The same as a lap joint, but typically used to refer to sheet metal, and sometimes plates, that are welded, soldered, or riveted at the overlapping joint.

lap siding Horizontal siding with each row covering part of the next, rather than butted up against each other.

lap splice In concrete construction, the simplest method for providing continuity of steel reinforcement. Ends of reinforcing bars are overlapped a specified number of diameters, usually no fewer than 30, and tied with wire.

lap weld pipe Pipe made by welding along a scarfed longitudinal seam in which one part is overlapped by another.

large quantity generator A classification of the Resource Conservation and Recovery Act (RCRA). One is considered a large quantity generator (LQG) for a particular month if 2,200 pounds or more of hazardous waste is generated, or more than 2.2 pounds of acute hazardous waste per calendar month.

larmier (corona, lorymer) A specific drip strip or molding that is part of a cornice. By projecting from the surrounding cornice, it catches rain and forces it to drip off away from the wall.

laser An acronym for light amplification by stimulated emission of radiation, a device that amplifies radiation in the visible or infrared parts of the spectrum.

laser level Surveying tool that employs a laser as a reference for measurements or verifying alignment.

laser level

last in, first out (LIFO) A method of accounting for inventory in which it is assumed that goods bought last are sold first. This allows automatic updating of inventory values.

late dates Calculated in the backward pass of time analysis, late dates are the latest dates on which an activity can start and finish without delaying a successor activity.*

latent defect A defect in materials or equipment that would not be revealed under reasonably careful observation. A patent defect, on the other hand, is one that may be discovered by reasonable observation.

lateral buckling The failure of any structural column, wall, or beam that has undergone excessive side-to-side (lateral) deflection, movement, or twist.

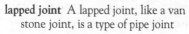

L

179

lateral sewer A sewer that discharges into another sewer or branch, but is engineered without any other common tributary to it.

lateral support Any bracing, temporary or permanent, that provides greater support in resisting side-to-side (lateral) forces and deflections. Floor and roof members typically provide lateral support for walls, columns, and beams. Vertical pilasters or secondary walls may also provide support.

lateral thrust Any force applied at a right angle to a member.

latest event occurrence time In the CPM (critical path method) of construction scheduling, the final deadline by which a particular item of work must be completed in order to avoid delaying the entire project.

latest finish date In the critical path method of scheduling, a completion deadline for a particular activity. Work performed after this deadline will result in project delay.

latest start date In the critical path method of scheduling, the deadline for starting a particular activity. A late start will throw off the schedule and delay the project.

EARLY START		EARLY FINISH
Account No.		
ACTIVITY DESCRIPTION		
DURATION		FLOAT
LATE START		LATE FINISH
	KEY	

latest start date

latex caulk A semisolid caulk containing water, ground calcium carbonate, plasticizers, mineral spirits, ethylene, glycol, surfactants, and pigments. Used to close joints or cracks between material.

latex paint A paint with a latex binder, usually a polymeric compound, characterized by its ability to be thinned or washed from applicators with water.

latex patching compound A compound used to fill voids, large gaps or penetrations in a floor, especially a subfloor before application of a floor covering.

lath Strips of wood or metal used as a base for plaster.

lathe A machine used to shape circular pieces of wood, metal, or other material. The stock is rotated on a horizontal axis while a stationary tool cuts away the unwanted material or creates ornamental turned work.

lath hammer A hammer used chiefly for cutting and nailing wood lath, designed with a nail-driving hammerhead, as well as a hatchet blade with a lateral nick that is used for pulling out nails.

lattice Typically, a diagonal network or grid of strips of material. Lattice is often used as ornamental screening, or to provide privacy.

lattice truss (lattice beam, lattice girder) A structural truss in which the web is a latticework of diagonal members.

lavas Natural stone used in building as a facing, veneer, and decoration. Pink, purple, and black are typical shades.

lattice

lavatory 1. A basin, with running water and drainage facilities, used for washing the face and hands. **2.** A room with a washbasin and a toilet, but no bathtub. **3.** A room containing a toilet.

lay A term used to define the direction of twist of the strands or wires in wire rope. The strands or wires have either a right-hand or left-hand lay.

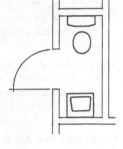

lavatory

lay-in Describes tile or panels that are installed into metal channels in suspended ceiling systems.

laying line (laying guide) Lines printed on felt or roll roofing as a guide for the amount of lap required.

laying out 1. Marking of materials in preparation for work, showing where they are to be cut. **2.** Marking the location for the placement of building members.

layout A design scheme or plan showing the proposed arrangement of objects and spaces within and outside a structure.

lay-up 1. The reinforcing material that is placed into position in the mold during the manufacturing of reinforced plastics. Also, the resin-impregnated reinforcement. **2.** The method or process of assembling veneers to manufacture plywood.

L-beam A beam whose section has the form of an inverted L, usually occurring in the edge of a floor, of which a part forms the top flange of the beam.

L-column The portion of a precast concrete frame composed of the column, the haunch, and part of the girder.

L/D ratio The relationship of the span (L) of a beam, column, slab, or truss to its depth (D).

leachate Liquid from rainfall, groundwater, or other sources that has percolated through a landfill mass and that contains biological and/or chemical wastes and dissolved or suspended materials. This waste is often hazardous and may need to be collected and treated to avoid groundwater contamination.

leaching The process of separating liquids from solid materials by allowing them to percolate into surrounding soil.

leaching well (leaching pit) A pit with porous walls that retain solids but permit liquids to pass through. It is not used to treat raw sewage but may be used to allow septic tank effluent to be absorbed into the surrounding soil.

lead 1. (Pb) A soft, dense heavy metal easily formed and cut. Historically, lead was used for flashing and for the joints in stained glass windows. **2.** End sections of a masonry wall, usually at the corners, which are built up, in steps, before the main part of the wall is begun. Also, a string stretched between these end sections that serves as a guide for the rest of the wall. **3.** In electricity, conduction of electric current from the electric source to point of contact such as a welding lead. **4.** Time that an activity precedes the start of its successor(s). Lead is the opposite of lag.*

lead-based paint (LBPs) Toxic oil-based paint containing lead-based pigments often found on old buildings. Prohibited in residential construction by the federal government starting in 1978, as exposure poses a health hazard.

lead chromate One of a number of opaque pigments that range from orange to yellow in color and have strong tinting properties.

lead-covered cable (lead-sheathed cable) An electric cable protected from damage and excess moisture by a lead covering.

leaded light A window whose diamond-shaped or rectangular glass panes are set in lead frames called cames.

leader 1. In a hot-air heating system, a duct that conveys hot air to an outlet. 2. A downspout.

leader head That part of a drainpipe assembly that is placed at the top of a leader and serves as a catch basin to receive water from the gutter.

lead foil tape An acrylic, weather-resistant, highly malleable tape with good tack, bond and resistance to solvents and heat. Used as a moisture barrier, a maskant in electroplating and chemical milling, and in thermopane window sealing. Also used in alarm systems, with conductive tape attached to a window to detect breakage.

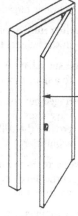

leading edge The vertical edge of a hinged swinging door or window that is opposite the hinged edge and in proximity to the knob or latch.

leading edge

lead joint A joint in a water pipe, such as a bell-and-spigot joint, into which molten lead is poured.

lead-lag ballast A ballast that serves to cut down on the stroboscopic effect of two fluorescent lamps, one of which is on a leading current and the other on a lagging current.

lead-lined door A door with lead sheets lining its internal core to prevent the penetration of x-ray radiation.

lead-lined frame A frame that is used with lead-lined doors and is itself lined internally with lead sheets to prevent penetration of x-ray radiation.

lead-lined door

lead-lined sheetrock Sheetrock internally lined with sheet lead to provide protection from x-ray radiation.

lead plug 1. A cylinder of lead placed inside a hole in a masonry, plaster, or concrete wall. A screw or nail driven into it will then be firmly held in place. 2. In stonemasonry, a piece of lead that holds adjacent stones together. Grooves are cut in both rock surfaces to be joined, and molten lead is poured into it.

lead roof A flat roof with a surface of sheet lead.

leads A collective term for short sections of electric conductors, generally insulated.

lead shield A lead sleeve used to provide anchorage for expansion bolts or screws. Similar to a *lead plug*.

lead shield

lead slate (copper slate, lead sleeve) A cylinder of sheet lead or sheet copper that surrounds a pipe at the point where it passes through a roof, ensuring a watertight intersection.

lead soaker A piece of lead sheeting that forms a weathertight joint at the intersection of a roof and of any vertical wall that passes through the roof at a hip or valley.

lead spitter The tapered part of a drainpipe assembly that connects a lead gutter with a downpipe.

lead tack 1. A lead strip used to attach a lead pipe to some means of support. 2. A lead strip placed along the edge of metal flashing. One side of the strip is attached to the structure, while the other side is folded over the free edge of the flashing.

leakage current Unwanted electric current that remains in a circuit or device. Can occur with improperly grounded electrical equipment. Leakage current can be observed by wire or semiconductor detectors.

lean construction Eliminating the waste associated with the delivery of construction projects.

lean mix (lean mixture) 1. A mixture of concrete or mortar with a relatively low cement content. 2. A plaster with too much aggregate and not enough cement, which thus renders it unworkable. 3. A mixture of gasoline and oil in which the gasoline portion is on the high side in relation to the oil.

lean mortar A mortar with a low cement content, which makes it sticky, overly adherent to the trowel, and difficult to apply.

lean-to roof A roof with one pitch, supported at one end by a wall extending higher than the roof.

learning curve A term applied to the time required for a new craftsman or crew to attain the productivity level of an experienced craftsman or crew.

ledge 1. A molding that projects from the exterior wall of a building. 2. A piece of wood nailed across a number of boards to fasten them together. 3. An unframed structural member used to stiffen a board or a number of boards or battens. 4. Bedrock.

ledger 1. A horizontal framework member that carries joists and is supported by upright posts or by hangers. 2. A slab of stone laid flat, such as that over a grave. 3. A horizontal scaffold member, positioned between upright posts, on which the scaffold planks rest.

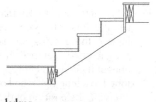

ledger

ledger board 1. One of multiple boards attached horizontally across a series of vertical supports, as in the construction of a fence. 2. A ribbon strip.

ledger strip On a beam that carries joists flush with its upper edge, the strip of wood attached along the bottom edge of the beam that serves to seat the joists and to support them.

leech field A designated area of a lot where sewage is permitted to be filtered and discharged into. Leech fields are most commonly found in areas not accessible to a municipal sewer system.

L

LEED® Leadership in Energy and Environmental Design (LEED), a U.S. Green Building Council rating system for single-family, commercial, institutional, and high-rise residential buildings. Used to evaluate environmental performance from a "whole building" perspective over a building's life cycle.

left-hand stairway A stairway on which the handrail is positioned on the left-hand side in the direction of ascent.

left on the table The dollar difference between the low bid and the next bid above.

legal notice A covenant, often incorporated in the language of an agreement between two or more parties, that requires communication in writing, serving notice from one party to the other in accordance with terms of the agreement.

length The longest dimension of an object.

lessons learned A project team's learning, usually defined during close out. Should be limited to capturing/identifying work process improvements.*

let in In joinery, to fasten a timber securely in place by inserting or embedding it in another.

let-in brace A diagonal brace inserted or let in to a stud.

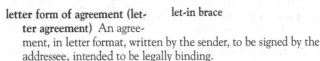

let-in brace

letter form of agreement (letter agreement) An agreement, in letter format, written by the sender, to be signed by the addressee, intended to be legally binding.

letter of credit Letter from a lender that promises to pay a beneficiary a specified sum of money for performance of a specific set of conditions, occasionally used in lieu of a payment bond.

letter of intent A letter that states the intent to enter into a formal agreement. Terms of the anticipated agreement may be stated in a general way.

level **1.** A term used to describe any horizontal surface that has all points at the same elevation and thus does not tilt or slope. **2.** In surveying, an instrument that measures heights from an established reference. **3.** A spirit level, consisting of small tubes of liquid with bubbles in each. The small tubes are positioned in a length of wood or metal that is handheld and, by observing the position of the bubbles, used to find and check level surfaces.

level A–D clothing The EPA's classification of personal protective equipment. Level A clothing offers the greatest amount of protection and should be worn whenever there is a risk to skin, eyes and the respiratory system. It utilizes an encapsulated chemical protective suit. Level D offers little protection and should only be worn on a site with no hazards.

level control Benchmarks or other devices used to identify points of known elevation on a project site.

leveling The procedure used in surveying to determine differences in elevation.

leveling plate A bearing plate set to an elevation used for setting structural steel.

leveling rod (leveling staff) A graduated straight rod used in construction with a leveling instrument to determine differences in elevation. The rod is marked in feet and fractions of feet, and may be fitted with a movable target or sighting disc.

leveling plate

leveling rule A long level used by plasterers to detect irregularities in the height of horizontal surfaces measured at various points.

level spreaders A stormwater management device installed parallel to a slope that changes concentrated flow to sheet flow.

leveraged buyout Any acquisition of a company where the buyer uses the company's internal assets and cash flows to provide the collateral for the loans required to make the acquisition.

lever arm In a structural member, the distance from the center of the tensile reinforcement to the center of action of the compression.

lever tumbler In a lock, a type of pivoted tumbler.

lewis (lifting pin) A metal device used to hoist heavy units of masonry. A lewis is equipped with a dovetailed tenon that is made in sections and fitted into a corresponding recess cut into the piece of masonry to be moved.

lewis bolt (lewis pin) **1.** A bolt shaped into a wedge at its end, which is inserted into a prepared hole in a heavy unit or stone and secured in place with poured concrete or melted lead. **2.** An eyebolt inserted into heavy stones and used in the manner of a lewis to lift and move the stones.

lewis bolt (lewis pin)

L-head The top of a shore formed with a braced horizontal member projecting from one side and forming an inverted L-shaped assembly.

liability A situation in which one party is legally obligated to assume responsibility for another party's loss or burden. Liability is created when the law recognizes two elements: the existence of an enforceable legal duty to be performed by one party for the benefit of another, and the failure to perform the duty in accordance with applicable legal standards.

liability insurance Insurance designed to safeguard the insured from liability resulting from injury to another person or another person's property.

license The permission by competent authority to do an act that, without such permission, would be illegal.

licensed contractor An individual or a firm that has, where required by law, obtained certification from a government office, to practice construction contracting.

lien A legal means of establishing or giving notice of a claim or an unsatisfied charge in the form of a debt, obligation, or duty. A lien is filed with government authorities against title to real property. Liens must be adjudicated or satisfied before title can be transferred.

life 1. That period of time after which a machine or facility can no longer be repaired in order to perform its design function properly. 2. The period of time that a machine or facility will satisfactorily perform its function without a major overhaul.*

life cycle A term often used to describe the period of time that a building can be expected to actively and adequately serve its intended function.

life-cycle assessment The determination of the environmental burdens associated with a product or process, including materials used and wastes released. The assessment considers the extraction and manufacturing of the materials, transportation and distribution, use and maintenance, and final disposal or reuse.

life-cycle costing The determination of the value of a system, such as roof covering, amortized over the projected life of the system, as opposed to the value determined by the initial cost only. Life-cycle costs include such costs as service and maintenance.

life safety code Developed by the NFPA Committee on Safety to Life, an international organization dedicated to saving lives from fire.

lift 1. The concrete placed between two consecutive horizontal construction joints, usually consisting of several layers or courses, such as in slip forming. 2. The amount of grout, mortar, or concrete placed in a single pour. 3. A British term for elevator.

lifting block A combination of pulleys or sheaves that provide a mechanical advantage for lifting a heavy object.

lift joint A surface at which two successive concrete lifts meet.

lifts (tiers) The number of frames of scaffolding erected one above each other in a vertical direction.

lift slab A method of concrete construction in which floor and roof slabs are cast at ground level and hoisted into position by jacking. Also, a slab that is a component of such construction.

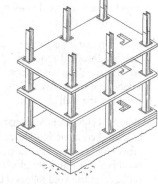

lift slab

light 1. A man-made source of illumination, such as an electric light. 2. A pane of glass.

light-framed construction Building construction that features vertical and horizontal structural components made up of wood or light-gauge steel framing.

lighting cost The total of the cost of a system's lamps, operating energy, and lamp replacement.

light

lighting outlet An electrical outlet that serves to accommodate the direct connection of a lighting fixture or of a lamp holder and its pendant cord.

lighting panel An electric panel housing fuses and circuit breakers, which serves to protect the branch circuits of lighting fixtures.

light loss factor In illumination calculations, an adjustment factor that estimates losses in light levels over time due to aging of the lamp, dirt on the room surfaces, and other causes. Losses corrected by lamp replacement or cleaning are termed recoverable. Nonrecoverable losses would be due to deterioration of the fixture or voltage drops.

lightning arrester A device that connects to and protects an electrical system from lightning and other voltage surges.

lightning conductor (lightning rod) A cable or rod built of metal that protects a building from lightning by providing a direct link to ground.

light pollution The glare from inefficient outdoor lights, especially around highly populated areas, making it difficult to discern the features of the night sky.

light shelves A daylighting system based on sun path geometry, used to bounce light off a ceiling, project light deeper into a space, distribute light from above, and diffuse it to produce a uniform light level below.

light shelves

light to solar gain ratio (LSG) The ratio of visible light transmittance to the solar heat gain coefficient (SHGC). LSG measures the ability of glazing to provide light without excess solar heat gain.

lightweight block A cement masonry unit manufactured using lightweight aggregate and often used to reduce the weight of partitions.

lightweight concrete Concrete of substantially lower unit weight than that made using gravel or crushed-stone aggregate.

light well An inside shaft with an open top through which light and air are conveyed from the outdoors to windows opening on the shaft.

lignin 1. A substance that occurs naturally in wood and joins with cellulose to make up the chief constituents of wood tissue. 2. A crystalline byproduct of paper pulp, used in manufacturing plastics, wood chipboard, and protective chemical coatings that prevent corrosion.

like-kind property In a 1031 exchange, property that is of the same nature.

lime Specifically, calcium oxide (CaO). Also, a general term for the various chemical and physical forms of quicklime, hydrated lime, and hydraulic hydrated lime.

lime-and-cement mortar A lime, cement, and sand mortar used in masonry and cement plaster. In addition to imparting a favorable consistency to the mix, the lime also increases the flexibility of the dried mix, thus limiting cracks and minimizing water penetration.

L

lime concrete A lime, sand, gravel, and concrete mix made without Portland cement. Lime concrete is found in older structures, but is no longer in general use.

lime putty A thick lime paste used in plastering, particularly for filling voids and repairing defects.

limestone A sedimentary rock composed mostly of calcium carbonate, calcium, or dolomite. Limestone can be used as a building stone, with its commercial grades being A. Statuary, B. Select, C. Standard, D. Rustic, E. Variegated, and F. Old Gothic. It is also crushed into aggregate, crushed for agricultural lime, or burned to produce lime.

limit control A safety device for a variety of mechanical systems that detects unsafe conditions, sounds an alarm, and shuts off the system.

limit design Any of a number of structural design methods based on limits related to stability, elasticity, fatigue, deformation, and other structural criteria.

limited combustible material A material that is noncombustible, as defined by the National Fire Protection Agency, yet fails to meet the NFPA's definition of combustible material. Must not exceed a potential heat value of 3500 Btu per pound.

limit of liability The greatest amount of money that an insurance company will pay in the event of damage, injury, or loss.

limit switch **1.** An electrical switch that controls a particular function in a machine, often independently of other machine functions. **2.** A safety device, such as a switch that automatically slows down and stops an elevator at or near the top or bottom terminal landing.

line A marked or defined limit or border.

lineal foot A straight-line measurement of one foot, as distinguished from a cubic foot volume or a square foot area.

linear diffuser An elongated diffuser with parallel slots with deflectors to divert airflow in various directions.

linear heat detector A cable-like component of a fire detection system that detects heat anywhere along its length.

linear measurement (long measure) A unit or system of units for measuring length: 12″ = 1′, 3′ = 1 yard, 1 yard = 0.9144 meters, 1 mile = 5,280′.

linear prestressing Prestressing as applied to linear structural members, such as beams, columns, etc.

line drilling In the blasting of rock, the boring of a series of holes along the desired line of breakage. Holes are spaced several inches apart to create a plane of weakness.

line drop A decrease in voltage caused by the resistance of conductors in an electric circuit.

line item Any item specifically called for on a plan or specification price-out sheet and listed with all of the quantities, unit prices, and extensions.

line level A spirit level used in excavation and pipe laying. Each end of the level has a hook used to hang it from a horizontal line.

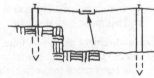

line level

line of credit A fixed amount of credit granted to cover a series of transactions.

line of levels A series of differences in elevation as measured and recorded by surveyors.

line of sight (sight line) The line extending from a telescope or other long-distance sighting device, along which distant objects can be viewed.

line pipe A welded or seamless pipe typically used to convey gas, oil, or water.

lining Any sheet, plate, or layer of material attached directly to the inside face of formwork to improve or alter the surface texture and quality of the finished concrete.

lining panel A strip that secures the lower edges of sheet metal roofing sheets along the eaves of a roof.

lining paper A waterproof or water-resistant building paper placed under siding and roofing shingles.

link The circuit that connects two points.

linked switch A series of mechanically connected electrical switches designed to act simultaneously or sequentially.

link fuse A type of exposed fuse attached to electrically insulated supports.

linoleum A form of resilient floor covering that is manufactured of ground cork and oxidized linseed oil. Linoleum is applied to a coarse fabric backing and possesses a low resistance to staining, dents, and abrasion.

linseed oil A drying oil processed from flaxseed and used in many paints and varnishes.

lintel A horizontal supporting member, installed above an opening such as a window or a door, that serves to carry the weight of the wall above it.

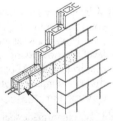

lintel

lintel block (U-block) A special U-shaped concrete block used with other blocks to form a continuous-bond beam or lintel. Reinforcing steel is placed in the void followed by mortar or grout.

lintel block (U-block)

lintel course A course of stonemasonry set level to a lintel, but different from the rest of the wall by virtue of size or finish.

lip A rounded, overhanging edge or member.

liquefaction The sudden failure of a loose soil mass due to total loss of shearing resistance. Typical causes are shocks or strains that abruptly increase the water pressure between soil particles, causing the entire mass to behave similarly to a liquid.

liquid adsorption A natural process in which molecules of a liquid are physically attracted to and held at the surface of a solid.

liquid asphaltic material An asphaltic product too soft to be measured by a penetration test at normal temperatures. The material is principally used for cement surface treatments.

liquidated damages An amount, specified in a contract for construction, to cover damages incurred by the owner as a result of the contractor's failure to complete the work within the time frame set forth in the contract.

liquid indicator A device located in the liquid line of a refrigerating system and having a sight port through which flow may be observed for the presence of bubbles.

liquid limit The water content at which soil passes from the plastic to the liquid state under standard test conditions. The limit is expressed as a percentage of the dry weight of the soil.

liquid line In a refrigeration system, the pipe transporting refrigerant away from the condenser.

liquid membrane forming compound A material sprayed or rolled on a fresh concrete surface to restrict the loss of moisture from the concrete.

liquid petroleum gas (LPG) A general term referring to propane, butane, and other similar hydrocarbons stored as liquids and used as fuel.

liquid roofing Any of a number of different liquid or semiliquid roofing materials used to create a seamless waterproof membrane.

liquid volume measurement A measurement of grout on the basis of the total volume of solid and liquid constituents.

listed Refers to approved equipment or material that has been evaluated to meet appropriate testing and standards.

liter A metric measure of capacity equal to 61.022 cubic inches, or 2.113 American pints.

litigation The process by which parties submit their disputes to the jurisdiction and procedures of federal or state courts for resolution.

litmus A piece of test paper containing a chemical indicator that changes color when exposed to liquids. Litmus paper is used to determine acidity or alkalinity values expressed as pH variations.

live 1. Descriptive of a wire or cable connected to a voltage source. 2. A descriptive term for a room with a very low level of sound absorption.

live load The load superimposed on structural components by the use and occupancy of the building, not including the wind load, earthquake load, or dead load.

live part Any part or component of an electrical device or system that is engineered to function at a voltage level different from that of the earth.

live steam Steam that has not condensed and still retains its energy, such as steam issuing from a boiler or radiator.

load 1. The force, or combination of forces, that act upon a structural system or individual member. 2. The electrical power delivered to any device or piece of electrical equipment. 3. The placing of explosives in a hole.

load-bearing tile A form of tile used in masonry walls that is capable of supporting loads superimposed on the wall structure.

load-bearing wall A wall specifically designed and built to support an imposed load in addition to its own weight.

load-bearing wall

load binder A device used to tighten chains that are holding loads in place on a truck bed.

loader A construction machine used to push or transport earth, crushed stone, or other construction materials. The bucket, or scoop, is located on the front of the vehicle and can be raised, lowered, or tilted.

loader

load factor 1. In structural design, the factor applied to the working load to determine the design's ultimate load. 2. In a drainage system, the percentage of the total flow that occurs at a particular location in the system. 3. A ratio of the average air-conditioning load on a system to the maximum capacity.

loading dock leveler Typically, an adjustable mechanized platform built into the edge of a loading dock. The platform can be raised, lowered, or tilted to accommodate the handling of goods or material to or from trucks.

loading dock seal A flexible pad installed around the door of a loading dock to form a tight seal between the receiving doors and the opening of a truck that is backed into the dock.

loading platform (loading dock) A platform adjoining the shipping and receiving door of a building, usually built to the same height as the floor of the trucks or railway cars on which shipments are delivered to and from the dock.

loading ramp A fixed or adjustable inclined surface that adjoins a loading platform and is installed to ease the conveyance of goods between the platform and the trucks or railway cars that transport goods.

loam Soil consisting primarily of sand, clay, silt, and organic matter.

local area network (LAN) A type of distance-limited communications network used for data transfer, text, facsimile, and video applications.

local buckling In structures, the failure of a single compression member. The local failure may cause the failure of the whole structural member.

location factor An instantaneous overall total project factor for translating the summation of all project cost elements of a defined construction project scope of work, from one geographical location to another. Location factors include given costs, freights, duties, taxes, field indirects, project administration, and engineering and design. Location factors do not include the cost of

land, scope/design differences for local codes and conditions, and the cost for various operating philosophies.*

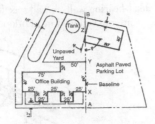

location survey The establishment of the position of points and lines on an area of ground, based on information taken from deeds, maps, and documents of record, as well as from computation and graphic processes.

location survey

lock corner A corner held together by interlocking construction of adjacent members, such as the dovetail joint on the front panel of a drawer.

lock face The surface of a mortise lock, which remains visible in the edge of a door when the lock is installed.

locking device In scaffolding, a device used to secure a cross brace to the frame or panel.

locking pliers Pliers with clamping jaws that can be tightened with a screw attachment on one handle.

lock keeper The box on a doorjamb that accommodates the extending bolt on a lock.

locknut 1. A special nut that locks when tightened so that it will not come loose. 2. A second nut used to prevent a primary nut from loosening.

lock rail On a door, the horizontal structural member situated between the vertical stiles at the same height as the lock.

lock rail

lock reinforcement A metal plate installed inside the lock stile or lock edge of a door and designed to receive a lock.

locksaw A saw used for cutting the seats for locks in doors. The saw is designed with a tapering blade that can be flexibly maneuvered.

lock seam (lock joint) In sheet metal roofing, a joint or seam formed by bending the two adjoining edges over in the form of hooks, which are interlocked. The hooks are then pressed down tightly to form a seam.

lockset A complete system including all the mechanical parts and accessories of a lock, such as knobs, reinforcing plates, and protective escutcheons.

lock stile (closing stile, locking stile, striking stile) On a door or a casement sash, the vertical member that closes against the jamb of the frame that surrounds it. The stile is located on the side away from the hinges.

lock-strip gasket (structural gasket) Typically, a thick and stiff black neoprene glazing gasket that holds and attaches panes of glass to each other or to the surrounding structure. During installation the gasket is tightened by the insertion of a wedge-like strip (the lock-strip) along the entire length of the gasket.

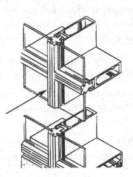

lock-strip gasket (structural gasket)

lock washer A circular washer with a break that helps it keep a nut tightly in position.

lock washer

locus A small map included on a site plan that shows the general location of a project with respect to local highways, roads, and recognized landmarks.

loess Silty material that is deposited by the wind, but maintains significant cohesion due to the presence of clay or other cementitious materials.

loft 1. Space beneath a roof of a building, most commonly used for storage of goods. 2. In a barn, the upper space at or near the ceiling with an elevated platform on which hay and grains are stored. 3. The upper space in a church or auditorium, sometimes enclosed and cantilevered, which accommodates a pipe organ or area for a choir. 4. The space between the grid and the upper part of the proscenium in a theater stagehouse. 5. Within a loft building, the unpartitioned upper spaces visible from the floor immediately below.

logic panel Any electronic control panel that is designed to perform a specific control sequence.

log mean Refers to the value of insulation thickness for curved piping in order to produce the same resistance to heat flow as a straight, flat area.

long column In structural design, a column of sufficient slenderness to necessitate a reduction of its load-bearing capacity.

long float A concrete finishing float designed to be handled by two men.

long header A header that runs a wall's full depth.

longitudinal bond In masonry, a bond in which a number of courses are laid only with stretchers and used principally for thick walls.

longitudinal bracing Bracing that extends lengthwise or runs parallel to the center line of a structure.

longitudinal joint Any joint parallel to the long dimension of a structure or pavement.

longitudinal reinforcement Steel reinforcement placed parallel to the long axis of a concrete member.

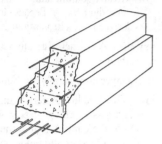

longitudinal reinforcement

long lead items Those components of a system or piece of equipment for which the times to design and fabricate are the longest and for which an early commitment of funds may be desirable or necessary in order to meet the earliest possible date of system completion.*

long-life lamp An incandescent lamp that has a lower luminous output than standard lamps of equal wattage, but a longer design life than the value set for lamps of its general class.

long-radius elbow In plumbing, a pipe elbow with a larger radius than is standard. The elbow is designed to mitigate losses from friction and to facilitate the flow of liquids through the pipe.

long-radius elbow

long screw A nipple, usually measuring 6″ long, with one thread that is longer than average.

long span The distance between supports in a structure, usually spanned by a truss or heavy timber.

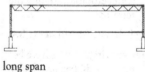

long span

long span lintel A lintel used in light-gauge metal stud framing that has a steel channel for extra support over an opening.

long span steel joist A structural framing joist that provides large open areas within a building. It is usually a very deep beam due to the nature of the building system.

long span structure A building that uses long span roof joist systems in its design to create large unobstructed areas within the structure, e.g., a domed stadium.

long ton A unit of weight equal to 2,240 pounds (1,016 kilograms).

lookout A short wooden brace or block that supports an overhanging portion of a roof.

loop Sometimes called a local line or subscriber loop, the local circuit between a subscriber station and the exchange.

looping in In interior electrical wiring, the connection of an outlet by two conductor cables, one to and one from the outlet. Splices (junction boxes) are thus avoided, but more wire is used.

loop vent In plumbing, a venting configuration for multiple fixtures, as in a public restroom. The vent pipe is connected to the waste branch in only two places, before the first and last fixtures. The fixtures are not individually vented. The two vents are connected together in a loop, and the loop is then connected to the vent stack.

loose cubic yard (meter) A unit of measure with which to express the volume of loose soil, rock, or blasted earth material.

loose-fill insulation Any of several thermal insulation materials in the form of granules, fibers, or other types of pieces that can be poured, pumped, or placed by hand.

loose knot A knot on a piece of lumber that is not fixed firmly and is likely to fall out.

loose lintel A lintel that is placed across a wall opening during construction to support the weight of the wall above, but which is not attached to another structural member.

loose material Soil, rock, or earth materials in loose form, whether blasted or broken by artificial or natural means.

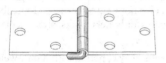

loose-pin hinge

loose-pin hinge A hinge, usually for a door, that can be separated by the removal of a vertical pin.

loose yards A term defining the cubic measurement of earth or blasted rock after excavation, as when loaded on a truck. Equal to a volume of 27 cubic feet.

Los Angeles Abrasion Test A test for abrasion resistance on concrete aggregates.

loss of prestress In prestressed concrete, the reduction in prestressing force that results from the combined effects of strain in the concrete and steel, including slip at anchorage, relaxation of steel stress, frictional loss due to curvature in the tendons, and the effects of elastic shortening, creep, and shrinkage of the concrete.

loss on ignition The percentage loss in weight of an ignited sample to constant weight at a specified temperature, usually 900°C to 1,000°C.

lot A parcel of land that is established by a survey or delineated on a recorded plot.

lot line The limit or boundary of a land parcel.

louver A framed opening in a wall, fitted with fixed or movable slanted slats. Though commonly used in doors and windows, louvers are especially useful in ventilating systems at air intake and exhaust locations.

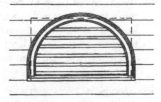

louver

louver board (luffer board) One of multiple narrow boards or slats on a louver door, window, or ventilator. The boards are installed at an angle.

louver door A door or louver, usually assembled with its blades in a horizontal position, which allows air to pass through the door when it is closed.

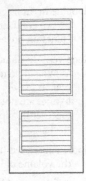

louver door

low-alkali cement A cement containing smaller-than-usual amounts of sodium and/or potassium. Its use is necessary with certain types of aggregate that would otherwise react with high levels of alkali.

low-alloy steel Steel composed of less than 8% alloy.

low bed (low boy) A flatbed trailer used to carry heavy equipment.

low bed (low boy)

low bid In bidding for construction work, the lowest price submitted for performance of the work in accordance with the plans and specifications.

low-carbon steel (mild steel) Steel with less than 0.20% carbon. This type of steel is not used for structural members, due to its ductility. It is good for boilers, tanks, and objects that must be formed.

low-density concrete Any concrete with a unit weight lower than 50 pounds per cubic foot.

low-density polyethylene (LDPE) A widely used thermoplastic that is inexpensive and easy to process. Low-density polyethylene is the softest and most flexible version of this material, and is commonly used in packaging and injection molding.

low emissivity glass (low e glass) Insulating glass with a coating or special component that keeps heat in during winter and out during summer. Low e glass reflects internal long-wave radiation, as that from a heat source, back into a home while keeping the short-wave radiation of the sun out.

lowest responsible bidder (lowest qualified bidder) The bidder who has submitted the lowest legitimate bid. The owner and architect must agree that this person (or firm) is capable of performing the work covered by the bid proposal.

lowest responsive bid The lowest bid that meets the requirements set forth in the bid proposal.

low-flow toilet An economically and environmentally efficient toilet that uses less water per flush than a conventional model.

low-hazard contents Building contents with such an exceptionally low level of combustibility that they are unable to propagate or sustain a fire in and of themselves.

low-heat cement (type IV cement) A special cement that minimizes the amount and rate of heat generation during hydration (setting). Strength is also achieved at a slower rate. Use is limited to structures involving large masses of concrete, such as dams, where the heat generated would be excessive if normal cement were used.

low-iron glass Glass with a low iron content that has a higher visible transmittance and thus a greater ability to collect solar energy.

low-lift grouting The common and simple method of unifying concrete masonry, in which the wall sections are built to a height of not more than 4' (1.2 meters) before the cells of the masonry units are filled with grout.

low-pressure mercury lamp A mercury-vapor lamp, including germicidal and fluorescent lamps, whose partial pressure during operation is no more than 0.001 atmosphere.

low-pressure sodium lamp A lamp that produces light via radiation from sodium vapor. Because it renders most colors gray, it is considered a monochromatic light source.

low steel A characteristically soft steel that contains less than 0.25% carbon.

low-temperature supply air Supply air below 50°F.

low VOC Building materials and finishes that exhibit low levels of off-gassing, the process by which volatile organic compounds (VOCs) are released from the material, impacting health and comfort indoors and producing smog outdoors. Low (or zero) VOC is an attribute to look for in an environmentally preferable building material or finish.

low-voltage lamp A lamp, typically compact halogen, that operates at 12V and requires a transformer. Popular lamps include MR11, MR16, and PAR36.

low-voltage lighting control A remote-control system that controls a number of lighting circuits. Lights are turned on or off automatically by electronic controls located within the low-voltage lighting control panel.

L runner The fastener used at the base of solid gypsum lath.

lubricant A substance used to minimize friction between two areas.

luffing boom (live boom) A crane boom with the ability to move vertically while slewing 360 degrees.

lug 1. Any of several types of projections on a piece of material or equipment. Such projections are used during handling and installation. **2.** A connector for fastening the end of a wire to a terminal.

lug bolt A bolt with a flat iron bar welded to it.

lumber Timbers that have been split or processed into boards, beams, planks, or other stock that is to be used in construction and is generally smaller than heavy timber.

lumber core (stave core) Wood door core made up of narrow strips of lumber glued together at the edges and commonly held together by a veneer, which is glued to both faces with its grain at 90° to that of the core wood.

lumber core (stave core)

lumber, matched Lumber whose side or end edges are cut to form tongue-and-groove joints that fit together when laid side by side or end to end.

lumen A unit of luminous flux that defines the quantity of light.

lumen-hour A measurement of light equal to one lumen for one hour.

lumen method A simple way to calculate the luminaires required to achieve a desired lighting intensity for an area.

luminaire A lighting fixture, with or without the lamps in it.

luminaire efficiency In lighting calculations, a special ratio of the light emitted by a light fixture to the light emitted by the lamps inside the fixture.

luminaire shielding Any device such as a louver, lens, or baffle that controls or directs light from a lamp.

luminous ceiling An area lighting system, mounted on a ceiling, that has a surface of light-transmitting materials with light sources installed above it.

luminous paint A phosphorescent or fluorescent paint that glows in the dark after exposure to direct light. Has many safety applications, including use on light switch plates, exit signs, stair edges, and fuse boxes.

lump sum An item or category priced as a whole rather than broken down into its elements.

lute 1. A long-handled scraper used to level asphalt or wet concrete. **2.** A straightedge used to strike off clay from a brick mold.

lute

lux A measure of illumination striking a surface. One lux is equal to one lumen per square meter.

L

Abbreviations

m meter

M thousand, bending moment (on drawings)

M&V measurement (or monitoring) and verification

ma milliampere

MA mechanical advantage, mixed air

MAC media access control; message authentication code; moves, adds and changes

mach machine, machinist

mag magazine, magneto

MAN manual, metropolitan area network

manuf manufacture

mas masonry

MAT mixed air temperature

mat, matl material

max maximum

mb millibar

MBF 1,000 board feet

MBH 1,000 Btus per hour

MBM, mbm thousand feet board measure

MBMA Metal Building Manufacturer's Association

MC moisture content, metal-clad, mail chute

MDF medium-density fiberboard

MDO medium-density overlay

me marbled edges

ME mechanical engineer

meas measure

mech mechanic, mechanical

med medium

memb member

MEMS micro-electro-mechanical systems

mep mean effective pressure

MEP mechanical, electrical, plumbing

MER mechanical equipment room

MERV minimum efficiency reporting value

met metallurgy

mezz mezzanine

mf mill finish

MF04 MasterFormat 2004

mfg manufactured

Mg magnesium

MG motor generator, mixed grain

mgt management

MH manhole

MHW mean high water

mi mile

mid middle

min minimum, minor, minute

MIS Management Information System

misc miscellaneous

mix, mixt mixture

mks meter-kilogram-second

ml, ML material list

mldg, MLDG molding

MLS multiple listing service

MLW mean low water

MMF magnetomotive force

Mn manganese

MN magnetic North, main

Mo molybdenum

MO month

mod, modif modification

MOD model

MOE modulus of elasticity

MOL maximum overall length

MOT motor

MOU memorandum of understanding

mp melting point

mpg miles per gallon

mph miles per hour

MPOE minimum point of entry

MPS master production schedule, materials and resources

mr moisture-resistant

MRL machine room-less (elevator systems)

MRO maintenance, repair and operations

MRP manufacturing resource planning, materials resource planning

MRT mean radiant temperature

MSDS Material Safety Data Sheet

MSF per 1,000 square feet

MSG Model Support Group of the International Alliance for Interoperability

msl mean sea level

MSR machine stress rated

MTBF mean time between (system or device) failures

mtg, mtge mortgage

MTTR mean time to restore or repair (a system or device)

mult multiple, multiplier

mun, munic municipal

MUTOA multiuser telecommunications outlet assembly

mxd mixed (lumber industry)

Definitions

macadam A method of paving in which layers of uniformly graded, coarse aggregate are spread and compacted to a desired grade. Next, the voids are completely filled by a finer aggregate, sometimes assisted by water (water-bound), and sometimes assisted by liquid asphalt (asphalt-bound). The top layers are usually bound and sealed by some specified asphaltic treatment.

machine bolt A threaded straight bolt usually specified by gauge, thread, and head type.

machine bolt

machine burn Any darkening or burning of a material during milling or other machining.

machined A term used to describe a smooth finish on a metal surface.

machine excavation Digging or scooping performed by a machine, as opposed to that performed by hand.

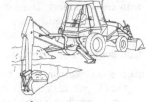

machine excavation

machine finish A finish on a stone surface produced by a smooth-edged planer.

machine rating A standard describing the power that an electric motor is designed to produce.

machine room-less elevator systems Elevators that do not require an area to house equipment.

machine stress-rated lumber Lumber that is rated mechanically by a machine that evaluates its structural properties.

made ground Land or ground created by filling in a low area with rubbish or other fill material. Often, such created land is not suitable for building without the use of a pile foundation.

magnesia Finely processed magnesium oxide.

magnesite flooring A finished surface material consisting of magnesium oxide, sawdust, and sand combined in various proportions, and subsequently applied to integral concrete floors.

magnesium A lightweight silver-colored metal that is highly flammable and immune to alkalies and is usually used in an alloy. A scale-forming element found in some boiler feed water.

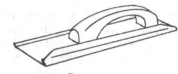

magnesium float A hand tool with a flat magnesium base used to finish concrete flatwork.

magnesium float

magnetic bearing The horizontal angle from magnetic north for a given survey line.

magnetic catch A door catch that uses a magnet to hold it in the closed position.

magnetic overload relay An overcurrent protection switch or fuse.

magnetic switch An electric switch using an electromagnet for operation.

magnetite A naturally occurring, black iron oxide used as an iron ore and as a high-density aggregate in concrete.

mahogany A straight-grained, medium-density wood originating in the West Indies, Central America, and South America, and used principally in interior plywoods and cabinetry.

mail slot A slot in a wall or door for receiving incoming mail. The slot usually has a cover to prevent draft.

main 1. In electricity, the circuit that feeds all subcircuits. 2. In plumbing, the principal supply pipe that feeds all branches. 3. In HVAC, the main duct that feeds or collects air from the branches.

main bar (main reinforcement) A reinforcing bar in a concrete member designed to resist stresses from loads and moments, as opposed to those designed to resist secondary stresses.

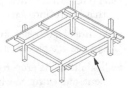

main beam A structural beam that transmits its load directly to columns, rather than to another beam.

main beam

main cross-connect The connecting point (in a structured cabling system) between entrance cables, equipment cables, and inter-building backbone cables.

main couple The main truss in a timber roof.

main office expense A contractor's main office expense consists of the expense of doing business that is not charged directly to the job. Depending on the accounting system used, and the total volume, this can vary from 2% to 20%, with the median about 7.2 percent of the total volume.

main rafter A structural roofing member that extends from the plate to the ridge pole at right angles.

main runner In a suspended ceiling system, one of the main supporting members.

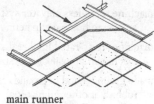
main runner

main sewer In a public sanitary sewer system, the trunk sewer into which branch sewers are connected.

main stack In plumbing, a vent that runs from the building drains up through the roof.

maintenance The process of implementing measures to conserve a site, building, structure, or object over an extended period of time to prevent deterioration.

maintenance bond A contractor's bond in which a surety guarantees to the owner that defects of workmanship and materials will be rectified for a given period of time. A one-year bond is commonly included in the performance bond.

maintenance curve For a light source, a plot of lumens vs. time.

maintenance factor In lighting calculations, the ratio of illumination of a light source or lighted surface at a given time to that of the initial illumination. This factor is used to determine the depreciation of a lamp or a reflective surface over a period of time.

main tie In a roof truss, the bottom straight member that connects the two feet.

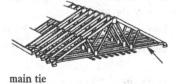
main tie

makeup air unit A unit to supply conditioned air to a building to replace air that has been removed by an exhaust system or by combustion.

makeup water Water that is added to a system to replace water that has been lost through evaporation or leaking.

male connector An electrical connector with contacts that fit into a female connector.

male nipple A short length of pipe with threads on the outside of both ends.

male plug An electrical plug that inserts into a receptacle.

male thread A thread on the outside of a pipe or fitting.

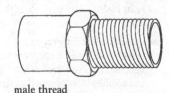
male thread

malleability The property of a metal that enables it to be hammered, bent, and extruded without cracking.

malleable iron Cast iron that has been heat-treated to reduce its brittleness.

mallet A small wooden hammer used to drive another tool, such as a chisel or a gouge.

management host computer A computer that serves as a management-machine interface in a building automation system.

mandrel A retractable insert for driving a steel pile.

manganese An alloy that is added to most steels as a hardener and deoxidizer.

manhole A vertical access shaft from the ground surface to a sewer or underground utilities, usually at a junction, to allow cleaning, inspection, connections, and repairs.

manhole cover A removable cast iron cover for a manhole. As many manholes are in paved areas, the cover must be strong enough to bear the weight of traffic.

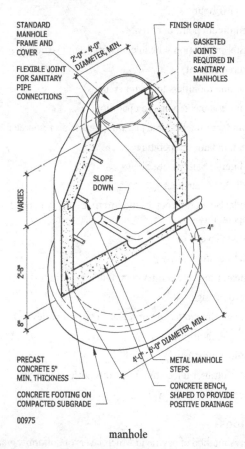
manhole

manhole frame The cast iron frame into which a manhole cover fits.

manhole invert In a sewer manhole, the elevation or grade of the inlet or outlet pipes.

manhole step A preformed metal or fiberglass step that is permanently fixed to the inside of a manhole or catch basin.

manifest 1. A list of the contents or the cargo of any shipment.
2. A specific form used by a generator of hazardous waste to track the waste from the site of generation to the site of final treatment or disposal.

M

manifold A distribution or collection pipe or chamber having one inlet and several outlets, or one outlet and several inlets.

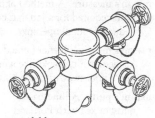

manifold

man lift A motorized scaffold characterized by a bucket or platform. Motorized buckets, also known as "cherry pickers," can hold up to three people. Motorized platform capacity is limited only by weight.

man lock A chamber in which personnel pass from one environmental pressure to another, such as when entering or leaving a caisson.

manometer A U-shaped tube filled with a liquid used to measure the differential pressure of gasses.

mansard roof

mansard roof A roof with a double pitch on all four sides, the lower level having the steeper pitch. *See also* **curb roof.**

mantel (mantelpiece) The shelf above and the finished trim or facing around a fireplace.

manual A system of controls that can be operated by hand.

manual batcher A batcher with gates and scales that can be operated by hand.

manual fire alarm box A device that initiates an alarm signal when activated by hand.

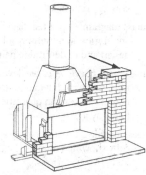

mantel (mantelpiece)

manual fire pump A pump for water supply to a sprinkler or standpipe system that must be activated by hand.

manufactured homes (mobile homes) Manufactured homes are factory-finished inside and out, and usually have wheeled chassis.

manufactured sand A fine aggregate that is produced by crushing stone, gravel, or slag.

manufactured wood A joist, truss, beam, or other product made from smaller pieces of wood, usually glued together. Manufactured wood can provide greater strength and has the environmental benefit of using waste material rather than virgin lumber.

manufacturer's specifications Documented installation or maintenance instructions produced by a product manufacturer. Often these instructions must be carefully followed in order to maintain a product warranty.

maple A hardwood that grows in North America and Europe and has a dense uniform texture. Maple is used primarily in flooring and furniture.

marble A metamorphic rock, chiefly calcium carbonate, with various impurities that give it distinctive colors. Marble is used in the architectural facing of both interior and exterior walls.

margin 1. The amount added to the cost of materials as a markup. 2. An edge projecting over the gable of a roof. 3. The space between a door and the jambs. 4. The measurement of the exposure of overlapped shingles.

marginal bar A glazing bar that separates a large glazed area in the middle of a window from smaller panes around the outside.

margin draft In stonemasonry, a dressed border on the edge of the face of a hewn stone.

margin strip In wood flooring, a narrow strip that forms a border.

margin trowel A plasterer's hand trowel on which the edges are turned up to finish plaster in corners.

marine glue A waterproof glue used on exterior plywoods and other wood-gluing applications where water may be encountered.

marine paint A paint containing elements to withstand exposure to sunlight, salt, and fresh water.

marine plywood A high-grade plywood especially adaptable to boat hull construction. All inner plies must be B grade or better.

marked face The front or veneer side of a wood building product.

market value The monetary price upon which a willing buyer and a willing seller in a free market will agree to exchange ownership, both parties knowing all the material facts but neither being compelled to act. The market value fluctuates with the degree of willingness of the buyer and seller and with the conditions of the sale.*

marking gauge A carpenter's hand tool for scribing a line parallel to an edge. The gauge has a scribe on a rod whose distance is adjustable at the head, and rides along the edge of the material.

mark out To lay out the locations where cuts are to be made on lumber.

markup A percentage of other sums that may be added to the total of all direct costs to determine a final price or contract sum. In construction practice, the markup usually represents two factors important to the contractor. The first factor may be the estimated cost of indirect expense often referred to as *general overhead.* The second factor is an amount for the anticipated profit for the contractor.

marl A silty clay, found in the bottom of lake beds or swamps, with a high percentage of calcium carbonate.

marquee A canopy extending out from an entrance for protection from the weather.

marquetry Mosaics of inlaid wood and sometimes ivory and mother of pearl.

Martin's cement Similar to Keene's cement and used in plaster. This type of cement contains potassium carbonate as an additive in place of the alum used in Keene's cement.

mash hammer A heavy, short-handled mason's hammer with two striking faces. Used with a chisel to remove old mortar between bricks in preparation for repointing. This term is also used regionally as another name for a sledgehammer.

masking The temporary covering of areas adjacent to those to which paint is to be applied. Masking is applied either by sticking something on, as with masking tape, or by covering with a firm mask.

masking tape An adhesive-backed tape used for masking that comes in rolls and various widths. The tape is applied to the surface that is to be left unpainted and removed after the painting has been done, leaving a clean, straight line.

M

mason A workman skilled in the trade of masonry and/or the finishing of concrete floors.

masonry Construction composed of shaped or molded units, usually small enough to be handled by one man and composed of stone, ceramic brick, or tile, concrete, glass, adobe, or the like. The term *masonry* is sometimes used to designate *cast-in-place concrete*.

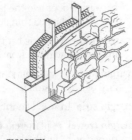

masonry

masonry anchor A metal device attached to a door or window frame that is used to secure it to masonry construction.

masonry bonded hollow wall A hollow masonry wall in which the inner and outer wythes (thicknesses) are tied together with masonry units rather than metal ties.

masonry cement A mill-mixed mortar to which sand and water must be added.

masonry fill Insulation material used to fill the voids in masonry units.

masonry filler unit Masonry units that are placed between joists or beams prior to placing the concrete for a concrete slab. The filler unit is used to reduce the amount of concrete required and the weight of the slab.

masonry insulation Sound and thermal insulation used in masonry walls. The material can be either rigid insulation, or an expanded aggregate such as perlite.

masonry nail A hard steel nail with a fluted shank that can be driven into masonry or concrete.

masonry panel A prefabricated masonry wall section that is constructed on the ground or in a shop and erected by crane.

masonry pointing Troweling mortar into a masonry joint after the masonry units have been laid.

masonry reinforcing Refers to both the lateral steel rods or mesh laid between the courses of masonry units and the vertical rods that are grouted into the voids.

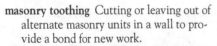

masonry reinforcing

masonry toothing Cutting or leaving out of alternate masonry units in a wall to provide a bond for new work.

masonry unit Natural or manufactured building units of burned clay, stone, glass, gypsum, concrete, etc.

masonry veneer A single wythe of masonry for facing purposes.

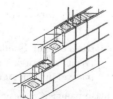

masonry unit

mason's hammer A steel hammer having one square face for striking and one curved chisel face for trimming masonry units.

mason's level Three separate levels set in a straight bar of wood or metal for determining level or plumb lines. The level is usually about 4' long.

mason's lime Lime used in preparing plaster or mortar.

mason's line A heavy string or cord used by masons to align courses of masonry.

mason's measure A method of making a quantity survey of masonry units required for a job that counts corners twice and does not deduct for small openings.

mason's miter A corner formed out of a solid masonry unit, the inside of which looks like a miter joint. The joints are actually butt joints away from the corner.

mason's putty A lime-based putty mixed with Portland cement and stone dust, and used in ashlar masonry construction.

mason's scaffold A self-supporting scaffold for the erection of a masonry wall. The scaffold must be strong enough to support the weight of the masons, the masonry units, and the mortar tubs during construction.

mason's scaffold

mass Property of a body that resists acceleration and produces the effect of inertia. The weight of a body is the result of the pull of gravity on its mass.

mass diagram A plotted diagram of the cumulative cuts and fills at any station in a highway job. The diagram is used in highway design and to determine haul distances and quantities.

mass haul curve A curve developed from the mass diagram to display haul distances and quantities in a highway job.

mass profile A road profile graphically showing volumes of cut and fill between stations.

mass shooting The simultaneous detonation of explosives in blast holes, as opposed to detonation in sequence with delay caps.

mast The vertical member of a tower crane that carries the load lines.

mast arm The bracket attached to an exterior lamppost that supports a light.

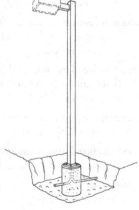

mast arm

master A term applied to the third and highest level of achievement for a tradesman or mechanic, who by supervision, experience, and examination has earned a master's license attesting that he is a master of the trade and no longer requires supervision of his work, as is the case with the journeyman and apprentice levels.

Master Builder A term applied to one who performs the functions of both design and construction. The Master Builder approach to building construction has been a practice commonplace in much of the world for many centuries. In the United States, design and construction are traditionally seen as two separate and distinct functions.

master clock system An electrical system that synchronizes all the clocks in a building.

MasterFormat The name owned and created by the Construction Specifications Institute (CSI) of the United States and

Construction Specifications Canada (CSC) denoting a numerical system of organization for construction-related information and data, based on a 16-division format.

master key A key that operates all the locks in a master-keyed series.

master-keyed lock A locking system intended for use in a series, each lock of which may be actuated by two different keys, one capable of operating every lock of the series, and the other capable of operating only one or a few of the locks.

master plan A zoning plan of a community classifying areas by use, or zoning code used as a guide for future development.

master plumber A plumber licensed to install and assume responsibility for contractual agreements pertaining to plumbing, and to secure needed permits. A journeyman plumber is licensed to install plumbing only under the supervision of a master plumber.

master schedule The most complete schedule for a project, it covers not only the construction portions, but also items that are not strictly construction-related, such as financing deadlines and community board reviews. The master schedule includes all of the details of the project, but can be presented in a summary or executive-level format, with the ability to "drill down" into specific parts to get more detailed information, as needed.

MASTERSPEC® A widely used master specification system, developed by the American Institute of Architects, for design professionals and members of the construction and building industries.

master switch An electrical switch that controls two or more circuits.

mastic 1. A thick bituminous-based adhesive used for applying floor and wall tiles. 2. A waterproof caulking compound used in roofing that retains some elasticity after setting.

mat 1. A heavy, flexible cover for retaining blasted rock fragments that is usually made of wire, chain, or cordage. 2. A grid of reinforcing bars.

matched boards Boards having been worked with a tongue on one edge and/or end, and a groove at the opposite edge and/or end to provide a tight joint when two pieces are fitted together.

matched boards

material cost The cost of everything of a substantial nature that is essential to the construction or operation of a facility, both of a direct or indirect nature. Generally includes all manufactured equipment as a basic part.*

material safety data sheet (MSDS) A form published by manufacturers of hazardous materials to describe the hazards thereof.

materials cage The platform on a hoist used for transporting materials to upper floors.

materials lock The chamber through which materials are passed from one environmental pressure to another.

mat foundation A continuous thick-slab foundation supporting an entire structure. This type of foundation may be thickened or have holes in some areas and is typically used to distribute a building's weight over as wide an area as possible, especially if soil conditions are poor.

matrix In concrete, the mortar in which the coarse aggregate is embedded. In mortar, the cement paste in which the fine aggregate is embedded.

matte A dull surface finish with low reflectance.

matte-surfaced glass Glass that has been etched, sandblasted, or ground to create a surface that will diffuse light.

mattock A heavy digging tool with a hoe blade on one side of the head and a pick or ax on the other.

mattress A grade-level concrete slab used to support equipment, such as transformers and air conditioning units, outside a building.

maturing The curing and hardening of construction materials such as concrete, plaster, and mortar.

maul A long-handled heavy wooden mallet.

maximum demand 1. The greatest anticipated load on an electrical system during a given period of time. 2. The greatest anticipated load on a sanitary waste system during a given period of time.

maximum density The largest unit weight to which a material may be compressed.

maximum power point (MPP) In photovoltaic systems, the point at which the most possible current is drawn from a cell, and the voltage subsequently drops off. The MPP changes slightly with temperature and intensity of sunlight. Most photovoltaic (PV) systems have power conditioning electronics, called maximum power point trackers (MPPT), that constantly adjust the voltage in order to maximize power output. Simpler systems operate at a fixed voltage close to the optimal voltage.

maximum rated load The greatest live load, plus dead load, which a scaffold is designed to carry, including a safety factor.

meager lime A low-purity lime commonly used in plaster. Contains at least 15% impurities.

mean roof height The distance from average grade to average roof elevation.

means of egress Any continuous exit path from a building to the outside.

measurement standard Any standard set to ensure that measurements are recorded in a reliable, uniform manner.

mechanic 1. A person skilled in the repair and maintenance of equipment. 2. Any person skilled in a particular trade or craft.

mechanical advantage The ratio of the weight lifted by a machine divided by the force applied.

mechanical application The placing of plaster or mortar by pumping or spraying, as opposed to placement by hand with a trowel.

mechanical bond (mechanical connection) A bond formed by keying or interlocking as opposed to a chemical bond by adhesion, as plaster bonding to lath or concrete bonding to deformed reinforcing rods.

mechanical completion Unit is essentially complete for start-up operation and test run. All major work is completed. Minor work not interfering with operation may not be completed, such as punch list and minor touchup work.*

mechanical drawing 1. A graphic representation made with drafting instruments. 2. Plans showing the HVAC and plumbing layout of a building.

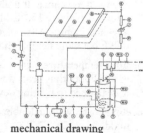
mechanical drawing

mechanical equivalent of heat A measure of mechanical energy that equates work (measured in Joules) and thermal energy (measured in calories). Equal to one Btu of heat and 778 foot-pounds.

mechanical joint A plumbing joint that uses a positive clamping device to secure the sections, such as a flanged joint using nuts and bolts.

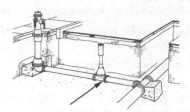

mechanical joint

mechanical plan In construction documents, the print that shows piping, ductwork, HVAC equipment, and fire safety systems.

mechanical properties The properties of a material defining its elasticity and its stress-strain relationships.

mechanical room A room or space that houses HVAC, plumbing, and electrical equipment, as well as controllers for a building automation system.

mechanical trowel A machine with interchangeable metal or rubber blades used to compact and smooth plaster.

media filtration A physical process that removes suspended solids from an aqueous waste stream by forcing the fluid through a porous medium.

medium 1. Any material used for the transmission of signals (radio, light, and sound waves). A medium could be cable or wire (radio); optical fiber (light); or water, air, or free space (sound). **2.** In paint, the liquid in which the other ingredients are suspended or dissolved.

medium carbon steel Steel with a carbon content from 0.3 to 0.6 percent.

medium curing asphalt Liquid asphalt composed of asphalt cement and a kerosene-type dilutent (thinner) of medium volatility.

medium-density fiberboard (MDF) A fiberboard made of compressed wood fibers glued together. The smooth and stable surfaces of MDF provide an excellent substrate for painting or the application of decorative lamination or wood veneers. Commonly used to manufacture furniture, cabinets and flooring systems.

medium-density overlay (MDO) An exterior-grade plywood or fiberboard with a fiber resin veneer on both sides. An ideal base for paint, MDO is commonly used for signs.

medium-density polyethylene (MDPE) A widely used, inexpensive thermoplastic that is easy to process and has good to excellent chemical resistance. It is also soft and cannot be used in temperatures much above 150°F. MDPE combines the characteristics of low- and high-density polyethylene. It is less translucent than LDPE but more flexible than HDPE.

medium-temperature water-heating system A water-heating system with a boiler that heats water to between 250°F and 350°F before it is supplied to heating devices.

meeting posts With a double gate, the stiles that meet in the middle.

meeting rail With a double-hung window, the horizontal rails that meet in the middle.

megabit Approximately one million bits.

megahertz One million hertz (Hz).

melamine A plastic laminate made by fusing a resin-impregnated surface material, under heat and pressure, into a dense, ~5EF″ board. Has superior scratch and water resistance and is not prone to delamination.

member A general term for a structural component of a building, such as a beam or column.

membrane curing A process of controlling the curing of concrete by sealing in the moisture that would be lost to evaporation. The process is accomplished either by spraying a sealer on the surface or by covering the surface with a sheet film.

membrane fireproofing A lath and plaster layer applied as a fireproofing barrier.

membrane roofing A term that most commonly refers to a roof covering employing flexible elastomeric plastic materials from 35 to 60 mils thick, that is applied from rolls and has vulcanized joints. The initial cost of an elastomeric-membrane roof covering system is higher than a built-up roof, but the life-cycle cost is lower.

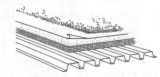

membrane roofing

membrane theory In thin-shell design, the assumption that a shell has no strength in bending because of deflection, and, that the only stresses in any section are in tension, compression, and shear.

membrane waterproofing The application of a layer of impervious material, such as felt and asphaltic cement, to a foundation wall.

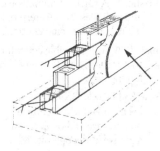

membrane waterproofing

mending plate A steel strap with predrilled screw holes used to span and strengthen wood joints.

mensuration The determination of length, area, and volume.

merchant pipe A lightweight pipe that is up to 8% lighter than standard pipe.

mercury switch An electrical switch that contains mercury in a vial to make a silent contact.

mesh 1. A network of wire screening or welded wire fabric used in construction. **2.** The number of openings per lineal inch in wire cloth.

mesothelioma A rare form of cancer linked in almost all cases to asbestos exposure.

metal-clad fire door A flush door with a wood core or stiles and rails and heat-insulating material covered with sheet metal.

metal crating Open metal flooring for pedestrian or vehicular traffic used to span openings in floors, walkways, and roadways.

metal curtain wall A metal exterior building wall that is attached to the structural frame but does not support any roof or floor loads.

metal deck Formed sheet-metal sections used in flat-roof systems.

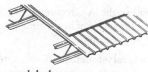
metal deck

metal framing Metal framed partitions commonly used for fire-rated construction

metal gutter Typically, a preformed aluminum or galvanized steel trough attached at the eaves of a sloped roof.

metal halide lamp A high-intensity discharge (HID) lamp that produces light from a metal vapor such as mercury or sodium.

metal lath Any of a variety of metal screening or deformed and expanded plate used as a base for plaster. The metal lath is attached to wall studs or ceiling joists.

metal framing

metallic insulation A heat shield of thin metal applied over insulating board or sheathing.

metallic paint A paint containing metal flakes that reflect light.

metal nosings Metal enclosures over the cut ends of acoustic lining sections in ductwork.

metal pan A form used for placing concrete in floors and roofs. A metal pan may also be made of molded fiberglass.

metal primer The first coat of paint on a metal surface. Primer usually contains rust inhibitors and/ or agents to improve bonding.

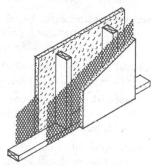
metal lath

metal roof panel A metal roof sheet that interlocks and has a weather exposure after installation less than 3 SF per sheet.

metal roof panel

metal sash block A concrete masonry unit with a groove in the end into which a metal sash can fit.

metal track In metal-stud construction, a U-shaped channel member used along floors, ceilings and walls as a brace for metal studs.

metal trim Grounds, angle beads, picture rails, and other metal accessories that are attached prior to plastering.

metal valley A roof valley gutter lined with sheet metal flashing.

metal wall ties The prefabricated metal strips that secure a masonry veneer to a structural wall.

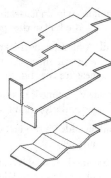
metal wall ties

metal window A solid metal-framed window such as those used in factories.

metamorphic rock A rock mass whose crystalline structure, composition, or texture has been altered by great heat or pressure during its formation.

meter 1. The base unit of length in the metric system of measurement. A meter is equivalent to 39.37″. **2.** A device for measuring the flow of liquid, gas, or electrical current.

meter stop A valve in a water service line that cuts off the flow of water before it reaches the meter.

metes and bounds The limits or boundaries of property, as identified by measured distances and compass bearings.

methyl methacrylate (MMA) A rigid, transparent material widely used in the manufacture of acrylic resins and plastics, as well as in surface-coating resins, emulsion polymers, and impact modifiers.

metric ton A weight equal to 1,000 kilograms or 2,205 pounds.

mezzanine A suspended floor, usually between the first floor and the ceiling, that covers less area than the floor below.

mica A naturally occurring, clear silicate used in thermal and electrical insulation, paint suspensions, and composite roofing materials.

microcell A cell within a cellular phone network—with a low-power base station that covers a facility such as a hotel, transportation hub, or mall.

microlam Refers to high-strength, construction-grade engineered wood beams constructed of wooden strands bonded with adhesive under pressure. Also known as laminated veneer lumber (LVL).

micrometer 1. Instrument used for accurately measuring extremely small distances. **2.** Unit of linear distance equal to one-millionth of a meter.

micron A metric unit of measure equal to one-millionth of a meter, commonly used in particle measurement. One micron is approximately 1/25,400 (0.00004) of an inch.

microwave A very short electromagnetic wave with a wavelength range from one millimeter to one meter.

microwave motion detector A device that detects motion by sending out a microwave signal and looking for a return. A moving object will reflect back some microwave energy, triggering a response. Applications include security, lighting and automatic door systems.

middle lap joint A T-shaped joint formed when the end of one member and the middle of another are joined. Each member is halved along the area of the joint to form a flush connection.

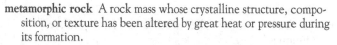

M

middle strip In flat-slab framing, the slab portion that occupies the middle half of the span between columns.

mil A measure of thickness typically used to describe materials such as plastic sheeting, trash bags, or vinyl, equal to 0.001″. Used to measure coating, wire, and material thickness.

mildew A fungus that grows on damp fabric and other materials, particularly when there is a lack of air circulation.

mildewcide A product used to retard or prevent the growth of mildew.

mild steel A steel having a low carbon content, and therefore being relatively soft and ductile. This type of steel is sometimes used for the manufacture of boilers and tanks, but not for structural beams, columns, or lintels. Mild steel is not as resistant to corrosion as wrought or cast iron and corrodes faster if not regularly maintained.

mile A distance measure equal to 5,280 feet, 1,760 yards, or 8 furlongs.

mileage tax A license tax levied on intrastate transportation business to compensate for use of the state's public roads.

milestone A specific, predetermined measurement of the completion of a project typically used for progress payment purposes.

milkiness A whitish haze caused by moisture and often occurring in a varnish finish.

milk of lime A hydrated lime slaked in water to form a lime putty.

mill To shape metal or wood to a desired dimension by a machine that removes excess material.

mill construction Historically, a type of construction used for factories and mills, and consisting of masonry walls, heavy timbers, and plank floors.

Miller Act A federal labor law that requires general contractors working on federally funded construction projects to obtain performance bonds and labor and material payment bonds to protect the interests of subcontractors and suppliers. The Miller Act applies to all U.S. government construction contracts valued at more than $25,000.

mill finish The type of finish produced on metal by the extrusion or cold rolling of sections.

milling 1. In metal, the process of shaping an item by rotary cutting machines. **2.** In stonework, the shaping of a stone to the desired dimensions.

mill length (random length) Refers to length of pipe, usually for power plant or oil field use, often made in double random lengths of 30′ to 35′. (The usual run-of-the-mill pipe is 16′ to 20′ in length.)

mill run Products from a mill, such as a sawmill, that have not been graded or sized.

mill scale A thin, loose coat of iron oxide that forms on iron or steel when heated.

millwork All the building products made of wood that are produced in a planing mill, such as moldings, door and window frames, doors, windows, blinds, and stairs. Millwork does not include flooring, ceilings, and siding.

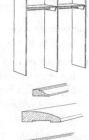

millwork

millwright A carpenter skilled in the layout, installation, and alignment of heavy equipment such as that used in manufacturing.

mineral admixture A substance added to concrete as both a filler, improving the physical structure by occupying the spaces between the cement particles, and as a "pozzolan," reacting chemically to impart far greater strength and durability.

mineral dust Aggregate passing the No. 200 screen, usually a by-product of crushed limestone or traprock.

mineral fiber insulation Insulation primarily composed of rock, slag, or glass fibers.

mineral fiber tile A preformed ceiling tile composed of mineral fiber and a binder with good acoustical and thermal properties.

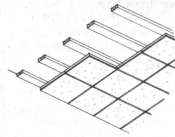

mineral fiber tile

mineral-filled asphalt Asphalt with mineral dust in suspension to improve its body and plasticity.

mineral-insulated cable Seamless copper tubing carrying one or more conductors that is embedded in refractory minerals and used in areas that may be subjected to high heat.

mineral spirits A liquid paint thinner and solvent obtained from a petroleum distillation process.

mineral-surfaced felt A roofing felt used on flat or sloped roofs that has a mineral-aggregate surface that improves its wearing and heat-reflecting properties.

mineral wool Fibers formed from mineral slag, the most common being glass wool, which is used in loose or batt form for thermal and sound insulation and for fireproofing.

minimum acceptable pressure The lowest pressure in an air and water distribution system that still allows for safe and efficient operation.

Minimum Efficiency Reporting Value (MERV) A classification used for air filters, determined by a standard ASHRAE test. A higher number MERV indicates a filter that traps more contaminants.

minimum point of entry (MPOE) The point at which carrier lines terminate, usually 12" inside of a building's foundation wall.

minimum wage law Common term used to describe the Fair Labor Standards Act enacted by Congress in 1938. This act established a minimum wage for workers and the 40-hour workweek.

minor change A job change requiring field approval only. No change order is necessary.

minor diameter The smallest diameter of a screw thread.

minority business enterprise (MBE) A business that is at least 51% owned and operated by African Americans, Asian Americans, Hispanic Americans and/or Native Americans.

minute An angle measurement equal to 1/60 of a degree.

mission tile A clay roofing tile shaped like a longitudinal segment of a cylinder. The tile is used on sloped roofs with the concave side alternately up, then down.

miter box A device used by a carpenter or cabinetmaker to cut the bevels for a mitered joint.

miter clamp A clamping device that holds a mitered joint during fastening and gluing.

miter cut The beveled cut, usually 45°, made at the end of a piece of molding or board that is used to form a mitered joint.

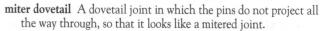

mission tile

miter dovetail A dovetail joint in which the pins do not project all the way through, so that it looks like a mitered joint.

mitered hip A roofing hip that has been close cut.

mitered valley A roofing valley that has been close cut.

miter gauge A gauge that measures the angle of a miter.

miter joint A joint, usually 90°, formed by joining two surfaces beveled at angles, usually 45° each.

miter knee The miter joint formed when the horizontal handrail at a landing is joined to the sloping handrail of the stairs.

miter plane A carpenter's planing tool used for preparing the surfaces for miter or butt joints.

miter rod A flat metal plate with one end cut at a 45° angle that plasterers use to finish inside corners.

miter square A carpenter's square with one edge having an angle of 45° for laying out miter joints.

miter valve A valve with a disk that sits at a 45° angle to the valve axis.

mitigation of damages A duty that the law imposes on an injured party to make a reasonable effort to minimize his or her damages after an injury.

mix design The selection of specific materials and their proportions for a concrete or mortar batch, with the goal of achieving the required properties with the most economical use of materials.

mixed in place An asphalt course of mineral aggregate and emulsified asphalt mixed at the site by special road-mixing equipment.

mixer A machine for blending the ingredients of concrete, mortar, or grout. Mixers are divided into two categories: batch mixers and continuous mixers. Batch mixers blend and discharge one or more batches at a time, whereas continuous mixers are fed the ingredients and discharge the mix continuously.

mixer

mixer truck See **transit mix**.

mixing box In HVAC systems, a chamber, usually located upstream of the filters, that collects outside air and return air.

mixing cycle The elapsed time between discharges of a batch mixer.

mixing speed In mixing a batch of concrete, the rate of rotation of a mixer drum or of the mixing paddles expressed in revolutions per minute (RPM). The rate can also be expressed as the distance traveled, in feet per minute (FPM), of a point on the circumference of a mixer drum at its maximum diameter.

mixing valve A valve that mixes two liquids or a liquid and a gas, such as steam with water or hot water with cold water.

mixing water The water used in mixing a batch of concrete, mortar, or grout, exclusive of water previously absorbed by the aggregates. As a general rule, the water should be clean enough to drink.

mix proportions The quantities of cement, coarse aggregate, fine aggregate, water, and other additives in a batch of concrete by weight or volume.

mobile crane Any crane mounted on wheels or tracks. They are classified by lifting capacity.

mobile gantry A movable framework housing a work platform or means to support equipment.

mobile hoist A personnel or material platform hoist that can be towed to and around the site on its own wheels.

mobile scaffold A scaffold that can be moved on wheels or casters.

model 1. A scale representation of an object, system, or building used for structural, mechanical, or aesthetic analysis. **2.** A compilation of parameters used in developing a system.

model codes Professionally prepared building regulations and codes, regularly attended and revised, designed to be adopted by municipalities and appropriate political subdivisions by ordinance. Model codes are used to regulate building construction for the welfare and safety of the general public.

model server See **BIM server**.

modem Short for *modulator/demodulator*. An electronic device that transmits data to or from a computer via telephone lines. A modem translates between digital signals (used by data processing equipment) and analog signals (used in voice switching equipment). Once the data is sent through the telephone lines, it is retranslated back into digital form in the computer at the receiving end.

modification (to construction contract documents) 1. A change to a contract that is made after the contract has been signed by both parties. **2.** A change order.

modified accelerated cost recovery system (MACRS) The method currently required for determining the calculation of asset depreciation for tax purposes. MACRS comprises two subsystems: the General Depreciation System (GDS) and the Alternative Depreciation System (ADS).

modified asphalt Asphalt whose binder has been modified by additives such as rubber or polymers for specific applications.

modified bitumen A heavy roofing material employing multiple layers of asphalt and reinforcers around a core of plastic or rubber modifiers. Installed with a special torching apparatus, a cold adhesive, or hot-mopped into place using methods of asphalt application.

modular masonry unit A brick or block manufactured to a modular dimension of 4″.

modular ratio The ratio of the modulus of elasticity of steel to that of elasticity of concrete, denoted by n in the formula: $n = E_s/E_c$.

M

modulation The tendency of a control to adjust by increments and decrements, rather than in an "on/off" fashion.

modulus of compression The measure of a material's resistance to inward pressure.

modulus of elasticity The unit stress divided by the unit strain of a material that has been subjected to a strain below its elastic limit.

modulus of resilience The measure of the elastic energy absorbed by a unit volume of a material when it has reached its elastic limit in tension.

modulus of rigidity The measure of a material's resistance to shear. The ratio of unit shearing stress to unit shearing strain.

modulus of rupture The measure of a beam's maximum load-carrying capacity. The ratio of the rupture's bending moment to the beam's section modulus.

modulus of toughness The measure of energy per unit volume that is absorbed by a material when subject to impact, up to the point of fracture.

mogul base A screw-in type base for a large incandescent lamp usually of 300 watts or more.

Mohs scale An arbitrary scale devised to determine the relative hardness of a mineral by its resistance to scratching by another mineral. Talc is rated No. 1, and diamond is rated No. 10.

moist room An enclosure maintained at a given temperature and relative humidity and used for curing test cylinders of concrete or mortar.

moisture barrier A dampproof course or vapor barrier, but not necessarily waterproof. *See also* **vapor barrier**.

moisture content The weight of water in materials such as wood, soil, masonry units, or roofing materials, expressed as a percentage of the total dry weight.

moisture gradient The difference in moisture content between the inside and the outside of an object, such as a wall or masonry unit.

moisture migration The movement of moisture through the components of a building system such as a floor or wall. The direction or the movement is always from high-humidity areas to low-humidity areas.

moistureproofing The application of a vapor barrier.

moldboard The curved shape or blade of a plow, bulldozer, grader, or other earth-moving equipment.

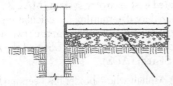

moistureproofing

molded brick Brick that has been cast rather than pressed or cut, often with a distinct design or shape.

molded-case circuit breaker An electrical circuit breaker manufactured in a molded insulated housing.

molded insulation Thermal insulation premolded to fit plumbing pipes and fittings. Common materials are fiberglass, calcium silicate, and urethane foams, with or without protective coverings.

molded plywood Plywood that has been permanently shaped to a desired curve during curing.

molding An ornamental strip of material used at joints, cornices, bases, door and window trim, and the like, and most commonly made of wood, plaster, plastic, or metal.

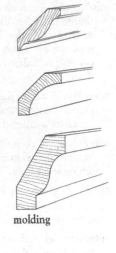

molding

mole An excavation machine used to bore tunnels.

molecular sieve A device for selective collection, by adsorption, of a substance in a gas or a liquid.

moler brick Brick, made from moler earth or diatomite, that has better insulation properties than common brick.

moment An applied load or force that creates bending in a structural member. It is numerically expressed as the product of the force times the length of the lever arm, and given in units such as foot-pounds.

moment connection A rigid connection between structural members that transfers moment from one member to the other, and thus resists the moment force. A pinned connection cannot resist moment forces, only shear forces.

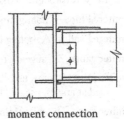

moment connection

moment of inertia In a structural member, the product of each element of mass times the square of the distance from an axis.

momentum The mass times the velocity of a moving object.

money damages A monetary award that a party who has breached a contract is ordered by a court to pay as compensation to the non-breaching party.

monitor **1.** A raised section of a roof, often along the ridge of a gable roof, with louvers or windows in the side for ventilation or light. **2.** In closed-circuit television, a video display device used to check the quality of a picture or image transmitted by a camera.

monkey wrench A wrench with a fixed jaw and an adjustable jaw.

monolithic In glazing, a window containing only one light or pane.

monolithic concrete Concrete that has been cast continuously with no joints other than construction joints.

monolithic construction The pouring of concrete grade beam and floor slab together to form a building foundation.

monolithic surface treatment A concrete finish obtained by shaking a dry mixture of cement and sand on a concrete slab after strike-off, then troweling it into the surface.

monolithic terrazzo Terrazzo applied directly over a concrete surface instead of over a mortar underbed.

monopost An adjustable metal column that supports a bearing point or beam.

Monotube® pile A cold-processed steel pile with a fluted cross section for use in deep foundations.

mop plate A protective plate at the bottom of a door, such as a kickplate.

mop sink A low, deep sink used by janitors.

mop sink

moratorium A temporary denial of permission to develop, used by local government to create an opportunity to formulate permanent growth policies and plans for an area.

mortar 1. A plastic mixture used in masonry construction that can be troweled and hardens in place. The most common materials that mortar may contain are Portland, hydraulic, or mortar cement, lime, fine aggregate, and water. 2. The mixture of cement paste and fine aggregate that fills the voids between the coarse aggregate in fresh concrete.

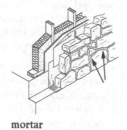

mortar

mortar aggregate (mortar sand) Natural or manufactured fine aggregate, usually washed screened sand.

mortar bed A layer of fresh mortar into which a structural member or flooring is set.

mortarboard A board about 3 feet square on which mortar is placed for use by a mason on a scaffold.

mortar box A shallow box in which mortar or plaster is mixed by hand.

mortar mill (mortar mixer) A machine with paddles in a rotating drum for mixing and stirring mortar.

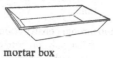

mortar box

mortise 1. A recess cut in one member, usually wood, to receive a tenon from another member. 2. A recess such as one cut into a door stile to receive a lock or hinge.

mortise and tenon joint A joint between two members, usually wood, which incorporates one or more tenons on one member fitting into mortises in the other member. Used on joints such as door stiles, door rails, window sashes, and cabinetry.

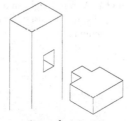

mortise and tenon joint

mortise chisel A carpenter's tool for cutting mortises.

mortise gauge A carpenter's tool for scribing the location on mortises. It is similar to a marking gauge, but scribes two parallel lines.

mortise machine A power-driven machine for cutting rectangular or round mortises in a wood member.

mortise pin A pin that secures a mortise and tenon joint by being driven through either the extension of the tenon or through the whole joint.

mosaic 1. An aerial photographic map pasted up using the center portion of overlapping vertical photographs. 2. A design created by inlaying pieces of stone, glass, or tile in a mortar bed. 3. A design of inlaid pieces of wood.

motion-detection equipment A device that detects motion by sending out a fixed electromagnetic or ultrasonic signal. A moving object will disrupt the frequency of the signal, triggering a response. Applications include security, lighting and automatic door systems.

motor controller A device that controls the power delivered to a motor or motors.

mouse 1. A device with a piece of curved lead and string for pulling a sash cord over a pulley. 2. A handheld device that is moved around on a desk or surface to control the cursor in a computer and select functions.

motor controller

movable form A large prefabricated concrete form of a standard size that can be moved and reused on the same project. The form is moved either by crane or on rollers to the next location.

movable partition A non-load-bearing demountable partition that can be relocated and can be either ceiling height or partial height.

moving ramp A continuously moving belt or other system designed for carrying passengers on a horizontal plane or up an incline.

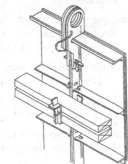

movable form

M-roof Any roof that incorporates two adjacent pitched roofs in resemblance to the letter "M."

muck 1. A soil high in organic material, often very moist. 2. Any soil to be excavated.

mucking Rearward removal of heading excavation.

mud 1. Soil containing enough water to make it soft and plastic. 2. A slang expression for joint compound.

mudjacking A method of raising a depressed concrete highway slab or slab-on-grade by boring holes at selected locations and pumping in grout or liquid asphalt.

mud pan Rectangular, angle-sided pan, shaped like a bread pan, used by joint finishers to handle portions of joint compound. The straight-cut lip of the pan ensures that the taping knife can be cleaned regularly.

mud rotary drilling method A drilling method that uses mud as the lubricant for the drill.

mudsill A plank or beam laid directly on the ground, especially for posts or shores for formwork or scaffolding.

mud slab A base slab of low-strength concrete from 2″ to 6″ thick placed over a wet subbase before placing a concrete footing or grade slab.

muffle A layer of grout over a plaster mold used to rough in the plaster. The muffle is chipped off when the final coat of plaster is applied.

mulch Organic material such as straw, leaves, or wood chips spread on the ground to prevent erosion, to control weeds, to minimize evaporation as well as temperature extremes, and to improve the soil.

mullion The vertical member separating the panels or glass lights of a window or door system.

multiconductor An electrical cable containing multiple wires inside a single outer jacket.

multimedia filter A water-supply system filter that incorporates several different filtering elements or materials.

M

multiple glazing Any glazing that incorporates two or more sheets of glass, as in a thermal- or sound-insulating application.

multiple-layer adhesive An adhesive film used as a bond between dissimilar materials.

multiple of direct salary expense An accounting method used to pay for professional services. It is based on direct salary expenses multiplied by a factor that accounts for the cost of benefits that are linked to direct salary as well as indirect expenses, other direct expenses, and profit.

multiple prime contract A contract used when one or more constructors are employed under separate contracts to perform work on the same project, either in a sequence or coincidentally.

multiple surface treatments A term applied to successive pavement surface treatments of asphaltic materials and aggregate.

multiplex To send signals from more than one source simultaneously over a single channel.

multiplier A factor used to adjust costs for modifications, such as for location, time, or size of project.

multistage stressing The prestressing of concrete members in stages as the construction progresses.

multiunit wall A masonry wall with two or more wythes.

multiuser telecommunications outlet assembly A connecting point within a structured cabling system; for multiple users for a work area.

multizone system **1.** An air-conditioning system that is capable of handling several individual zones simultaneously. **2.** A heating or HVAC system having individual controls in two or more zones in a building.

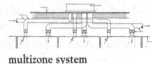

multizone system

multizone units Air-handling units with parallel heating and cooling air paths providing individual mixing of air-distribution circuits into a single duct for each zone.

municipal lien A claim or lien filed by a municipality against a property owner for collection of the property owner's proportionate share of a public improvement made by the municipality that also improves the property owner's land.

muntin A short vertical or horizontal bar used to separate panes of glass in a window or panels in a door. The muntin extends from a stile, rail, or bar to another bar. *See also* **vertical bar**.

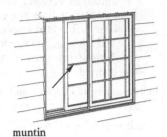

muntin

muriatic acid An acid mixed with water to clean masonry. Muriatic acid is also commonly used to etch smooth concrete surfaces for painting.

mushroom The spreading out of concrete at the top of a caisson pier, causing it to be wider than the thickness of the foundation wall.

mushroom construction A system of flat-slab concrete construction, with no beams, in which columns are flared at the top to resist shear stresses near the column head.

mylar A thin, tough, smooth polyester film used as a drafting medium.

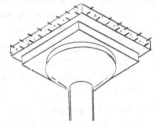

mushroom construction

M

Abbreviations

n, N noon, number, North, nail, nitrogen, normal

Na sodium

NAAMM National Association of Architectural Metal Manufacturers

NAIOP National Association of Industrial Office Properties

NAT natural

NBC National Building Code

NBFU National Board of Fire Underwriters (now merged into the American Insurance Association)

NBIMS National Building Information Model Standard

nbm net board measure

NBS National Bureau of Standards

NC noise criterion

NCM noncorrosive metal

NCX fire-retardant treated wood

NDT nondestructive testing

NEC National Electrical Code

NECA National Electrical Contractors Association

NELMA Northeastern Lumber Manufacturers Association

NEMA National Electrical Manufacturers Association

NESC National Electrical Safety Commission

NESHAPS National Emission Standards for Hazardous Air Pollutants

NFC National Fire Code

NFIP National Flood Insurance Program

NFoPA National Forest Products Association

NFPA National Fire Protection Agency

NFRC National Fenestration Rating Council

NHLA National Hardwoods Lumber Association

NHPMA Northern Hardwood and Pine Manufacturers Association

Ni nickel

NIBS National Institute of Building Sciences

NIC not in contract

NIOSH National Institute of Occupational Safety and Health

NLRA National Labor Relations Act

NLRB National Labor Relations Board

NM nonmetallic sheathed

NOC network operations center

NOM nominal

NOP not otherwise provided for

norm normal

NPS nominal pipe size

NPV net present value

nr near, noise reduction

NRC noise reduction coefficient

NS not specified

ntp normal temperature and pressure

NTS not to scale

nt. wt., n.wt. net weight

num numeral

NwFA National Wood Flooring Association

NWWDA National Wood Window and Door Association

N1E nosed one edge (lumber industry)

N2E nosed two edges (lumber industry)

Definitions

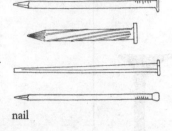

nail A slender piece of metal with a point on one end that is driven into construction materials by impact. Nails are classified by size, shape, and usage.

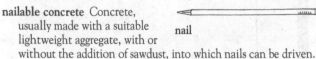
nail

nailable concrete Concrete, usually made with a suitable lightweight aggregate, with or without the addition of sawdust, into which nails can be driven.

nailer A strip of wood or other fitting attached to or set in concrete, or attached to steel to facilitate making nailed connections.

nailing block (nog) A wood block set into masonry or steel and used to facilitate fastening other structural members by nailing.

nailing ground A nailing strip to which trim is attached.

nailing schedule A specific pattern or configuration of nails, most often used for fastening components of a structural assembly.

nail pop The protrusion of a nail from a wall or ceiling, usually attributed to the shrinkage of or use of improperly cured wood framing.

nail puller One of a variety of hand tools for pulling nails. The shape and size depends on the nails to be pulled.

nail puller

nail set A handheld, tapered steel rod specifically designed to drive nail heads below the surface of wood. The rod is used specifically in finish carpentry work.

naphtha-based oil Petroleum oil used as an additive in herbicides. Naphtha oil alone has herbicidal properties on some weeds and grasses.

narrow light door A door with a narrow vertical light near the lock stile.

narrow-ringed timber Lumber with fine-grained, closely spaced growth rings.

National Building Information Model Standard A standard developed by a National Institute of Building Sciences Facility Information Council committee. It integrates life-cycle information for use by facility management and AEC professionals.

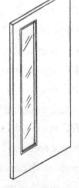
narrow light door

National Electric Code® (NEC) A nationally recognized code that addresses proper installation of electrical systems and equipment. Published by the National Fire Protection Association, the code is revised and reissued every three years.

National Emission Standards for Hazardous Air Pollutants (NESHAPS) Federal air pollution regulations instituted by the Clean Air Act and enforced by the Environmental Protection Agency (EPA).

National Fenestration Rating Council (NFRC) An organization that provides ratings for energy performance for windows and skylights, doors, and accessories.

National Fire Protection Association (NFPA) A nonprofit organization that publishes the National Electrical Code®, the Life Safety Code®, the Fire Prevention Code™, the National Fuel Gas Code®, and the National Fire Alarm Code®.

National Institute of Building Sciences A nonprofit organization that works with the government and private sector on "solutions for the built environment," including standards, new technologies, and dissemination of information.

National Labor Relations Act An act of Congress sometimes known as the Wagner Act, enacted in 1935. This act mandated a framework of procedure and regulation by which management-labor relations are to be conducted.

National Lumber Grades Authority One of seven regional grading agencies in North America that are authorized to write and publish grading rules for lumber.

native landscape Plantings that are selected because they have adapted to thrive in the local environment without irrigation, fertilizer, or pesticides, and that provide storm-water management.

natural asphalt Asphalt occurring in nature through natural evaporation of petroleum. This type of asphalt can be refined and used in paving materials.

natural cement A hydraulic cement produced by heating a naturally occurring limestone at a temperature below the melting point, and then grinding the material into a fine powder.

natural convection The movement of air resulting from differences in density usually caused by differences in temperature.

natural draft Refers to the movement of air through a cooling tower by the force of air density differential.

natural gas A naturally occurring combustible gas used for industrial and domestic heating and power.

natural grade The undisturbed elevation of a property before any excavation operations.

naturally durable wood As defined for the purpose of building codes, the heartwood of certain species of decay-resistant wood, including black walnut, cedar, redwood, and black locust.

natural resin Any naturally occurring thermoplastic substance. Used in paints and varnishes, among other applications.

natural seasoning In the lumber industry, a curing process using natural air convection.

natural stone Stone shaped and sized by nature as opposed to stone that has been quarried and cut.

natural stone

NCX fire-retardant treated wood Treated wood generally used on exterior balconies, steps, and roof systems. Carries the Underwriters' Laboratories rating of FRS.

neat 1. Idiom for exact dimensions; e.g., excavation to the designed width of the footing. 2. A term referring to a process by which a material is prepared for use without addition of any other materials except water. Examples include neat cement or neat plaster.

neat cement A cement mortar or grout made without addition of sand or lime.

neat line The line or plane defining the limits of work, particularly in excavation of earth or rock. Excavation beyond the neat line is usually not a pay item in a unit price contract.

neat plaster Plaster mixed with no aggregate.

needle 1. In underpinning, the horizontal beam that temporarily holds up the wall or column while a new foundation is being placed. 2. In forming or shoring, a short beam passing through a wall to support shores or forms during construction. 3. In repair or alteration work, a beam that temporarily supports the structure above the area being worked on.

needle nose pliers Pliers that are outfitted with long thin jaws for use in narrow spaces.

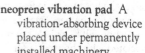

needle nose pliers

needle valve A type of globe valve in which a long pin or needle, tapered at the end, moves in and out of a conical seat to regulate the flow of liquid.

negative cash flow On a construction project, when the periodic payments a contractor receives from an owner for the performance of the contract's construction work amount to less than what he will pay out during the same period for expenses associated with the work. (The amount of money coming in is less than the amount going out.)

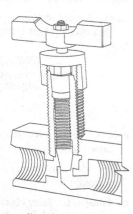

needle valve

negative float The amount of delay that a task on the critical path of the schedule has incurred, expressed in days.

negative friction The additional load placed on a pile by the settling of fill placed around it. The effect of negative friction is to pull the pile down.

negative pressure ventilation system A method of providing low-velocity airflow from uncontaminated areas into contaminated areas by means of a portable exhaust system equipped with HEPA filters.

negligence The failure of a party to conform its conduct to the standard of care required by law. The law requires that a person exercise that degree of care which a reasonable person would exercise under the same or similar circumstances.

negotiated procurement A procedure used by the U.S. government for contracting whereby the government and potential contractor negotiate on both price and technical requirements after submission of proposals. Award is made to the contractor whose final proposal is most advantageous to the government.

negotiation A process used to determine a mutually satisfactory contract sum, and terms to be included in the contract for construction. In negotiations, the owner directly selects the constructor and the two, often with assistance of the design professional, derive by compromise and a meeting of the minds the scope of the project and its cost.

neon lamp A lamp that gains its illumination by electric current passing through neon gas.

neoprene A synthetic rubber with high resistance to petroleum products and sunlight. Neoprene is used in many construction applications, such as roofing and flashing, vibration absorption, and sound absorption.

neoprene roof A roof covering made of neoprene sheet material with heat-welded joints, which can be either ballasted or nonballasted. This type of roof covering has good elastic and durability properties over a long time span.

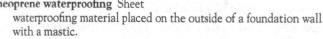

neoprene roof

neoprene vibration pad A vibration-absorbing device placed under permanently installed machinery.

neoprene vibration pad

neoprene waterproofing Sheet waterproofing material placed on the outside of a foundation wall with a mastic.

nestable joist Standard shape for a light-gauge framing joist.

nested nails Nails having a crescent-shaped piece missing in the head to allow them to be fitted tightly together.

nested studs Two studs placed together for additional support in framing an opening.

net cross-sectional area In defining a masonry unit, the gross cross-sectional area minus the area of the ungrouted cores or cellular voids.

net cut In excavation, the total cut, minus the compacted fill required, between particular stations.

net fill In excavation, the compacted fill required, minus the cut material available, between particular stations.

net floor area The occupied area of a building not including hallways, elevator shafts, stairways, toilets, and wall thicknesses. The net floor area is used for determining rental space and fire code requirements.

net load In heating calculations, the heating requirement, not considering heat losses, between the source and the terminal unit.

net metering Allowing electric meters of power-generating facilities, such as solar or wind power, to turn backward when more

energy is produced than customers consume. Net metering allows customers to use the excess energy their own system generates to offset their consumption over an entire billing period, not just at the time the electricity is produced.

net savings (NS) A measure of long-run profitability of an alternative relative to a base case. It can be calculated as an extension of the life-cycle costing (LCC) method as the difference between the LCC of a base case and the LCC of an alternative.

net site area The area of a building site less streets and roadways.

net weight The weight of an article, cargo, or other load minus the transporting vehicle.

network In CPM (critical path method) terminology, a graphic representation of activities showing their interrelationships.

network schedule A method of scheduling the construction process where various related events are programmed into a sequential network on the basis of starting and finishing dates.

neutral 1. An electrical conductor used as the primary return path for current during normal operation of an electrical device. **2.** The center tap of a three- or four-wire transformer. Neutral cable will carry current when there is an unbalanced load between wires.

neutral axis In structural design, an imaginary line in a structural member where no tension, compression, or deformation exist. If holes are to be drilled through a structural member for conduits or pipes, they should be drilled at the neutral axis.

neutral cash flow On a construction project, when the periodic payments a contractor receives from an owner for the performance of the contract's construction work equal what will be paid out during the same period for expenses associated with the work. (The amount of money coming in equals the amount going out.) This is virtually impossible to achieve.

neutralize To reduce the pH of an alkali, or to raise the pH of an acid, to approximately 7.0.

neutralizing The treatment of concrete, plaster, or masonry surfaces with an acid solution in order to neutralize the lime before application of paint.

neutralizing basin In a building drainage system, a device that neutralizes acid-bearing wastes before their entry into the system.

newel The post supporting a handrail at the top and bottom of a flight of stairs. Also, the center post of a spiral staircase.

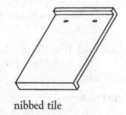
newel

N-grade wood 1. In molding, stock intended for natural or clear finishes. The exposed face must be of one single piece. **2.** In plywood, cabinet quality panels for natural finishes.

nibbed tile A small lug at the upper end of a roofing tile that hooks over a batten.

nib grade (nib guide, nib rule) A wooden straight edge nailed on the ceiling's base plaster coat as a guide for a cornice molding.

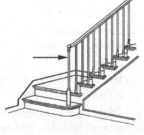
nibbed tile

nickel A silver-colored, hard ductile metal used in alloys, batteries, and electroplating. Nickel is used extensively on plumbing fixtures, where its anticorrosive properties are important.

nickel silver A silver-colored class of nickel alloys containing copper and zinc but not silver.

nickel steel A steel alloy with 3% to 5% nickel content, that gives it greater strength, ductility, and anticorrosive properties.

nidge (nig) To dress or shape the edge of a masonry stone with the sharp point of a hammer, as opposed to doing so with a chisel and mallet.

night cycle program In a building automation system, an energy-saving computer program that maintains a reduced temperature in the heating season and an elevated temperature during the cooling season during unoccupied periods.

night purge program In a building automation system, a computer program that replaces building air with cooler outdoor air during the early morning hours of the cooling season.

nighttime ventilation Cooling buildings with outside air at night to minimize the cooling load during the day. Ventilation can be achieved naturally via temperature drops, wind, cross-ventilation, and stack effect, or through use of wind towers and mechanical ventilation.

nippers A hand tool for cutting wire and small rods.

nipple A piece of pipe less than 12″ long and threaded on both ends. Pipe over 12″ long is regarded as a cut measure.

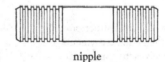
nipple

nobble The process of shaping building stones to the roughly desired dimensions while the stone is still at the quarry.

noble A technical term describing a metal's degree of resistance to corrosion.

no damages for delay clause A contract provision that states that a contractor's sole remedy for delay is an extension of time without any monetary compensation even if the delay is caused by the owner or its agents.

node 1. In critical path method scheduling, a junction of arrows containing the i-j number, early start, and late finish. **2.** In electrical wiring, a junction of several conductors.

no-fines concrete A concrete mixture with little or no fine aggregate.

no-hub pipe Cast iron pipe that is fabricated without hubs for coupling.

noise In closed-circuit television, interference detected on a cable circuit that reduces or destroys the clarity of the signal. Sometimes called snow.

noise absorption The reduction of noise in an enclosure by introducing sound-absorbing materials or methods of construction that restrict the transmission of sound.

noise criterion curve Defines the limits that the octave band spectrum of a noise source must not exceed if a certain level of occupant acceptance is to be achieved.

noise insulation Sound-absorbing materials installed in partitions, doors, windows, ceilings, and floors.

noise reduction **1.** The difference, expressed in decibels, between the noise energy in two rooms when a noise is produced in one of the rooms. **2.** The difference in noise energy from one side of a partition to another when a noise is produced on one side.

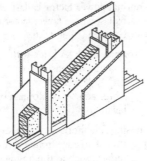
noise insulation

noise reduction coefficient The average sound absorption coefficient of a material to the nearest 0.05 at four frequencies: 250, 500, 1,000, and 2,000 cycles.

nominal Term indicating that the full measurement is not used. Usually slightly less than the full net measurement, as with 2″ × 4″ studs that have an actual size when dry of 1½″ × 3½″. Pipe size designations are also described as nominal. For example, 2″ nominal is 2~3⅛″ O.D.

nominal dimension **1.** The size designation for most lumber, plywood, and other panel products. In lumber, the nominal size is usually greater than the actual dimension, thus, a kiln-dried 2″ × 4″ ordinarily is surfaced to 1½″ × 3½″. In panel products, the size is generally stated in feet for surface dimensions and increments of ¹⁄₁₆″ for thickness. Product standards

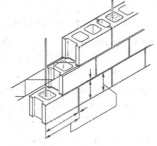
nominal dimension

permit various tolerances for the latter, varying according to the type and nominal thickness of the panel. **2.** In masonry, a dimension larger than the one specified for the masonry unit by the thickness of a joint.

nominal mix The proportions of the constituents of a proposed concrete mix.

nonagitating unit A truck-mounted container for transporting central-mixed concrete, not equipped to provide agitation (slow mixing) during delivery.

non-air-entrained concrete Concrete in which neither an air-entraining admixture nor air-entraining cement has been used.

nonbearing partition A partition that is not designed to support the weight of a floor, wall, or roof.

nonbearing wall A wall designed to carry no load other than its own weight.

noncohesive soil A soil in which the particles do not stick together, such as sand or gravel.

noncollusion affidavit A notarized statement by a bidder that the bid was

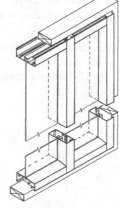
nonbearing wall

prepared without any kind of secret agreement intended for a deceitful or fraudulent purpose.

noncombustible Any material that will neither ignite nor actively support combustion in air at a temperature of 1,200°F when exposed to fire.

noncompensable delay A delay for which the contractor receives a time extension only and may not recover its delay-related costs from the owner.

nonconcordant tendons In a statically indeterminate structure, tendons that are not coincident with the pressure line caused by the tendons.

nonconductor A material that does not easily conduct electric current. Such materials are used as insulators.

nonconforming work Work that does not fulfill the contractually agreed upon requirements.

noncorrosive flux A flux used in soldering operations that does not chemically attack the base metal.

nondestructive testing The examination of an object with technology that does not affect the object's future usefulness. Applications include the evaluation of the strength of concrete or a weld using ultrasonic measures.

nondrying A term applied to a material containing oils that do not oxidize or evaporate, and therefore do not form a surface skin. The term is often applied to glazing compounds.

nonelectric delay blasting cap A detonating cap with a delay device built in so that it detonates at a designated time after receiving an impulse or signal from a detonating cord.

nonevaporable water The water that is chemically combined during cement hydration and that is not removable by specified drying.

nonexcusable delay A delay that is the fault of the contractor for which the contractor will receive neither a time extension nor compensation.

nonferrous A term referring to any metal or alloy that does not contain iron, such as brass or copper.

nonflammable A material that will not burn with a flame.

nonfreeze sprinkler system A fire protection sprinkler system that is designed to operate in freezing temperatures.

nonionic surfactant An adjuvant with no electrical charge. Nonionic surfactants are compatible with all types of pesticides.

nonmetallic sheathed cable Two or more electrical conductors enclosed in a nonmetallic, moisture-resistant, flame-retardant sheath.

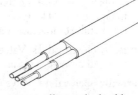
nonmetallic seathed cable

nonnailable decks In built-up roofing, a deck or substrate requiring the base sheet to be adhered rather than mechanically fastened.

non-point-source pollution Runoff contamination from an overall site or land use and not discharged from a single pipe, such as sediment from construction sites, oils from parking lots, or fertilizers and pesticides washed from farm fields.

N

nonpotable substitution system A system that uses byproduct water to replace potable water for systems that do not require fresh water.

nonpressure pipe A pipe with no pressure rating, and therefore only suitable for conveying liquid by gravity.

nonprestressed reinforcement Reinforcing steel in a prestressed, concrete structural member that is not subjected to prestressing or posttensioning.

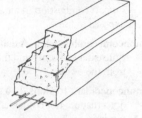

nonprestressed reinforcement

nonproduction costs Overhead costs, or the cost of supporting actual production activities.

nonrestrictive specification A type of specification that is written so as not to restrict the product to a particular manufacturer or material supplier.

nonreturn valve A check valve that allows flow in only one direction.

nonsimultaneous prestressing The posttensioning of tendons individually rather than simultaneously.

nonsiphon trap A plumbing fixture that creates a water seal that cannot be siphoned, but still allows the free flow of liquids.

nonskid floor A concrete floor surface treated with carborundum powder, iron filings, or other material to improve its traction qualities, especially when wet. The surface may also be brushed before the concrete sets to create a rough finish.

nonstaining mortar A mortar, with low free-alkali content, that avoids efflorescence or staining of adjacent masonry units by migration of soluble materials.

nontilting mixer A rotating drum concrete mixer on a horizontal axis. Concrete is discharged by inserting a chute that catches the concrete from the rotating fins.

normal consistency 1. The degree of wetness exhibited by a freshly mixed concrete, mortar, or neat cement grout when the workability of the mixture is considered acceptable for the purpose at hand. 2. The physical condition of neat cement paste determined with the Vicat apparatus in accordance with a standard method of testing.

normalizing The heating of steel or other ferrous alloys to a specified temperature above the transformation range, followed by cooling in ambient air. The process reduces the brittleness and strength of the metal, but increases its ductility.

normally closed valves Valve ports are closed to flow when external power or pressure is not being applied.

normally open valves Valve ports are open to flow when external power or pressure is not being applied.

normal power factor ballast A ballast whose power factor is between 0.4 and 0.6.

Norman brick A brick with nominal dimensions of 2¾″ × 4″ × 12″. Three courses of Norman brick lay up to 8″.

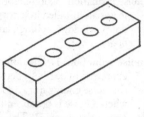

Norman brick

nosing The horizontal projection of an edge from a vertical surface, such as the nosing on a stair tread.

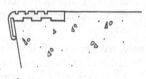

nosing

no-slump concrete Fresh concrete with a slump of 1″ or less.

notching A timber joint in which one or both of the members have a section or notch cut out.

notch joist A joist with a section cut out to fit a ledger or girder.

notice to bidders A notice included in the bidding documents that informs prospective bidders of the bidding procedures and the opportunity to submit a bid.

notice to creditors During bankruptcy proceedings, the formal notification to creditors of a meeting or the granting of an order for relief.

notice to proceed A written notice from the project owner to the contractor in which the contractor is authorized to proceed with the work on a specified date.

novation One party's agreement to release another party from a contract in exchange for a new party as substitute.

nozzle An attachment to the outlet of a pipe or hose that controls or regulates the flow.

N-truss A Pratt truss.

nuisance alarm An alarm triggered by a malfunction or incorrect information.

nurses call system An electrically operated system for use by patients or personnel for summoning a nurse.

nut A short metal block with a threaded hole in the center for receiving a bolt or threaded rod.

nylon fiber A synthetic fiber used extensively in floor coverings, wall coverings, drapery, and other furnishings.

nut

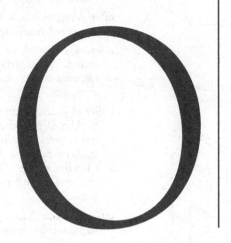

Abbreviations

O oxygen

OA overall; outside air

O/A on approval

OAI outside air intake

OA&M operation, administration, and maintenance

OAS outside air supply

OASIS Organization for the Advancement of Structured Information Standards

OAT outdoor air temperature

oBix open Building Information Xchange

OBM ordinance benchmark

OBS obsolete; open back strike

OC, oc on center

OCT octagon

OD outside diameter

ODS ozone-depleting substance

OEM original equipment manufacturer

OFCI owner-furnished and contractor-installed

OFE owner-furnished equipment

OFF office

OFOI owner furnished and owner installed

OG, og ogee

O/H, OVHD, OH overhead

OHS oval-headed screw

OJT on-the-job training

OLED organic light-emitting diode

O&M operations and maintenance

OO object-oriented; owners and operators

Opex operating expense

opp opposite

opt optional

OR outside radius, owner's risk

ord order, ordinance

ORIG original

OS operating system

OS&Y outside screw and yoke

OSCR Open Standards Consortium for Real Estate

OSHA Occupational Safety and Health Administration; Occupational Safety and Health Act

OSS operations support systems

OZ, oz ounce

Definitions

oak A hard, dense wood used for heavy framing, flooring, interior trim, plywood, and furniture. The two types available from mills are white oak and red oak.

oakum A caulking material made from hemp fibers that are sometimes saturated with tar. Oakum is commonly used with a bell-and-spigot joint in cast iron pipe. Oakum is packed into the joint with a hammer and chisel before molten lead is poured into the joint to seal it.

obligation A result of custom, law, or agreement by which an individual is duty bound to fulfill an act or other responsibility.

oblique drawings One face of an object drawn directly on the picture plane. Projected lines are drawn at a 30° or 60° angle.

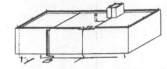

oblique drawings

oblique photograph An aerial photograph taken with the camera axis inclined away from the vertical. A high oblique photograph is one that shows the horizon; a low oblique does not include the horizon.

obscure glass Glass that transmits light but does not allow a view of objects on the other side, such as ground glass or frosted glass; translucent glass.

observation of the work The architect, during construction-phase visits to a project, observes the progress and quality of the work. The purpose of the visits is to verify that work is proceeding as specified in the contract documents.

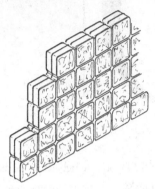

obscure glass

obsolescence 1. The condition of being out of date. A loss of value occasioned by new developments that place the older property at a competitive disadvantage. A factor in depreciation. **2.** A decrease in the value of an asset brought about by the development of new and more economical methods, processes, and/or machinery.

occupancy The designed or intended use of a building. Also, the ratio of present space being used or rented to the designed full use, expressed as a percent.

occupancy management Identifying the presence of building users in different parts of a facility, for example by using occupancy sensors, in order to save energy by automatically turning off lights in unoccupied spaces.

occupancy rate The number of persons per building, unit, room, or floor.

occupancy sensor A heat-sensitive electronic sensor that reacts to moving heat sources within its monitored field. Used as a component of security and lighting systems.

occupant load The maximum number of persons in an area at a given peak period.

occupant unit cost estimate An early-stage planning estimate based on the projected number of facility users. Costs are expressed per common unit, which may be desks for schools, seats for auditoriums, beds for hospitals, or rental rooms for hotels. Also referred to as *end product unit, end unit,* or *capacity* estimates.

Occupational Safety and Health Act The federal law governing the safety of workers in the workplace.

odorless mineral spirit A thinner used in interior paints because of its odorless qualities.

odorless paint A water-based paint, or a base using odorless mineral spirit, that produces very little odor.

off-center (out-of-center) A load that is applied off the geometric center of a structural member, or a structural member that is placed off the geometric center of an applied load.

offer A proposal, as in a wage and benefits package, to be accepted, negotiated, or rejected.

off-gassing The release of airborne particulates, often from installed construction materials such as carpeting, cabinetry, or paint, that can cause allergic reactions and other health problems in building occupants.

Office of Federal Contract Compliance (OFCC) An agency of the U.S. Department of Labor that requires contractors for federal projects to maintain affirmative action in providing equal rights for employees under the provisions of Title IV of the Civil Rights Act of 1964, which forbids discrimination by an employer on the basis of race, color, religion, sex, or national origin.

off-road hauler A heavy-duty rear dump truck or bottom dump wagon capable of routinely hauling over rough haul roads.

offset 1. In surveying, a line or point placed at a given distance from a control line or point used to reestablish the original location. **2.** In plumbing, an assembly of fittings on a pipeline that takes one section of pipe out of line but parallel to a second section. **3.** Any bend in a pipe.

offset bend An intentional distortion from the normal straightness of a steel reinforcing bar in order to move the center line of a segment of the bar to a position parallel to the original position of the centerline. The offset bend is commonly applied to vertical bars that are used to reinforce concrete columns.

offset screwdriver A screwdriver whose head is set at 90° to its shaft.

offset stakes Stakes placed by the excavator to mark the corners of a building after the surveyor's stakes have been removed.

offset stakes

ogee 1. A double curve much like the letter S. **2.** The union of concave and convex lines.

ogee molding Cornice molding (may be multiple moldings) utilizing ogee and/or reverse ogee curves to provide definition.

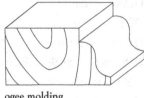

ogee molding

ohm A unit of electrical resistance in a conductor that produces a decrease in voltage of 1 volt with a constant current of 1 ampere.

ohmmeter An instrument for measuring the electrical resistance in a conductor or appliance in ohms.

Ohm's law A scientific law stating that the current in an electrical circuit is directly proportional to the voltage and inversely proportional to the resistance. Stated as an equation: I (current) = E (voltage) / R (resistance).

oil-based paint Commonly known as oil paint, any paint that has an oil base or binder along with coloring pigments. Becomes brittle and yellow with age.

oilborne preservative One of two general classifications for wood preservatives, the other being waterborne salts. Examples of oilborne preservatives include creosote and various chlorinated phenols, such as pentachlorophenol.

oil burner A fuel-oil-burning unit installed in a furnace or boiler.

oil burner

oiled and edge sealed A process used to resist moisture and preserve plywood concrete form panels. Oiling and sealing increases the number of times a panel can be used and makes the panel easier to release from the concrete.

oiler The second person assigned to a piece of equipment such as a crane. An oiler's principal job is to lubricate the equipment.

oil-filled transformer A transformer having its core and coils immersed in an insulating oil such as mineral oil.

oil paint A paint with drying oil as a base, as opposed to a water base or other base.

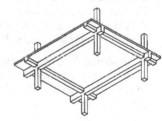

oil-filled transformer

oil separator In a refrigeration system, a device for purging the refrigerant of oil and oil vapor.

oil stain A stain, with an oil base containing dye or pigment, used to penetrate and permanently color wood or other porous materials.

oilstone A stone with a fine-grained, oil-lubricated surface used to sharpen the cutting edge of tools.

oil switch (oil-immersed switch) A switch immersed in oil or in another insulating fluid.

oil varnish A high-gloss varnish that contains a blend of drying oil and gum or resin, principally used for interior finishes.

oil/water separator A device that allows oils mixed with water to become trapped in a holding section for removal, while the water is allowed to pass through for disposal.

old wood Reused wood that has previously been worked.

olive knuckle hinge (olive butt, olive hinge) A single-pivot paumelle hinge with knuckles that join to form an oval shape.

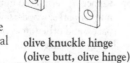

olive knuckle hinge (olive butt, olive hinge)

OmniClass A construction classification system that includes prior systems (MasterFormat, UniFormat, and EPIC).

on center **1.** A measurement of the distance between the centers of two repeating members in a structure. **2.** A term used for defining the spacing of studs, joists, and rafters.

on-demand bond A monetary form of contractor performance assurance, typically provided by a bank, where the bond is payable to the owner simply on his demand, and usually without the owner having to provide the bank any evidence or details of the contractor's failure to perform.

one by A slang term for lumber with a nominal thickness of 1″, as in a 1″ × 3″ board.

one-hour rating A measure of fire resistance, indicating that an object can be exposed to flame for an hour without losing structural integrity or transmitting excessive heat.

one-line system A graphic representation of a power distribution system using lines and symbols to indicate its major components.

one-on-two (one-to-two) A slope in which the elevation rises one foot in two horizontal feet.

one-piece toilet A toilet whose tank and bowl were manufactured as a single fixture.

one-pipe system **1.** In drainage systems, two vertical pipes with waste and soil water flowing down the same pipe and all branches connected to the same antisiphon pipe. **2.** A heating circuit in which all the flow and return connections to the radiators come from the same pipe. The radiator at the far end is therefore much cooler than the radiator that is nearest the heat source.

one-way joist construction A concrete floor or a roof framing system with monolithic parallel joists cast with the slab. The joists are supported on girders, which in turn are supported by columns.

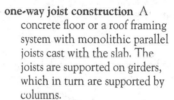

one-way joist construction

one-way slab A slab panel, bound on its two long sides by beams and on its two short sides by girders. The dead and live loads acting on the slab area may be considered to be entirely supported in the short or transverse direction by the beams; hence, the term *one-way*.

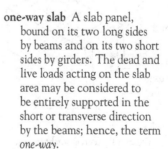

one-way slab

one-way system The arrangement of steel reinforcement within a slab that is designed to bend in only one direction.

on grade A concrete floor slab resting directly on the ground.

on-off sprinkler A fire safety system sprinkler that activates like a traditional sprinkler but ceases operation when room temperature reaches a safe level. On-off sprinklers can reduce the amount of water discharged during a fire, thus reducing cleanup and water damage.

opacity In painting, the ability of a paint to cover or hide the original background.

opalescent glaze A smooth surface with a milky appearance.

open assembly time The time required between the application of glue to veneer or to joints and the assembly of the pieces.

O

open bid An offer to perform a contract in which the bidder reserves the right to reduce his bid to compete with a lower bid.

open bidding A bidding procedure wherein bids or tenders are submitted by and received from all interested contractors, rather than from a select list of bidders privately invited to compete.

open Building Information Xchange (oBix) A standard for Web communications between building electrical and mechanical systems and applications.

open cell A cell that interconnects with other cells in foam rubber, cellular plastics, and similar materials.

open-cell foam Cellular plastic with a great many open and interconnected cells.

open-cell process A process for fixing preservative in wood under pressure. The chemical used as a preservative is retained in the cell walls only, with the cells left empty.

open circuit The absence of a direct connection between two points in an electrical network.

open construction A description of any building element that is temporarily left unfinished or exposed to allow for easy inspection.

open cut An excavation in the ground that is open to the sky at its surface, as opposed to a tunnel or horizontal mine shaft.

open cut

open defect Any hole or gap in lumber, veneer, or plywood that has not been filled and repaired.

open-end block 1. An A-block or H-block of concrete masonry. 2. A block of standard material and size built with recessed end webs.

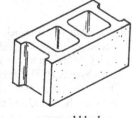
open-end block

open excavation Excavation in which no shores, piles, or sheeting are used to hold back the soil at the edge of the excavation.

open-frame girder An open-web girder or truss built with verticals connected rigidly to top and bottom chords, but with no diagonals.

open-graded aggregate 1. An aggregate that contains almost no mineral filler. 2. A compacted aggregate with relatively large void spaces.

open grain 1. Lumber that is not restricted by the number of rings per inch or rate of growth. 2. Timber with a coarse texture and open pores.

open-hole inspection Inspection by a building inspector or engineer of an open excavation, to identify, based on soil type and conditions, the type of foundation that should be installed.

opening light An operable pane or sash in a window that may be open or shut, as opposed to a fixed light.

opening protective A device installed over an opening to guard it against the passage of smoke, flame, and heated gases.

opening size The size of a door opening as measured from jamb to jamb and from the threshold or floorline to the head of the frame, allowing for the size of the door and for the necessary clearance to open and close it freely.

open joint 1. A joint in which two pieces of material that are joined together are not entirely flush. 2. A joint that is not tight.

open loop control A hydraulic control system that does not have a direct link between the valve of the controlled variable (temperature, pressure) and the controller.

open mortise A mortise or notch that is open on three sides of the piece of timber into which it is cut.

open plan A building plan that has relatively few interior walls or partitions to subdivide areas for different uses.

open plumbing Plumbing that is exposed to view beneath its fixtures, with ventilated drains and traps readily accessible for inspection and repairs.

open riser The space between two successive treads of a staircase built without solid risers.

open-riser stair A stair built without solid risers.

open roof (open-timbered roof) A style of roof with exposed rafters, sheathing, and supporting timbers visible from beneath, and no ceiling.

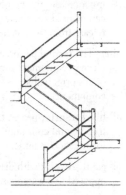
open-riser stair

open shaft A vertical duct or passage in a building that is used to ventilate interior spaces, drawing in outside air from its open top.

open sheeting (open sheathing, open timbering) Vertical or horizontal supporting planks and timbers that are set along an excavated surface at intervals. Open sheeting is used whenever the ground is firm and dry enough to be effectively shored up without closed sheeting.

open shelving Shelving that is exposed to view, not concealed by doors or cabinets.

open shelving

open shop (merit shop, nonunion shop) A term describing a firm whose employees are not covered by collective bargaining agreements.

open space A term used in urban planning for parks, woods, lawns, recreation spaces, and other areas on which no building stands.

open stairway A stairway with one side or both sides open to the room in which it is located.

open string (open stringer, cut string) A stairway string with its upper edge cut or notched to conform with the treads and risers of the stairs.

open system A piping system for water or fuels in which the conduits that

open stairway

circulate the liquid are attached to an elevated tank or tower, with open vents to facilitate storage, access, and inspection.

open-timbered floor A style of floor construction in which the joists and other supporting timbers are exposed and visible from the underside.

open-timbering Description of structures built with their timberwork or wooden framework exposed to view, or without plaster or another covering; exposed timberwork.

open time The time that is required for the completion of the bond after an adhesive has been spread.

open-top mixer A mixer consisting of a trough or a segment of a cylindrical mixing compartment within which paddles or blades rotate about the horizontal axis of the trough.

open valley A type of roof valley on which shingles and slates are not applied to the intersection of two roof surfaces, leaving the underlying metal or mineral-surfaced roofing material exposed along the length of the valley.

open web A form of construction on a truss or girder in which multiple members, arranged in zigzag or crisscross patterns, are used in place of solid plates to connect the chords or flanges.

open-web steel joist A steel truss with an open web, constructed of hot-rolled structural shapes or shapes of cold-formed, light-gauge steel.

open-web steel joist

open-web studs Light-gauge steel framing construction of wire rods welded to flanges for use as backing lath.

open wiring A network of electrical wiring that is not concealed by the structure of a building, but is protected by cleats, flexible tubing, knots, and tubes, which also support its insulated conductors.

operable partition A partition made of two or more large panels suspended on a ceiling track, and sometimes also supported on a floor track, which may be opened by sliding the panels so that they overlap. The panels form a solid partition when closed.

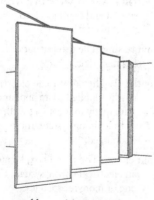

operable partition

operable window A window that may be opened and shut to accommodate ventilation needs, as opposed to a fixed light or fixed sash.

operating engineer The worker or technician who operates heavy machinery and construction equipment.

operating expense (opex) Expenses attributed to

operable window

operations, generally with a useful life of one accounting period or less. These costs are not subject to amortization or depreciation.

operating pressure The pressure registered on a gauge when a system is in normal operation.

operations control center May refer to a building area where building systems operations are managed and tracked—or to those functions themselves. Also called a *network operations center*.

opposed-blade damper A damper installed over an air passage to regulate the volume of air entering an enclosure. The damper is operated by means of two sets of blades linked so that adjacent blades can open and turn in opposite directions.

optical access control A security device, often featuring turnstiles, that allows only properly identified people to pass through an access port.

optical coatings Very thin coatings applied to glass or other transparent materials to increase the transmission of or reduce the reflection of sunlight.

optical fiber cable A medium through which light pulses are transmitted, consisting of a glass core surrounded by a protective sheath. Light pulses are introduced into the optical fiber by a laser or light-emitting diode.

optics 1. The study of light and vision. 2. A system of lenses, filters, prisms, or mirrors used in electronics to direct, disperse, reflect, or otherwise control light rays.

optimum moisture content The percentage or degree of moisture in soil at which the soil can be compacted to its greatest density. Optimum moisture content is used in specifications for compacting embankments.

optimum start program A building automation system that delays the startup of a heating or cooling system until the last possible moment while still maintaining building comfort levels during occupancy.

orangeburg 1. A standard paneling pattern used in decorative plywood. Orangeburg panels have random-width grooving, with the width between each panel following a pattern of 4-8-4-7-9-6-4-6 inches. 2. A common term for bituminous fiber drainage pipe.

orange peel (orange peeling) 1. A surface flaw on paint, resulting from poor flow or application, that leaves the finish pocked with tiny holes like citrus skins. 2. A wavy surface defect on porcelain enamel. 3. A segmented hemispherical bucket that resembles a peel of half an orange, equipped with self-opening and closing capabilities and used to excavate earth. 4. A distinctive texture applied to drywall.

orbital sander A handheld electrical sander whose base, covered with sandpaper or abrasive material, moves rapidly in an elliptical pattern and is used chiefly for coarse work.

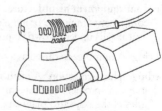

orbital sander

order-of-magnitude estimate An estimate made without detailed engineering data based on a unit such as bells in a hospital or classrooms in a school.

O

213

ordinary-hazard contents Building contents that burn at moderate speed and give off smoke, but release no poisonous fumes or gases that would cause an explosion under fire conditions.

or-equal clause A clause within the technical specification section of the project manual that allows a contractor to propose a product or method that is equal to that which has been specified by the design professional.

organic Descriptive of materials or compounds produced from vegetable or animal sources.

organic clay Clay containing a high volume of composted animal or vegetable materials.

organic light-emitting diode (LED) An LED with an emissive layer made of organic compounds.

organic silt Silt composed in large part of organic substances.

organic soil A highly compressible soil with heavy organic content, considered generally undesirable for construction because of its inability to bear sizable loads.

oriel In architecture, a projecting bay, frequently outfitted with one or more windows, that is corbeled out from the wall of a structure or supported by brackets and that serves both to expand interior space and to enhance the appearance of the building.

oriel window A window housed in an oriel. An oriel window's projection does not extend to the ground, differentiating it from a bay window.

oriel window

orientation The siting of a building relative to compass direction and, therefore, to the sun, which can impact heating, lighting, and cooling costs.

oriented-core barrel A surveying instrument that takes and marks a core to show its orientation, and at the same time records the bearing and slope of the test hole.

oriented strand board (OSB) Panels made of narrow strands of wood fiber oriented lengthwise and crosswise in layers, with a resin binder. Depending on the resin used, OSB can be suitable for interior or exterior applications.

orifice meter A device used to measure the amount of liquid or gas flowing through a pipe.

original equipment manufacturer (OEM) A company that uses components from one or more other companies to build a product that it sells under its own company name.

O ring A round gasket used as a sealant in pipe joints and valves.

orthographic projection A method of representing the exact shape of an object by dropping perpendiculars from two or more sides of the object to planes, generally at right angles to each other. Collectively, the views on these planes describe the object completely. The term *orthogonal* is sometimes used for this system of drawing.

orthotropic A contraction of the terms *orthogonal* and *anistropic* as in the phrase "orthogonal anistropic plate"; a hypothetical plate consisting of beams and a slab acting together with different flexural rigidities in the longitudinal and transverse directions, as in a composite beam bridge.

oscillating saw A power saw with a straight blade that oscillates in short strokes.

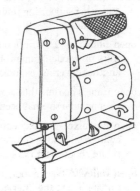

oscillator Any electronic component that generates alternating voltage.

oscilloscope An electronic device that uses a cathode ray tube or a fluorescent-coated tube to produce visual displays corresponding to changes in electrical signals.

Ottawa sand Silica sand produced by processing material obtained by hydraulic mining of massive orthoquartzite situated in deposits near

oscillating saw

Ottawa, Illinois. The sand is composed almost entirely of naturally rounded grains of nearly pure quartz and is used in mortars for testing of hydraulic cement.

outband A masonry jamb stone that serves as a stretcher and is cut to accommodate a frame.

outcrop A segment of an underground rock stratum or a formation that breaks through the surface of the earth and forms a visible protuberance.

outer string On a stairway, the string that stands away from the wall on the exposed outer edge of the stair.

outfall The final receptacle or depositing area for sewage and drainage water.

outfall sewer A sewer that receives the sewage from the collecting system and conducts it to a point of final discharge or to a disposal plant.

outgassing The driving out or freeing of gases from fabrics and building materials.

outlet 1. The point in an electrical wiring circuit at which the current is supplied to an appliance or device. 2. A vent or opening, principally in a parapet wall, through which rainwater is released. 3. In a piping system, the point at which a circulated liquid is discharged.

outlet box The metal box, located at the outlet of an electrical wiring system, that serves to house one or more receptacles.

outlet ventilator An opening covered with a louvered frame, serving as an outlet from an enclosed attic space to the outside.

outline specifications A listing of shortened specification requirements (normally part of schematic or design development documents).

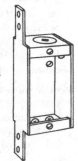

outlooker In roof construction, a projecting member that supports the portion of the roof beyond the face of a gable.

outlet box

out-of-plumb Deviating from a true vertical line of descent, as determined by a plumb line.

out-of-true Descriptive of a structural member that is twisted or otherwise out of alignment.

output 1. The net volume of work produced by a system. 2. The maximum capacity or performance that a system is capable of under normal conditions of operation.

outrigger 1. A beam that projects beyond a wall in order to support an overhanging roof or extended floor, built in a direction perpendicular to the joists of the structural member it serves. 2. An extended beam that supports scaffolds or hoisting tackle as work is being performed on or near a building's wall. 3. A beam that gives stability to a crane by widening its base.

outrigger scaffold A scaffold suspended from outrigger beams or brackets fixed to the outer wall of a building.

outrigger shore (horsing) A bracket installed temporarily to support an outrigger beam or another projecting member.

outside air Fresh, unconditioned air from outside of a building.

outside-air intake (fresh-air intake) An opening or inlet to the outside of a building, through which fresh air is introduced to the boiler room or to an air-conditioning system.

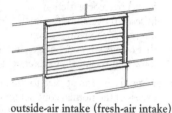

outside-air intake (fresh-air intake)

outside caliper A measuring instrument set on adjustable legs and used especially to measure the outside circumference and diameter of round or cylindrical objects and structures.

outside casing (outside architrave, outside facing, outside lining, outside trim) The supporting members of the jamb or head on a cased window that face the outside and have the appearance of trim.

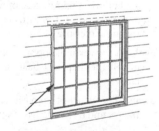

outside corner The outward-projecting corner where two walls meet.

outside corner molding A molding that covers and protects the projecting outside angle of two intersecting surfaces, as in wood veneer. *See also* **corner bead**.

outside casing (outside architrave, outside facing, outside lining outside trim)

outside foundation line A line that indicates where the outer side of a foundation wall is located.

outside glazing External windows or glass doors installed in a building from the outside.

outside screw and yoke A valve configuration where the valve stem, with exposed external threads supported by a yoke, indicates the open or closed position of the valve.

outside studding plate The soleplate or double top plate in the construction of a wood-frame wall or partition, usually built with stock of a size equal to the studding.

outstanding leg A leg of a structural angle member, generally unconnected to any other member.

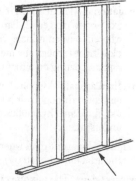

outside studding plate

outsulation 1. The placing of insulation to the exterior of a wall. 2. The elimination of all thermal bridges between the inner and outer surfaces of a wall.

out-to-out In measurements, a term meaning that the dimensions are overall.

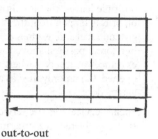

out-to-out

oven dry The condition resulting from having been dried to essentially constant weight in an oven at a temperature that has been fixed, usually 221°F and 230°F (105°C and 115°C).

oven dry wood (bone dry wood) Wood that gives off no moisture when subjected to a temperature of 212°F (100°C).

overall (overall dimension) The total external dimension of any building material, including all projections.

overbreak Excavation performed beyond the work limits established by the neat line.

overburden 1. A mantle of soil, rock, gravel, or other earth material covering a given rock layer or bearing stratum. 2. An unwanted top layer of soil that must be stripped away to open access to useful construction materials buried beneath it.

overcloak The portion of a metal roofing sheet that overlaps the edge of an adjacent sheet set underneath it.

overcurrent protection Safety provisions within an electrical system, such as would be furnished by ground fault circuit interrupters, that guard against damage and injury resulting from excessive current by shutting off the flow of current when it reaches a certain level.

overdesign A term used to describe adherence to structural design requirements beyond service demands, as a means of compensating for statistical variation, anticipated deficiencies, or both.

overdig The amount of clean soil that is removed below or beyond the extent of contamination at a site to make sure all waste has been taken away.

overflow (overflow pipe) 1. A pipe installed to prevent flooding in storage tanks, fixtures, and plumbing fittings, or to remove excess water from buildings and systems. 2. An outlet fitted to a storage tank to set the proper level of liquid and to prevent flooding.

overhand work The process by which bricklayers install brick in an external wall while standing on a scaffold or on the floor inside a building or structure.

overhang The extension of a roof or an upper story of a building beyond the wall/story situated directly beneath.

over-haul The distance excavated material is transported beyond that given as the stated hauling distance.

overhead (indirect expense, overhead expense) The costs to conduct business other than direct job costs; included in bidder's markup.

overhead balance A tense steel coil or spring installed in the head jamb of a window frame to serve as a balance for the sash.

overhead concealed closer A door closer, installed out of view in the head of the door frame, designed with a hinged arm that connects the door with the top rail of the frame.

O

overhead door (overhead-type garage door) A door, constructed of a single leaf or of multiple leaves, that is swung up or rolled open from the ground level and assumes a horizontal position above the entranceway it serves when opened. Commonly used as a garage door.

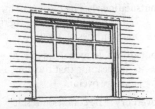

overhead door (overhead-type garage door)

overhead service Electric service that is delivered to a structure by lines that are above the ground.

overhead shovel A tractor loader that digs at one end, swings the bucket overhead, and dumps at the other end.

overhead traveling crane A lifting machine generally power-operated at least in its hoisting operation. The crane is carried on a horizontal girder, reaching between rails above window level at each side of a shop, and consists of a hoisting cab that can travel from end to end on the girder. The whole area between the rails can thus be traversed by the cab.

overlaid plywood (overlay) Plywood with a surfacing material added to one or both sides. The material usually provides a protective or decorative characteristic to the side, or a base for finishing. Materials used for overlays include resin-treated fiber, resin film, impregnated paper, plastics, and metal.

overlapping astragal (wraparound astragal) A molding that is attached lengthwise along the masting edge on one of a pair of doors to close the gap between them, providing a weather-resistant seal and stopping the transmission of light or smoke from one side of the doors to the other.

overlapping astragal (wraparound astragal)

overlay 1. A layer of concrete or mortar, seldom thinner than 1″ (25 mm), placed on and usually bonded to the worn or cracked surface of a concrete slab either to restore or to improve the function of the original surface. **2.** The surfacing of a plywood face with a solid material other than wood. *See also* **overlaid plywood.**

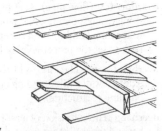

overlay flooring Finish flooring of maple, mahogany, oak, or other hardwood cut into narrow tongue-and-groove strips.

overlay flooring

overload 1. A load exceeding that for which the bearing structure was designed. **2.** Excess power, current, or voltage in an electrical device or circuit that is not designed to accommodate it.

overload capacity The limit of excess power, current, or voltage that an electrical device or circuit can accommodate before it is damaged.

overload relay A relay in a motor circuit that disconnects the motor from its power source if the current that feeds the motor surpasses a certain predetermined level.

overrun brake (overriding brake) A brake fitted to a towed vehicle, such as a concrete mixer to a trailer. It operates as soon as the towing truck slows down and the towed vehicle tends to push into it. Movement of the towed vehicle applies the overrun brake, making safe high-speed towing possible.

oversailing course 1. A course of masonry that extends beyond the face of the wall in which it is set. **2.** A string course.

oversize brick A brick measuring greater than 2½″ × 3½″ × 7½″.

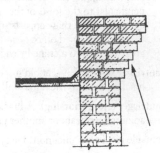

oversailing course

overstretching Stressing of tendons to a value higher than designed for the initial stress to: (a) overcome friction losses, (b) overstress the steel temporarily to reduce creep that occurs after anchorage, and (c) counteract loss of prestressing force that is caused by subsequent prestressing of other tendons.

overtime 1. A term applied to the number of hours worked in excess of the normal contract for one day or one week. **2.** A term applied to the payment for this time, frequently 1½ or double the normal rate of pay.

overturning The failure of a retaining wall as the result of hydraulic or earth pressure on one side. Overturning occurs when walls are built on a narrow base or with materials too light to withstand surrounding pressure.

ovolo A convex molding approximately the shape of a quarter circle.

ovolo

owner The owner of a project, that is also party to the owner-contractor and owner-designer agreements.

owner–architect agreement Contract between owner and architect for professional design services.

owner–contractor agreement The contract formed between owner and contractor describing performance of the construction work for a project (or a portion thereof).

owner's inspector A party hired by an owner to inspect the work.

owner's representative The designated official representative of the owner (may be an architect, engineer, or contractor) to oversee a project.

oxidation 1. The reaction of a chemical compound mixed or exposed to oxygen. **2.** Part of the asphalt-refining process, wherein oxygen is incorporated in hot, bituminous liquids by blowing it through the melted substance. **3.** The hardening of asphalt coating on a roof under exposure to sun and air.

oxidized asphalt Asphalt that has been specially treated by having air blown through it at high temperatures, making it suitable for use in roofing, hydraulics, pipe coating, membrane envelopes, and undersealing.

oxidizer An agent that, when acting on another substance, causes the attachment of an oxygen atom thereto.

oxyacetylene The mixture of oxygen and acetylene, forming a highly combustible gas used for cutting and welding metal.

oxychloride cement (sorel cement) A hard, durable cement made up of calcined magnesia and magnesium chloride, occasionally blended with fillers.

oxygen cutting A process of metal cutting in which the separation of the metal is effected by its chemical reaction with oxygen at high temperatures.

ozone Triatomic oxygen (O_3), an unstable form of oxygen that is produced by ultraviolet activity and electrical discharges and is used as an oxidizing agent, a deodorizer in air-conditioning and cold storage systems, and as an agent for stemming the growth of bacteria, fungus, and mildew. Excessive concentrations of ozone are poisonous to humans.

O

Abbreviations

p part, per, pint, pipe, pitch, pole, post, port, power

P phosphorus, pressure, pole, page, projection

P & T pressure and temperature

P & I purchase and install

P & SDS piping and surface decontamination solvents

P1E planed one edge

P1S planed one side

P1S2E planed one side and two edges

P4S planed four sides

P & T posts and timbers, pressure and temperature

PA particular average, power amplifier, preliminary assessment, professional association, purchasing agent, public address system

PAH polycyclic aromatic hydrocarbons

pan panel

Pape paperhanger

PAPI precision approach path indicator

PAPR powered air purifying respirator

PAR. paragraph, part, partition, parabolic reflector

par. parapet

par planed all round

PARP full-facepiece air purifying respirator

part, partn partition, partial

partbd particleboard

PASS passenger, passage

Patt pattern

pat. patent

PAX private automatic (telephone) exchange

Pb lead

Pb pushbutton

PC Portland cement, power connector, personal computer

pc piece

PCB polychlorinated biphenyl

PCCP prestressed concrete cylinder pipe

PCE pyrometric cone equivalent

pcf pounds per cubic foot

pckg packaged

PCM phase contract microscopy

pc/pct percent

pcs pieces

PD per diem, potential difference

pd paid

Pd palladium

PE professional engineer, probable error, plain end, polyethylene, porcelain enamel

pe plain-edged

pecky cyp pecky cypress

ped pedestal, pedestrian

pegbrd pegboard

pelec photoelectric

pend pendant

PEP Public Employment Program

per perimeter, by the, period

PERF perforate, perforated

PERM permanent

PERP perpendicular
Pers personal
PESB preengineered steel building(s)
PERT project evaluation and review technique
PET precision end trimmed
PF power factor, profile
PFA pulverized fuel ash
PFD preferred
P&G post and girder
ph phase, phot
PH phase, Phillips head
pH hydrogen-ion concentration
ph phase
phos phosphate
photo photograph
Ph phenyl
PI pressure injected
PIB polyisobutylene
PID photoionization detector
pil pilaster
pile pile driver
PITI principal, interest, taxes, and insurance
piv pivoted
PIV post indicator valve
Pjtn projection
Pjtr projector
pk park, peak, plank
pkd packed
pkfr plank frame
pkg package
pkgng packaging
pkng packing
pkt pocket
pkwy parkway
pl plain, plate
plah plasterer helper
plas plaster
P/L plastic laminate
PL pile, plate, plug, power line, pipe line, private line
pl place, plate
platf platform
PLC programmable logic controller
Plf pounds per linear foot
PLG piling
plh production labor-hour

Pll pallet
PLM polarized light microscopy
plmb, plb, PLMB plumbing
plstc plastic
pluh plumber's helper
plum plumber
ply plywood
PLYWD plywood
PM post meridiem
pmh production man-hour
pmp pump(s)
PNEU pneumatic
PNL panel
Pntd painted
Pntg painting
PO purchase order
POL polish, petroleum, oil and lubricants, polished
polarogr polarographic
polthn polyethylene
polyest polyester
polyiso polyisocyanurate
polyprop polypropylene (also PP and PL)
PORC porcelain
Pord painter, ordinary
Port portable
PORT CEM portland cement
pos, POS positive
posn position
pot. potential
POTW publicly owned treatment works
pp ponderosa pine, pages, piping
PP-AC air-conditioning power panel
Ppd prepaid
PPE personal protective equipment
PPGL polished plate glass
PPH parts, per hour
ppm parts per million
PPSD package, power supply
ppt, pptn precipitate, precipitation, parts per thousand
PR payroll, pair
Prcs process(es)
Prscg processing
prcst precast
pre-assm preassembled
preb prebend

prec preceding

precp precipitation

pre-eng preengineered

prefab prefabricated

prefin prefinished

prelim preliminary

prem premium

prep preparation, prepared

press pressure

pretreat pretreated, pretreatment

prfcn purification

prgm program

pri primary

prin principal

pris prismatic

prl parallel

prod. production, product

prog progressive

proj project, projection, projecting

prom programmable read-only memory

prop property, propelled, propeller, proportional

prot protection, protective

prov provisional

prox proximity

PRP potentially responsible party, purpose

prs pairs

prt particle

PRV pressure-regulating valve, pressure relief valve

ps pieces, power shift

PS polystyrene

p.s.e. planed and square-edged

psf pounds per square foot

psi pounds per square inch

psig pounds per square inch

psj planed and square-jointed

PSP plastic sewer pipe

Pspr painter, spray

Psst painter, structural steel

PT pipe thread, potential transformer, part, point, packed tower

pt paint, pint, payment, port, point

ptfe polytetrafluoroethylene

ptg planed, tongued, and grooved

PTN partition

PTTU packed tower treatment unit

PU pickup, plutonium, ultimate load

PUD pickup and delivery

pur purlins

PUR polyurethane

PVA polyvinyl acetate

PVC polyvinyl chloride

pvmt pavement

pvntr preventer

PW paper wrapped

PWA Public Works Administration

pwr power

pwred powered

PWR pressurized water reaction

pwt pennyweight

1PH single phase

Definitions

3PH three phase

pace A landing in a staircase.

pache Color coding used on drawings to aid in quantity takeoffs for estimating.

Pacific silver fir *Ables ammilis*. This species is found in British Columbia, Washington, and Oregon. The name comes from the silvery appearance of the underside of the tree's needles. Its wood is classed in the hem-fir group.

Pacific yew *Taxus brevifolia*. This species, generally small in size, is not a commercially important tree. Its wood is heavy and strong and is used for such purposes as archery bows. Yews are usually found growing in the shade of larger trees.

pack The bundling in which shakes and shingles are shipped. In shakes, the most prevalent pack is a 9/9. This describes a bundle packed on an 18″ wide frame with nine courses, or layers, at each end. The most common pack for shingles is 20/20. Because of their smoother edges, shingles can be packed tighter than shakes, a bundle of shakes usually contains a net of about 16″ of wood across the 18″ width of the frame.

packaged air conditioner A factory-assembled air-conditioning unit ready for installation. The unit may be mounted in a window, an opening through a wall, or on the building roof. These units may serve an individual room, a zone, or multiple zones.

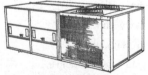

packaged air conditioner

packaged boiler A factory-assembled water or steam heating unit ready for installation. All components, including the boiler, burner, controls, and auxiliary equipment, are shipped as a unit.

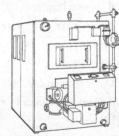

packaged boiler

packer A device inserted into a hole in which grout is to be injected, which acts to prevent return of the grout around the injection pipe. A packer is usually an expandable device actuated mechanically, hydraulically, or pneumatically.

packer-head process A method of casting concrete pipe in a vertical position in which concrete of low water content is compacted with a revolving compaction tool.

packing 1. Stuffing of shaped elastic material to prevent fluid leakage at a shaft, valve stem, or joint. **2.** Small stones, usually embedded in mortar, used to fill cracks between larger stones.

packing gland A protective sleeve used over cable or piping in applications where pressure or other factors pose a threat.

pack set The condition where stored cement will not flow from a container such as a rail car or silo. It is caused by interlaced particles or electrostatic charges on particles.

pad 1. A plate or block used to spread a concentrated load over an area, such as a concrete block placed between a girder and a load-bearing wall. **2.** A shoe of a crawler-type track.

paddle wheel scraper A heavy-duty excavation machine with a wheel that scrapes soil into a large bowl section.

pad foundation A thick slab-type foundation used to support a structure or a piece of equipment.

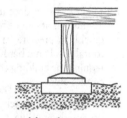
pad foundation

padlock A unit lock with a U-shaped bar that is passed through a staple of a hasp or link in a chain, and the bar pressed into the body to lock.

pad out To add shims to framing so that a finished wall or ceiling will be correctly aligned.

pailing Sheathing constructed of vertical boards that is used in concrete formwork.

paint 1. A mixture of a solid pigment in a liquid vehicle that dries to a protective and decorative coating. **2.** The resultant dry coating.

paint base The liquid vehicle into which a pigment is mixed to produce a paint.

paint grade A description of a wood product that is more suitable for painting than for a clear finish.

paint remover A liquid solvent applied to dry paint to soften it for removal by scraping or brushing.

paint roller A tube with a fiber surface that is mounted on a roller and handle, and used to apply paint.

paint system A specific combination of paints applied in sequence. A paint system consists of a combination of some of the following coats: sealer or primer, stain, filler, undercoat, and one or more topcoats.

paint thinner A liquid compatible with the vehicle of a paint, used to make a paint flow easier. Paint thinner lowers the viscosity of paints, adhesives, etc.

pale (paling) 1. One of the stakes in a palisade. **2.** A picket in a fence.

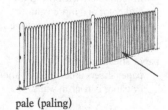
pale (paling)

palisade A fence of poles driven into the ground and pointed at the top.

palladian window A tripartite or three-light window that features a central arched window with balanced rectangular lights on each side.

palladian window

pallet 1. A platform used for stacking material and arranged to be handled by a forklift truck. **2.** A wood insert in a brick wall used for support of a surface system.

palletized A term used frequently in the shingle and shake industry. Both items are often shipped on pallets from the mill for ease in handling while in transit. These shipments are referred to as palletized loads.

palm sander A handheld electrical sander with a vibrating or orbital base to which sandpaper is attached for finer sanding.

pan 1. A prefabricated form unit used in concrete joist floor construction. **2.** A container that receives particles passing the finest sieve during mechanical analysis of granular materials. **3.** A structural panel.

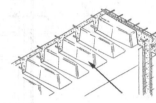
pan

pan and roll roofing tile A roofing tile system consisting of two types of tile: A flat or slightly curved tile with a flange on each side, and curved tile that fits the flanges and closes the joints.

pan construction A type of concrete floor or roof in which pan forms are used to create intersecting ribs, resulting in a waffle-like undersurface.

pane A flat sheet of glass installed in a window or door. The installed sheet is also referred to as a light.

panel 1. A section of form sheathing, constructed from boards, plywood, metal sheets, etc., that can be erected and stripped as a unit. **2.** A concrete member, usually precast, rectangular in shape, and relatively thin with respect to other dimensions. **3.** A sheet of plywood, particleboard, or other similar product, usually of a standard size, such as 4' × 8'.

panelboard A board on which electric components and/or controls are mounted.

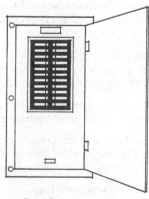
panelboard

P

panel box A box in which electric switches and fuses are mounted.

panel clip A specially shaped metal device used in joining panels in roof construction. The clip substitutes for lumber blocking and helps to spread the load from one panel to the next one.

panel construction A general term used to describe construction where building components are assembled elsewhere before being brought to a site for placement.

panel door A door constructed with panels, usually shaped to pattern, installed between the stiles and rails, which form the outside frame of the door.

panel door

paneling The material used to cover an interior wall. Paneling may be made from a 4′ × 4′ *select* milled to a pattern and may be either hardwood or softwood plywood, often prefinished or overlaid with a decorative finish, or hardboard, and usually prefinished.

panel insert A metal unit used instead of glass in a panel door.

panel molding A decorative molding, originally used to trim raised panel wall construction.

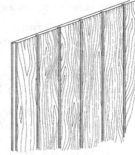

paneling

panel point Point of intersection of the members of a truss.

panel product Any of a variety of wood products such as plywood, particleboard, hardboard, and waferboard, sold in sheets or panels. Although sizes vary, the board size for most panel products is 4′ × 8′.

panel strip 1. A strip extending across the length or width of a flat slab for design purposes. 2. A narrow piece of wood or metal used to hide a joint between two sheathing boards forming a panel.

pan form stair A metal stair assembly with metal sheet pans at the treads to hold precast or cast-in-place masonry or stone treads.

pan fraction 1. The reported results of mechanical analysis of granular materials. 2. The weight of the material retained on any one sieve divided by the initial weight of the sample.

pan head A head of a screw or rivet shaped like a truncated cone.

panic bolt The bolt in panic hardware that is released by pressure on a horizontal bar.

panic exit hardware, mortise type Panic exit hardware in which the lock mechanism is concealed within the door or set into a rectangular cavity (called a *mortise*) that has been cut in the edge of the door.

panic exit hardware, rim type Panic exit hardware in which the lock mechanism is located on the inside face of the door.

panic exit hardware, vertical rod type Panic exit hardware with latches at the top and/or bottom of the door. The latches are connected to a vertical rod that, in turn, is connected to a crossbar. In an emergency, the latches are released simply by pressing the crossbar.

panic hardware A door-locking assembly that can be released quickly by pressure on a horizontal bar. Panic hardware is required by building codes on certain exits.

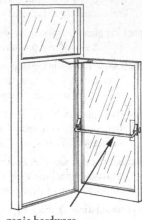

panic hardware

pants Steel plates attached to the hammer of a pile driver to aid in driving sheet piling.

pan-type humidifier A pan with water placed in a flow of air to increase humidity. A heating element may be placed in the pan for greater evaporation.

paper-backed lath Any lath with building paper attached. A paper-backed lath serves as formwork and reinforcing for a concrete floor over open web joists.

paperboard (pasteboard, cardboard) A stiff cardboard composed of layers of paper, or paper pulp, compressed into a sheet.

paper form A heavy paper mold used for casting concrete columns and other structural shapes.

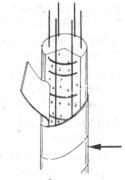

paper form

paperhanger A tradesperson experienced and trained in preparing surfaces for and hanging wall coverings.

paper overlay Paper prepared for application to the face of a panel after first being printed in four colors with the grain and color of a more valuable wood, or in a decorative design.

paper sheathing Felt or heavy paper sheets used as an air and/or vapor barrier in walls.

paper sheathing

papreg A paper product produced by impregnating sheets of high-strength paper with synthetic resin and then laminating the sheets to form a dense, moisture-resistant product.

parabolic aluminized reflector (PAR) lamp A lamp that utilizes a light reflector shaped so as to project light from a small source in an approximately parallel beam, as in a spotlight.

parabolic luminaire A fluorescent fixture with a louver of parabolic-shaped baffles that provide excellent light control and reduce glare.

paraform Paraformaldehyde, an additive used with wood flour as a hardener in adhesives. *See also* **resorcinal resin adhesive**.

parallel **1.** The condition in which two lines or planes are an equal distance apart at all points. **2.** Electric blasting caps arranged so that the firing current passes through all of them at the same time.

parallel activities Two or more activities than can be done at the same time. Allows a project to be completed faster than if activities were arranged sequentially.*

parallel chord truss An engineered structural component, composed of a combination of members, with its top and bottom members positioned flat and parallel to each other.

parallel circuit An electrical circuit that has at least two paths for electricity to flow, with loads parallel to each other as a result.

parallel connection A connection at which a flow is diverted to two or more parallel conduits.

parallel flow An arrangement of a heat exchanger where the hot and cold materials enter at the same end and flow to the exit.

parallel-laminated veneer A product in which the veneers have been laminated with their grains parallel to one another. Parallel laminated veneer is used in furniture and cabinetry to provide flexibility over curved surfaces, and in the production of laminated veneer structural products.

parallel siding (square-edged siding) **1.** Siding that is not beveled. **2.** Siding having edges of the same thickness.

parallel welding The joining of metal parts by fusion, with the electric current that produces the weld divided and routed through the electrode and metal along similar paths.

parallel-wire unit A posttensioning tendon composed of a number of wires or strands that are approximately parallel.

parameter A variable in a mathematical expression.

parametric estimate Estimating algorithms or cost estimating relationships that are highly probabilistic in nature (i.e., the parameters or quantification inputs to the algorithm tend to be abstractions of the scope). Typical parametric algorithms include, but are not limited to, factoring techniques, gross unit costs, and cost models (i.e., algorithms intended to replicate the cost performance of a process of system).*

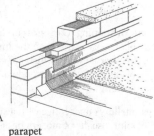

parapet **1.** That part of a wall that extends above the roof level. **2.** A low wall along the top of a dam. **parapet**

parapet gutter A gutter built or placed behind a parapet.

paretta Cast masonry with a surface of protruding pebbles.

parge coat A coat of masonry cement applied to masonry for resistance to penetration of moisture.

pargetting **1.** Lining of a flue to aid in smooth flow and increase fire resistance. **2.** Application of a dampproofing masonry cement. **3.** Ornamental, often elaborate, facing for plaster walls.

pargetting

paring chisel A long-handled chisel used to shape wood by hand without the use of a mallet.

paring gouge A long, thin woodworking gouge, the cutting edge of which is beveled on the concave side.

parliament hinge An H-shaped hinge.

parquet flooring A floor covering composed of small pieces of wood, usually forming a geometric design.

parquet flooring

parsing A thin coat of plaster or masonry cement.

partial cover plate A cover plate attached to the flange of a girder but not extending the full length of the girder.

partially air-dried (PAD) Wood seasoned to some extent by exposure to the atmosphere without artificial heat, but still considered green or unseasoned.

partial occupancy An owner's occupation and use of a project before final completion.

partial prestressing Prestressing to a stress level such that, under design loads, tensile stresses exist in the precompressed tensile zone of the prestressed member.

partial release Release in a prestressed concrete member of a portion of the total prestress initially held wholly in the prestressed reinforcement.

particle size **1.** Minimum particle diameter that will be removed by an air filter. **2.** Diameter of a pigment particle in paint. **3.** Diameter of a grain of sand in a mechanical analysis test.

particle size distribution A tabulation of the result of mechanical analysis expressed as the percentage by weight passing through each of a series of sieves.

particleboard A generic term used to describe panel products made from discrete particles of wood or other ligno-cellulosic material rather than from fibers. The wood particles are mixed with resins and formed into a solid board under heat and pressure.

parting bead A narrow strip between the upper and lower sashes in a double-hung window frame.

parting slip A thin piece of wood in the cased frame of a sash window separating the sash weights. Also called a *parting stop*.

parting tool A turning tool with a narrow blade and V-shaped gouge used for cutting recesses or grooves in wood.

partition A dividing wall within a building, usually non-load-bearing.

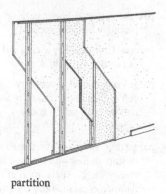

partition

partition block Light concrete masonry unit with a nominal thickness of 4″ to 6″.

partition plate The top horizontal member of a partition, which may support joists or rafters.

partition stud A steel or wood upright in a partition.

partition tile A hollow, clay unit for use in interior partitions. The surface of a partition tile is often grooved for plastering.

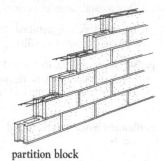

partition block

partnering A set of collaborative processes (as opposed to a relationship only); a commitment to achieve mutually identified objectives between two (or more) organizations, based on cooperation, open communication, and heightened effectiveness through continuous improvement.

partnership The joining of two or more individuals for a business purpose whereby profits and liabilities are shared.

party wall A common wall between two living units.

Pascal's law A principle that states that pressure applied to a confined fluid at any point is transmitted equally in all directions.

pass 1. One-direction application of a substance, such as paint or a layer of shotcrete, placed in one movement over the field of operation. **2.** A single progression of a welding operation along a joint, resulting in a weld bead.

passage (passageway) A horizontal space for moving from one area of a building to another.

passenger elevator An elevator mainly used for people.

passive solar energy system A solar energy system that collects and distributes thermal energy through a structure via natural means, without using pumps or fans.

pass-through An opening in a partition for passing objects between adjacent areas.

pass-through clause Contract language that allows a general contractor to pass risk and responsibilities to subcontractors by reference.

paste content (of concrete) Proportional volume of cement paste in concrete, mortar, or the like, expressed as volume percent.

of the entire mixture. **paste paint** A pastelike mixture of pigment and solvent, usually requiring additional solvent for use.

pat A specimen of neat cement paste about 3″ (76 mm) in diameter and ½″ (13 mm) in thickness at the center, and tapering into a thin edge on a flat glass plate for indicating setting time.

patch 1. A piece of wood or synthetic material used to fill defects in the plies of plywood. Also called a *plug*. **2.** A compound used in stonemasonry to replace chips and broken corners or edges in fabricated pieces of cut stone or to fill natural voids. The patch is applied in plastic form.

patch board (patch panel) A board with jacks and plugs for terminals of electric circuits. The circuits may be temporarily interconnected by patch cords.

patch gun 1. A hand tool that "shoots" a premixed material for patching and repairing exterior finishes like stucco. **2.** A hand tool used to apply joint compound to drywall.

patent defect A defect present in materials, equipment, or completed work detectable by reasonably careful observation. A patent defect is distinguished from a latent defect, which could not be discovered by reasonable observation.

patent glazing Any of a number of devices, usually preformed neoprene gaskets, for securing glass in frames without putty.

patent-hammered Stonework finish applied to the face of building stone.

patent knotting A solution of shellac and benzine or similar solvent used to seal knots in wood.

patent stone (artificial soil) Stone chips embedded in a binder of mortar, cement, or plaster. The surface may be ground and/or polished.

patina Color and texture added to a surface as a result of oxidation or use, such as the green coating on copper or its alloys.

patten The base of a column.

pattern 1. A plan or model to be a guide in making objects. **2.** A form used to shape the interior of a mold.

pattern cracking Fine openings on concrete surfaces in the form of a pattern, resulting from a decrease in volume of the material near the surface and/or an increase in volume of the material below the surface.

patternmaker's saw A small hand saw with fine teeth used to make intricate cuts.

pattern staining Dark areas on finished plaster, particularly on the interior of external walls, which are caused by different thermal conductance of backings.

paumelle A door hinge with a single joint, usually of modern design.

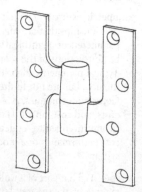

paumelle

paved invert In piping, the lower portion of a corrugated metal pipe whose corrugations have been filled with smooth bituminous material to resist scour and erosion and improve flow.

pavement base The layer of a pavement immediately below the surfacing material and above the subbase.

pavement, concrete A layer of concrete over roads, sidewalks, canals, playgrounds, and those areas used for storage or parking.

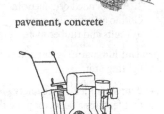
pavement, concrete

pavement saw A self-propelled machine with a circular saw blade for cutting control joints.

pavement sealer A bituminous coating used to seal and renew the surface of asphalt paving.

paver 1. A block or tile used as a wearing surface. 2. A machine that places concrete pavements.

pavement saw

pavilion roof 1. A roof composed of equally hipped areas. 2. A pyramid-shaped roof.

paving The hard surface covering of areas such as walks, roadways, ramps, waterways, parking areas, and airport runways.

paving aggregate The various solid materials, such as sand, gravel, or slag, used in construction of a pavement.

paving asphalt A sticky residue from the refining of crude oil. Paving asphalt is used in built-up roofing systems, as the binder in asphaltic concrete, or as a waterproofing agent.

paving breaker (chipper) A handheld, pneumatic tool for cutting pavements.

paving brick A vitrified clay brick with good resistance to abrasion.

paving unit A fabricated or shaped unit used in a pavement surface.

payback method A technique of economic evaluation that determines the time required for the cumulative benefits from an investment to recover the investment cost and other accrued costs.*

payback period (PB) Measures the length of time until accumulated savings are sufficient to pay back the initial cost. Discounted payback (DPB) takes into account the time value of money by using time-adjusted cash flows. If the discount rate is assumed to be zero, the method is called simple payback (SPB).

payment bond A form of security purchased by the contractor from a surety, which is provided to guarantee that the contractor will pay all costs of labor, materials, and other services related to the project for which he is responsible under the contract for construction.

payment schedule An arrangement for payments to the contractor, typically based on amounts of work completed.

payments withheld A provision of AIA General Conditions of the Contract for Construction, which provides that the owner may withhold payments to the contractor if, in the opinion of the design professional, the work falls behind the schedule of construction, or in the event that the work deviates from the provisions of the contract documents.

payout time The time required to recover the original fixed investment from profit and depreciation. Most recent practice is to base payout time on an actual sales projection.*

p bar A heavy steel bar, shaped like a chisel at one end, used for prying.

pea gravel Screened gravel, most of the particles of which will pass a ⅜″ (9.5 mm) sieve and be retained on a No. 4 (4.75 mm) sieve.

peak joint The joint of a roof truss that is at the ridge.

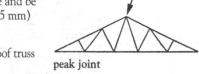
peak joint

peak levels In measuring ambient air contamination, above average levels of ambient contamination due to the sudden release of a contaminant into the air. Usually occurs for a short period of time immediately following the release.

peak load The maximum demand or design load of a device, system, or structure over a designated time period.

peak-load controller An electrical controller used to limit the maximum power demands to a device or system.

peat Fibrous organic matter in various stages of decomposition, found in swamps and bogs, and used to enrich soil for plantings.

peavey A long-shafted tool with a hook that is used by loggers to roll logs, break jams, pry rocks, tighten chains, and push over trees.

pebble dash (rock dash) An exterior finish in which crushed rock or pebbles are embedded in mortar, plaster or stucco.

pecan *Carya illinoensis.* One of the largest native hickories, its wood is used in furniture and flooring.

peck 1. Channeled or pitted areas or pockets sometimes found in cedar or cypress, the decay resulting from fungus in isolated spots. 2. A dry measure equal to 2 gallons.

pedestal 1. An upright compression member whose height does not exceed three times its average least lateral dimension, such as a short pier or plinth used as the base for a column. 2. Utility boxes that house connections and/or switches for telephone, electrical, or cable television service.

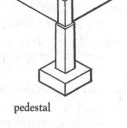
pedestal

pedestal floor A flooring system that is elevated from the subfloor to accommodate cabling, piping, ductwork, or other building systems.

P

pedestal lavatory Lavatory supported by a pedestal rather than wall hung. Supply and waste lines are enclosed by the pedestal.

pedestal pile A cast-in-place concrete pile constructed so that concrete is forced out into a widened bulb or pedestal shape at the foot of the pipe, which forms the pile.

pedestrian control device Any device, especially turnstiles, but including gates, railings, or posts, used to control the movement of pedestrians.

pediment A decorative unit, often triangular in shape, above a doorway.

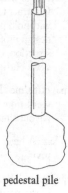

pedestal pile

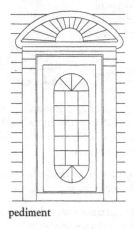

pediment

peel To produce veneer from a log through rotary cutting.

peeler A log from which veneer is peeled on a lathe, for the production of plywood. A peeler-grade log most frequently is from an old-growth tree, with a high proportion of clear wood.

peeling 1. A process in which thin flakes of mortar are broken away from a concrete surface, such as by deterioration or by adherence of surface mortar to forms as they are removed. **2.** Separation of paint or other coating from the surface to which it was applied, often caused by poor adhesion due to inadequate pre-painting preparation, or effects of moisture.

peen The end of a hammer, other than a claw hammer, opposite the hammering face. A peen may be pointed or ball- or cone-shaped. A peen is used for chipping, indenting, and metalworking.

peen-coated nail A mechanically galvanized nail coated by tumbling in a container with zinc dust and glass balls.

peg 1. A pointed pin of wood used to fasten wood members together. **2.** A short, pointed wooden stick used as a marker by surveyors.

peg-and-plank A rustic wood floor of planks and exposed pegs.

pegboard A hard fiberboard sheet, usually ¼″ thick with regular rows of holes for attaching pegs or hooks.

pelmet A valance or cornice, sometimes decorative, at the head of a window to conceal a drapery track or other fittings.

pelmet board A board at the head of a window, acting as a pelmet.

pelmet lighting Lighting furnished by sources that are concealed by a pelmet.

penalty clause A clause in a contract specifying a charge against the contractor for failure to complete the work by a prearranged date.

penciling Painting mortar joints, usually white.

pencil rod Plain metal rod of about ¼″ (6 mm) diameter.

pendant 1. An electric device suspended from overhead. **2.** A suspended ornament in Gothic architecture, used in vaults and timber roofs.

pendant luminaire A suspended lighting unit.

pendant luminaire

pendent sprinkler A fire safety system sprinkler with a head that features a deflector plate that directs discharged water downward.

penetrant An additive that increases a liquid's ability to penetrate a surface or enter the pores of a substrate. Penetrants are typically used as wetting agents.

penetrating finish A low-viscosity oil or varnish that penetrates into wood with only a film of material at the surface.

penetration 1. A test of the hardness of an asphalt utilizing a weighted needle at standard conditions. **2.** The cut-off depth of piles or sheet piling. **3.** The depth of a caisson below ground level. **4.** The intersection of two surfaces of vaulting.

penetration macadam Pavement made from layers of coarse, open-graded aggregate (crushed stone, slag, or gravel) followed by the spray application and penetration of emulsified asphalt.

penetration test A test to estimate the bearing capacity of soil by recording the number of blows required to drive a standard tool into soil.

penitent post A short post placed against the wall and supporting an arch or tie beam.

penny A measure of the length of a nail. The larger the number is, the longer the nail.

pentachlorophenol A chemical used in wood preserving, usually applied under pressure so that it will penetrate the wood.

percentage fee A fee paid to the contractor or the architect that is a percentage of the total construction cost.

percentage humidity The ratio, expressed as a percentage, of the weight of water vapor in a pound of dry air to the weight of water vapor if the same weight of air were saturated.

percentage void The ratio, expressed as a percentage, of the volume of voids to the gross volume of material.

percent complete An estimate of the percentage complete for an activity as of a particular data date. Percent complete may be based on time expended, cost or resources employed, or measurement of work in place.*

percent fines 1. Amount, expressed as a percentage, of material in aggregate finer than a given sieve, usually the No. 200 sieve. **2.** The amount of fine aggregate in a concrete mixture expressed as a percent by absolute volume of the total amount of aggregate.

percent saturation The ratio, expressed as a percentage, of the volume of water in a soil sample to the volume of voids.

percolation The movement of a fluid through a soil.

percolation test A test to estimate the rate at which a soil will absorb waste fluids, performed by measuring the rate (percolation rate) at which the water level drops in a hole full of water.

percussion drill A pneumatic or electric tool that drills holes by applying a rapid series of blows.

percussion drill

perfections Shingles 18″ long and 0.45″ thick at the butt.

perforated drain A subsurface draining system that uses pipes with holes in the bottom to allow water or other liquid to percolate into the soil.

perforated facing A perforated sheet or board used as a finished surface and allowing a fraction of sound to penetrate the surface to an absorbent layer.

perforated metal pan A unit that forms the exposed surface of a type of acoustical ceiling. The perforated pan contains a sound-absorbent material, usually in pad form.

perforated tape A special paper tape used to reinforce the material covering joints between gypsum boards.

performance 1. A term meaning fulfillment of a promise made by one party to a contract or agreement in return for compensation. 2. The manner in which or the efficiency with which something acts or reacts in the manner in which it is intended.

performance-based fee A fee structure that rewards a consultant's effort to meet or exceed the clients goals, such as minimizing a project's life-cycle cost. The designer's fee is based on a measurement, such as energy use or operating cost of the completed facility.

performance bond 1. A guarantee that a contractor will perform a job according to the terms of the contract, or the bond will be forfeited. 2. A bond procured by the contractor which shows that a surety guarantees (to the owner) that the work will be performed in accordance with the contract documents. Unless prohibited by statute, the performance bond can be combined with the labor and material payment bond. *See also* **surety bond**.

performance measurement baseline The time-phased budget plan against which contract performance is measured. It is formed by the budgets assigned to scheduled work elements and the applicable indirect budgets.*

performance measurement system 1. An organization's defined processes for monitoring and updating project and/or organization progress at a detailed level over time. 2. A quantitative tool (for example, rate, ratio, index, percentage) that provides an indication of an organization's performance in relation to a specified process or outcome.*

performance specification A description of the desired results or performance of a product, material, assembly, or piece of equipment with criteria for verifying compliance.

perimeter drain A drainage system around a house usually comprised of perforated plastic pipe.

perimeter grouting Injection of grout, usually at relatively low pressure, around the periphery of an area that is subsequently to be grouted at greater pressure. Perimeter grouting is intended to confine subsequent grout injection within the perimeter.

perimeter heating system A system of warm-air heating in which outlets for air ducts are located near the outside walls of rooms and are close to the floor. The returns are near the ceiling.

perimeter installation A method of floor installation where adhesive is only used along the outside edges and seams of the flooring material, allowing for faster installations and easier repairs

perimeter isolation A method of installing a building material, such as concrete or wallboard, so that it is separated from structural elements. This isolation helps reduce the cracking that can result when structural members shift. Perimeter isolation is also used as a sound control measure.

perlite A volcanic glass having a perlitic structure, usually having a higher water content than obsidian when expanded by heating. Perlite is used as a lightweight aggregate in concretes, mortars, and plasters.

perlite composite board A rigid insulation board formed of expanded perlite, fibers, and a sizing material. Often recommended as a product for sustainable design because it is commonly manufactured with postconsumer paper.

perm A unit of water vapor transmission through a material, expressed in grains of vapor per hour per inch of mercury pressure difference.

permanent bracing Bracing that forms part of a structure's resistance to horizontal loads. Permanent bracing may also function as erection bracing.

permanent form Any form that remains in place after the concrete has developed its design strength. A permanent form may or may not become an integral part of the structure.

permanent load The load, including a dead load or any fixed load, that is constant through the life of a structure.

permanent shore An upright used to support dead loads during alterations to a structure and left in place.

permeable paving Paving materials that allow rainwater to pass through into the ground to replenish the water table.

permeability 1. The property of a material that permits passage of water vapor. 2. The property of soil that permits the flow of water.

permeance The resistance, measured in perms, to the flow of water vapor through a given thickness of material.

permit A document issued by a governing authority such as a building inspector approving specific construction. Among the types of permits are the building permit, demolition, zoning, grading, septic, plumbing and electrical permits.

perpend 1. A stone that extends completely through a wall and is exposed on each side of the wall. 2. A joint between masonry units.

personal protective equipment (PPE) Special clothing, devices and equipment worn by workers for protection against documented or potential site hazards. Examples include respirators and eye, ear and skin protection. *See also* **level A–D clothing**.

P

PERT (project evaluation and review technique) Along with CPM, PERT is a probabilistic technique for planning and evaluating progress of complex programs. Attempts to determine the time required to complete each element in terms of pessimistic, optimistic, and best-guess estimates.*

pervious soil A soil that allows relatively free passage of water.

pessimistic time estimate The maximum time required for an activity under adverse conditions. It is generally held that an activity would have no more than one chance in a hundred of exceeding this amount of time.*

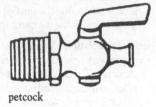

petcock

petcock A small valve installed on equipment or piping for drainage of liquids or air.

petroleum asphalt Asphalt refined directly from petroleum, as opposed to asphalt from natural deposits.

petroleum hydrocarbon Petroleum products are complex mixtures of hydrocarbon compounds, ranging from light, volatile, compounds to heavy, long-chained, branched compounds. The composition of petroleum hydrocarbons varies depending upon the source of the crude oil and the refining practices used. A number of solvents are refined from cured petroleum and used to lower the viscosity of oils and resins.

petroleum spirit A thinner, for paints and varnishes, having a low-aromatic hydrocarbon content, obtained in petroleum distillation.

petrology The science of rocks, dealing with their origin, structure, composition, etc., from all aspects and in all relations.

pew A benchlike seat used in a church.

pH A measure of hydrogen ion concentration expressed as an exponential number used to determine the relative acidity or alkalinity of a liquid.

phase A major period in the life of an asset or project. A phase may encompass several stages.

phase converter An electrical device that converts single-phase power to the smooth, continuous and universally adaptable three-phase power.

phasing Working on a construction project in parts and at specified intervals. Usually phasing is required when construction activity is not permitted at certain locations or on particular dates during the project.

phenol A product of the petroleum industry used in the production of phenolic resin, epoxy resins, plasters, and exterior plywood glue. Phenol is made from benzene. It exists naturally in coal tar and wood tar. However, phenol from these sources is rarely used in the production of glue.

phenolic insulation A rigid, closed cell foam insulation product made with phenolic plastics. No longer used in the roofing industry because when wet it will chemically attack steel decks. However, it has not posed problems when used over concrete or wood decks, or gypsum boards sealed with a vapor retarder.

phenolic resin glue An adhesive used for bonding exterior plywood. Phenolic resin is produced in a reaction between phenol and formaldehyde. An extender is usually added to the phenolic resin prior to use in the plywood-manufacturing process.

Philadelphia leveling rod A leveling rod in two sliding parts with color-coded graduations. The rod can be used as a self-reading leveling rod.

Philippine mahogany The wood of several types of trees found in the Philippines. The wood resembles mahogany in grain. Density varies from very light to quite heavy. The heavier, darker woods are durable and strong and used like mahogany. The lighter-weight colored woods are used for interior plywood.

Phillips head screw A screw with a recessed head and an X-shaped driving indentation.

phosphatizing A chemical treatment process used on steel to prevent corrosion.

phosphor mercury-vapor lamp A high-pressure mercury-vapor lamp with a phosphor-coated glass cover over the lamp proper. The phosphor in the cover adds colors not generated by the lamp.

photoelectric cell An electronic device for measuring illumination level or detecting interruption of a light beam. The electric output or resistance of the device varies according to the illumination.

photoelectric control An electric control that responds to a change in incident light.

photometer A device that measures luminous intensity, light distribution, color, and other qualities of a luminaire.

photo-oxidation Oxidation resulting from exposure to sunlight.

photovoltaics (PV) Devices that convert sunlight directly into electricity. PVs generate power without noise, pollution, or fuel consumption, and are useful where utility power is not available, reliable, or convenient.

physical percentage complete Percentage of work content of an activity or project achieved as of a particular date. Physical completion of any activity represents the most accurate, unbiased measure or appraisal in accordance with the accept method of measurement, tempered with judgment and experience. Physical completion is not linked to work hours budgeted or expended.*

physical progress The status of a task, activity, or discipline based on preestablished guidelines related to the amount or extent of work completed.*

physical restraint A situation in which a physical activity or work item must be completed before the next activity or work items in the sequence can begin (e.g., concrete must harden before removing formwork).*

phytoremediation A low-cost option for site cleanup when the site has low levels of contamination that are widely dispersed. Phytoremediation (a subset of bioremediation) uses plants to break down or uptake contaminants.

piano hinge A continuous strip hinge used in falling doors, etc.

pick A hand tool, consisting of a steel head pointed at one or both ends and mounted on a wooden handle, for loosening and breaking up compacted soil or rock.

picked finish A surface finish for stonemasonry in which the surface is covered with small pits made by striking it perpendicularly with a pick or chisel.

P

picket A sharpened or pointed stake, post, or pale, usually used as fencing.

picket fence A fence consisting of vertical piles, often sharpened at the upper end, supported by horizontal rails.

pickled A metal surface that has been treated with strong oxidizing agents to remove scale and provide a tough oxide film.

pickup 1. The amount by which an estimate for an item is higher than the actual cost; savings; underrun. 2. Unwanted adherence of solids to the open surface of a sealant. 3. Common term for a small open-body truck with a half-ton to one-ton capacity.

pickup load The heat consumption required to bring piping and radiators to their operating temperature when a heating system is first turned on.

picture window A large window, usually a fixed sheet of plate or insulating glass.

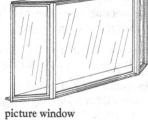

picture window

pieced timber 1. A timber made from two or more pieces of timber fitted together. 2. A damaged timber patched with a fitted piece of wood.

pien check In a stair constructed of stone, a rabbet cut in the front edge of a tread that fits over the riser below it.

pier 1. A short column to support a concentrated load. 2. Isolated foundation member of plain or reinforced concrete. 3. Means of support for bridge spans or the ends of a lintel or arch. 4. A marine dock or breakwater structure.

pierced louver A louver set in the face sheets or panels of a door.

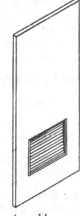

pierced louver

pigment A coloring matter, usually in the form of an insoluble fine powder, dispersed in a liquid vehicle to make paint.

pigment/binder ratio Amount of pigment in paint relative to the amount of binding agent.

pig spout A sheet metal flashing system that directs water out through the face of a gutter instead of through a downspout.

pigtail 1. An antisiphon piping device used to protect a pressure gauge. 2. A flexible conductor attached to an electric component or appliance to connect it to a circuit.

pigtail splice A connection of two electric conductors, made by placing the ends of the conductors side by side and twisting the ends about each other.

pike pole A long pole, with a spear-type point and a hook on one end, used to move logs around in a mill pond.

pilaster A column built within a wall, usually projecting beyond the wall.

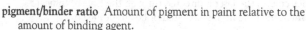

pilaster

pilaster block Concrete masonry units designed to form plain or reinforced concrete masonry pilasters of the projecting type.

pile A slender timber, concrete, or steel structural element, driven, jetted, or otherwise embedded on end in the ground for the purpose of supporting a load.

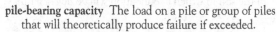

pile

pile-bearing capacity The load on a pile or group of piles that will theoretically produce failure if exceeded.

pile bent Two or more piles driven in a row transverse to the long dimension of the structure and fastened together by capping and (sometimes) bracing.

pile cap 1. A structural member placed on, and usually fastened to, the top of a pile or a group of piles and used to transmit loads into the pile or group of piles and, in the case of a group, to connect them into a bent. Also known as a *rider cap* or *girder*. 2. A masonry, timber, or concrete footing resting on a group of piles.

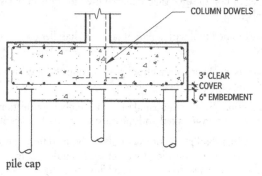

COLUMN DOWELS

3" CLEAR COVER

6" EMBEDMENT

pile cap

pile core The mandrel used to drive the shell of a cast-in-place concrete pile.

pile driver A machine for driving piles, usually by repeated blows, from a free-falling or driven hammer. A pile driver consists of a framework for holding and guiding the pile, a hammer, and a mobile plant to provide power.

pile extractor A machine for loosening piles in the ground by exerting upward striking blows. The actual removal is by a crane.

pile friction The friction forces on an embedded pile limited by the adhesion between soil and pile and/or the shear strength of the adjacent soil.

pile hammer A weight that strikes a pile to drive it into the ground. The weight may fall freely or be assisted by steam or air pressure.

pile height The height of piles in a carpet measured from the top surface of the backing to the top of the pile.

pile load test A static load test of a pile or group of piles used to establish an allowable load. The applied load is usually 150% to 200% of the allowable load.

pile point Hardened steel tip affixed to the end of a pile to reduce damage.

pile shoe A pointed or rounded device on the foot of a pile to protect the pile while driving.

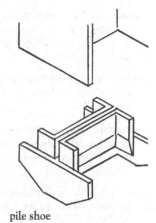

pile shoe

P

pile tolerance The permitted deviation of a pile in the horizontal and vertical planes.

pillar 1. A post or column that supports part of a structure. 2. A column of ore left in a mine to support the ground overhead.

pilot boring A preliminary boring or series of borings to determine the nature of the soil in which a foundation will be dug or a tunnel driven.

pilot hole A guiding hole for a nail or screws, or for drilling a larger hole.

pilot light 1. A small, constantly burning flame used as an ignition source in a gas burner. 2. A low-wattage light used to indicate that an electric circuit, control, or device is active.

pilot nail A temporary nail used to align boards until permanent nails are driven.

pin A peg or bolt of some rigid material used to connect or fasten members.

pincers A joined tool with a pair of jaws and handles used to grip an object.

pinch bar A steel bar with a chisel point at one end used as a lever for lifting or moving heavy objects.

pin connection In structural analysis, any member connection designed to transfer axial and shear forces, but not moments.

pine oil A high-boiling-point essential oil obtained from the steam distillation of pine needles, twigs, etc. Used industrially as a solvent. Used in paint to provide good flow properties and as an antiskinning agent.

pine shingles Shingles made from pine wood.

pine tar A blackish-brown liquid distilled from pine wood. Pine tar is used as an antiseptic externally and an expectorant internally and also to make the grips of tools sticky.

pin hinge A butt hinge with a pin for the pivot.

pin hinge

pinhole (pin hole) 1. In wood, a small, round hole made by a beetle or worm in the standing timber. 2. In plaster, a surface defect caused by trapped air. 3. In painted surfaces, a defect usually caused by impurities or dirt. 4. In glazed ceramic surfaces, a small round hole.

pin knot A knot with a diameter no larger than ½″.

pinnacle 1. The highest point. 2. A turret or elevated portion of a building. 3. A small ornamental body or shaft terminated by a cone or pyramid.

pintle A vertical pin fastened at the bottom and serving as a center of rotation.

pin tumbler A lock mechanism having a series of small pins that must be properly aligned by a key to open.

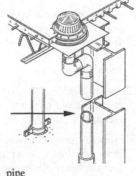

pipe 1. A hollow cylinder or tube for conveyance of a fluid. 2. From ASTM B 251–557: Seamless tube

pipe

conforming to the particular dimensions commonly know as "standard pipe size."

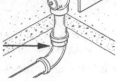

pipe bend (pipe elbow) A pipe fitting used to change direction.

pipe bend (pipe elbow)

pipe chase A vertical space in a building reserved for vertical runs of pipe.

pipe column A column made of steel pipe and often filled with concrete.

pipe coupling A fitting used to connect two lengths of pipe in a direct line.

pipe covering Any wrapping on a pipe that acts as thermal insulation and/or a vapor barrier.

pipe cross A fitting used to connect four lengths of pipe in the same plane with all lengths at right angles to each other.

pipe cutter A hand tool for cutting pipe or tubing, consisting of a frame with a cutting wheel and drive wheels. Cutting is accomplished by forcing the cutting wheel into the pipe material and rotating the tool around the pipe a number of times.

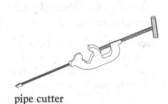

pipe cutter

pipe die An adjustable tool for cutting threads on or in a pipe.

pipe fitting Bends, tees, and other connectors used in assembling pipe.

pipe gasket A fabricated packing to seal joints in pipe.

pipe hanger A device or an assembly to support pipes from a slab, beam, or other structural element that is above the pipe.

pipe laser 1. A self-leveling instrument used in surveying that employs a laser as a reference for measurements or verifying alignment. 2. A plumber's tool used to simplify pipe alignment and installation. Once a hole is drilled in a stud or beam for pipe placement, the laser is directed through the hole to indicate where the hole in the next stud or beam should go.

pipelayer 1. A tradesperson skilled and trained in laying and joining pipes of glazed clay, concrete, iron, or steel in a trench. 2. An attachment for a tractor or other machine consisting of a winch and side boom for placing lengths of pipe in a trench.

pipeline heater A heater, usually a wrapping, with an electric element used to prevent the liquid in the pipe from freezing or to maintain the viscosity of the liquid.

pipe pile A cylinder, usually 10″ to 24″ in diameter, generally driven with open ends to form a friction pile. This pile may consist of several sections from 5′ to 40′ long joined by special fittings, such as cast-steel sleeves. A pipe pile is sometimes used with its lower end closed by a conical steel shoe.

pipe plug A pipe fitting with outside threads and a projecting head used to close the opening in another fitting.

pipe reducer A pipe fitting used to connect two lengths of pipe of different diameters.

pipe ring A circular-shaped metal part used to support a pipe from a suspended rod.

pipe run Any path taken by pipe in a distribution or collection system.

pipe saddle An assembly to support a pipe from the underside.

pipe scaffolding A flexible scaffold anchored against a building. This prevalent type of scaffold is also referred to as "metal tube" or "coupler scaffold."

pipe scaffolding

pipe sleeve A cylindrical insert cast in a concrete wall or floor for later passage of a pipe. Also a larger pipe usually used in embankments that allows a smaller pipe to be installed later or to protect it from heavy loads.

pipe stop A stopcock in a pipe.

pipe strap A thin metal strip used as a pipe hanger.

pipe tee A T-shaped fitting to connect three lengths of pipe in the same plane with one length at right angles to the other two.

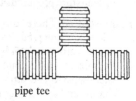

pipe tee

pipe thread A V-cut screw thread cut on the inside or outside of a pipe or fitting. The diameter of the thread tapers.

pipe wrench (alligator wrench) A heavy hand tool with adjustable serrated jaws for gripping, screwing, or unscrewing metal pipe.

pipe wrench (alligator wrench)

piping 1. An assembly of lengths of pipe and fittings, i.e., a run of pipe. **2.** Movement of soil particles by percolating water that produces erosion channels.

piping loss The heat lost from piping between the heat source and the radiators.

piston A solid cylinder that fits inside a larger cylinder and moves as a result of the power it receives. It can be used to transmit power to or from a connecting rod. Found in reciprocating engines, pumps, and compressors.

pit 1. An excavation, quarry, or mine made or worked by the open cut method. A pit seldom goes below the ground water level. **2.** The area between the stage and the first row of seats in a theater. **3.** A small hole or cavity on a surface.

pitch 1. An accumulation of resin in the wood cells in a more or less irregular patch. Pitch is classified for grading purposes as light, medium, heavy, or massed. **2.** The angle or inclination of a plane such as a roof, which varies according to climate, design, and materials used, and is expressed as a ratio of rise per run. **3.** The set, or projection, of teeth on alternate sides of a saw to

provide clearance for its body. **4.** The ratio of rise to run of stairs. **5.** The reciprocal of the number of threads per inch.

pitched roof A roof having one or more surfaces with a slope greater than 10° from the horizontal.

pitched roof

pitch pocket 1. A flanged metal device used to provide a watertight seal around columns or other roof penetrations. **2.** An opening between growth rings which usually contains or has contained resin, or bark, or both. A pitch pocket is classified for grading purposes as very small, small, medium, large, closed, open, or through.

pitch select D select or better, except that the grade admits any amount of medium to heavy pitch. Massed pitch is admissible but limited to half the area of an otherwise high line piece. Dimensions are 4″ and wider, 6′ and longer in multiples of one foot.

pith knot A minor defect in lumber, a pith knot is a knot whose only blemish is a small pith hole in the center.

pitot tube A device, used with a manometer or other pressure-reading device, to measure the velocity head of a flowing fluid.

pit-run gravel Ungraded gravel used as taken from a pit.

pitting Development of relatively small cavities in a surface due to phenomena such as corrosion, cavitation, or, in concrete, localized disintegration.

pivot A short shaft or pin about which a part rotates or swings.

pivoted door A door that swings on pivots, rather than a door hung on hinges.

pivoted window A window with a sash that rotates about fixed horizontal or vertical pivots.

pivot hinge A hinge with a short fixed shaft or pin upon which a part rotates or swings. Commonly used for cabinet doors.

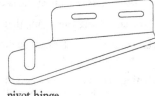

pivot hinge

placement 1. The process of placing and consolidating concrete. **2.** A quantity of concrete placed and finished during a continuous operation.

placing The deposition, distribution, and consolidation of freshly mixed concrete in the place where it is to harden. Also, inappropriately referred to as *pouring*.

placing drawings Detailed drawings used to position steel reinforcement in concrete construction.

plain bar A reinforcing bar without surface deformations, or one having deformations that do not conform to the applicable requirements.

plain concrete 1. Concrete without reinforcement. **2.** Reinforced concrete that does not conform to the definition of reinforced concrete. **3.** Used loosely to designate concrete containing no admixture and prepared without special treatment.

P

plain end (P.E.) Used to describe the ends of pipe that are shipped from the mill with unfinished ends. These ends may eventually be threaded, beveled, or grooved in the field.

plain rail A meeting rail in a double-hung window that is the same thickness as the other members of the frame.

plain-sawn Wood sawn from logs so that the annual rings intersect the wide faces at an angle less than 45°.

plan 1. A two-dimensional overview of the design, location, and dimensions of a project (or a portion of a project). 2. Formalized, written method of accomplishing a project task. 3. An intended future course of action. 4. The basis for project controls. 5. A generic term used for a statement of intentions whether they relate to time, cost, or quality in their many forms. 6. A predetermined course of action over a specified period of time which represents a projected response to an anticipated environment in order to accomplish a specific set of adaptive objectives.*

planar frame A structural frame with all members in the same plane.

plane 1. A flat surface. 2. A tool used to smooth or shape wood. 3. To run sawn wood through a planer to smooth its surface.

planed all round A piece that has been surfaced on all four sides.

planed lumber Lumber that has been run through a planer to finish one or several sides.

plane of weakness The plane along which a body under stress will tend to fracture. The plane of weakness may exist by design, by accident, or because of the nature of the structure and its loading.

planer A machine used to surface rough lumber.

planer heads Sets of cutting knives mounted on cylindrical heads that revolve at high speed to dress lumber fed through them. Top and bottom heads surface or pattern the two faces, while side heads dress or pattern the two edges or sides.

plane surveying Surveying that neglects the curvature of the earth.

planimeter A mechanical device that measures plane areas on a map or drawing.

planing The process of smoothing a surface by shaving off small chips.

plank A piece of lumber 2″ or more thick and 6″ or more wide, designed to be laid flat as part of a load-bearing surface, such as a bridge deck.

planking Material used for flooring, decking, or scaffolding.

planking

planking and strutting The temporary timbers supporting the soil at the side of an excavation.

planned cost The approved estimated cost for a work package or summary item. This cost when totaled with the estimated costs for all other work packages results in the total cost estimate committed under the contract for the program or project.*

planned value Measure of the value of work planned to have been performed so far.*

planner In project control, a team member with the responsibility for planning, scheduling and tracking of projects. They are often primarily concerned with schedule, progress, and manpower resources.*

planning The process of developing a scheme of a building or group of buildings by studying the layout of spaces within each building, and of building and other installations in an open space.

plan room A service provided by construction industry organizations or service companies, sometimes available to interested constructors, materialmen, vendors, and manufacturers. Plan rooms provide access to contract documents for projects currently in the process of receiving competitive or negotiated bids.

planting In masonry, laying the first courses of a foundation on a prepared bed.

plant mix 1. A mixture of aggregate and asphalt cement or liquid asphalt, prepared in a central or traveling mechanical mixer. 2. Any mixture produced at a mixing plant.

plant overhead Those costs in a plant that are not directly attributable to any one production or processing unit and are allocated on some arbitrary basis believed to be equitable. Includes plant management salaries, payroll department, local purchasing and accounting, etc.*

plan view A drawing that depicts an object, assembly, or floor plan from above.

plaster 1. A cementitious material or combination of cementitious material and aggregate that, when mixed with a suitable amount of water, forms a plastic mass or paste. When applied to a surface, the paste adheres to it and subsequently hardens, preserving in a rigid state the form or texture imposed during the period of elasticity. 2. The placed and hardened mixture created as in definition 1 above.

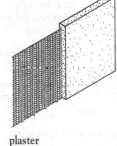

plaster

plaster base

plaster aggregate Graded mineral particles and mineral, vegetable, or animal fibers to be used with gypsum or cement-base plasters to produce a plaster mix.

plaster base Any working ground to receive plaster, including wood, metal, or gypsum lath, insulating board, or masonry.

plaster bead An edging, usually metal, to strengthen applied plaster at corners.

plasterboard (sheetrock, drywall) Any prefabricated board of plaster with paper facings. Plasterboard may be painted or used as a base for a finish coat of applied plaster.

plasterboard nail A nail for fastening plasterboard to a supporting system. The nails are galvanized with a flat head and a deformed shank.

P

plaster bond The mechanical or chemical adhesion of plaster to a surface.

plasterer's putty A hydrated lime with just enough water added to make a thick paste for use as a hole or crack filler.

plaster ground A wood strip or metal bead used as a guide for application of a desired thickness of plaster or for attaching trim.

plaster lath A supporting structure for plaster, such as a wood lath, metal lath, or lath board.

plaster of paris Gypsum, from which three-quarters of the chemically bound water has been driven off by heating. When wetted, it recombines with water and hardens quickly. *See also* **hemihydrate**.

plaster ring A metal collar attached to a base and used as a guide for thickness of applied plaster and a fastener for trim.

plaster set The initial stiffening of a plaster mix that may be reworked without the addition of water.

plastic bond fire clay 1. A fire clay of sufficient natural plasticity to bond nonplastic material. 2. A fire clay used as a plasticizing agent in mortar.

plastic cement A synthetic cement used in the application of flashing.

plastic consistency 1. Condition of freshly mixed cement paste, mortar, or concrete that allows that deformation to be sustained continuously in any direction without rupture. 2. In common usage, concrete with slump of 3″ to 4″ (80 to 100 mm).

plastic cracking Cracking that occurs in the surface of fresh concrete soon after it is placed and while it is still plastic.

plastic curtains Curtains or strip doors that reduce infiltration and exfiltration within a building. These barriers typically consist of several strips of heavy plastic (often transparent or translucent) that form a fairly tight seal, yet allow easy passage.

plastic deformation Deformation that does not disappear when the force causing the deformation is removed.

plastic glue Resin bonding materials used in joining wood pieces. These materials include: 1. Thermosetting resins such as phenol-formaldehyde, urea-formaldehyde, and melamine resin. 2. Thermoplastics such as acryl polymers and vinyl polymers. 3. Casein plastics. 4. Natural resin glues.

plasticity 1. The capability of being molded, or being made to assume a desired form. 2. A property of a material that allows it to retain its form when bent. 3. A complex property of a material involving a combination of qualities of mobility and magnitude of yield value. 4. That property of freshly mixed cement paste, concrete, or mortar that determines its resistance to deformation and ease of molding.

plasticizer 1. A material that increases plasticity of a cement paste, mortar, or concrete mixture. 2. Various substances added to organic compounds to create a more flexible finished product. These additives are frequently used in roofing materials and concrete.

plasticizing 1. Producing plasticity or becoming plastic. 2. Softening wood by hot water, steam, or chemicals to increase its moldability.

plastic laminate A thin board used as a finished surfacing, made from layers of resin-impregnated paper fused together under heat and pressure.

plastic limit The water content at which a soil will just begin to crumble when rolled into a thread approximately ⅛″ (3 mm) in diameter. *See also* **Atterberg limits**.

plastic lumber An alternative building material that is usually manufactured with recycled plastics. In contrast to wood, it will not splinter, rot, or warp. Used in decking and pilings, among other applications.

plastic skylight A molded unit of transparent or translucent plastic that is set in a frame for use as a skylight.

plastic soil Any soil that can be molded or deformed by moderate pressure without crumbling.

plastic wood A quick-drying putty of nitrocellulose, wood flour, resins, and solvents used as a filler for holes and cracks.

plastic skylight

plate 1. In formwork for concrete, a flat, horizontal member at the top and/or bottom of studs or posts. If on the ground, a plate is called a mudsill. 2. In structural design, a member, the depth of which is substantially smaller than its length and width. 3. In framing, the top plate (horizontal) connects with the top of wall studs. The floor joists, rafters, or trusses rest on it. The sole plate is at the bottom of wall studs. The still plate (horizontal) rests on and is anchored to the foundation. 4. A flat rolled iron or steel product.

plate anchor An anchor bolt used to fasten a plate or sill to a foundation.

plate girder A girder fabricated from plates, angles, or other structural shapes, welded or riveted together.

plate glass High-quality glass of the same composition as window glass but thicker, up to 1¼″, with ground and polished faces, usually used for large areas in a single sheet.

plate stock Component that makes up the bottom and top of a typical wood framed wall. Usually the same dimension as the wall framing stock but may be a lesser grade.

plate-type tread A stair tread fabricated from metal plate and/or floor plate. The riser may be integral.

plate girder

plate vibrator A self-propelled, mechanical vibrator used to compact fill.

platform 1. A floor or surface raised above the adjacent level. 2. A landing in a stairway. 3. A working space for persons, elevated above the surrounding floor or ground level such as a balcony or platform for the operation of machinery or equipment.

platform framing A framing system in which the vertical members are only a single story high, with each finished floor acting as a platform upon which the succeeding floor is constructed. Platform framing is the common method of house construction in North America.

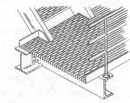
platform

plenishing nail A large nail used to fasten planks to joists.

plenum 1. A closed chamber used to distribute or collect warmed or cooled air in a forced air heating/cooling system. **2.** The space between the suspended ceiling and the floor above. **3.** The space between a raised floor and the floor below. **4.** A closed chamber used to collect or distribute fluids in a distribution or collection system.

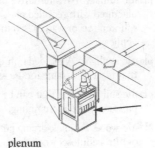

plenum barrier A barrier, erected in a plenum ceiling, used to reduce sound transmission between rooms or over a large area.

plenum

plenum chamber *See* **plenum.**

plenum fans (plug fans) Single-inlet, single-width centrifugal fans without the scroll, permitting 360° air delivery from the fan wheel.

pliers A pincer-like hand tool with opposing jaws for gripping, cutting, and bending.

plinth 1. A block or slab supporting a column or pedestal. **2.** The base course of an external masonry wall when of different shape from the masonry in the wall proper. **3.** The base of a monument or statue, often with inscription.

plinth course 1. The masonry course that forms the plinth of a stone wall. **2.** The final course of a brick plinth in a brick wall.

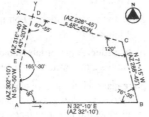

plot

plot 1. A measured and defined area of land. **2.** A ground plan of a building and adjacent land.

plot plan A diagram showing the proposed or existing use of a specified parcel of land.

plow 1. In molding, a rectangular slot of three surfaces cut with the grain of the wood. **2.** In carpentry, a tool that cuts grooves. **3.** A drywall tool with a bent trowel used to finish corners.

plug 1. A wood peg driven into a wall for support of a fastener. **2.** A stopper for a drain opening. **3.** A male-threaded fitting used to seal the end of a pipe or fitting. **4.** A fixture for connection of electric wires to an outlet socket. **5.** A fibrous or resinous material used to fill a hole and close a surface. **6.** Material that stops or seals the discharge line of a channel or pipe.

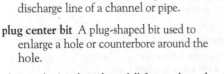

plug center bit A plug-shaped bit used to enlarge a hole or counterbore around the hole.

plug cock A valve where full flow is through a hole in a tapered plug. Rotating the plug 90° completely stops the flow.

plug cock

plug fuse A fuse contained in an insulated container with a metal screwbase. There is a small window on the face of the container for checking the condition of the fuse element.

plugged lumber Lumber in which a defect has been filled by material to provide a smooth paint surface.

plugging chisel A steel rod, with a star-shaped point, used for drilling holes in masonry by striking with a hammer.

plug tenon A short tenon that projects from the material into which it is fitted, the free end fitting into a mortise. A plug tenon is used to provide lateral stability for a wood column.

plug weld A weld made through a circular hole in one of the members to be connected.

plumb Vertical, or to make vertical.

plumb bob A cone-shaped metal weight, hung from a string, used to establish a vertical line or as a sighting reference to a surveyor's transit.

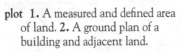

plumb bob

plumb bond Any bond in masonry in which the vertical joints are in line.

plumb cut A vertical cut, as in the cuts in a rafter at the top ridge where it meets the ridge plate.

plumber's rasp A coarse rasp used to file lead.

plumber's round iron A specially shaped soldering iron used to solder seams in tanks.

plumber's solder An alloy with a low melting point used for joining metal pieces such as copper pipes.

plumbing 1. The work or practice of installing in buildings the pipes, fixtures, and other apparatuses required to bring in the water supplies and to remove water-borne wastes. **2.** The process of setting a structure or object truly vertical.

plumbing boot Metal support installed to reinforce wall studs in cases where a cut has been made for a plumbing drain line.

plumbing fixture A receptacle in a plumbing system, other than a trap, in which water or wastes are collected or retained for use and ultimately discharged to drainage.

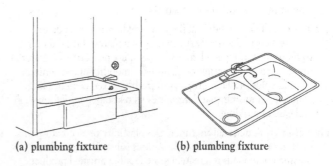

(a) plumbing fixture (b) plumbing fixture

plumbing ground Drain and waste lines underneath a basement floor.

plumbing jack A sleeve surrounding a drain or vent pipe on a roof.

plumbing, rough Advance work done by a plumbing contractor. Includes installation of waste and supply piping, shower pans and tubs, and gas piping.

plumbing stack A vent pipe installed through the roof materials.

plumbing system Arrangements of pipes, fixtures, fittings, valves, and traps in a building that supply water and remove liquid-borne wastes.

plumbing trim Last stage of the plumbing contractor's work prior to final inspection. It includes pipe connections to fixtures and appliances.

plumbing waste line Plastic piping for removal of sewage.

plumb joint A sheet metal joint made by lapping the edges and soldering them together flat.

plumb level A level that is set in a horizontal position by placing it at a right angle to a plumb line.

plumb rule A board or metal rule, fitted with one or more leveling bubbles, used to establish horizontal and vertical lines.

plume 1. The effluent mixture of heated air and water vapor discharged from a cooling tower. 2. An identifiable and definable stream of pollutants in an otherwise clean volume of air or water.

plunger (plumber's friend) A tool, consisting of a large rubber suction cup on a wood handle, for clearing plumbing traps of minor obstructions.

ply 1. A single layer or sheet of veneer. 2. One complete layer of veneer in a sheet of plywood.

plymetal Plywood covered on one or both sides with sheet metal.

plywood A flat panel made up of a number of thin sheets (veneers), of wood. The grain direction of each ply, or layer, is at right angles to the one adjacent to it. The veneer sheets are united under pressure by a bonding agent. Interior-grade plywood is suitable for indoor use or for outside use when subject only to occasional, temporary moisture, while exterior-grade plywood uses a weather-resistant adhesive.

plywood grade The widely recognized grading system administered by the American Plywood Association that rates plywood on the quality of its veneer (from A to D) and on its exposure durability.

pneumatic control system A system in which control is effected by pressurized air.

pneumatic drill A reciprocating drill actuated by compressed air.

pneumatic feed Delivery equipment in which material is conveyed by a pressurized air stream.

pneumatic hammer An air-powered tool with a linear driven shaft fitted with a chisel or hammer.

pneumatic structure A fabric envelope supported by an internal air pressure slightly above atmospheric pressure. The pressure is provided by a series of fans.

pneumatic water supply A water supply system for a building in which water is distributed from a tank containing water and compressed air.

pocket 1. A recess in a wall to receive an end of a beam. 2. A recess in a wall to receive part or all of an architectural item, such as a curtain or folding door. 3. The slot on the pulley stile of a double-hung window through which the sash weight is placed in the sash weight channel.

pocket channel A U-shaped opening in a window frame or sash where the glazing is inserted.

pocket chisel A chisel with a wide blade that is sharpened on both sides.

pocket door A door that opens by sliding into and "hiding" in a wall recess.

pocket piece A small piece of wood that closes the *pocket* in the pulley stile of a double-hung window.

podium 1. A stand for a speaker. 2. An elevated platform for a conductor. 3. The masonry platform on which a classical temple was built.

point 1. A fee equal to 1% of the principal amount of a loan. Charged by the lender when the loan is made. 2. A tooth for a saw. 3. A mason's tool. 4. A thin, triangular or diamond-shaped piece of metal used in glazing to hold glass in a wooden frame. 5. A piece of equipment that is monitored or controlled by a building automation system.

pointed work The rough finish on the face of a stone that is made by a pointed tool.

pointing 1. The finishing of joints in a masonry wall. 2. The material with which joints in masonry are finished.

pointing trowel A diamond-shaped trowel used in pointing or repointing masonry joints.

point load A term used in structural analysis to define a concentrated load on a structural member.

point of inflection (point of contraflexure) 1. The point on the length of a structural member subjected to flexure where the curvature changes from concave to convex or conversely, and at which the bending moment is zero. 2. Location of an abrupt bend in a plotted locus of points in a graph.

point of support A point on a member where part of its load is transferred to a support.

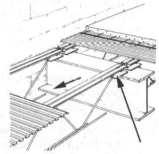

point of support

point source 1. A light source, the dimensions of which are insignificant at viewing distance. A fluorescent lamp is a point source at a large distance. 2. A pollution source that is discharged at an identifiable point such as a single pipe or smoke stack.

Poisson's ratio The ratio of transverse (lateral) strain to the corresponding axial (longitudinal) strain resulting from uniformly distributed axial stress below the proportional limit of the material.

polarity The direction of electric current flow in a DC circuit.

polarized receptacle An electric receptacle with contacts arranged so a mating plug must be inserted in only one orientation.

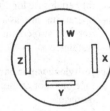

polarized receptacle

pole 1. A long, usually round piece of wood, often a small diameter log with the bark removed, used to carry utility wires or for other purposes. A pole is often treated with preservative. 2. Either of two oppositely charged terminals, as in an electric cell or battery. 3. Either extremity of an axis of a sphere.

pole-frame construction A construction system using vertical poles or timbers.

pole plate A horizontal board or timber that rests on the tie beams of a roof and supports the lower ends of the common rafters at the wall, and also raises the rafters above the top plate of the wall.

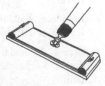

pole sander A sanding block on the end of a long pole that makes the sanding of drywall ceiling joints easier.

pole sander

polish 1. To give a sheen or gloss to a finish coat of plaster. **2.** The operation in which fine abrasives are used to hone a finished surface to a desired smoothness.

polished finish A finish so smooth that it forms a reflective surface, usually produced by mechanical buffing and chemical treatment of a surface with no voids.

polishing varnish A hard varnish that can be polished by rubbing with abrasive and mineral oil without dissolving the resin.

polycarbonate A transparent thermoplastic with a high impact strength and a high modulus of elasticity. Its excellent insulating qualities make it ideal for many electrical applications.

polychlorinated biphenyls (PCBs) A group of hydrocarbon-containing chlorine compounds that, before being banned as pollutants, were used in hundreds of industrial and commercial applications, including electrical and hydraulic equipment, sealants, rubber, paints, and plastics. PCBs are not readily biodegradable, and the United States stopped producing them in 1977.

polychloroprene (neoprene) An oil-resistant, synthetic rubber. In roofing, used for membranes and flashing. The common name for polychloroprene is neoprene.

polyester resin A synthetic resin that polymerizes during curing and has excellent adhesive properties, high strength, and good chemical resistance.

polyethylene A thermoplastic high-molecular-weight organic compound. In sheet form, polyethylene is used as a protective cover for concrete surfaces during the curing period, a temporary enclosure for construction operations, and as a vapor barrier. It is also commonly used for culvert pipes and in other piping systems.

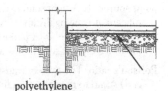

polyethylene

polyethylene vapor barrier A plastic film used to prevent the passage of vapor or moisture into areas where it could collect and do damage.

polyisobutylene (PIB) A synthetic rubber derived from the polymerization of isobutylene. In roofing, used for membranes and flashing.

polyisocyanurate (polyiso) A polymer with a high R-value commonly used as insulation. It is available as a liquid, sprayed foam or rigid foam boards, usually faced with a foil paper. Increasingly, polyiso board products are also being used for sheathing. Often specified for applications where increased fire resistance is desired.

polymer 1. The product of polymerization. Some polymers are elastomers, while others are plastics. **2.** A rubber or resin consisting of large molecules formed by polymerization.

polymer-cement concrete A mixture of water, hydraulic cement, aggregate, and a monomer or polymer. The cement is polymerized in place when a monomer is used.

polymethyl-methacrylate (PMMA) A transparent thermoplastic that offers good weather resistance and a high strength-to-weight ratio.

polypropylene A tough plastic with good resistance to heat and chemicals. Polypropylene is a polymer of propylene, and is found in everything from packaging to molded automobile parts.

polystyrene foam A low-cost, foamed plastic weighing about 1 lb. per cu. ft., with good insulating properties and resistance to grease.

polystyrene resin Synthetic resins, varying in color from water-white to yellow, formed by the polymerization of styrene on heating, with or without catalysts. These resins may be used in paints for concrete, for making sculptured molds, or as insulation.

polysulfide coating A protective coating system prepared by polymerizing a chlorinated alkyl polyether with an inorganic polysulfide. This coating exhibits outstanding resistance to ozone, sunlight, oxidation, and weathering.

polytetrafluoroethylene (PTFE) A fluorocarbon-based polymer with high chemical and weather resistance, low friction, and electrical and thermal insulation. The most common brand name is Teflon®. PTFE's many construction applications include usage in hydraulic machinery and in plumber's joint tape.

polyurethane Reaction product of an isocyanate with any of a wide variety of other compounds containing an active hydrogen group. Polyurethane is used to formulate tough, abrasion-resistant coatings, and is also used for foam insualtion products.

polyurethane insulation (PUR insulation) Any of a number of insulation products made of polyurethane. Forms include rigid boards, spray foams, and pourable mixes.

polyvinyl acetate (PVA) Colorless, permanently thermoplastic resin, usually supplied as an emulsion or water-dispersible powder, which may be used in paints for concrete. Polyvinyl acetate is characterized by flexibility, stability towards light, transparency to ultraviolet rays, high electric strength, toughness, and hardness. The higher the degree of polymerization, the higher the softening temperature.

polyvinyl chloride (PVC) A thermoplastic resin derived from the polymerization of vinyl chloride. Plasticizers have been added to give it flexibility. Widely used in piping products, it is also used for siding, floor covering, window housing, and fencing. In roofing, it is used for membranes and flashing. Also used in the manufacture of nonmetallic waterstops for concrete. PVC materials are to be avoided in green construction because it has been linked to cancer, birth defects, and groundwater contamination.

P

pommel **1.** A knob at the top of a conical or domelike roof. **2.** A rounded metal block on an end of a handle, raised and dropped by hand to compact soil.

ponderosa pine (western white pine, western yellow pine) *Pinus ponderosa.* A pine species found in a wide range that reaches from British Columbia to Mexico, and from the Pacific coast to the Dakotas. The wood is widely used in general construction, most often as boards, but is more valued for its uses in millwork and in cuttings for remanufacture.

ponding **1.** The process of flooding the surface of a concrete slab by using temporary dams around the perimeter in order to satisfactorily cure the concrete. **2.** The accumulation of water at low points in a roof. The low points may be produced or increased by structural deflections.

pop (blow, blister) A delaminated area in a plywood panel.

popcorn concrete No-fines concrete containing insufficient cement paste to fill voids among the coarse aggregate so that the particles are bound only at points of contact.

poplar A member of the willow family. Its wood is used in furniture core stock, crates, and plywood. In North America: *Populus tretnula,* aspen; *P. balsamifera,* cottonwood; and *P. tacamahaca, balsam poplar.*

popout The breaking away of small portions of a concrete surface due to internal pressure, leaving a shallow, typically conical, depression.

popping Shallow depressions ranging in size from pinheads to ¼″ in diameter, immediately below the surface of a lime-putty finish coat. Popping is caused by expansion of coarse particles of unhydrated lime or of foreign substances.

pop rivet A fastener installed with a rivet gun to connect metal pieces.

pop valve A safety valve made to open immediately when the fluid pressure is greater than the design force of a spring.

porcelain A hard glazed or unglazed ceramic used for electrical, chemical, mechanical, or thermal components.

porcelain enamel A silicate glass bonded to metal by fusion at a temperature above 800°F (427 °C). Porcelain enamel is not a true porcelain.

porcelain tile A dense, usually impervious, fine-grained, smooth-surfaced, ceramic mosaic tile or paver.

porch A structure attached to a building, usually roofed and open-sided, and often at the entrance. Sometimes screened or glass-enclosed.

porch

porcupine boiler A vertical, cylindrical boiler with many projecting, closed stubs to provide an additional thermal surface.

porosity The ratio, usually expressed as a percentage, of the volume of voids in a material to the total volume of the material, including the voids.

porous paving Paving surfaces designed to allow storm water infiltration and reduce runoff.

portico **1.** A covered walk consisting of a roof supported on columns. **2.** A colonnaded (continuous row of columns) porch.

portico

portlandite A mineral, calcium hydroxide, that occurs naturally in Ireland. Portlandite is a common product of hydration of Portland cement.

Portland stone A limestone, from the island of Portland off the coast of England, used as a building stone.

position **1.** A trader's open contracts in the futures market. **2.** A reference to a shipping period, as in "Feb/March position."

positioned weld A weld on a joint that has been oriented to facilitate the welding.

position indicator A device that shows the position of an elevator in its hoistway. Also called a *hall position indicator* if at a landing, or a *cab position indicator* if in the cab.

positive cutoff A below-ground wall that extends to an impervious lower stratum to block subsurface seepage.

positive displacement Moving a fluid by capturing and then discharging a fixed amount of fluid. A piston pump is one example of a positive displacement pump.

positive moment A condition of flexure in which, for a horizontal simply supported member, the deflected shape is normally considered to be concave downward and the top fibers subjected to compression stresses. For other members and other conditions, consider positive and negative as relative terms.

positive float Amount of time available to complete noncritical activities or work items without affecting the total project duration.*

positive pressure Pressure that is greater than atmospheric pressure.

positive reinforcement Reinforcement for positive moment.

post **1.** A member used in a vertical position to support a beam or other structural member in a building, or as part of a fence. In lumber, 4 × 4s are often referred to as posts. Most grading rules define a post as having dimensions of 5″ × 5″ or more in width, with the width not more than 2″ greater than the thickness. **2.** Vertical formwork member used as a brace. Also called a shore, prop, and jack. **3.** A secondary column located at the end of a building to support its girts.

post and beam framing A structural framing system in which beams rest on posts rather than bearing walls.

post and pane A type of construction in which timber framings are filled in with brick or plaster panels, leaving the timbers exposed.

postbuckling strength The load that can be carried by a structural member after it has been subjected to buckling.

postconstruction services 1. Services rendered after the release of the final invoice for payment or over 60 days from the date of substantial completion of the project. **2.** Any services necessary to allow the owner to use and/or occupy the facility.

postconsumer recycled content Material that has been used by consumers, such as used newspaper, and has been diverted or separated from waste management systems for recycling.

post shore, adjustable timber single-post Individual timber used with a fabricated clamp to obtain adjustment and not normally manufactured as a complete unit.

post shore, adjustable timber single-post

post shore, fabricated single-post Type 1: Single all-metal post, with a fine-adjustment screw or device in combination with pin-and-hole adjustment or clamp. Type 2: Single or double wooden post members adjustable by a metal clamp or screw and usually manufactured as a complete unit.

post-tensioned concrete Concrete that has the reinforcing tendons tensioned after the concrete has set.

potable water Water that satisfies the standards of the responsible health authorities as drinking water.

potato masher A simple hand tool used to mix joint compound. The tool has a wire mixer similar to that of the kitchen implement of the same name.

potentiometer An instrument for controlling electrical potential. Measures an unknown voltage by comparing it to a standard voltage.

pot life Time interval, after preparation, during which a liquid or plastic mixture is usable.

pound-calorie The amount of heat required to raise one pound of water 1°C.

pour coat (top mop) The top coating of asphalt on a built-up roof, sometimes including embedded gravel or slag.

pour point The lowest temperature at which a lubricant flows under specified conditions.

pour strip In concrete formwork, a narrow guide placed inside the form to direct the concrete.

powder post A condition in which wood has decayed to powder or been eaten by borers that leave holes full of powder.

power The rate of performing work or the rate of transforming, transferring, or consuming energy. Power is usually measured in watts, Btu/hour, or horsepower.

power buggy A wheelbarrow-sized machine powered by a gasoline engine or an electric motor.

power cable A usually heavy cable, consisting of one or more conductors with insulation and jackets, for conducting electric power.

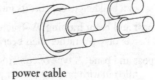

power cable

power consumption The rate at which power is consumed by a device or unit (such as a building), usually expressed in kilowatt-hours, Btu/hour, or horsepower-hours.

power drops Electrical power outlets to serve specific pieces of equipment.

power factor A calculation, expressed as a percentage, that relates the volt amperes of an AC circuit, or the apparent power, to the wattage, or the true power. In essence, it provides a measurement of how effectively electrical power is being used. A higher power factor means a more effective use of electrical power.

power panelboard A panelboard used for circuits supplying motors and other heavy power-consuming devices, as opposed to a panelboard used for lighting circuits.

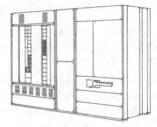

power panelboard

power sander An electric-powered hand tool used for smoothing and/or polishing.

power shovel A self-propelled, power-operated machine used to excavate and/or load soils or debris.

power take-off On construction equipment, an attachment enabling the power from the prime mover to be used to drive an auxiliary machine or tool.

power transformer A device in an alternating-current electrical system that transfers electric energy between circuits, usually changing the voltage in the process.

power transformer

power vent A vent with a fan that boosts the flow of air.

pozzolan A siliceous, or siliceous and aluminous, material which, in itself, possesses little or no cementitious value but will, in finely divided form and in the presence of moisture, chemically react with calcium hydroxide at ordinary temperatures to form compounds possessing cementitious properties.

pozzolan cement A natural cement, used in ancient times, made by grinding pozzolan with lime.

Pratt truss A type of truss with parallel chords, all vertical members in compression, and all diagonal members in tension. The diagonals slant toward the center.

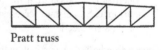

Pratt truss

practical coverage The actual coverage of a paint or other coating based on the intended dry film thickness and with an allocation (usually 15%) for material loss.

preaction sprinkler system A dry pipe sprinkler system in which water is supplied to the piping when a smoke or heat detector is activated.

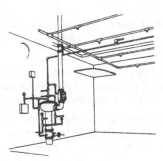

preaction sprinkler system

precast concrete Concrete structural components, such as piles, wall panels, beams, etc., fabricated at a location other than in place.

precise level An instrument similar to an ordinary surveyor's level but capable of finer readings and including a prism arrangement that permits simultaneous observation of the rod and the leveling bubble.

precompressed zone The area of a flexural member that is compressed by the prestressing tendons.

preconsolidation pressure The greatest effective pressure a soil has experienced.

pre-construction services Services provided by the construction contractor during the development of the design, which may include preliminary cost estimating, preparing a preliminary construction schedule, and constructability review.

precooling coil In an HVAC system, a cooling coil located at the air-entering side of the primary cooling coil.

precuring In plywood manufacturing, the premature curing of an adhesive due to press temperatures being too high, a too rapid resin-curing speed, or a malfunctioning press. Precuring can result in plywood delamination or a poor quality surface in particleboard.

precut A lumber item, usually a stud, that is cut to a precise length at the time of manufacture, so that it may be used in construction without further trimming at the job site.

precycling Proactive approach of selecting products and materials according to their potential for lessening the amount of material that goes into the waste stream and for future recycling. Precycling includes buying in bulk, avoiding one-time use products, and choosing products that are biodegradable and have the least amount of throwaway packaging, for example.

predesign services Services provided by the design professional that precede customary services. Predesign services include assistance of the owner in establishing the program, schedule, budget, and project limitations.

predrilled Materials, such as roof decking, that have been drilled at the mill to accommodate bolts or other hardware.

prefabricate To fabricate units or components at a mill or plant for assembly at another location.

prefabricated construction A construction method that uses standard prefabricated units that are assembled at a site along with site fabrication of some minor parts.

prefabricated flue A vent for fuel-fired equipment that is assembled from factory-made parts.

prefabricated joint filler A compressible material used to fill control, expansion, and contraction joints and may also be used alone, or as a backing for a joint sealant.

prefabricated masonry panel A wall panel of masonry units constructed at an assembly site and moved to a job site for erection.

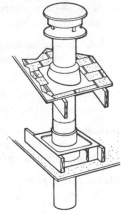

prefabricated flue

prefabricated modular units Units of construction that are preassembled at the factory and shipped as a complete unit to the job site. They usually can be installed with a minimum of adjustments.

prefabricated tie A manufactured assembly consisting of two heavy parallel wires tied together by welded wires. The tie is laid in masonry joints to tie two wythes together.

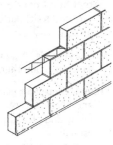

prefabricated tie

prefilled A particleboard panel whose surface has been made smooth by the application of a solvent-based filler before being shipped. Such panels have decorative overlays or laminates applied to them.

prefilter A filter placed before the main filter(s). A prefilter is coarser and is used to remove larger particles.

prefinished Products with a finish coating of paint, stain, vinyl, or other material applied before they are taken to the job site.

prefiring Raising the temperature of refractory concrete under controlled conditions prior to placing it in service.

preformed asphalt joint filler Premolded strips of asphalt, vegetable or mineral filler, and fibers for use as a joint filler.

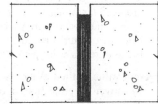

preformed asphalt joint filler

preformed foam Foam produced before it is mixed with other ingredients to make cellular foam.

preformed sealant A factory-shaped sealant that requires little field fabrication prior to installation.

preheat coil A coil, in an air-conditioning system, used to preheat air that is below 32°F (0°C).

preheater 1. A heat exchanger used to heat air that is to be used in the combustion chamber of a large boiler or furnace. 2. A heat exchanger used to heat a fluid flowing into a process or location.

An example would be to preheat outdoor air entering a heated space using the exhaust air.

preheat fluorescent lamp A fluorescent lamp, the electrodes of which must be preheated before the arc can be started. The preheating can be manual or automatic.

prehung door A packaged unit consisting of a finished door on a frame with all necessary hardware and trim.

preliminary site assessment The first phase in an environmental remediation operation, in which it is determined whether there is a reasonable probability that a hazardous waste exists at a site.

preliminary drawings Drawings prepared in the early phase of building design.

preliminary estimate A rough estimate made in an early stage of the design work, prior to receipt of firm bids.

premises wiring Strictly speaking, the external and internal wiring that runs from the service point of utility conductors into a structure and to its outlets. Also includes all associated hardware and fittings.

premium grade A general term describing the quality of one item as superior to another.

premolded asphalt panel A panel with a core of asphalt, minerals, and fibers, covered on each side with asphalt-impregnated felt or fabric and pressure-bonded. The outside is then coated with hot asphalt.

preposttensioning A method of fabricating prestressed concrete in which some of the tendons are pretensioned and a portion of the tendons are posttensioned.

prepreg In reinforced plastic, the reinforcing with applied resin before molding.

prequalification of bidders The investigation and subsequent approval of prospective bidders' qualifications, experience, availability, and capability regarding a project.

present value The value of a benefit or cost found by discounting future cash flows to the base time. Also, the system of comparing proposed investments, which involves discounting at a known interest rate (representing a cost of capital or a minimum acceptable rate of return) in order to choose the alternative having the highest present value per unit of investment.*

present value method A means of evaluating capital expenditures by converting projections of cash inflows and outflows over time to their present value, using an estimated discounting rate.

preservationist A term applied to one who objects to the use of natural resources because of a belief that such use will destroy basic values of the resource. The term is often used to refer to a member of various groups opposed to the expansion of industrial/commercial uses of public lands.

preservative Any substance applied to wood that helps it resist decay, rotting, or harmful insects.

preshrimmed tape sealant A preformed sealant that resists deformation when compressed.

preshrunk concrete 1. Concrete that has been mixed for a short period in a stationary mixer before being transferred to a transit mixer. 2. Grout, mortar, or concrete that has been mixed one to three hours before placing to reduce shrinkage during hardening.

press brake A machine used to bend and shape cold-form metal sheets and strips.

pressed brick Brick that is molded under mechanical pressure. The resulting product is sharp-edged and smooth, and is used for exposed surfaces.

pressed earthen block Blocks of earthen material made by compressing a mixture of soil and aggregate without the use of chemical additives. Application and usage is similar to that of adobe.

pressed wood A panel wood product manufactured by compressing wood fibers and adhesive under heat. Common examples include particleboard, hardwood plywood paneling, and medium-density fiberboard.

pressure 1. The force per unit area exerted by a liquid or gas on the walls of a container. 2. The force per unit area transferred between surfaces.

pressure bulb The zone in a loaded soil mass that is bounded by a selected stress isobar.

pressure cell An instrument used to measure the pressure within a soil mass or the pressure of the soil against a rigid wall.

pressure connector A mechanical device that forms a conductive connection between two or more electric conductors, or between one or more conductors and a terminal, without the use of solder.

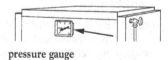

pressure connector

pressure differential valve A valve controlled by the pressure difference in the supply and return main to divert flow from the supply main to the return main.

pressure drop 1. The drop in pressure between two ends of a pipe or duct, between two points in a system, or across valves, fittings, etc., caused by friction losses. 2. In a water system, the drop caused by a difference in elevation.

pressure forming A thermoforming process for plastics in which pressure forces a sheet against a mold, as opposed to vacuum forming.

pressure gauge An instrument for measuring fluid pressure.

pressure gauge

pressure gluing A method used to glue wood that places the members under high pressure until the glue sets.

pressure-injected footing A cast-in-situ concrete pile that features a steel cylindrical shaft with an enlarged base that is driven into the ground. It can be used in nearly all soil conditions and can safely withstand very high compressive and tensile forces and substantial horizontal loads.

pressure-reducing valve A valve that maintains a uniform fluid pressure on its outlet side as long as pressure on the inlet side is at or above a design pressure.

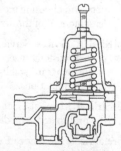

pressure-reducing valve

P

pressure-regulating valve A valve that automatically reduces water pressure to maintain a predetermined design pressure.

pressure-relief damper A damper that will open when pressure on the inside exceeds a design pressure.

pressure-relief device A device or valve that is designed to open or rupture when pressure on a designated side exceeds a design value.

pressure-relief hatch A roof hatch designed to open or blow off under pressure from an explosion in a building. Some smoke and heat vents are also designed as pressure-relief hatches.

pressure-relieving joint A horizontal expansion joint in panel wall masonry, usually below supporting hangers at each floor. These joints prevent the weight of higher panels from being transmitted to the masonry below.

pressure reset system A control system for boilers and furnaces that allows wide fluctuations in pressure. As a result, the burners can be shut off for longer periods and stay on for longer amounts of time, with fewer cycles. Avoiding short cycles increases the net system efficiency.

pressure-sensitive adhesive An adhesive material that remains tacky after the solvents evaporate and will adhere to most solid surfaces with the application of light pressure.

pressure treating A process of treating lumber or other products with various chemicals, such as preservatives and fire retardants, by forcing the chemicals into the structure of the wood using high pressure.

pressure wire connector Any device that maintains a mechanical connection between electrical conductors through the use of pressure.

prestressed concrete wire Steel wire with a very high tensile strength, used in prestressed concrete. The wire is initially stressed close to its tensile strength. Then some of this load is transferred to the concrete, by chemical bond or mechanical anchors, to compress the concrete.

prestressing Applying a load to a structural element to increase its effectiveness in resisting working loads. Prestressed concrete is a common example.

prestressing cable A cable or tendon made of prestressing wires.

prestressing steel High-strength steel used to prestress concrete, commonly seven-wire strands, single wires, bars, rods, or groups of wires or strands.

pretensioned concrete Concrete that has its reinforcing tendons stressed before the concrete is placed. Tension on the tendons is then released to provide load transfer where concrete has achieved strength.

pretensioned concrete

pretensioning A method of prestressing reinforced concrete in which the tendons are tensioned before the concrete has hardened.

prevailing wage Wage set by federal and state governments for construction work based on wages paid for similar work in the same local area.

preventive maintenance (PM) Periodic, scheduled work on selected equipment and building components, usually consisting of required inspection, cleaning, lubrication, and minor adjustment to help prevent systems failure.

price index A number which relates the price of an item at a specific time to the corresponding price at some specified time in the past.*

price out 1. The activity of applying dollar values to the items in a takeoff. **2.** The final estimate sheet showing all dollar values.

pricing In estimating practice, after costing an item, activity, or project, the determination of the amount of money asked in exchange for the item, activity, or project. Pricing determination considers business and other interests (e.g., profit, marketing, etc.) in addition to inherent costs. The price may be greater or less than the cost depending on the business or other objectives. In the cost estimating process, pricing follows costing and precedes budgeting.*

prick punch A pointed steel hand punch used to mark metal.

prick punch

prills Bulk, porable dry explosive consisting of pea-sized granules.

primacord A detonating fuse for high explosives, consisting of an explosive core in a strong, waterproof covering. Also called detonating cord.

primary air Air that is fed to a burner to be mixed with gas.

primary battery A battery of two or more primary cells.

primary blasting The blasting operation in which a natural rock formation is dislodged from its original location.

primary branch 1. A drain between the base of a soil or waste stack and a building drain. **2.** The largest single branch of a water supply line or an air supply duct in a building.

primary cell A cell that generates electric current by electrochemical action. In the process, one of the electrodes is consumed and the process cannot be effectively reversed. The cell cannot be recharged from an external source of electric power.

primary consolidation Soil compaction in saturated fine grain soils caused by the application of sustained loads, principally due to the squeezing out of water in the voids of the affected soil.

primary distribution feeder A feeder that operates at primary voltage supplying a distribution circuit.

primary excavation Excavating undisturbed soil.

primary light source 1. A source of light in which the light is produced by a transformation of energy. **2.** The most obvious source of light when several sources are present.

primary member One of the main load-carrying members of a structural system, generally columns or posts.

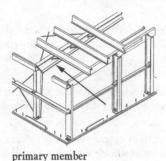

primary member

prime 1. A grade of finish lumber ranking below superior, the highest grade, and above E, the lowest grade of finish. Finish graded prime must present a fine appearance and is designed for application where finishing requirements are less exacting. 2. To supply water to a pump to enable it to start pumping. 3. In blasting, to place the detonator in a cartridge or charge of explosive. 4. The primary architect or contractor on a job that features several. 5. To seal a porous surface to prevent or reduce staining, shrinkage, and drying out.

prime bid A bid presented directly to the owner or his agent, rather than a subcontractor's bid to a general contractor.

prime coat 1. An application of low-viscosity liquid asphalt to an absorbent surface. 2. The first or preparatory coat in a paint system.

prime contract An agreement formed between the owner and the contractor for a major portion of the work on a construction contract.

prime contractor Any contractor on a project having a contract directly with the owner.

primer A base, preparatory paint that is applied to an uncoated surface to improve the adhesion and durability of the finish coat. Primer can be latex or alkyd (oil-based) paint.

primer/sealer A paint product manufactured to both prime a material for future paint application while sealing it from moisture.

priming 1. The application of a prime coat. 2. Filling a pump or siphon with fluid to enable flow. 3. The first or annual filling of a canal or reservoir with water.

princess post Subsidiary verticals, between queen posts or the king post, and walls, used to stiffen a roof truss.

principal 1. The principal authority or person responsible for a business such as architecture, engineering, or construction. 2. The capital amount of a loan or other obligation as distinguished from the interest. 3. In professional practice, any person legally responsible for the activities of that practice. 4. The person or entity under whose debt or obligation is the subject or a performance or payment bond issued by a surety.

principal beam The main beam in a structural frame.

principal-in-charge The professional individual within a design firm ultimately responsible for monitoring services in connection with a given project.

principal post A corner post in a framed building or a door post in a framed partition.

principal post

principal rafter One of the diagonals in a roof truss that support the purlins on which rafters are laid.

principal stress Maximum and minimum normal stresses at a point in a stressed body.

prism A surveying device used as a reference target when using electronic equipment to measure distance.

prismatic beam A beam with a uniform cross section. Both of its flanges run parallel along its longitudinal axis.

prismatic glass Glass with parallel prisms rolled into one face. The prisms refract light rays and change their direction.

private branch exchange (PBX) A telephone system, located on a customer's premises, that is owned and operated by the customer rather than a telephone company. A PBX system switches internal calls between the customer's users on local lines while also making external lines available. Using a PBX can save a customer money because it eliminates the need for direct lines from each user to the telephone company.

privy An outhouse serving as a toilet.

probabilistic design Method of design of structures using the principles of statistics (probability) as a basis for evaluation of structural safety.

processed shake A sawn cedar shingle that is textured on one surface to resemble a split shingle.

processed shake

Proctor compaction test A test to determine the moisture content of a soil at which maximum compaction can be obtained. The test establishes the density-moisture relationship in a soil.

Proctor curve Plot of the dry density of soil resulting from standard lab compactive effort versus moisture content.

product data Information furnished by the manufacturer to illustrate a material, product, or system for some portion of the work which includes illustrations, standard schedules, performance charts, instructions, brochures, diagrams, warranties.

production costs Parts of the actual, physical project, including materials, labor, tools, equipment, and subcontractor costs for each task or activity incorporated in the final structure.

production schedule A short-interval schedule used to plan and coordinate a group of activities.*

productivity Work performed per unit of time, or time to perform unit of work, such as square feet per hour.

products liability insurance Insurance for liability imposed for damages caused by an occurrence arising out of goods or products manufactured, sold, handled, or distributed by the insured or others trading under the insured's name. Occurrence must occur after product has been relinquished to others and away from premises of the insured.

P

product standard A published standard that establishes: **1.** dimensional requirements for standard sizes and types of various products; **2.** technical requirements for the product; and **3.** methods of testing, grading, and marking the product. The objective of product standards is to define requirements for specific products in accordance with the principal demands of the trade.

professional corporation A corporation created expressly for the purpose of providing professional practice and related services, which may have special requirements under the law, as opposed to requirements for corporations in general.

professional engineer A professionally qualified and duly licensed individual that performs engineering services such as structural, mechanical, electrical, sanitary, and civil engineering.

professional liability insurance Insurance coverage protecting against legal liability for damage claims sustained by others. Damage claims allege negligent acts, errors, or omissions in the performance of professional services.

professional practice The conduct and work of a design professional in which services are rendered within the framework of recognized professional ethics, standards, and applicable legal requirements.

profile A drawing showing a vertical section of ground, usually taken along the center line of a highway or other construction project.

profit Earnings from an ongoing business after direct and project indirect costs of goods sold have been deducted from sales revenue for a given period (gross profit).

profit margin Ratio of profit to either total cost or total revenue. Usage often varies depending on the type of company. Retail companies generally use the profit-to-revenue ratio. Wholesale companies and contractors generally use the profit-to-cost ratio.*

profit sharing Provisions in special agreements or contracts for construction where the contractor, as an incentive to save money for the owner, is paid, in addition to the final contract sum, some percentage of any net savings he may achieve if he is able to deliver the finished project to the owner's satisfaction at a total cost below a specified limiting amount.

proforma A project specific projection of future income and expenses used to evaluate the project's investment worthiness.

pro forma invoice An invoice sent before the order has been shipped in order to obtain payment before shipment.

program A written statement presenting design objectives, constraints, and project criteria, including space requirements and relationships, flexibility and expandability, special equipment, and systems and site requirements.

program manager An official in the program division who has been assigned responsibility for accomplishing a specific set of program objectives. This involves planning, directing and controlling one or more projects of a new or continuing nature, initiation of any acquisition processes necessary to get project work under way, monitoring of contractor performance and the like.

programming phase The design stage in which the owner develops and provides full information regarding requirements for the project, including a program.

progress chart A chart that shows various operations in a construction project, such as excavating and foundations, along with planned start and finish dates in the form of horizontal bars. Progress is indicated by filling in the bars.

progressive kiln A dry kiln in which green lumber enters one end and is dried progressively as it moves to the other end where it is removed.

progress payment A scheduled partial payment made during the work process to cover costs of work completed or materials delivered to date.

progress schedule A pictorial or written schedule (including a graph or diagram) that shows proposed and actual start and completion dates of the various work elements.

project The total construction activities, usually on one site. The work performed under the contract documents may be the whole or a part of the project.

project architect The party designated by the principal-in-charge to manage a given project for a firm.

project certificate for payment A statement to the owner confirming the amounts due individual contractors. Issued by the design professional where multiple contractors have separate necessary agreements with the owner.

project cost The total project cost, which includes the cost of construction, professional compensation, land, furnishings and equipment, financing, and other charges.

project designer (designer's office) The person assigned by the principal-in-charge to be responsible for guiding the overall direction of a project's design.

projected window A window with one or more sashes that swing either inward or outward.

projected window

project engineer An assistant to the contractor's project manager who is responsible for the documentation required for the management of the project. Often refers to an individual who uses engineering and management skills though is not necessarily a registered professional."

project evaluation and review technique (PERT) schedule A schedule that charts the activities and events anticipated in a work process.

project float The time that exists between the early finish of the last activity of a CPM network and the contractual completion date of the project. Project float can be internalized into the network and become network float.*

projecting belt course A course of masonry that projects beyond the face of the wall to form a decorative shelf.

projecting brick One of a number of bricks that project from a wall to form a pattern.

projection Any component member or part that extends out from a building for a relatively short distance.

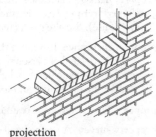

projection

project management **1.** The utilization of skills and knowledge in coordinating the organizing, planning, scheduling, directing, controlling, monitoring and evaluating of prescribed activities to ensure that the stated objectives of a project, manufactured product, or service, are achieved. **2.** The art and science of managing a project from inception to closure as evidenced by successful product delivery and transfer.*

project manager The individual designated by the principal-in-charge to manage a given project. Normally includes administrative and technical responsibilities.

project manual A bound booklet that contains the contract documents, with the exception of the drawings, specifically organized into bid requirements, contract information, general conditions for constructions and technical specifications.

project record documents The documents, certificates, and other information relating to the work, materials, products, assemblies, and equipment that the contractor is required to accumulate during construction and convey to the owner for use prior to final payment and project closeout.

project representative The architect's representative at the project site, who assists in the administration of the construction contract.

project team A collection of professional entities directed and otherwise coordinated to perform work or services for a project.

project web site An Internet site created for those involved in a specific construction project with the goal of increasing efficiency by improving access to information. Such sites can make a variety of information available—from a file of project team members and their e-mail addresses, to the project manual, field reports, meeting notes, and the shop drawing log.

promissory estoppel An equitable doctrine that prevents subcontractors from withdrawing their bids prior to acceptance by the prime contractor, making a subcontractor's quote binding if it meets specific criteria.

promissory note A legal instrument, agreement, or contract made between a lender and a borrower by which the lender conveys to the borrower a sum or other consideration known as principal for which the borrower promises repayment of the principal plus interest under conditions set forth in the agreement.

proof stress **1.** Stress applied to materials sufficient to produce a specified permanent strain. **2.** A specific stress to which some types of tendons are subjected in the manufacturing process as a means of reducing the deformation of anchorage, reducing the creep of steel, or ensuring that the tendon is sufficiently strong.

property damage insurance Insurance covering legal liability for claims for injury to or destruction of tangible property (including loss of use). See also **care, custody, and control**.

property insurance Insurance that compensates the insured for loss of property (real or personal) resulting from direct physical damage. Damages include fire, lightning, extended coverage perils, vandalism, and malicious mischief, among other damages.

property line A recorded boundary of a plot.

property survey A professional examination of documents and land to determine specific boundaries.

proportional control In an HVAC system, the controlled device (valve or damper) is positioned proportionally in response to slight changes in the controlled variable (temperature, pressure).

proportional limit The greatest stress that a material is capable of developing without any deviation from proportionality of stress to strain (Hooke's Law).

proprietary specification A specification that describes a product, material, assembly, or piece of equipment by trade name and/or by naming the manufacturer or manufacturers who may produce products acceptable to the owner or design professional.

protected corner Corner of a slab with adequate provision for load transfer, so that at least 20% of the load from one slab corner to the corner of an adjacent slab is transferred by mechanical means or aggregate interlock.

protected noncombustible construction Noncombustible construction in which bearing walls (or bearing portions of walls), whether interior or exterior, have a minimum fire resistance rating of two hours and are stable under fire conditions. Roofs and floors, and their supports, have minimum fire resistance ratings of one hour. Stairways and other openings through floors are enclosed with partitions having minimum fire resistance ratings of one hour.

protected opening An opening, in a rated wall or partition that is fitted with a door, window, or shutter having a fire resistance rating appropriate to the use of the wall.

protected ordinary construction Construction in which roofs and floors and their supports have a minimum fire resistance rating of one hour, and stairways and other openings through floors are enclosed with partitions that have minimum fire resistance ratings of one hour. Such construction must also meet all the requirements of ordinary construction.

protected wood-frame construction Construction in which roofs and floors and their supports have minimum fire resistance ratings of one hour, and stairways and other openings through floors are enclosed with partitions with minimum fire resistance ratings of one hour. Such construction must also meet all the requirements of wood frame construction.

protection board Asphalt-impregnated boards used in roofing installations to protect bituminous coatings from damage.

protocol A procedure or practice established by long or traditional usage and currently accepted by a majority of practitioners in similar professions or trades. Protocol represents the generally accepted method of action or reaction that may be expected to be followed in a transaction.

proving ring A device for calibrating load indicators of testing machines, consisting of a calibrated elastic ring and a mechanism or device for indicating the magnitude of deformation under load.

proximate cause The cause of an injury or of damages which, in natural and continuous sequence, unbroken by any legally recognized intervening cause, produces the injury, and without which the result would not have occurred. Existence of proximate cause involves both **1.** causation in fact, i.e., that the wrongdoer actually produced an injury or damages, and **2.** a public policy determination that the wrongdoer should be held responsible.

proximity switch A sensor that is activated by the intrusion of objects into an area.

psychrometer An instrument for measuring water vapor in the atmosphere, utilizing both wet- and dry-bulb thermometers.

psychrometric chart A chart that graphically represents the interrelation of air temperature and moisture content. Commonly used by building engineers and designers.

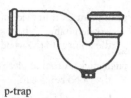

p-trap P-shaped trap that provides a water seal in a waste or soil pipe, used mostly at sinks and lavatories.

p-trap

public liability insurance Insurance covering liability for negligent acts causing bodily injury, disease, or death of persons other than employees of the insured. Also covers liability for property damage.

public sewer A common sewer controlled completely by a public authority.

public space 1. An area within a building to which the public has free access, such as a foyer or lobby. **2.** An area or piece of land legally designated for public use.

public utility A service for the public such as water, sewers, telephone, electricity, or gas.

public water main A water supply pipe controlled by public authority.

public way A street, alley, or other parcel of land open to the outside air and leading to a public street. A public way is deeded or otherwise permanently appropriated for public use. A minimum width is usually specified by code.

puddle 1. To settle loose soil by flooding and turning it over. **2.** To vibrate and/or work concrete to eliminate honeycomb. **3.** Clay that has been worked with some water to make it homogeneous and increase its plasticity so it can be used to seal against the passage of water.

puddle weld A weld used to join two sheets of light-gauge metal. A hole is burned in the upper sheet and filled with weld metal to fasten the two together.

puff pipe A short vent pipe on the outlet side of a trap; used to prevent siphoning.

pug box 1. A box, with a removable cover, placed in an electric raceway to facilitate the pulling of conductors through the raceway. **2.** A manual activator for a fire alarm system.

pugging A layer of clay, mortar, sawdust, or felt used for soundproofing purposes.

pug mill 1. A machine for mixing and tempering clay. **2.** The part of an asphaltic concrete plant where the heated batch materials are mixed.

pull 1. A handle used for opening a door, drawer, etc. **2.** To loosen rock at the bottom of a hole by blasting.

pull-chain operator A chain or control used to open or close a device such as a damper.

pulldown handle A handle fixed to the bottom rail of the upper sash of some double-hung windows.

pulley (pulley sheave) A wheel, with a grooved rim, that carries a rope or chain, and turns on a frame.

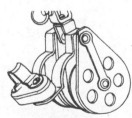

pulley (pulley sheave)

pulley block A frame that contains one or more pulleys.

pulley stile The upright in a window frame on which the sash pulleys are supported and along which the sash slides.

pulling 1. The drag on a paintbrush caused by high paint viscosity. **2.** Installing and connecting wires in an electric system.

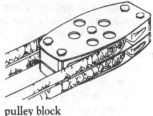

pulley block

pull scraper A hand scraper, consisting of a steel blade at approximately right angles to the handle, that is used to remove old finishes or for smoothing wood.

pulpboard A solid board composed of wood pulp.

pulsed arc transfer An arc welding method that involves operating a power source between low and high current levels. The high current level, called a "pulse," forces an electrode drop to the workpiece, while the low current level, or "background," maintains the arc between pulses. Major benefits of the process include better arc characteristics and greatly reduced spatter and fume generation.

pulse tube boiler A boiler using a sealed combustion system in which residual heat from the initial cycle ignites all subsequent air-gas mixtures, and flue gas condensation takes place in the heat exchanger.

pumice A highly porous lava, usually of relatively high silica content, composed largely of glass drawn into approximately parallel or loosely entwined fibers.

pumice concrete A lightweight concrete in which pumice is used as the coarse aggregate and which has good thermal insulation value.

pumice stone A solid block of pumice used to rub or polish surfaces.

pump A machine, operated by hand or a prime mover, used to compress and/or move a fluid.

pumped concrete Concrete that is transported through a hose or pipe by means of a pump.

pump head The pressure differential produced by an operating pump.

pumping (of pavements) The ejection of water, or water and solid materials, such as clay or silt, along transverse or longitudinal joints and cracks, and along pavement edges. Pumping is caused by downward slab movement activated by the passage of loads over the pavement after the accumulation of free water on or in the base course, subgrade, or subbase.

pump jack An adjustable support used to raise and lower scaffolding.

pump mix Concrete formulated for placement using a pump.

puncheon 1. Roughly dressed, heavy timber used as flooring, or as a footing for a foundation. **2.** Short timbers supporting horizontal members in a cofferdam.

punching shear 1. Shear stress calculated by dividing the load on a column by the product of its perimeter and the thickness of the base or cap, or by the product of the perimeter taken at one half the slab thickness away from the column and the thickness of the base or cap. **2.** Failure of a base when a heavily loaded column punches a hole through it.

punch list A list of items within a project, prepared by the owner or his representative, and confirmed by the contractor, which

P

remain to be replaced or completed in accordance with the requirements of the contract for construction at the time of substantial completion.

punitive damages (exemplary damages) Damages are awarded by a judge to a plaintiff not merely to compensate the plaintiff for losses incurred, but to punish the defendant for wrongful conduct and to use the plight of the defendant as an example to potential wrongdoers.

purchase order A formal written authorization to a vendor to provide certain goods or services and to bill the buyer for them at the specified price. The purchase order becomes a contract when it is accepted by the vendor.

purge To remove unwanted liquid, sediment, air or gas from a ductline, pipeline, container, space, or furnace.

purge pump A compressor that removes noncondensibles from a refrigeration system.

purlin One of several horizontal structural members that support roof loads and transfer them to roof beams.

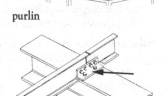

purlin

purlin cleat A shaped metal fastener used to secure a purlin to its support.

purlin plate A purlin, in a curb roof, located at the curb and supporting the ends of the upper rafters.

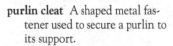

purlin cleat

purlin post A strut that supports a purlin to reduce sag.

purlin roof A roof in which purlins are supported directly on walls rather than rafters.

push bar A heavy bar across a glazed door, screen door, or horizontally pivoted window sash, used to open or close the door or window.

push drill A hand drill that is operated by pushing on its handle. A spiral ratchet rotates the bit.

push hardware A fixed bar or plate operated by pushing.

push plate A metal plate used to protect a door while it is pushed open.

push point A thin piece of metal used in glazing to hold glass in a wooden frame. A bent section of the point makes it easy to push the point into the wood.

push stick A tool or piece of wood used with a table saw to guide the wood being cut.

putlog Short pieces of timber that support the planks of a scaffold. One end of the timber is supported by the scaffold, the other is inserted in a temporary hole left in the masonry.

putlog hole The temporary hole in masonry that supports one end of a putlog.

putty A doughlike mixture of pigment and vehicle, used to set glass in window frames and fill nail holes and cracks.

putty coat Final smooth coat of plaster.

putty knife A knife, with a broad flexible blade with a flat end, used to apply putty.

pycnometer A vessel for determining the specific gravity of liquids or solids.

pylon 1. A steel tower used to support electrical high-tension lines. 2. A movable tower for carrying lights.

pyramidal light A skylight shaped like a pyramid.

pyramidal light

pyrolyze To bring about a chemical change through the action of heat.

pyrometer An apparatus for measuring high temperatures.

pyrometric cone A small, slender, three-sided oblique pyramid made of ceramic or refractory material for use in determining the time-temperature effect of heating and in obtaining the pyrometric cone equivalent of refractory material.

P

Abbreviations

q quart

Q quantity heat flow

qa quality assurance

qc quality control

QC quick coupling

QA/QC quality assurance/quality control

qda quantity discount agreement

QF quick firing

qr quarter

QR quarter-round

Qrtz quartz

qs quarter-sawn

qt quart

qty quantity

QTR quarry-tile roof; on drawings, quarter

quad quadrant

QUAD quadrangle

QUAL quality

QUAN quantity

quar quarterly

Definitions

quadrangle (quad) An open rectangular courtyard surrounded on all sides by buildings.

quadrant 1. One quarter of the circumference of a circle; an arc of 90°. **2.** An angle-measuring instrument.

quad-tube lamp A compact fluorescent lamp with a double twin tube configuration.

qualifications and assumptions Items that are not completely defined in the project documents for which the estimator is required to use judgment in developing the estimate.*

qualification submittals Data pertaining to a bidder's qualifications that must be submitted as set forth in the instructions to bidders.*

qualitative analysis The analysis of a sample (solid, liquid, or gas) to identify its components.

quality Characteristics of a product, material, or installation/workmanship that may determine its durability, longevity, appearance, safety, efficiency, and/or other attributes. A grade of Idaho white pine equivalent to D *select* in other species.

quality acceptance criteria Specified limits placed on characteristics of a product, process, or service defined by codes, standards, or other requirement documents.*

quality appraisal Quality activities employed to determine whether a product, process, or service conforms to established requirements, including: design review, specification review, other documentation review, constructability review, materials inspection/tests, personnel testing, quality status documentation, and post project reviews.*

quality assurance A system of procedures for selecting the levels of quality required for a project or portion thereof in order to perform the functions intended, and for assuring that these levels are obtained.

quality audit A formal, independent examination with intent to verify conformance with the acceptance criteria. An audit does

not include surveillance or inspection for the purpose of process control or product acceptance.*

quality conformance Quality management activities associated with appraisal, training, and prevention adapted to achieve zero deviations from the established requirements.*

quality control A system of procedures and standards by which a constructor, product manufacturer, materials processor, or the like, monitor the properties of the finished work.

quality corrective action Measures taken to rectify conditions adverse to quality and, where necessary, to preclude repetition. Corrective action includes rework and remedial action for non-conformance deviations.*

quality management Concerns the optimization of the quality activities involved in producing a quality product, process or service. As such, it includes appraisal, training, and prevention activities. *

quality performance tracking system A management tool providing data for the quantitative analysis of certain quality-related aspects of projects by systematically collecting and classifying costs of quality.*

quantification In estimating practice, an activity to translate project scope information into resource quantities suitable for costing. In the engineering and construction industry, a take-off is a specific type of quantification that is a measurement and listing of quantities of materials from drawings.*

quantitative analysis The analysis of a sample (solid, liquid or gas) to determine the proportions of its various components.

quantity overrun/underrun The difference between estimated quantities and the actual quantities in the completed work.

quantity survey A detailed analysis of material and equipment required to construct a project.

quarry An open excavation in the surface of the earth for mining stone.

quarry run Indiscriminate building stone as it comes from the pit without regard to color or structure.

quarrystone bond In masonry, a term applied to the manner in which stone is arranged in rubblework.

quarry (promenade) tile Machine-made, unglazed tile.

quarry tile Clay tile used for flooring or walls.

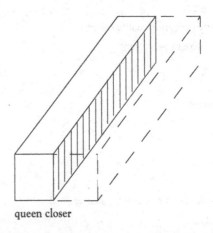
quarter bend

quarter bend A 90° bend, as in piping.

quarter closer (quarter closure) A brick cut to one-fourth of its normal length; used either as a spacer, or to complete a course.

quartered lumber Lumber that has been quarter-sawn approximately radially from the log.

quartered veneer Veneer that has been sliced in a radial direction, that is, at right angles to the growth rings. The term *quartered* comes from the use of blocks that have been cut into quarters before slicing. Quarter slicing brings out the presence of medullary rays; quarter-sliced veneer appears striped.

quarter hollow A concave molding formed by a 90° arc; the opposite of a quarter round.

quarter round A type of molding used as a base shoe; presenting the profile of a quarter circle.

quarter round

quartersawn (rift-sawn) Lumber sawn so that the annual rings form angles from 45° to 90° with the surface of the piece.

quarter-turn Descriptive of a stair that turns 90° as it progresses from top to bottom.

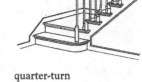
quarter-turn

quartz glass (silica glass) Glass consisting of pure, or nearly pure, amorphous silica. Of all glass, qua6eriors often incorporated turrets, dormers, hip and gable roofs, bay windows, and extensive decorative moldings or carvings. Wood, stone, or stucco was used in addition to brick.

queen closer A half-brick of normal thickness, but half-normal width. Used in a course of brick masonry to prevent vertical joints from falling above one another.

queen closer

queen-post truss (queen truss) A pitched roof support using two vertical tie posts connected between the tie beam and the rafters.

quenching Immersing hot, solid items in a cool liquid. Used as a means of tempering metals.

quick condition Soil that is weakened by the upward flow of water. Minute channels are created that significantly reduce the bearing capacity of the soil.

quick-load test Compression testing of piles that applies loads of increasing weight for brief time periods.

quick response sprinkler A sprinkler very sensitive to high temperatures, that reacts quickly to a fire, helping to limit smoke damage by arresting the fire at an earlier stage.

quick test A test performed on a cohesive soil to measure shear, before the sample has drained.

quirk 1. A narrow groove or bead located at or near the intersection of two surfaces or next to a molding, so as to reduce the possibility of uncontrolled cracking. **2.** Acute angle between adjoining pieces of molding. **3.** Thin groove on the bottom of a drip cap to keep water away from the joint.

quirk molding 1. Trim piece with a narrow groove. **2.** Trim with both inward and outward curves.

quoin (coign, coin) 1. A right-angle stone in the corner of a masonry wall to strengthen and tie the corner together. **2.** Keystone in an arch.

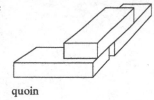

quoin

quotation A price for materials or services provided by a contractor, subcontractor, supplier, or vendor.

quote To make an offer at a guaranteed price.

Q

R

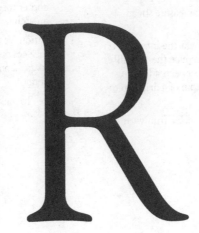

Abbreviations

r rain, range, rare, red, river, roentgen, run

R radius, right, symbol for capacity of a pile

Ra radium

RA remedial action, return air, registered architect

R&A (facility) repair and alteration

rab rabbeted

RAB rabbet

rad radiator

raft rafter

RAG return air grate

RAM rapid access memory, risk assessment methodology

RAMCAP Risk Analysis and Management for Critical Assets Protection

RAT return air temperature

RBM reinforced brick masonry

RC, R/C reinforced concrete, red cedar

RC asphalt rapid-curing asphalt

RCM reliability-centered maintenance

RCP reinforced concrete pipe

RCR room cavity ratio

RCRA Resource Conservation and Recovery Act

RCx retro-commissioning (of existing buildings)

1/4 RD quarter-round

1/2 RD half-round

rd road, rod, round

RD roof drain, round, remedial design

rdm random

rebar reinforcing bar

recap recapitulation

recd received

recip reciprocal

RECP receptacle

rec room recreation room

rect rectangle, rectified

red reduce, reduction

ref reference, refining

REF refer, reference

REFR refractory, refrigerate

refrig refrigeration

reg registered

Reg regular

REG register, regulator

rein, reinf reinforced

REINF reinforce, reinforcing

REIT real estate investment trust

REM removable

remod remodel

rent rental

rep, REP repair

repl, REPL replace, replacement

REPRO reproduce

reqd, REQD required

res resawn

ret retain, retainage

RET return

RETS real estate transaction standard

rev revenue, reverse, revised

REV revise

rf roof

RF roof, radio frequency, return fan

RFA RCRA Facility Assessment

RFC request for comment

RFD reduced function device

Rfg roofing

RFI radio frequency interference, request for information

RFI/CMS RCRA Facility Investigation/Corrective Measures Study. *See also* RCRA.

RFP request for proposal

RFQ request for quotation, request for qualifications

rgh, Rgh rough

RG 6 type of coaxial cable

Rh rhodium, Rockwell hardness

RH relative humidity

RHN Rockwell hardness number

RI refractive index, remedial investigation

RIBA Royal Institute of British Architects

rib gl ribbed glass

RI/FS Remedial Investigation/Feasibility Study

RIS Redwood Inspection Service

riv river

Rj road junction

RKI rotary kiln incineration

R/L, RL random lengths

rm ream, room

RM room

r mld raised mold

rms root mean square

ROD record of decision

ROP record of production

ROPS rollover protection system

rot rotating, rotation

rpm revolutions per minute

RPP remote power panel

RRGCP reinforced rubber gasket concrete pipe

RRM rapidly renewable materials

RRS railroad siding

R/S resawn

RSF rentable square feet

RSJ rolled steel joist

RSPA Research and Special Program Administration

rt right

RT raintight, real temperature, real time

RTD resistance temperature detector

RTN return to normal

Rub, rub Ruberoid, rubble

rw redwood, roadway, right-of-way

RW, R/W random widths, right-of-way

R/W&L, RW/L random widths and lengths

rwy, ry railway

rab A rod or stick used to mix concrete or mortar.

Definitions

rabbet A cut or groove along or near the edge of a piece of wood that allows another piece to fit into it to form a joint.

rabbet depth The depth of a glazing rabbet.

rabbeted lock (rebated lock) A lock that is fitted into and flush with the rabbet on a rabbeted doorjamb.

rabbet joint A longitudinal edge joint formed by fitting together rabbeted boards.

rabbet plane A plane for cutting a rabbet or fillister in wood, open on one side, with the blade extending to the open side.

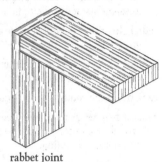

rabbet joint

rabbet size The actual size of a rabbeted glass opening, equal to the glass size plus two edged clearances.

race 1. A channel intended to contain rapidly moving water. 2. A groove in a machine part in which an object moves.

raceway 1. Any furrow or channel constructed to loosely house electrical conductors. These conduits may be flexible or rigid, metallic or nonmetallic, and are designed to protect the cables they enclose. 2. A man-made channel for directing water flow as in a hydroelectric plant.

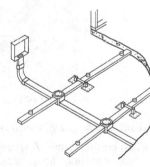

raceway

rack A framework of metal bars used to prevent waterborne trash from entering a waterway.

rack-and-pinion elevator A platform or enclosure in which linear motion is created by the use of a rotating gear pinion mated to, and revolving on, a vertical stationary rack.

racked A temporary support used to brace and prevent deformation.

racking (racking back) 1. A method entailing stepping back successive courses of masonry in an unfinished wall. 2. The movement of structural elements out of level or plumb by forces such as stress, wind, or material shrinkage or expansion. 3. The installation of roof shingles in a straight line across the roof instead of on

R

the diagonal. This method is quick, but is not often used because it can produce shingles incorrectly nailed and color variations.

rack saw A saw with wide teeth.

radial Radiating from, or converging to, a common center.

radial arch roof A roof supported by arches that radiate out from a central point.

radial-arm saw (radial saw) A circular saw suspended above the saw table on a cantilevered arm. The material remains stationary while the saw is free to move along its projecting beam.

radial-curved fans Backward-curved centrifugal fans with flat blades.

radial grating Grating in which the *bearing bars* extend radially from a common center, and the *cross bars* form concentric circles.

radial grating

radially cut grating Rectangular grating cut into panels shaped as segments of a circle to fit annular or circular openings.

radial shrinkage The shrinkage of wood across its diameter or growth rings as it dries.

radian The standard metric unit for a plane angle, as compared to customary units of degree (°), minute ('), and second (″).

radiance Light emitted or reflected from a designated area.

radiant barrier A reflective sheet or spray-on material applied to attic floors, rafters, or roof decking to reduce the flow of heat in and out of a building.

radiant cooling An efficient cooling system that typically involves running cool water through a building's floors, walls, or ceilings.

radiant cooling

radiant energy Energy traveling in electromagnetic waves.

radiant heating system A system with heating terminals that deliver heat by radiation from a hot surface, such as those heated by the flow of hot water or electric current.

radiation The transmission of energy by means of electromagnetic waves of very long wavelength. The energy travels in a straight line at the speed of light and is not affected by the temperature or currents of the air through which it passes.

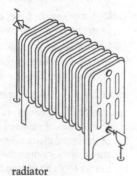

radiator A visually exposed heat exchanger consisting of a series of pipes that allows the circulation of steam or hot water. The heat from the steam or hot water is given up to the air surrounding the pipes.

radiator

radio frequency interference (RFI) Disturbance of radio frequency bands by other high-frequency equipment nearby. Fluorescent lighting creates RFI.

radiological screening A soil test than can determine approximate areas of local contamination.

radius gauge (fillet gauge) A gauge that measures the radius of curvature of small concave and convex surfaces.

radius of curvature A measure of the dimensions of a curve, as in the radius of curvature of a directional wall.

radius of gyration An imaginary distance from an axis to a point such that, if an object's mass were concentrated at the point, the moment of inertia would not change.

radon An odorless, colorless radioactive gas given off by some soils and rock with a short half-life (less than four days). If found in homes in sufficient concentrations, radon exposure can cause health problems.

rafter One of a series of sloping parallel beams used to support a roof covering.

rafter plate A plate used to support the lower end of rafters and to which they are fastened.

rafter

rafter table A carpenter's square with a table of values for determining the lengths and angles of cut for roof rafters.

rafter tail That part of a rafter overhanging the wall.

raft foundation A continuous slab of concrete, usually reinforced. Employed as a foundation laid over soft ground, or where heavy loads must be supported.

rag felt An asphaltic felt consisting of rags saturated with asphalt, creating a waterproof composition for use on roofs, i.e., roofing paper and asphalt shingles.

ragwork Courses of irregular stone masonry laid in a random pattern without parallel surfaces. The stones are roughly shaped with nonuniform joints.

rail 1. A horizontal member supported by vertical posts, e.g., a handrail along a stairway. 2. A horizontal piece of wood, framed into vertical stiles, such as in a paneled door. 3. Track used to direct or control the path of a vehicle or device.

rail

rail fence A barrier or boundary constructed of rails and their supporting posts.

railing 1. A solid wood band around one or more edges of a plywood panel. 2. A *balustrade*.

rails 1. The horizontal members that form the outside frame of a door, including pieces used as a cross bracing between the top and bottom rails. 2. The horizontal members of a fence between posts. 3. The side pieces of a ladder to which rungs or steps are attached.

rail steel reinforcement Reinforcing bars hot-rolled from standard T-section rails.

Raimann patch (football patch) A patch, elliptical in shape, used to fill voids caused by defects in the veneers of plywood panels.

The Raimann patch is one of the two patch designs allowed in A-grade faces.

rain cap A protective cover installed at the top of a roof vent to prevent rain from entering. Sometimes screened to keep out birds as well.

rain catchment system Connected gutters, downspouts, and barrels or cisterns that capture and store rainwater for irrigation and indoor use.

rainscreen A method of constructing walls in which the cladding is separated from a membrane by an airspace that allows pressure equalization to prevent rain from being forced in. Often used for high-rise buildings or for buildings in windy locations.

rain sensor An electronic device that shuts off an automatic sprinkler system when it rains, conserving water when irrigation is not needed.

rainwater harvesting On-site rainwater collection and storage systems used to offset potable water needs for a building and/or landscape. Systems can take a variety of forms, but usually consist of a surface for collecting precipitation (roof or other impervious surface) and a storage system.

rainwet 1. Lumber that has excess moisture content because of exposure to rain after it was dried. **2.** Surface lumber that has been stained or weathered by exposure to rain.

raised flooring system A floor constructed of removable panels supported by stringers allowing easy access to the space below.

raised girt (flush girt, raised girth) A stiffening member, parallel to and level with the floor joists, used to strengthen and protect.

raised molding A molding not on the same level or plane as the wood member or assembly to which it is applied.

raising hammer A hammer with a long head and a rounded face that is used in sheet metal work.

rake 1. To slant or incline from the vertical or horizontal. **2.** A board or molding that is placed along the sloping edge of a frame *gable* to cover the edges of the siding. **3.** A tool used to remove mortar from the face of a wall.

rake

raked joint A joint in a masonry wall that has the mortar raked out to a specified depth while it is still soft.

raker 1. A sloping brace for a *shore head*. **2.** A tool for raking out decayed mortar from the joints of brickwork.

rake trim Flashing used to close gap between the roof and the end walls.

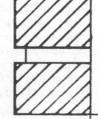

raked joint

raking course A course of bricks laid diagonally across faces of a brick wall for added strength.

raking flashing A parallel flashing used at the intersection of a chimney or other projection, with a sloping roof.

raking riser A riser that is not vertical, but slopes away from the *nosing* for added foot room on the lower riser.

raking stretcher bond A bond in brickwork in which the vertical joint of each brick is displaced a small, fixed distance from the vertical joint of the brick below.

ram 1. A cylinder that contains a plunger instead of a piston and rod. **2.** The plunger in a hydraulic press.

rammed earth Earth formed into thick, durable monolithic building walls that are energy efficient and fire resistant. Typically used in hot, dry climates, rammed earth walls are composed of screened engineered soil and cement, formed to be 18″ or 24″ thick. If used in cooler climates, rammed earth may require supplemental insulation or additional heating or cooling.

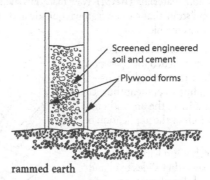

rammed earth

ramming A form of heavy tamping of concrete, grout, or the like by means of a blunt tool forcibly applied.

ramp A sloping surface to provide an easy connection between floors.

rampant vault A type of barrel vault exhibiting two quarters of different radii. The quarters have a common crown but the abutments are at different levels.

random ashlar (random bond) Constitutes ashlar masonry in which stones are set without continuous joints, and appear to follow a random pattern, although a large pattern may be repeated.

random ashlar (random bond)

random lengths (RL) Lumber of various lengths. A random length loading is presumed to contain a fair representation of the lengths being produced by a specific manufacturer.

random paneling Board paneling of varying widths of the same grade and pattern. The term refers to plywood paneling grooved to represent random width paneling.

random shingle One of a number of shingles of the same length, but varying widths.

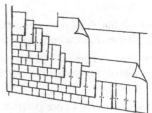

random shingle

random widths Boards, lumber, and shingles of varying widths.

range 1. A row or course of masonry. **2.** A straight line of objects such as columns. **3.** The difference between prices, costs, estimates, and bids.

range-in (wiggling-in) The term used by a surveyor for the trial-and-error procedure of finding a previously established line.

range lines North-south lines used as reference points in the rectangular survey system, originally created to survey land in the U.S. public domain.

range rod (range pole) A device that consists of a pole 7' to 8' long, equipped with a metal point. It is painted alternately red and white in bands 1' wide. The pole is used by surveyors for locating points of reference.

rapid-curing asphalt A liquid composed of asphalt, cement, and a highly volatile diluent such as naphtha or gasoline.

rapid-hardening cement A high-early-strength cement.

rapidly renewable materials (RRM) Raw materials such as cork, bamboo, and straw that can be regrown quickly and are therefore considered sustainable.

rapid-start fluorescent lamp A fluorescent lamp designed for operation with a ballast that provides a low-voltage winding to preheat the electrodes and initiate the arc without a starting switch or the application of high voltage.

ratchet

ratchet A mechanism with a hinged catch, or pawl, that slides over and locks behind sloped teeth on a gear or rod, allowing motion in one direction only.

ratchet brace A carpenter's clamp used in confined spaces where a full turn of the brace cannot be achieved. It is fitted with a pawl mechanism allowing the bit to be rotated while in the hole.

ratchet brace

ratchet drill A hand-operated drill that works with a hinged catch engaging on a wheel or chuck.

rated lamp life The average life of a particular lamp fixture.

rated load The overall weight that a piece of machinery is designed to carry.

rated speed The speed at which a piece of machinery is designed to operate with an appropriate load.

ratio The relative size of two quantities expressed as one divided by the other.

rat-tail file A file with a circular cross-section that tapers to a small diameter at the end opposite the handle.

raveling A term used for the progressive deterioration of asphalt pavement. The aggregate dislodges and becomes fragmented.

raw linseed oil Linseed oil that has been refined, but has not received further processing such as boiling, blowing, or bodying.

rawl plug A short fiber cylinder with a lead lining, driven into a hole in wood, masonry, glass, plaster, tile, concrete, or other materials to receive and hold a screw.

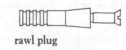
rawl plug

raw material A material used for the construction or manufacture of goods or products.

raw water Water that requires treatment before it can be used, such as water for steam generation.

raze To tear down, demolish, or level to ground.

RCRA Facility Assessment (RFA) A site investigation, regulated under RCRA, that is used to determine the nature and extent of contamination at a particular location.

RDX (cyclotrimethylenetrinitramine) An explosive commonly encountered in unexploded ordnance sites.

reactance Opposition to alternating current (AC) by capacitance or inductance or both. It is expressed in ohms, and symbolized by X.

reaction Resistive forces acting on a structural member that serve to hold it stable.

reactivated carbon Carbon that has been used for volatile organic compound adsorption, which is then treated for reuse.

reactive aggregate Aggregate containing substances capable of reacting chemically with the products of solution or hydration of the Portland cement in concrete or mortar, under ordinary conditions of exposure, resulting in harmful expansion, cracking, or staining.

reactive silica material Several types of materials that react at high temperatures with Portland cement or lime during autoclaving, including pulverized silica, natural pozzolan, and fly ash.

ready-mixed concrete Concrete manufactured for delivery to a purchaser in a plastic and unhardened state.

ready-mixed plaster A calcined gypsum plaster with aggregate added during manufacture. A powder product that requires the addition of water.

reagent A substance used in analysis and synthesis of chemical reactions.

realignment A change in the horizontal layout of a highway, may also affect vertical alignment.

real property (real estate) Land and anything growing on the land, or constructed on it, such as buildings, stone walls, fences, driveways, garages, underground swimming pools, trees, shrubs, and gardens. Real property also includes *fixtures*, or items that are permanently affixed to real property, such as light fixtures, oil burners, fuse boxes, and plumbing. Real property does not include things that are grown on the land for the purpose of being sold, such as crops, or trees from a tree farm. Such products that are to be severed from real property and sold are treated as *goods*.

reamer A bit with sharp, spiral, fluted, cutting edges along the shaft that may be slightly tapered. Used to enlarge drilled holes or remove burrs from the inside of pipe.

reamer

reasonable cost A cost that would be incurred by an ordinarily prudent person in the conduct of competitive business.

rebar Short for reinforcing bar.

rebound Aggregate and cement or wet shotcrete that bounces away from a surface against which it is being projected.

rebound hammer An apparatus providing a rapid indication of the strength of concrete based on the distance of rebound of a spring-driven missile.

rebutted and rejointed (R&R) Shingles with edges machine-trimmed to be exactly parallel and butts retrimmed at precisely right angles for use primarily in sidewall applications.

receipt of bids The official action of an owner in receiving sealed bids that have been invited or advertised in accordance with the owner's intention to award a contract for construction.

receiver A tank to hold fluids or gases, as on an air compressor.

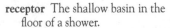

receptacle A contact device installed in an electric outlet box for the connection of portable equipment or appliances.

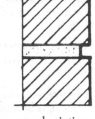

receptacle

receptor The shallow basin in the floor of a shower.

recessed fixture A lighting fixture the bottom edge of which is flush with the finish surface.

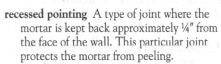

recessed fixture

recessed pointing A type of joint where the mortar is kept back approximately ¼" from the face of the wall. This particular joint protects the mortar from peeling.

recharging 1. The replenishment of ground water through direct injection or infiltration from trenches outside the area. 2. The replenishment of electric energy in a storage battery.

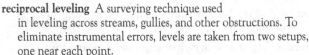

recessed pointing

reciprocal leveling A surveying technique used in leveling across streams, gullies, and other obstructions. To eliminate instrumental errors, levels are taken from two setups, one near each point.

reciprocating chiller (reciprocating compressor) Single-acting compressor using pistons that are driven by a connecting rod from a crankshaft.

reciprocating compressor An air compressor using a piston to compress the air, and employing valves for intake and discharge.

reciprocating engine A steam or internal-combustion engine with cylinders and pistons as opposed to a turbine.

reciprocating pump A water pump using a piston(s) for intake and discharge. Depending upon valve arrangement, the pump can be either single- or double-acting.

reciprocating saw A saw operated in a back and forth or up and down motion extending from an engine or other power source, e.g., a saber saw.

reciprocating saw

recirculated air Return air that is reconditioned and distributed once again, as opposed to makeup air.

reclaimed lumber Wood building materials removed from a deconstructed building in such a way that they can be reused for nonstructural purposes.

recloser Device that protects a power system from service interruption by automatically opening and closing as overcurrents (short circuits) are detected. They detect the fault, close and reset to open the line, often more than once. It clears the fault and reenergizes the distribution line.

reconstruction The process of depicting (by means of new construction) the form, features and detailing of a nonsurviving site, landscape, structure or object for the purpose of replicating its appearance at a specific period of time, and in its historic location.

record of decision (ROD) An agreement between the site owner(s) and other responsible parties and the regulators that stipulates the remedial action requirements, schedule, and cost sharing for cleanup.

recovery peg A temporary surveying marker of known location and elevation, used to reestablish a permanent marker that is to be replaced.

rectangular survey system A land survey system using longitude and latitude geographical coordinates. Established by the U.S. government, it is also called *government survey system*.

rectangular tie A piece of bent heavy wire in the shape of a rectangle used as a wall tie.

rectifier An electrical device that converts alternating current (AC) to direct current (DC).

recycle Revisiting partially or fully completed activities to perform additional work due to a change.*

recycled concrete Hardened concrete that is crushed to be used as aggregate.

recycled content Products such as steel, plastic lumber, and carpet cushion, fabricated with postconsumer materials or postindustrial byproducts.

recycling Reusing, reprocessing, or refabricating products after their initial use. Examples of recycled products include some types of tile, glass, asphalt paving, masonry, metal framing, insulation, toilet compartments, and carpet.

redesign Creating a plan to change an existing building to a new appearance or configuration.

red herring A proposed contract term that is not critical, but is offered for the purpose of distracting the other party.

red label A grade of shingle between blue label #1 and black label #5, graded by the Red Cedar Shingle and Handsplit Shake Bureau.

red lead A compound of lead used in paints to improve the anticorrosive properties of the paint used on steel and iron.

red oxide A pigment used in paints to prevent rusting and corrosion.

redress 1. Financial compensation for a loss caused by the actions of another. 2. To set right, remedy, or rectify.

red rot An early stage of decay.

redry To turn material to a dry kiln or veneer dryer for additional drying when the material is found to have a higher moisture content level than desired.

reduced function device (RFD) In building automation systems, a device that serves just one function, such as temperature monitoring. In contrast, a full-function device might control variable air volume.

reducer **1.** A solvent used in paints to reduce the viscosity of the paint. **2.** A pipe fitting, larger at one end than at the other. A reducing coupling. **3.** A substance that chemically reacts with another by removing an oxygen atom from the chemical structure of the second substance.

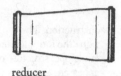
reducer

reducing joint The connecting joint of unequally sized electrical conductors.

reduct A small piece of material cut from a larger piece to make the larger piece uniform or symmetrical.

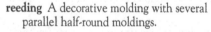
reducing joint

reduction of area The difference between the original cross-sectional area of a tensile specimen and the smallest area after rupture, expressed as a percentage of the original area.

redundant power An extra power source. It may be a complete second power system or a just an additional generator. Duplicate power service allows users to switch large amounts of a building's power load to another system.

Redwood Inspection Service One of seven regional grading agencies in North America that are authorized to write and publish grading rules for lumber.

reed switch Two or more highly conductive reeds inside a glass container. The reeds open or close together in response to a magnetic field.

reeding A decorative molding with several parallel half-round moldings.

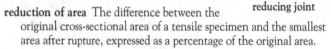
reeding

reentrant angle An internal angle of usually less than 90°.

reeving **1.** Threading a rope through the blocks to assemble a block and tackle. **2.** Threading or placing a pulling or control line.

reference line A series of two or more points in line to serve as a reference for measurements.

reference mark A supplementary mark close to a survey station. One or more such marks are located and recorded with sufficient accuracy so that the original station can be reestablished from the references.

reference standards Professionally prepared generic specifications and technical data compiled and published by competent organizations generally recognized and accepted by the construction industry. These standards are sometimes used as criteria by which the acceptability and/or performance of a product, material, assembly, or piece of equipment can be judged.

reference standard specification A type of nonproprietary specification that relies on accepted reference standards to describe a product, material, assembly, or piece of equipment to be incorporated into a project.

refill tube A tube in toilets that sends water from the ballcock into an overflow tube to refill the bowl after a siphon break.

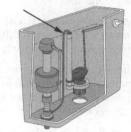
refill tube

reflectance The ratio of light reflected from a surface to the light incident on the surface. Reflectances are often used for lighting calculations. The reflectance of a dark carpet is around 20%, and a clean white wall is roughly 50% to 60%.

reflected plan A plan of an upper surface, such as a ceiling projected downward.

reflected sound Sound that has struck a surface and "bounced off."

reflection The return from a material's surfaces of light or sound waves.

reflection cracking The occurrence of cracks in overlays and toppings that coincide with the location of cracks in the base slab.

reflective insulation A thermal material having one or both faces metallically coated to reflect the radiant energy that strikes its surface.

reflective roof Roofing treated with a special coating that reflects the sun's heat away from the building, reducing the building's heat gain and prolonging the life of the roof.

reflector A device used to redirect light or sound energy by the process of reflection.

refraction The change in path of a light ray or energy wave as it passes obliquely from one medium to another.

refractories Materials, usually nonmetallic, used to withstand high temperatures.

refractory Resistant to high temperatures.

refractory brick A brick manufactured for high temperature use.

refractory concrete Concrete having refractory properties and suitable for use at high temperatures, generally about 315°C to 1,315°C, in which the binding agent is a hydraulic cement.

refrigerant The medium used to absorb heat in a cooling cycle.

refrigerant charge The quantity of refrigerant in a refrigeration system.

refrigerant compressor unit A unit consisting of a pump with various controls for regulating refrigerant flow.

refrigeration cycle A thermodynamic process whereby a refrigerant accepts and rejects heat in a repetitive sequence.

refrigeration system A system in which a refrigerant is compressed, condensed, and expanded as a means of removing heat from a cold reservoir. The heat is rejected elsewhere at a higher temperature.

refusal The depth at which a pile cannot be driven any deeper.

regenerative heating The use of heat rejected from one part of the heating cycle and transferred to another part.

register An opening to a room or space for the passage of conditioned air. The register has a grille and a damper for flow regulation.

reglaze To replace the glass in a window.

reglet A groove in a wall to receive flashing.

regulated-set cement A Portland cement with admixtures for the purpose of controlling its set and early strength.

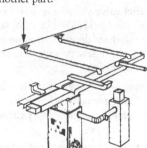
register

regulation The ability of a ballast to hold constant (or nearly constant) the output watts (light output) during fluctuations in the voltage feeding of the ballast. Normally specified as +/− percent change in output compared to +/− percent change in input.

regulator 1. A device that maintains an appropriate rate of release of pressurized substances. **2.** An electronically controlled device that regulates system voltage.

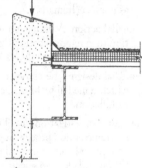
reglet

rehabilitation (rehab) The process of making possible a compatible use for a property through repair, alternations and additions, while preserving those portions or features that convey what have been determined to be important cultural or architectural values.

reheat coil A coil in an air supply duct, used to control the temperature of air being supplied to individual spaces or a group of spaces.

reheating The heating of return air in an air-conditioning system for reuse.

reinforced bitumen felt A light roofing felt saturated with bitumen and reinforced with jute cloth.

reinforced blockwork Masonry blockwork with reinforcing steel and grout placed in the voids to resist tensile, compressive, or shear stresses.

reinforced cames Lead bars with a steel core for reinforcing, used in leaded lights.

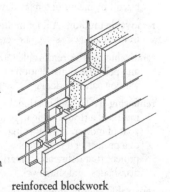
reinforced blockwork

reinforced concrete Concrete containing adequate reinforcement, prestressed or not prestressed, and designed on the assumption that the two materials (steel and concrete) act together in resisting forces.

reinforced masonry Unit masonry construction in which steel reinforcement is so embedded that the materials act together in resisting tensile, compressive, and/or shear stresses.

reinforced plastic Plastic that has components added to it, such as glass fiber, to increase its strength and stiffness.

reinforced T-beam A concrete T-beam strengthened internally with steel rods to resist tensile and/or shear stresses.

reinforcement Bars, wires, strands, and other slender members embedded in concrete in such a manner that the reinforcement and the concrete act together in resisting forces.

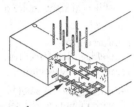
reinforcement

reinforcement, cold-drawn wire Steel wire made from rods that have been hot-rolled from billets and cold-drawn through a die used for concrete reinforcement of small diameter such as in gauges not less than 2 mm nor greater than 16 mm.

reinforcement, cold-worked steel Steel bars or wires that have been rolled, twisted, or drawn at normal ambient temperatures.

reinforcement displacement Movement of reinforcing steel from its specified position in the forms.

reinforcement, distribution bar Small-diameter bars, usually at right angles to the main reinforcement, intended to spread a concentrated load on a slab and to prevent cracking.

reinforcement, four-way A system of reinforcement in flat slab construction comprising bands of bars parallel to two adjacent edges and also to both diagonals of a rectangular slab.

reinforcement, helical Steel reinforcement of hot-rolled bar or cold-drawn wire fabricated into a helix, more commonly known as spiral reinforcement. Used in round columns.

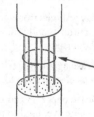
reinforcement, helical

reinforcement, high-strength Concrete reinforcing bars having a minimum yield of 60,000 psi or 414 MPa.

reinforcement, hoop A one-piece closed tie or continuously wound tie not less than #3 in size, the ends of which have a standard 135° bend with a 10-bar-diameter extension, that encloses the longitudinal reinforcement in a column.

reinforcement, lateral Usually applied to ties, hoops, and spirals in columns or column-like members.

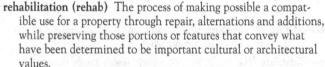
reinforcement, hoop

reinforcement, principal Elements or configurations of reinforcement that provide the main resistance of reinforced concrete to tensile loads borne by structural components.

reinforcement ratio Ratio of the effective area of the reinforcement to the effective area of the concrete at any section of a structural member.

reinforcement, secondary Reinforcement other than main reinforcement.

reinforcement, transverse Reinforcement at right angles to the longitudinal reinforcement. Transverse may be main or secondary reinforcement.

reinforcement, two-way Reinforcement arranged in bands of bars at right angles to each other.

reinforcement, welded Reinforcement joined together by welding.

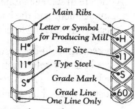
reinforcement, two-way

reinforcing bar A steel bar, usually with manufactured deformations, used in concrete and masonry construction to provide additional strength.

reinforcing plate An added plate used to strengthen a member or part of a member.

reinforcing bar

reinforcing tape A tape that is used to hide and strengthen joints between drywall sheets.

rejection of work The act of rejecting work that is defective, or that which does not conform to contractually agreed upon requirements.

relamping The replacement of lamps (bulbs) in light fixtures. May be done as each fails, or may be done on a schedule.

related trades The different building trades required to complete a project.

relative compaction The dry density of soil expressed as a percentage of the density of the soil after a standard compaction test.

relative humidity The ratio of the quantity of water vapor actually present to the amount present in a saturated atmosphere at a given temperature, expressed as a percentage.

relaxation of steel Decrease in stress in steel as a result of creep within the steel under prolonged strain, or as a result of decreased strain of the steel, such as results from shrinkage and creep of the concrete in a prestressed concrete unit.

relay An electromagnetic or electromechanical device using small currents and voltages to activate switches or other secondary devices.

release agent Material used to prevent bonding of concrete to a surface, such as to forms.

release paper A protective film, coated lightly on one side with adhesive, used to protect a surface during shipment and installation.

relief Carved or embossed decoration raised above a background plane.

relief angle Structural angle introduced to help support masonry over an opening or at specified elevations or multistory construction.

relief cut An initial cut made to keep the saw from binding when a curved section is about to be cut.

relief damper A damper in an air-conditioning system that opens automatically at a set pressure differential to balance pressures in a building.

relief valve A pressure-activated valve held closed by string tension and designed to automatically relieve pressure in excess of its setting.

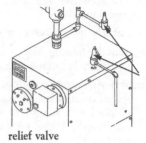
relief valve

relief vent An auxiliary vent that is supplementary to regular vent pipes. The primary purpose is to provide supplementary circulation of air between drainage and vent pipes.

relish In carpentry, the shoulder around or at the side of a tenon.

reluctance The opposition a material offers to magnetic force.

rem (rad equivalent man) These are units used to measure radiation as it affects humans.

remedial action A permanent remedy taken instead of, or in addition to, a removal action that is in response to a release or threatened release of hazardous substances.

remedial design The phase of the site remediation process in which remedial technologies and processes are designed prior to installation.

Remedial Investigation/Feasibility Study (RI/FS) The phase of the site remediation process in which the nature of contaminants is determined and a remedial remedy is selected. It is a site investigation and cleanup selection process that includes activities such as project scoping, data collection, risk assesment, treatability studies, analysis of alternatives, and remedy selection.

remediation The process of correcting a problem such as mold infestation.

remodeling Making alterations to an existing structure such that it will be better suited to current needs.

remoldability The readiness with which freshly mixed concrete responds to a remolding effort such as jigging or vibration, causing it to reshape its mass around reinforcement and to conform to the shape of the form.

remolding test A test to measure remoldability.

remote-control circuit An electrical circuit that controls another circuit by means of a relay device.

removable mullion A door mullion that can be removed from a door frame to allow passage of oversized objects.

removable stop A removable molding or trim used to stop motion and permit the installation of windowpanes or doors.

rendering 1. The application, by means of a trowel or float, of a coat of mortar. 2. A drawing of a project or a portion of a project that delineates materials, shades, and shadows.

rendering

renewable energy Energy from natural resources that replen-

ish themselves, such as the sun, wind, rain, tides, and geothermal sources.

renewable energy

renewable energy certificates (RECs) Also referred to as "green tags," RECs each represent one megawatt hour of renewable energy. RECs may be sold by those who produced the energy to offset the cost of generating it.

renovation Modernizing the elements within a structure to meet current functional and aesthetic requirements.

repeater A device that can receive a weak one-way or two-way communications signal, amplify or reshape it, and retransmit it. In digital transmission, a weak pulse train of discrete signals is received, amplified, retimed, and reconstructed for delivery back to the source. In fiber optics, a low-power light signal is converted electronically and retransmitted via a light-emitting diode or laser light source.

repeating theodolite A theodolite designed so that an angle can be turned several times; and successive angle measurement accumulated, with the final reading representative of the accumulations.

repetitive member One of a series of framing or supporting members such as joists, studs, planks, or decking that are continuous, or spaced not more than 24 inches apart; and are joined by floor, roof, or other load-distributing elements. In repetitive-member framing, each member is connected to, and receives some shared support from, the others.

repetitive member

replacement value The estimated cost to replace an existing building based on current construction costs.

reproductive toxin A substance that chemically or biologically interferes with the reproductive process.

request for information (RFI) A formal request for preliminary information for the purpose of evaluating potential bids for design services, construction services, or both. This request does not usually signify a firm commitment by an owner, but it provides information on potential bidders and their capabilities. The information gathered by RFIs is sometimes used to develop a list of architects, engineers, or contractors who will eventually receive a request for proposal (RFP).

request for job order (JO) proposal The owner's notification to the contractor of an upcoming project (JO). This document identifies and briefly describes the project, the desired or required dates and times for performing the work, and contact information of the owner's designated project representative. The request may include other project-specific information. Response times to the request are in accordance with contract requirements.

request for proposal (RFP) Material provided to potential contractors to communicate government requirements and solicit proposals.

requisition To apply for progress payments based on work completed at the end of the pay period (usually one month). Abbreviated term for application for payment.

reroofing Replacing or covering an existing roof.

resaw 1. To saw a piece of lumber along its horizontal axis. 2. A band saw that performs such an operation.

resawn board A piece of lumber, most commonly $^{11}/_{16}''$ thick, obtained by resawing a piece of 6/4 common. Used mostly for sheathing and industrial applications. Produced most often from ponderosa pine and white fir.

resealing trap A trap connected to a plumbing fixture drainpipe so constructed as to allow the rate of flow to seal the trap without causing self-siphonage.

reservoir A tank or receptacle used for the accumulation and retention of a fluid.

reset controls Controls for hot water systems that inversely monitor the hot-water loop set point as compared to the outdoor temperature. For example, the system may be set for 180°F when the outdoor temperature is 0°F, and, inversely, the loop temperature could be 120°F when the outdoor air temperature is 45°F.

resetting (of forms) Setting of forms separately for each successive lift of a wall to avoid offsets at construction joints.

reshoring The construction operation whereby the original shoring or posting is removed and replaced in such a manner as to avoid deflection of the shored element or damage to partially cured concrete.

resident engineer An engineer retained by the owner as a representative on the construction site. Frequently used on governmental projects.

residential occupancy Occupancy of a building in which sleeping accommodations are provided for normal residential purposes.

residual approach A methodology for determining the cost of an asset within a construction project.

residual deflection The amount of deviation that remains after an applied load has been removed.

residual stress The stress remaining in a member after an applied force has been removed.

resilience The work done per unit volume of a material in producing strain.

resilient channel A mounting device with flexible connectors used for fastening gypsum board to studs or joists. Helps reduce the transmission of vibrations.

resilient clip A flexible metal device for mounting gypsum board to studs or joists; used to reduce noise and vibrations.

resilient flooring A manufactured interior floor covering material, in either sheet or tile form, that is resilient.

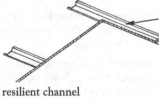

resilient channel

resilient clip

resin 1. A natural or synthetic solid, or semisolid organic material of indefinite and often high molecular weight having a tendency to flow under stress. It usually has a softening or melting range and fractures conchoidally. **2.** A natural vegetable substance occurring in various plants and trees, especially the coniferous species, used in varnishes, inks, medicines, plastic products, and adhesives.

resin chipboard A particleboard that uses a resin to ensure a uniform consistency.

resistance brazing A process in which coalescence is produced by the heat obtained from the resistance of the work to the flow of an electrical current.

resistance welding A process creating a metallurgical bond between metal by producing heat obtained from the resistance offered by the work to the flow of electric current, and by the application of an external pressure.

resistor A device included in an electric circuit to introduce a desired resistance.

resonance A condition reached in an electrical or mechanical system when the applied load frequency coincides and is equal to the natural undamped frequency of the system.

resonator 1. A device that intensifies vibration and volume, usually of sound waves. **2.** A hollow metal container that is utilized to create microwaves.

resorcinol resin adhesive A synthetic resin adhesive made with resorcinol-formaldehyde resin using a hardener, usually composed of wood flour and paraform. The adhesive provides an exterior quality glueline and will cure at room temperature. It is used in the manufacture of laminated beams or scarf joints.

Resource Conservation and Recovery Act (RCRA) The primary federal law that governs hazardous and toxic materials at operating facilities. It defines solid and hazardous waste, authorizes the EPA to set standards for facilities that generate or manage hazardous waste, and establishes a permit program for hazardous treatment, storage, and facilities.

respirator A device used to facilitate breathing in a hazardous atmosphere.

respond A support, usually a pilaster or corbel, attached to a wall and supporting one end of an arch, groin, or vault rib.

respondent The party defending a construction claim.

responsiveness The bid's conformance with the solicitation's salient requirements (price, quantity, quality, performance time).

restitution A court-ordered money award that is intended to restore the parties to their financial position as it existed before the contract was formed.

restoration Returning an existing property to a condition that depicts its form, features, and character as they appeared at a particular period of time. Restoration may involve removal of features from other periods in the property's history and reconstruction of missing features from the targeted period.

restraint (of concrete) Restriction of free movement of fresh or hardened concrete following completion of placing in formwork or molds, or within an otherwise confined space. This restraint can be internal or external and may act in one or more directions.

restricted Relative to electrical systems, it refers to areas classified hazardous; or a confined area.

retainage A specific portion of the contract sum that is withheld from progress payments as specified in the owner-contractor agreement.

retaining wall 1. A structure used to sustain the pressure of the earth behind it. **2.** Any wall subjected to lateral pressure other than wind pressure.

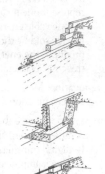

retardation Reduction in the rate of hardening or setting, i.e., an increase in the time required to reach initial and final set or to develop early strength of fresh concrete, mortar, or grout.

retarder An admixture that delays the setting of cement paste, and hence of mixtures such as mortar- or concrete-containing cement.

retempering Addition of water and remixing of concrete or mortar that has lost its workability and become too stiff. This is not usually recommended, as some of the strength is lost.

retaining wall

retention 1. A percentage, usually 10%, withheld from a periodic payment to a contractor, in accordance with the owner-contractor agreement, for work completed. The retention is held until all terms of the contract have been fulfilled. **2.** The amount of preservative, fire-retardant treatment, or resin retained by treated or impregnated wood.

retentivity The capacity of a material to hold magnetism.

reticulated Covered with crossing lines, netted.

reticuline bar Sinuously bent bars that interconnect adjacent bearing bars in a grating.

retract The mechanism on a dipper shovel bucket excavator that moves the bucket.

retro-commissioning (RCx) Commissioning of existing buildings.

retrofit To modify an existing structure or system within the structure to accommodate upgrading.

return The continuation, in a different direction, of a molding or projection, usually at right angles.

return-air Air returned from a conditioned room or space for processing and recirculation.

return-air intake An opening, usually with a control damper, through which return air reenters an air-conditioning system.

return bend A 180° bend in a pipe.

return bend

return corner block A concrete masonry unit with a plane surface at one end as well as the front face; used at outside corners.

return grille The grille on a return air intake, usually employing a damper.

return mains Pipes or ducts that reroute fluid or air back to the supply source.

return on investment (ROI) A calculation in investment analysis that expresses income as a percentage of capital. There are several specific varieties in general use, such as the internal rate of return (IRR), and the rate of return on equity capital (ROE).

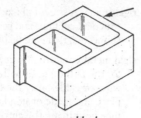

return corner block

return pipe The drain line that returns the water from condensed heating steam to the boiler for reuse.

return system A series of ducts, pipes, or passages that returns a substance, whether it be air or water, to the source for reuse.

reveal 1. The side of an opening in a wall for a window or door. **2.** Reveal is the depth of exposure of aggregate in an exposed aggregate finish.

reveal pin (reveal tie) A screw or clamp located inside a window opening used to secure scaffolding to the opening.

revent pipe An individual vent that connects directly to a waste fixture and the main or branch vent pipe.

reverberation The perpetuation of sound within a space after the source has ceased, such as an echo.

reverse-acting controllers Devices that decrease control pressure as the control variable (temperature, pressure) increases.

reverse-acting diaphragm valve A valve that opens on application of pressure on a diaphragm and closes when the pressure is released.

reverse board & batten A siding pattern in which the wider boards are nailed over the battens, producing a narrow inset.

reverse-trap water closet Toilet that has a siphonic trapway at the rear of the bowl and integral flushing action.

reversible lock A lock that can be adapted to fit a door of either hand.

reversible siding Resawn board siding that may be installed with either the surfaced or sawn side exposed.

reversion 1. The tendency of a processed material to return to its original state. **2.** Chemical reaction leading to deterioration of a sealant due to moisture trapped behind the sealant.

revertible flue A flue or chimney that momentarily redirects the gas flow path downward during its normal upward ascent.

revet To face a foundation or embankment with a layer of stone, concrete, or other suitable material.

revetment Facing, such as masonry, used to support an embankment.

revolving door An exterior door consisting of four leaves set at right angles to each other and revolving on a central pivot. Used to limit the passage of air through the opening and eliminate drafts.

R factor The measurement of a material's (such as insulation, windows, and roofing) resistance to the passage of heat. Also known as *R-value*.

rheology The science dealing with flow of materials, including studies of deformation of hardened concrete, the handling and placing of freshly mixed concrete, and the behavior of slurries, pastes, and the like.

rheostat An electric resistor, so constructed that its resistance may be varied without opening the circuit in which it is installed. Used to control the flow of electric current as in a light dimmer.

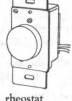

rheostat

rib 1. One of a number of parallel structural members backing sheathing. **2.** The portion of a T-beam that projects below the slab. **3.** In deformed reinforcing bars, the deformations or the longitudinal parting ridge.

ribbed panel A panel composed of a thin slab reinforced by a system of ribs in one or two directions, usually orthogonal.

ribbed vault A vault with ribs that support, or appear to support, the web of the vault.

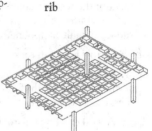

rib

ribbing (crimping, washboarding) A more or less regular corrugation of the surface of wood, caused by differential shrinkage.

ribbing up Circular joinery made by gluing several layers of veneer with a parallel grain structure.

ribbed panel

ribbon A narrow strip of wood or other material used in formwork.

ribbon board 1. A horizontal brace used in balloon framing where the board is applied to notches in the studs. **2.** A *let-in brace*.

ribbon loading Method of batching concrete whereby the solid ingredients, and sometimes the water, enter the mixer simultaneously.

ribbon strip A horizontal board set into the studs to help support the ends of rafters or joists.

riblath (rib lath) Stiffened expanded metal lath with ribs that provide greater strength and allow for increased spacing between members. It is not used for curved surfaces.

rice hull ash A byproduct from burning agricultural rice waste that can be used in applications similar to cement.

rich concrete Concrete of high cement content.

rich lime A pure lime that, when mixed with mortar, improves the plasticity or workability of the mortar.

riddle A coarse sieve for separating and grading granular material.

ridge The horizontal line formed by the upper edges of two sloping roof surfaces.

ridge beam A horizontal timber to which the tops of rafters are fastened.

ridgeboard (ridgepole, roof tree) The longitudinal board set on edge used to support the upper ends of the rafters.

ridgeboard (ridgepole, roof tree)

ridgecap (ridge cap) A layer of wood or metal topping the ridge of a roof.

ridge rib A projecting structural element following the ridge of a vault.

ridge shingles Shingles that cover a house's ridgeboard.

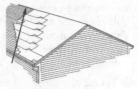

ridgecap (ridge cap)

ridge stop The flashing that forms a watertight barrier between the edge of the roofing membrane and the vertical wall or chimney rising above it.

ridge strap A metal connector that fastens opposing rafters together to resist suction forces caused by uplift at the ridge line.

ridge vent A vent installed along the top ridge of a roof to permit air to pass through to the attic or the peak of a cathedral ceiling.

ridging 1. An upward dislodgement of the roof membrane. **2.** Long narrow blisters in the surface of built-up roofing.

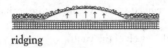

ridging

rift A narrow fissure in rock.

rig 1. To provide with equipment or gear for a special purpose. **2.** To assemble in a makeshift manner. **3.** An assembled piece of equipment, such as an oil-drilling rig.

rigger A worker who prepares heavy equipment or loads of materials for lifting.

rigging Lines or cables used to lift heavy loads.

riggot A trough or gutter for draining off rainwater.

right-hand rule 1. A rule for determining polarity produced by flowing electric current. **2.** Rule-of-thumb to envision the orientation of nut movement on a bolt to loosen/tighten it. Right thumb points along threaded shaft; direction of fist opening shows direction of nut loosening.

right-of-way A strip of land, including the surface and overhead or underground space, that is granted by deed or easement for the construction and maintenance of specific linear elements such as power and telephone lines, roadways and driveways, and gas or water lines.

right-to-know A right granted to workers by the Occupational Safety and Health Act (OSHA), by which they must be informed of the risks and hazards associated with the chemicals and substances that they are required to use in the workplace.

Right to Work Law State law providing, in general, that employees are not required to join a union as a condition of retaining or receiving a job.

rigid connection A connection between two structural members that prevents end rotation of one relative to the other.

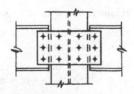

rigid connection

rigid foam insulation Panels of polyurethane foam that are used in commercial and residential construction as roof or outside wall insulation. The panels or boards include barrier board, and structural insulation panels (SIPs). The facings on the boards include metals, foil, glass fiber, kraft paper, or oriented strand board (OSB).

rigid frame A structural framing system in which all columns and beams are rigidly connected; there are no hinged joints.

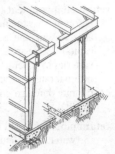

rigid frame

rigid metal conduit A raceway constructed for the pulling in or withdrawing of wires or cables after the conduit is in place and made of standard weight metal pipe permitting the cutting of standard threads.

rigid pavement Pavement that will provide high bending resistance and distribute loads to the foundation over a comparatively large area.

rim joist Perimeter joist for wood floor framing system. Usually referred to in conjunction with composite wood floor joists.

rim latch A surface-mounted latch.

rim lock A face-mounted door lock.

rime The rung of a ladder.

rimpull The capability of wheeled construction equipment to move a load.

ring-shank nail A nail with ring-like grooves around the shank to improve its grip.

rip To saw lumber parallel to the grain.

ripper A device with protruding claws that is pulled by a tractor. Used to penetrate and disrupt the earth's surface up to 3′ in depth.

ripping Fragmenting a rock formation by a tractor pulling a heavy tooth.

riprap Irregularly broken, large pieces of rock, used along stream banks and oceanfront property as protection against erosion.

ripsaw A coarse-toothed saw used for cutting wood in the direction of the grain.

rise 1. The vertical distance from the top of a tread to the top of the next higher tread. **2.** The height of an arch from springing to the crown. **3.** The vertical height from the supports to the ridge of a roof. **4.** Height difference between the crown of a road and its lowest point.

rise and run The angle of inclination or slope of a member or structure, expressed as the ratio of the vertical rise to the horizontal run.

riser 1. A vertical member between two stair treads. **2.** A vertical pipe extending one or more floors.

riser

riser height The vertical distance between the tops of two successive treads.

rising hinge (rising butt hinge) A door hinge fabricated with a slope on the knuckle, which causes the door to rise when opened. A rising hinge is used when there are obstructions along the door's bottom edge, e.g., carpeting.

risk management An approach to management and procedure designed to prevent occurrence of culpability, potential liability, contravention of law, or other potential risk that could bring about loss in the process of building construction.

risk probability A concept applied to assess a facility's security needs based on consideration of factors such as incidence history, trends in the surrounding environment, warnings or threats, and similar events occurring at other comparable schools.

rive Splitting wood along the grain. Shingles are cut in this fashion.

rivet A metal cylinder or rod with a head at one end that is inserted through holes in the materials to be fastened. The protruding end is flattened to tie the two pieces together.

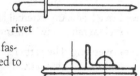
rivet

riveting

riveting The act of fastening or securing two or more parts with rivets.

R-number A number used by ASHRAE to indicate the molecular structure of chemical refrigerants.

road base A mixture, typically stone and sand, used as the base for road paving.

road forms Wood or steel forms set on edge to form the side of a concrete road slab, also used as screeds.

road heater A traveling machine that prepares a road surface for treatment by blowing a flame or hot air on it.

road oil A heavy petroleum oil, usually a slow-curing asphalt.

roadway The area of a highway including the surface over which vehicles travel as well as the land along the edges, such as slopes, ditches, channels, or other gradations necessary to ensure proper drainage and safe use.

rock anchor Steel rod anchored in rock by grouting used to hold temporary shoring or retaining walls around an excavation.

rock drill A high-powered pneumatic or electrically driven device for boring holes into rock.

rocker shovel Mechanical tunneling shovel.

rockers Slang term for drywall installers.

rock-faced A type of rough finish on stone. Essentially the same face that resulted when the stone was first struck from the quarry.

rock pocket A porous, mortar-deficient portion of hardened concrete, consisting primarily of coarse aggregate and open voids, caused by leakage of mortar from form, separation or segregation during placement, or insufficient consolidation.

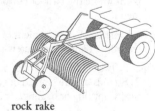

rock rake Tractor attachment with multiple teeth that is used to clear trees and rocks.

rock rake

rock saw A circular saw that removes a wide kerf on the upper surface of a log. A rock saw is used to remove stones or debris before a log enters the head rig.

Rockwell hardness A measure of the resistance of a material to indentation, expressed as an index number obtained by pressing a standard steel ball or diamond into the material under controlled conditions. A high index number indicates a hard material.

rock wool A type of mineral wool made by forming fibers from molten rock and slag. Used as insulation in walls and ceilings.

rod Sharp-edged cutting screed used to trim shotcrete to forms or ground wires.

rod bender A power-assisted device with adjustable rollers and supports that is used for bending steel reinforcing rods into usable shapes.

rod cutter Trench-type, hydraulically operated, wedgelike shear, used to cut steel reinforcing rods.

roddability The susceptibility of fresh concrete or mortar to compaction by means of a tamping rod.

rodding Compaction of concrete or the like by means of a tamping rod.

rod level A device used to ensure a leveling rod or stadia rod is in the vertical position before taking any instrument readings.

rod target (target rod) A metal disc that slides upon a track on a leveling rod, used for taking sights in surveying.

roll (noun) **1.** A quantity of sheet material wound in cylindrical form. **2.** A rounded strip of roofing fastened to and running along the ridge of a roof. **3.** Any type of rounded molding. **4.** Any heavy, metal cylinder used to flatten, smooth, or form material.

roll (verb) A slang term referring to the installation of floor joists or trusses.

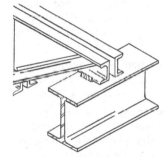
rolled beam (rolled steel beam)

rolled beam (rolled steel beam) A beam that is fabricated of steel and passed through a hot-rolling mill.

rolled glass Glass that is manufactured with a patterned surface that partially obscures vision and light.

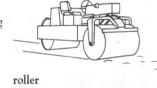
roller

roller 1. A heavy, self-propelled or towed device used to compact granular fill. **2.** A small hand tool used to smooth wall covering and flooring.

roller shade An automated window shade with a motor that adjusts the shade to control light, privacy, and solar heat gain.

rolling grille door A device similar to a rollup door, but with an open grille rather than slats; used as security protection.

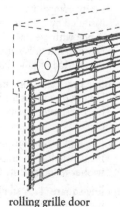
rolling grille door

rolling resistance The resistance of a substrate to turning wheels on top.

rolling scaffolding A reliable, inexpensive scaffold used on interiors up to 20′ in height. Suitable for close inspection and construction work on walls and ceilings.

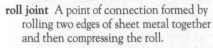

rolling scaffolding

roll joint A point of connection formed by rolling two edges of sheet metal together and then compressing the roll.

rollup door (rolling shutters) A device consisting of horizontal interlocking metal slats that ride along wall guides. When the door is opened the slats coil around a barrel assembly located above the door.

Roman brick Brick that measures 4″ × 2″ × 12″.

Romanesque architecture Style of building with massive walls, vaults, and rounded arches.

Romex The most familiar trade name for nonmetallic sheathed cable.

roof covering The covering material installed in a building over the roof deck. The type of covering used depends on the roofing system specified to weatherproof the structure properly.

roof deck The foundation or base upon which the entire roofing system is dependent. Types of decks include steel, concrete, cement, and wood.

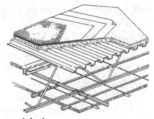

roof deck

roof drain A drain designed to accept rainwater on a roof and discharge it into a leader.

roof flange A collar that fits around a pipe penetrating through the roof, making the opening watertight.

roof framing A group of members fitted or joined together to provide support for the roof covering.

roof hatch (roof scuttle) A weathertight assembly with a hinged cover, used to provide access to a roof.

roofing bond A guarantee by a surety company that a roof installed by a roofer in accordance with specifications will be repaired if it fails within a certain period of time. Failure must be due to normal weathering.

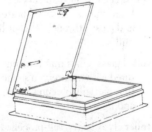

roof hatch (roof scuttle)

roofing bracket A bracket used in slope roof construction, having provisions for being fastened to the roof, or supported by ropes fastened over the ridge and secured to some suitable object.

roofing nail A special-purpose, short-threaded nail with a large head, usually galvanized or aluminum with a neoprene or plastic washer to aid in fastening roof coverings.

roofing square 1. A steel square used by carpenters. **2.** A measure of roofing material.

roofing tile A preformed slab of baked clay, concrete, cement, or plastic laid in rows as a roofing cover. Tiles have a variety of patterns, but fall into two classifications: roll and flat.

roofing tile

roof jack A sleeve installed around plumbing waste vent pipes.

roof joist A rafter on a flat roof that supports the roof loads; usually 2 × 10s or 2 × 12s.

roof live load Any external loads that may be applied to a roof deck, such as rain, snow, construction equipment, and personnel.

roof overhang The part of a roof that extends over the side wall or end wall of a building.

roof pitch The slope of a roof expressed as the ratio of the rise of the roof to the horizontal span.

roof plate A wall plate that supports the lower end of rafters.

roof principal A roof truss.

roof sheathing Any sheet or board material, such as plywood or particleboard, connected to the roof rafters to act as a base for shingles or other roof coverings.

roof slope The angle of a roof compared to a horizontal plane.

roof space The space between the roof and the ceiling of the highest room.

roof span The shortest distance between the seats of opposite common rafters.

roof structure Any structure on or above the roof of a building.

roof terminal The top point of the plumbing vent pipe as it exits through the roof, often 12″.

roof truss A truss used in the structural system of a roof.

roof vent A device used to ventilate an attic or roof cavity.

room cavity ratio (RCR) A ratio of room dimensions used to quantify how light will interact with room surfaces. A factor used in illuminance calculations.

room finish schedule Information provided on design drawings specifying types of finishes to be applied to floors, walls, and ceilings for each location.

root 1. The part of a tenon that widens at the shoulders. **2.** The point where the back or bottom of the weld meets the base metal.

root diameter A measurement of the major diameter on nuts and the minor diameter on screws.

rope 1. Twisted strands of fiber made into strong, flexible cord. **2.** Strands of wire braided or twisted together that are used for heavy hoisting or hauling.

rope caulk A preformed, ropelike bead of caulking compound that may contain twine reinforcement to facilitate handling.

rope diameter The largest diameter of the cross section of a wire rope. Fiber ropes are usually measured by circumference (3.14 × diameter).

rope lay The direction in which the wires or strands are twisted during manufacture.

rose The metal plate or escutcheon between a doorknob and the door.

rose bit A bit for countersinking holes in wood.

rose nail A wrought nail with a cone-shaped head.

rosette 1. A round pattern with a floral motif. 2. A circular or oval, ornamental, wood plaque that is used to terminate a wood piece such as a stair rail at a wall. 3. A decorative nailhead or screwhead.

rose window (wheel window) Circular window with a radial pattern resembling a wheel with spokes and a hub.

rosin paper A paper coated with resin. Rosin paper is laid between lead sheets and sheathing to allow for thermal movement in lead roofing.

rotary compressor An air compressor using a rotary impeller driving air through a curved chamber to compress the air.

rotary cutting A method of obtaining wood veneers by rotating logs against a flat knife and peeling the veneer off in a long continuous sheet. Peeling provides a greater volume and a more rapid production than does sawing or slicing.

rotary drill A machine for making holes in rock or earth by a cutting bit at the end of a metal rod, usually turned by a hydraulically—or pneumatically driven motor.

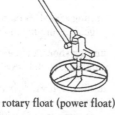

rotary float (power float) Motor-driven revolving blades that smooth, flatten, and compact the surface of concrete slabs or floor toppings.

rotary float (power float)

rotary hammer Electric boring tool that rotates with percussive action and cuts very hard materials.

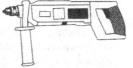

rotary hammer

rotary pump Any pump using gears, vaned wheels, or a screw mechanism to displace liquid, usually delivering large volumes at low pressure.

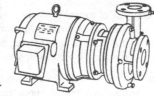

rotating laser A leveling and surveying instrument with a revolving head that emits a laser beam horizontally. A sensor then picks up the beam to determine an elevation.

rotary pump

rotor The rotating armature or member within an electrical device or motor.

rotunda A circular building or hall that is round inside and out and usually domed.

rough-cut joint (flat joint, flush joint, hick joint) A mortar joint that is flush with the face of the brickwork.

rough estimate An estimate made without detailed investigation.

rough floor A base or subfloor, consisting of a layer of boards or plywood nailed to the floor joists.

rough floor

rough grading Cutting and filling the earth for preparation of finish grading.

rough grind The initial operation in which coarse abrasives are used to cut the projecting chips in hardened terrazzo down to a level surface.

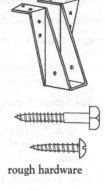

rough hardware Any fittings, such as screws, bolts, or nails that should be concealed for a finished product.

rough-in dimensions In plumbing installations, the distance from the center of a water supply, waste opening or mounting holes to the finished wall or floor.

rough hardware

roughing-in 1. The base coat in three-coat plasterwork. 2. Any unfinished work in a construction job. 3. Installing the concealed portion of electrical or plumbing systems to the point of connection for electrical outlets or appliances, or plumbing fixtures.

roughing-out A preliminary shaping operation in carpentry.

rough lumber Wood that has been sawn at the mill, but not trimmed.

rough opening An opening in a wall or framework into which a door frame, window frame, subframe, or rough buck is fitted.

rough rendering Applying a coat of plaster without removing the irregularities.

rough sill A horizontal member laid across the bottom of an unfinished opening to act as a base during construction of a window frame.

rough work The framing, boxing, and sheeting for a wood-framed building.

round 1. A molding that may be semicircular to full round, as in a closet rod. 2. A turn of wire rope around a drum.

rout 1. To deepen and widen a crack, preparing it for patching or sealing. 2. To cut out by gouging.

router An electrically driven device with various bits for cutting grooves or channels in wood.

router gauge A carpenter's tool consisting of a guide, a bar with a scale, and a narrow chisel as a cutter; used in inlaid work to cut out narrow channels in which colored strips are laid.

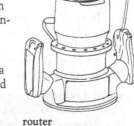

router

router patch (Davis patch) A wood patch in plywood with straight sides and rounded ends, similar to a tongue depressor; used to fill voids caused by defects. The router patch is one of two patch designs allowed in A-grade faces.

router plane (router plough, router plow) A woodworker's plane for cutting grooves or channels in wood.

rowlock course (bull header) Brick pattern in which bricks are set on their face edges with the ends visible in the wall face.

rowlock course (bull header)

R

royals Shingles with 24″ edges and a thickness of ½″ at the butt.

rubbed brick Brick that has one or more smooth faces.

rubbed finish 1. On woodwork, a dull finish obtained from hand rubbing with a rag or pad saturated with water or oil and pumice. **2.** A finish obtained by using an abrasive, often a carborundum stone, to remove surface irregularities from concrete.

rubbed joint A process for joining two narrow boards. Both boards are planed smooth, and then coated with glue and rubbed together until all air pockets and excess glue are expelled from the joint. No clamping is necessary, and the joint is extremely strong.

rubber 1. A highly resilient natural material manufactured from the juice of rubber trees and other plants. **2.** Any of various synthetically manufactured materials with properties similar to natural rubber. **3.** A cushioned backing for a carpet.

rubber silencer (bumper) A small round rubber device that attaches to a rabbeted doorjamb, which silences the noise caused by a slamming door.

rubber tile A soft and yielding floor covering that reduces the transmission of impact noises produced by walking or from other causes.

rubber-tired roller A machine for compacting and kneading soil using pneumatic-tired rollers.

rubbing brick A silicon-carbide brick used to smooth and remove irregularities from surfaces of hardened concrete.

rubble Rough stones of irregular shape and size, broken from larger masses by geological processes or by quarrying.

ruff-sawn A designation for plywood paneling or siding that has been saw-textured to provide a decorative, rough-sawn appearance.

rule 1. A straightedge with graduations used for measuring, laying out lengths, or drawing straight lines. **2.** A straightedge for working plaster to a plane surface.

rule of thumb A statement or formula that is not exact but is close enough for practical work.

run 1. In plumbing, a pipe or fitting that continues in the same straight line as the direction of flow. **2.** In roofing, the horizontal distance between the outer face of the wall and the roof ridge. **3.** In stairs, the horizontal distance from the face of the first riser to the face of the last riser.

runaround system A heat recovery system in which coils in an exhaust duct transfer a portion of the exhaust air heat to a fluid. The fluid is then circulated to coils that give up a portion of the fluid's heat to the air in a cold air duct.

rung A bar, usually of circular cross section, used as a step in a ladder.

run-in-place A method of producing ornamental plaster on-site, typically on a workbench. The plaster is attached to the wall after setting.

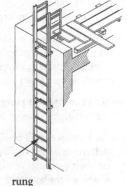

rung

run molding A formed molding of plaster or similar material formed by passing a template over the plastic material.

runner 1. The lengthwise horizontal bracing or bearing members. **2.** A cold-rolled channel used to support steel studs in a partition or ceiling tile.

running 1. Descriptive of a repeating design in a band having a smooth progression. **2.** Forming a cornice of plaster or similar material in place with a running mold. **3.** Operating a powered hand tool, particularly a drill.

running foot A linear foot. The term is a measurement of the actual length of a piece of lumber, without regard to the thickness or width of the piece.

running ground Earth in a semiplastic state that will not stand without support.

running inch A linear inch. The term is a measurement of the actual length of a piece of lumber. *See also* running foot.

running mold A template shaped to a desired cornice and mounted on a wood frame; used by plasterers to form a molding or cornice in place by applying plaster as the mold is moved along the ceiling line.

running screed A narrow strip of plaster used in place of a rule to guide a running mold.

running shoe A metal guide on a running mold to prevent wear and allow it to slide freely on a rule.

running trap A U-shaped pipe fitting installed in a drain line to prevent the backflow of sewer gases.

runway Decking over an area of concrete placement, usually of movable panels and supports, on which buggies of concrete travel to points of placement.

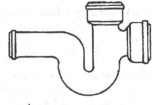

running trap

rust Any of various powdery or scaly reddish-brown or reddish-yellow, hydrated, ferric oxides formed on the surface of iron or steel that is exposed to moisture and air. Rust eventually will weaken or destroy the material if allowed to progress.

rustication Providing a continuous groove in vertical concrete at a preferred location by applying a form strip prior to placing concrete, and removing it after concrete has set.

rustic brick A brick that has a rough-textured finish produced by covering it with sand, wire brushing, or impressing it with a pattern.

rustication

rustic finish (washed finish) A type of terrazzo topping in which the matrix is recessed by washing prior to setting so as to expose the chips without destroying the bond between chip and matrix. A retarder is sometimes applied to the surface to facilitate this operation.

rustic joint A deeply sunk mortar joint that is accented by beveling the edges of the adjacent stone.

R

rustic stone Any rough, broken stone suitable for masonry with an uneven appearance, most commonly limestone or sandstone, usually set with the longest dimension horizontal.

rust joint A watertight pipe connection that uses iron filings as a catalyst to induce rusting in iron pipe joints.

rust pocket An area in the bottom of a ventilating pipe for the collection and removal of rust and debris.

R value A measure of a material's resistance to heat flow at a given thickness of material. The term is the reciprocal of the U value. The higher the R value, the more effective the particular insulation.

R

Abbreviations

S side, south, southern, seamless, subject, sulphur

S1E surfaced one edge

S1S surfaced one side

S1S1E surfaced one side and one edge

S1S2E surfaced one side and two edges

S2E surfaced two edges

S2S surfaced two sides

S2S1E surfaced two sides and one edge

S2S&CM surfaced two sides and center matched

S2S&SL surfaced two sides and shiplapped

S4S surfaced four sides

S4S&CS surfaced four sides and caulking seam

SA smart actuator, system air

S/A shipped assembled

SAB sound attenuation batt, blanket, or board

SAC security and access control

SAE Society of Automotive Engineers

SAF safety

SAFB sound attenuation fire batt

SAN sanitary

S&E surfaced one side and edge

S&G studs and girts

S&H staple and hasp

S&M surfaced and matched

sanit sanitation

SAP sampling and analysis plan, security access and parking

sar supplied air respirator

SAT supply air temperature

sat saturate, saturation

SB1S single bead one side

SBA Small Business Association

SBC Standard Building Code

SBS sick building syndrome

SC site closeout, substantial completion

scba self-contained breathing apparatus

sch school

SCH schedule

scp spherical candlepower

SCS structured cabling system

SD sea-damaged, standard deviation

S/D shop drawings

SDA specific dynamic action

Sdg siding

SDV switched digital video

Se selenium

S/E square-edged

sec second

SECT section

sed sediment, sedimentation

SEER Seasonal Energy Efficiency Ratio

sel select, selected

Sel select

sep, SEP separate

SERI Solar Energy Research Institute

SERV service

SE&S square edge and sound

SE Sdg square-edge siding

SEW sewer

sf, SF square foot, surface foot, specialty fabricator, supply fan

SFPE Society of Fire Protection Engineers

SFRC steel-fiber-reinforced concrete

Sftwd softwood

sfu supply fixture unit

SG slash grain

SGD sliding glass door

SH sheet, shower, single-hung

sh shingles

shf superhigh frequency

SHGC solar heat gain coefficient

shp shaft horsepower

SHSS site health and safety supervisor

sht sheet, sheath

SI system integrator

Si silicon

SIA Security Industry Association

SIC Standard Industrial Classification

sid siding

SIM similar

SJI Steel Joist Institute

SK sketch

Sk sack

SKU stock keeping unit

sky skylights

SL&C shipper's load and count

slid. sliding

S/L, S/LAP shiplap

SM standard matched, surface measure

SMACNA Sheet Metal and Air Conditioning Contractor's National Association

s mld stuck mold

SMS sheet-metal screw

so south

SO seller's option

SOC security operations center

SOHO small office/home office

soln solution

SOO sequence of operations

SOP standard operating (or operations) procedure

SOV shutoff valve

SOW statement of work

sp specific, specimen, spirit, single pitch (roof)

SP soil pipe, standpipe, self-propelled, single pole, static pressure

SPEC specification

sp gr, SP GR specific gravity

sp ht specific heat

SPKR loudspeaker

spl spline

SPL special

spr spruce

SPT standard penetration test

sp vol specific volume

sq square

sq e square edge

sq E&S square edge and sound

sq ft square foot

sq in square inch

sq yd square yard

SR sedimentation rate, service request

ss single strength (glass)

SS, S/S stainless steel, smart sensor, sustainable site

sst standing seam tin (roof)

SST stainless steel

st stairs, stone, street

ST steam, street

STC sound transmission class

std, STD standard

Std M standard matched

ST supply temperature

STG storage

STK stock, select tight knot

STL steel

STP standard temperature and pressure, shielded twisted pair

Stpg stepping

str stringers

Str structural

STR strike

STR1E straight line rip one edge

Struc structural

STS shared tenant services, static transfer switch

st sash steel sash

ST W storm water

sty story

S

sty hgt story height

SUB substitute

sub fl subfloor

subpar subparagraph

subsec subsection

sup supplementary, supplement

SUP supply

supp supplement

SUPSD supersede

supt, SUPT, SUPER superintendent

SUPV supervise

supvr supervisor

sur, SUR surface

surv survey, surveying, surveyor

svc service

SVE soil vapor extraction

sw switch

SW switch, seawater, southwest

SWBD switchboard

SWG, S.W.G. standard wire gauge

SY square yard

sy jet syphon jet (water closet)

SYM symmetrical

SYN synthetic

SYS system

syst system

Definitions

saber saw A handheld power saw with a reciprocating blade extending through the base of the saw.

sabin A unit of measure used to rate the sound absorption qualities of a material or assembly. Equal to one foot of a perfectly absorptive surface.

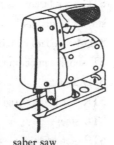

saber saw

sack A certain quantity of cement, typically 94 pounds (United States) and 87.5 pounds (Canada), for Portland or air-entraining Portland cement and as indicated for other cement types. See also **bag**.

sacking Correcting concrete surface defects by applying a mixture of cement and sand and rubbing with burlap or another coarse material.

sack joint A flush masonry joint that has been wiped or rubbed with a rag or an object such as a rubber heel or a burlap sack.

sack rub (sack finish) A finish for formed concrete surfaces, designed to produce even texture and fill all pits and air holes After dampening the surface, and before it dries, a mixture of dry cement and sand is rubbed over it with a wad of burlap or a sponge-rubber float to remove surplus mortar and fill the voids.

sacrificial protection The use of a metallic coating, such as zinc-rich paint, to protect steel. In the presence of an electrolyte, such as salt water, the metallic coating dissolves instead of the steel.

saddle 1. A fitted device used with hangers to support a pipe. 2. A series of bends in a pipe over an obstruction. 3. A short horizontal member set on top of a post as a seat for a girder. 4. Any hollow-backed structure with a shape suggesting a saddle, as a ridge connected to two higher elevations or a saddle roof.

saddle

saddle bar One of the horizontal iron bars across a window opening to secure leaded lights.

saddle bead A glazing bead or channel for securing two panes of glass together.

saddle bend A bend made in a conduit to provide clearance where it crosses another conduit.

saddle block The boom swivel block through which the stick of a dipper shovel slides.

saddle board A board used to cover the joint at the ridge of a pitched roof.

saddle fitting A type of gasketed fitting clamped around the exterior of a pipe; used when a connection to a previously installed pipe is required.

saddle flange A flange that is curved to conform to the surface of the boiler or tank to which it is welded, riveted, or otherwise attached, and designed to accept a threaded pipe.

saddle flashing Flashing installed over a cricket.

saddle joint 1. A joint in sheet-metal roofing, in which one end of one sheet is folded downward over the turned-up edge of the adjacent sheet. 2. A stepped joint in a projecting masonry course to prevent the penetration of water.

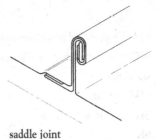

saddle joint

saddle scaffold A scaffold erected so as to bridge the ridge of a roof, usually used during chimney repair.

safe 1. A built-in or portable chamber used to protect materials or documents from fire and/or theft. 2. A pan or other collector placed beneath a pipe or fixture to collect leakage or overflow.

safe load The maximum load on a structure that does not produce stresses greater than those allowable.

safe room or shelter-in-place room A building space designed and constructed to provide protection in a natural or other disaster. The space may be structurally reinforced and may have limited air exchange.

safety belt A belt-like device worn around the waist and attached to a life-line or structure to stop a worker during a fall.

safety curtain A heavy iron or fiberglass curtain used in auditoriums to prevent fire spread. Also referred to as a fire curtain.

safety fuse A cord containing black powder or other burning medium encased in flexible wrapping and used to convey fire at a predetermined and uniform rate for firing blasting caps.

safety lintel A load-carrying lintel positioned behind a more decorative but somewhat less functional lintel, as in the aperture of a window or door.

S

safety nosing An abrasive, nonslip stair nosing whose surface is flush with the tread against which it is placed.

safety nosing

safety shutoff device A device in a gas burner that will shut off the supply of gas if the flame is extinguished.

safety stock The average amount of stock on hand when a replenishment quantity is received. Its purpose is to protect against the uncertainty in demand and in the length of the replenishment lead time. Safety stock and cycle stock are the two main components of any inventory.*

safety switch In an interior electric wiring system, a switch enclosed within a metal box that has a handle protruding from the box to allow switching to be accomplished from outside the box.

safety time In a time series planning system, material is frequently ordered to arrive ahead of the forecast requirement date to protect against forecast error. The difference between the forecast requirement date and the planned in-stock date is safety time.*

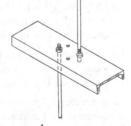

safety switch

safety tread A tread on a stair that has a roughened surface or roughened inserts to improve traction.

safe working pressure The maximum working pressure at which a vessel, boiler, flask, or cylinder is allowed to operate, as determined by the American Society of Mechanical Engineers Boiler Code; usually so identified on each individual unit.

safing 1. Noncombustible material used as a fire barrier around the perimeter of a floor or around protrusions or penetrations. 2. In ductwork, a type of barrier or similar device installed around a component to ensure that air flows through that component and not around it.

safing off Installation of fire safety insulation around floor perimeters, between floor slab and spandrel panels. Insulation helps retain integrity of fire-resistance ratings.

sagging 1. Subsidence of shotcrete material from a sloping, vertical, or overhead placement. 2. The condition of a horizontal structural member bending downward under load. 3. A paint defect where paint runs after it has been applied.

sag rod A tension member used to limit the deflection of a girt or purlin in the direction of its weak axis or to limit the sag in angle bracing.

sag rod

sailor A rectangular masonry unit, such as a brick, that is set upright with its largest side facing out.

salamander A portable source of heat, customarily oil-burning, used to heat an enclosure around or over newly placed concrete to prevent the concrete from freezing.

sal ammoniac (ammonium chloride) A material used in soldering flux and iron cement.

salient Descriptive of a projecting part of an object or member, as a salient corner.

sally A projection, such as the end of a rafter beyond the notch, which has been cut to fit a plate or beam.

salmon brick (chuff brick, place brick) Comparatively soft and underburnt brick, usually because of its high placement in the kiln, and so named because of its color.

salt-glazed brick (brown-glazed brick) Brick whose surface faces have a lustrous glazed finish resulting from the thermochemical reaction between silicates of clay and vapors of salt or other chemicals during firing.

salt-glazed tile Facing tile with a lustrous glazed finish, obtained by thermochemical reaction between silicates of clay and vapors of salt or other chemicals during firing.

salvage value The value assigned to the piece of equipment at the end of the depreciation period.

samples Examples of workmanship establishing the standards against which the rest of the work will be measured.

sampling approach A methodology for determining the cost of an asset within a construction project. Uses a model to analyze identical types of construction and is most popular for redundant types of projects such as retail chain stores.

sand 1. Granular material passing the ⅜″ sieve, almost entirely passing the No. 4 (4.75-millimeter) sieve, and predominantly retained on the No. 200 (75-micrometer) sieve, and resulting from natural disintegration and abrasion of rock or processing of completely friable sandstone. 2. That portion of an aggregate passing the No. 4 (4.75-millimeter) sieve, and resulting from natural disintegration and abrasion of rock or processing of completely friable sandstone. *Note:* The definitions are alternatives to be applied under differing circumstances. Definition 1 is applied to an entire aggregate either in a natural condition or after processing. Definition 2 is applied to a portion of an aggregate. Requirements for properties and grading should be stated in the specifications. Fine aggregate produced by crushing rock, gravel, or slag commonly is known as manufactured sand.

sand asphalt A mixture of ungraded sand and liquid asphalt used for an economical base or wearing surface for pavements.

sandbag A canvas bag that is filled with sand and used as a counterweight or for emergency damming of water flow.

sandblast A system of cutting or abrading a surface, such as concrete, by a stream of sand ejected from a nozzle at high speed by compressed air. Sandblasting is often used for cleanup of horizontal construction joints or for exposure of aggregate in architectural concrete.

sand box (sand jack) A tight box filled with clean, dry sand, on which rests a tight-fitting timber plunger that supports the bottom of posts used in centering. A plug from a hole near the bottom of the box permits the sand to run out when it is necessary.

sand clay A naturally occurring sand that contains about 10% clay, or just enough to make the material bind tightly when compacted.

sand-coarse aggregate ratio Ratio of fine to coarse aggregate in a batch of concrete, by weight or volume.

sand-dry Descriptive of a stage in the drying process of paint where sand will not adhere to the surface.

sanded-face shingle A shingle with retrimmed edges and butts that has been sanded to remove saw marks, etc., and is to be applied to a wall as part of a decorative effect.

S

sander A machine designed to smooth wood and remove saw or lathe marks and other imperfections. Sanders range in size from hand-held to large drums or belts capable of surfacing a full-size panel.

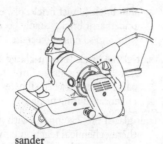
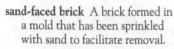

sander

sand-faced brick A brick formed in a mold that has been sprinkled with sand to facilitate removal.

sand filter A bed of sand laid over graded gravel, used as a filter for a water supply.

sand filter trenches A network of sewage-effluent-filtering trenches incorporating perforated pipe or drain tiles surrounded by fine sand sandwiched between coarse aggregate, and equipped with an underdrain to remove whatever material has passed through. The

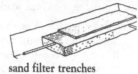

sand filter trenches

trenches are used to remove solid or colloidal material that cannot be removed by sedimentation.

sand finish 1. In plastering, a textured final coat, usually containing sand, lime putty, and Keene's cement. **2.** A smooth finish derived from rubbing and sanding the final coat.

sand grout (sanded grout) Any grout in which fine aggregate is incorporated into the mixture.

sanding machine A stationary machine having a moving belt, disk, or spindle with an abrasive surface, usually a sandpaper; used for smoothing surfaces.

sand interceptor (sand trap) A watertight device that can be directly or indirectly connected to a drainage system to intercept and prevent sand or other solids from passing into the system.

sandpaper Strong, tough paper coated on one side with glue or other adhesive material, into which an abrasive such as flint, silica, or aluminum oxide has been embedded. The resulting product is used primarily for resurfacing or cutting wood, metal, plastic, or glass and is available in an extensive range of coarseness or grit.

sandpile A foundation formed by ramming sand into a hole left by a pile that was driven and removed.

sand pine *Pinus clausa*. A type of pine found almost exclusively in north central Florida, where it grows in very sandy soil. The wood is used mostly for pulp.

sand plate A flat steel plate or strip welded to the legs of bar supports for use on compacted soil.

sand pocket A zone in concrete or mortar containing sand without cement.

sand seal Emulsified asphalt covered with fine aggregate applied to pavement to improve skid resistance and to prevent air and water penetration.

sandstone A cemented or otherwise compacted sedimentary rock composed predominantly of sand grains. It has been used as cut stone for carved ornament.

sand streak A streak of exposed fine aggregate in the surface of formed concrete caused by bleeding.

sand trap *See* **sand interceptor.**

sandwich beam *See* **flitch beam.**

sandwich construction Composite construction usually incorporating thin layers of a strong material bonded to a thicker, weaker, and lighter core material, such as rigid foam or paper honeycomb, to create a product that has high strength-to-weight and stiffness-to-weight ratios.

sandwich panel A panel formed by bonding two thin facings to a thick, and usually lightweight, core. Typical facing materials include plywood, single veneers, hardboard, plastics, laminates, and various metals, such as aluminum or stainless steel. Typical core materials include plastic foam sheets, rubber, and formed honeycombs of paper, metal, or cloth. **2.** A prefabricated panel that is a layered composite formed by attaching two thin facings to a thicker core. An example is precast concrete panels, which consist of two layers of concrete separated by a nonstructural insulating core.

sandwich panel

sanitary code Municipal regulations established to control sanitary conditions of establishments that produce and/or distribute food, serve food, or provide medical services.

sanitary cove A piece of metal used in a stair between the surface of the tread and the face of the riser to facilitate cleaning.

sanitary cross In a soil pipe system, a cross pipe, whose 90° transitions are curved to direct the flow from the branches toward the direction of the main flow.

sanitary engineering That part of civil engineering related to public health and the environment, such as water supply, sewage, and industrial waste.

sanitary sewage (domestic sewage) Sewage containing human excrement and/or household wastes that originate from sanitary conveniences of a dwelling, business, building, factory, or institution. Does not include storm water.

sanitary sewer 1. A sewer line designed to carry only liquid or waterborne waste from the structure to a central treatment plant. **2.** The conduit or pipe that carries sanitary sewage.

sanitary tee A T-fitting for pipe, having a slight curve in the 90° transition so as to channel flow from a branch line toward the main flow.

sanitary ware Devices of porcelain enamel, stainless steel, or other material, such as bathtubs, sewer pipes, toilet bowls, and washbasins.

sanitary tee

Santorin earth A volcanic tuff originating on the Grecian island of Santorini and used as a pozzolan.

sap grade 1. A grade of southern yellow pine export lumber. **2.** KD saps.

saponification Gray residue that has collected on the inside surface of a concrete sub-grade wall or floor. It is the result of moisture leaching chemicals from the concrete.

sapwood The wood just beneath the bark of a tree, normally lighter in color than the rest of the wood, but usually not as strong as the rest of the wood.

sarking (sarking board) A thin board employed in sheathing applications, as under the tiles or slates of a roof.

S

Sarnafill A roofing manufacturer known for a product used as a substitute for copper roofing because of its lower cost and green color.

sash (window sash) The framework of a window that holds the glass.

sash adjuster *See* casement stay.

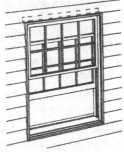

sash (window sash)

sash and frame A preassembled unit consisting of a cased frame and a double-hung window.

sash balance A spring-loaded device, usually a spring balance or tape balance, used as a counterbalance for a sash in a double-hung window. A sash balance replaces sash weights, cords, and pulleys.

sash block *See* jamb block.

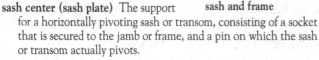

sash and frame

sash center (sash plate) The support for a horizontally pivoting sash or transom, consisting of a socket that is secured to the jamb or frame, and a pin on which the sash or transom actually pivots.

sash chain A metal chain used in place of a sash cord to connect a vertically hung sash with its counterweight.

sash chisel A chisel having a wide blade honed on both sides, used for deep cutting, such as cutting the mortises in pulley stiles.

sash cord A rope connecting a sash with its counterweight in a double-hung window.

sash-cord iron A metal holder used to connect a sash cord or chain to the window.

sash counterweight *See* sash weight.

sash door *See* glazed door.

sash fast (sash fastener, sash holder) Any fastening device that holds two window sashes together to prevent their opening or rattling. A sash fast is usually attached to the meeting rails of a double-hung window.

sash fillister 1. A rabbet cut in a glazing bar to receive the glass and glazing compound. **2.** A special plane for cutting such rabbets.

sash hardware All accessories used to balance a vertically hung sash, including chains or cords, weights, and pulleys.

sash holder *See* sash fast.

sash lift *See* window lift.

sash lift and hook A sash lift having a locking lever that holds the window fixed by contact with a strike in the frame. Raising the sash automatically releases the strike.

sash line *See* sash cord.

sash lock A sash fast that is controlled by a key. *See also* sash fast.

sash plane A carpenter's plane having a notched cutting blade for trimming the inside of a door frame or window frame.

sash plate (sash center) One of the pair of plates constituting the pivoting mechanism for a horizontally pivoting sash or transom.

sash pocket *See* pocket.

sash pull A plate, with a recess for fingers, set in a sash rail, or a handle attached to a rail to use in raising or lowering a window.

sash pulley A pulley mortised into the side of the frame of a double-hung window. The sash chord or chain passes over the pulley to the sash weight.

sash ribbon A metal tape used in place of a sash chord or chain.

sash saw 1. A small miter saw used to cut the tenons of sashes. **2.** A saw fitted in a frame that moves vertically.

sash pull

sash schedule Information provided on design drawings regarding the selection and installation of window sashes.

sash spring bolt *See* window spring bolt.

sash stop (window stop) A small strip fastened to a cased frame to hold a sash of a double-hung window in place.

sash stuff Wood cut to standard sizes and shapes for use in making window frames.

sash tool A round brush used in painting items such as window frames and glazing bars.

sash weight A weight, usually cast iron, used to balance a vertically hung window.

sash window Any window, but usually a double-hung window, having a vertically or horizontally sliding sash.

saturant 1. A substance, usually a diluted encapsulant, added to water to increase the amount of solute that can be dissolved at a certain temperature. A wetting agent used to improve penetration. **2.** A bituminous material with a low softening point, used for impregnating felt in asphalt-prepared roofing.

saturated air Air that contains the maximum amount of water vapor it can hold at its temperature and pressure.

saturated surface dry Condition of an aggregate particle or other solid when the voids between the particles are filled with water and no water is on the exposed surfaces.

saturated vapor pressure The pressure at a given temperature in a closed container that contains a liquid and the vapor from that liquid, once equilibrium conditions have been reached.

saturation temperature The temperature at which vapor and liquid coexist in stable equilibrium.

saucer dome

saucer dome A dome that has a rise less than its radius.

sauna A room in which a person bathes in steam produced when water is sprayed or poured over heated rocks or another heated surface.

savings-to-investment ratio (SIR) A dimensionless measure of performance that expresses the ratio of savings to costs, recommended for establishing priorities among projects.

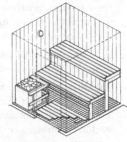

sauna

S

The numerator of the ratio contains the operation-related savings; the denominator contains the increase in investment-related costs.

saw **1.** To cut by means of a hand or powered tool having a thin, flat metal blade, band, or stiff plate with cutting teeth along the edge. **2.** A toothed steel device used to cut construction materials.

saw bench A bench on which a circular saw is mounted.

saw cut (control joint) A cut made in hardened concrete by diamond or silicone-carbide blades or discs.

sawdust concrete Concrete in which the aggregate consists mainly of sawdust from wood.

sawed finish Descriptive of the surface of any stone that has been sawn.

sawed joint A joint cut in hardened concrete by special equipment to less than the full depth of the member.

sawhorse (sawbuck) A four-legged bench, usually used in pairs, made primarily to hold wood while being sawed.

saw kerf A slot or kerf that is cut into wood with a saw.

sawn veneer Veneer that has been cut from a block with a saw, rather than peeled on a lathe or sliced off by a blade. Sawn veneer is sometimes said to be more solid than sliced or peeled veneer. Because of saw kerf waste, it is more costly to produce.

saw set **1.** The angle at which the teeth of a saw are set. **2.** A tool used to set the teeth of a saw at a desired angle.

saw table The table or platform of a power saw.

saw texture A texture put on a piece of siding or paneling by a saw or knurled drum to give it a textured, rough, and/or resawn appearance.

saw-tooth roof (sawtooth roof) A roof with a profile similar to the teeth in a saw, composed of a series of single-pitch roofs, whose shorter or vertical side has windows for light and air. This roof shape is found primarily on industrial buildings.

sawyer In a sawmill, a worker who operates the head rig, or main saw, to make the initial cuts on a log. Also refers to a concrete saw operator.

scab A short piece of wood fastened to two formwork members to secure a butt joint.

scabble To dress stone with a pick, scabbling hammer, or broad chisel, leaving prominent tool marks so that a rough surface is left. Finer dressing usually follows.

scabbling hammer (scabbing hammer) A hammer, used for rough dressing stone, with one end pointed for picking.

scabbling hammer

scaffold **1.** Any temporary, elevated platform and its supporting structure used for supporting workmen and/or materials. **2.** Any raised platform.

scaffold board A board used in forming a work platform on a scaffold.

scaffold height That height of a wall under construction that necessitates the addition of another section of scaffold so that construction of the wall can continue.

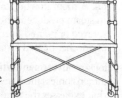

scaffold

scagliola Plasterwork in imitation of ornamental marble, consisting of ground gypsum and glue colored with marble or granite dust.

scalability The issue of how well a system behaves as the data it uses grows in size. Some applications operate well only with small datasets. File-based systems tend to have file size limitations while systems that use a database tend to be much less dependent on file size.

scale **1.** A draftsman's tool with proportioned, graduated spaces. **2.** A system of proportioned drawing in which lengths on a drawing represent larger or smaller lengths on a real object or surface. **3.** The flaky material resulting from corrosion of metals, especially iron or steel. **4.** A heavy oxide coating on copper or copper alloys resulting from exposure to high temperatures and an oxidizer. **5.** Any device for measuring weight.

scale drawing A drawing in which all dimensions are reduced proportionally according to a predetermined scale.

scaling Local flaking or peeling away of the near-surface portion of hardened concrete or mortar, or of a layer from metal

scaling hammer A hand or power hammer used in a chiseling fashion in masonry work.

scaling, light Does not expose coarse aggregate.

scaling, medium Involves loss of surface mortar to 5 to 10 mm in depth and exposure of coarse aggregate.

scaling, severe Involves loss of surface mortar to 5 to 10 mm in depth with some loss of mortar surrounding aggregate particles 10 to 20 mm in depth.

scaling, very severe Involves loss of coarse aggregate particles, as well as mortar, generally to a depth greater than 20 mm.

scalper A sieve for removing oversized particles.

scalping The removal of particles larger than a specified size by sieving.

scant Less than standard or required size.

scantling **1.** A small piece of lumber, ordinarily yard lumber, 2″ thick and less than 8″ wide, or lumber not more than 5″ square. **2.** The dimensions, especially width and thickness, of construction materials such as stone or timber. **3.** A stud or similar upright framing timber. **4.** Any hardwood that has been squared, but is not of standard dimensions.

scarf joint A joint made by chamfering, or beveling, the ends of two pieces of lumber or plywood to be joined. The angled cut on each is made to correspond to the other so that the surfaces of the two pieces being joined are flush.

scarify **1.** To break up or scratch a surface such as earth or pavement. **2.** To roughen a surface by sanding or other means to improve adhesion of paint.

scarp A steep slope, natural or man-made.

schedule A chronological itemization, often in chart form, of the sequence of project tasks.

schedule contingency Duration added to a schedule activity to allow for the probability of possible or unforeseen events. Use in this manner is not recommended as the contingency is hidden and may be misused.

schedule number Schedule numbers are American Standards Association designations for classifying the strength of pipe.

Schedule 40 is the most common form of steel pipe used in the mechanical trades.

schedule of values (cost breakdown) A listing of elements, systems, items, or other subdivisions of the work, establishing a value for each, the total of which equals the contract sum. The schedule of values is used for establishing the cash flow of a project.

schema As applied to databases, the abstract representation or model of data for some use. SQL is a popular database language for creating and operating on database schemas.

schematic design phase (schematic drawing) The phase of design services in which the design professional consults with an owner to clarify the project requirements. The design professional prepares schematic design studies with drawings and other documents illustrating the scale and relationship of the project's components to the owner. A statement of estimated construction cost is often submitted at this phase.

schist A metamorphic rock, the constituent minerals of which have assumed roughly parallel beds, used principally for flagging.

Schmidt hammer An instrument used to measure the compressive strength of concrete. Testing involves gauging the distance of the hammer's rebound after it strikes the material.

scissor lift An electric or gas-powered platform mounted on folding arms and used to provide elevated work areas or to raise or lower unit loads.

scissors truss A roof truss with tension members extending from the foot of each principal rafter to the upper half of its opposite member.

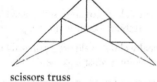
scissors truss

scope creep Gradual progressive change (usually additions to) of the project's scope such that it is not noticed by project management team or customer. Typically occurs when the customer identifies additional, sometimes minor, requirements that, when added together, may collectively result in a significant scope change, resulting in cost and schedule overruns.*

scope of work An accurate, detailed, and concise description of the work to be performed by the contractor, the owner, and third parties in a construction contract.

scoping A primary component process in response to a request for a JO proposal, which includes a site visit by the contractor's representative(s) and the owner's representative(s) to identify and document various site characteristics and job conditions that will be impacted by the project's design and/or owner's intent.

score 1. In concrete work, to modify the top surface of one pour, as by roughening, so as to improve the mechanical bond with the succeeding pour. **2.** To tool grooves in a freshly placed concrete surface to reduce cracking from shrinkage. **3.** To scratch or otherwise roughen a surface to enhance the bond of plaster, mortar, or stucco that will be applied to it. **4.** To groove, notch, or mark a surface for practical or decorative purposes.

S corporation A corporation that has elected to be taxed like a partnership, in accordance with the provisions of subchapter S of the Internal Revenue Code.

scotia A deep, concave molding more than ¼ round in section, especially as found in classical architecture at the base of a column.

scour 1. Erosion of a concrete surface by water movement exposing the aggregate. **2.** Erosion of a river bottom by water movement.

scouring Smoothing freshly applied mortar or plaster by working it in circular motions with a cross-grained wooden float.

scraper 1. A digging, hauling, and grading machine having a cutting edge, carrying bowl, a movable front wall or apron, and a dumping or ejecting mechanism. **2.** A machine that is towed behind a vehicle to level the ground surface.

scraping Removal of a top layer of paint with a sharp instrument, such as a scalpel.

scratch To score or groove a coat of plaster to provide a better bonding surface for a successive coat.

scratch-brushed finish (satin finish) A surface finish rendered by mechanical wire brushing or abrasive buffing.

scratch coat The first coat of plaster or stucco applied to a surface in three-coat work and usually cross-raked or scratched to form a mechanical key with the brown coat.

SCR brick A patented brick with nominal dimensions of 22/3″ × 6″ × 12″, as designated or classified by the Structural Clay Research (trademark of the Structural Clay Products Institute). It will render three courses in 8″ and will render a wall whose nominal thickness is 8″.

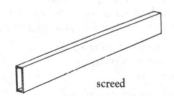
SCR brick

screed 1. To strike off concrete lying above the desired plane or shape. **2.** A tool for striking off the concrete surface, sometimes referred to as a *strikeoff*.

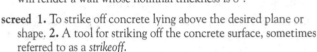
screed

screed coat The plaster coat made flush with the screeds.

screed guide Firmly established grade strips or side forms for unformed concrete that will guide the strikeoff in producing the desired plane or shape.

screeding The operation of forming a surface by the use of screed guides and a strikeoff. *See also* **strike off**.

screed strip One of a series of long narrow strips of plaster, carefully leveled to serve as guides for the application of plaster to a specified thickness.

screen façade An architectural facing used to disguise the shape or size of a building.

screenings That portion of granular material that is retained on a sieve.

screen mold A molding, originally used in the construction of screens and now used extensively in cabinetry and finished carpentry, where a clear strip is required, as on the edge of a shelf made of plywood or particle board.

screw A fastener with an external thread.

screw anchor A type of molly whose metal, plastic, or fiber shell is inserted into a hole in masonry, plaster, or concrete, and expanded when the screw is driven in.

screw clamp A woodworking tool consisting of a pair of opposing jaws that can hold pieces of wood and are adjusted by two screws.

screw conveyor A helical screw shaft turning on the concentric axis of a pipe. Conveys fine-grained or liquid material.

screwdriver A hand tool with the shanks tip shaped to fit the recess in the head of a screw; used to drive or remove the screw.

screwdriver

screwed joint A pipe joint consisting of threaded male and female parts joined together.

screwnail *See* **drivescrew**.

scribe To mark and cut the edge of a member for an irregular cut that will allow it to fit a similarly irregular space.

scriber A pointed tool used to mark guidelines on wood, metal, and plastic.

scrim A coarse, meshed material, such as wire, cloth, or fiberglass, that spans and reinforces a joint over which plaster will be applied.

scroll compressor A rotary positive displacement compressor with a fixed and a rotating scroll, in which compression takes place by confining gas volume by the meshing of the scrolls.

scroll molding An ornamental molding, semicircle in cross section, with a slight overhang on the rounded surface that faces outward.

scroll saw A handsaw consisting of a thin blade in a deep U-shaped frame and a handle. A scroll saw is used for cutting thin boards, veneers or plates, and is especially good for cutting curves.

scrub plane A wood plane having a blade with a convex cutting edge, used in rough carpentry work.

scrub sink A plumbing fixture equipped to enable medical personnel to scrub their hands prior to a surgical procedure. The hot and cold water supply is activated by a knee-action mixing valve or by wrist or foot control.

scrub sink

scum 1. A deposit, sometimes formed on the surface of clay bricks, caused by soluble salts in the clay that accumulate on the surface during drying, or by the formation of deposits during kiln firing. (2) A mass of organic matter that floats on the surface of sewage. **3.** A film of impure matter that forms on the surface of a body of water.

scupper Any opening in a wall, parapet, bridge curb, or slab that provides an outlet through which excess water can drain.

S curve A rudimentary method of illustrating the progress of billings as a function of time as measured against the schedule.

scupper

scutching Dressing stone with a special hammer whose head contains several steel points.

s-dry A description of lumber seasoned to a moisture content of 19% or less prior to surfacing.

seal A legal term used to describe the signature or other representation of an individual agreeing to the terms and conditions of an agreement or contract.

sealed bearing A conventional bearing that has been provided with seals on its sides so that the bearing can be used for longer periods without greasing.

sealed bid A bid, based on contract documents, that is submitted sealed for opening at a designated time and place.

sealed bidding A basic method of procurement that involves the solicitation of bids and the award of a contract to the responsible bidder submitting the lowest responsive bid. This type of bidding is commonly used on public works projects.

sealer 1. Any liquid applied to the surface of wood, paper, or plaster to prevent it from absorbing moisture, paint, or varnish. **2.** A liquid coating applied over bitumen or creosote to restrict it from bleeding through other paints. **3.** A final application of asphalt or concrete to protect against moisture. **4.** Any liquid coating used to seal the pores of the surface to which it is applied.

sealing joints Filling pavement joints with a flexible material.

seal slab A concrete slab placed along the bottom of a trench to stabilize bedding material.

seal weld A weld used primarily to seal a joint against leakage.

seam A joint between two sheets of material, such as metal.

seaming The process of joining metal sheets by bending over or doubling the edges and pinching them together.

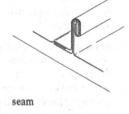
seam

seamless flooring Fluid- or trowel-applied floor surfaces that do not contain aggregates.

seamless pipe An extruded seamless tube having certain standardized sizes of outside diameter and wall thickness, commonly designated by nominal pipe sizes and American National Standards Institute's schedule numbers.

seam weld A resistance weld made in overlapping parts.

seasonal commodities Commodities that are normally available in the marketplace only in a given season of the year.*

seasonal energy efficiency ratio (SEER) A measure of cooling efficiency for air conditioning products. The higher the SEER rating, the more energy efficient the unit. The government's established minimum rating for air conditioning is 10.

seasoned 1. Timber that is not green, having a moisture content of 19% or less, and is air- or kiln-dried. **2.** Cured or hardened concrete.

seasoning check A small split that occurs in the grain of wood when moisture is extracted too rapidly.

seat The cut made at the lower end of a rafter that enables it to fit securely over the top wall plate

S

seat angle A short leg on a steel angle fastened to a column or girder to support the end of a beam.

seating pressure The pressure generated by the action of a spring and control air to close the automatic control valve plug against its seat.

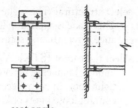
seat angle

second 1. A piece of secondary quality or one not meeting specified dimensions. **2.** A unit measure of time.

secondary air 1. Air that is introduced into a burner above or around the flames to promote combustion, in addition to the primary air that is premixed with the fuel or forced as a blast under a stoker. **2.** Air already in an air-conditioned space, in contrast to primary air that is supplied to the space.

secondary beam A flexural member that is not a portion of the principal structural frame of a building.

secondary branch In the plumbing of a building drain or water-supply main, any branch that is not the primary branch.

secondary combustion The burning of combustible gases and smoke that are not burned during primary combustion. Such combustion can be either by design to create added energy or an unintentional, undesirable event.

secondary consolidation (secondary compression, secondary time effect) The reduction in volume of a soil mass caused by the application of a sustained load to the mass and due principally to the adjustment of the internal structure of the soil mass after most of the load has been transferred from the soil water to the soil solids.

secondary containment A chamber for the collection of oil that has leaked from a fuel oil tank.

secondary feeder Any electrical conductor that runs from a building's service entrance to one of its distribution centers.

secondary light source A light source that is not itself a luminaire or otherwise intrinsically light-producing or light-emitting, but that instead receives light from another source and simply serves to redirect it, as by reflection or transmission.

secondary member A structural member that carries a load to a primary member. Examples include purlins, headers, and jambs.

secondary moment In statically indeterminate structures, the additional moments caused by deformation of the structure due to the applied forces. In statically indeterminate prestressed concrete structures, the additional moments caused by the use of a nonconcordant prestressing tendon.

secondary reinforcement Reinforcing steel in reinforced concrete, such as stirrups, ties, or temperature steel. Secondary steel is any steel reinforcement other than main reinforcement.

secondary sheave A supplementary sheave or block used to enable double wrapping of hoisting ropes and increase traction and/or lifting power.

secondary subcontractor A subcontractor employed by the contractor to

secondary reinforcement

complete minor portions of the work, or a subcontractor other than those identified as primary subcontractors.

secondary winding The winding on the output side of a transformer.

second coat The second coat of plaster, which is the brown coat in three-coat work or the finish coat in two-coat work.

second-growth timber Wood from trees grown after a virgin forest has been cut down.

seconds A grade of lumber used in the hardwood industry and sometimes in the overseas trade, as in firsts and seconds.

secret dovetail (miter dovetail) A joint whose external appearance implies a simple miter joint, but having dovetailing concealed within it.

section 1. A topographical measure of land area, equal to one mile square or 640 acres. One of the 36 divisions in a township. **2.** The most desired pieces of veneer, clipped to standard widths of 54″ and 27″, because of the ease of using them in assembling a panel. The actual width may vary from 48″ to 54″, or 24″ to 27″. **3.** A drawing of an object or construction member cut through to show the interior makeup. **4.** A segment of the project specifications that cover a work item.

sectional insulation Insulation that is manufactured to be assembled in the field, such as pipe insulation molded in two parts to fit around a pipe in the field.

sectional overhead doors Doors made of horizontally hinged panels that roll into an overhead position on tracks, usually spring-assisted.

sectional overhead doors

sectional valve In a piping system, a control valve that is used to isolate a section for repair or other reason.

section modulus A term pertaining to the cross section of a flexural member. The section modulus with respect to either principal axis is the moment of inertia with respect to that axis divided by the distance from that axis to the most remote point of the tension or compression area of the section, as required. The section modulus is used to determine the flexural stress in a beam.

security screen A heavy screen in a special frame used as a barrier against escapes or break-ins.

security screw A screw with a tamper-proof head design.

security system A system that monitors and may also document aspects of access control, various types of alarms, perimeter security, and closed circuit TV and video surveillance.

security window A steel window used in stores, warehouses, and similar commercial buildings to provide protection against burglary.

security screen

sediment (silt) Material that settles to the bottom of a liquid.

sedimentary rock Rock such as limestone or sandstone that is formed from deposits of sediment consolidated by cementitious material and/or the weight of overlying layers of sediment.

sediment basin A dug-out or depressed area used for retention of debris or sediment on a construction site.

S

seepage **1.** The slow percolation of water through a soil. **2.** The quantity of water that has moved slowly through a porous material.

seepage bed A trench at least a yard wide into which coarse aggregate and a system of distribution pipes are placed so as to allow the treated sewage that passes through them to seep into the surrounding soil.

segmental arch An arch where the curve through the length of the arch is less than a semicircle.

segmental member A structural member made up of individual elements prestressed together to act as a monolithic unit under service loads.

segment saw A large-diameter circular saw consisting of pie-shaped sections and whose narrow kerf makes it especially suitable for cutting veneer.

segregation The differential concentration of the components of mixed concrete, aggregate, or the like, resulting in nonuniform proportions in the mass.

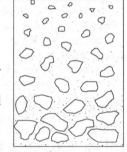

segregation

seismic bracing Structural reinforcement that is designed to protect against wind and earthquake.

seismic code A building code pertaining to earthquakes and other seismic activity that governs structural design and defines the requirements of preventive measures.

seismic load (earthquake load) The assumed lateral load an earthquake might cause to act upon a structural system in any horizontal direction.

seizing **1.** Friction damage to a metal surface by another metal surface. **2.** Restriction or prevention of motion of a metal component caused by fusion or coherence with another component.

select A high-quality piece of lumber graded for appearance. Select lumber is used in interior and exterior trim, and cabinetry.

selected bidder The bidder selected by the owner to consider the award of a construction contract.

selective bidding A process of competitive bidding for award of the contract for construction whereby the owner selects the constructors who are invited to bid to the exclusion of others, as in the process of open bidding.

selective digging Separating two or more types of soil while excavating, such as loam from sandy soil.

selective glazing Window material, such as glass, that screens out the infrared and ultraviolet portions of the solar spectrum, but allows visible light to pass. Selective glazing is recommended if a clear appearance is desired, or if a high visible transmittance is required to meet daylighting goals.

select material Excavated pervious soil suitable for use as a foundation for a granular base course of a road, or for bedding around pipes.

select merchantable **1.** A grade of boards intended for use where knotty-type lumber of fine appearance is required. **2.** An export grade of sound wood with tight knots and close grain, suitable for high-quality construction and remanufacture.

select structural The highest grade of structural joists and planks. This grade is applied to lumber of high quality in terms of appearance, strength, and stiffness.

select tight knot (STK) A grade term frequently used for cedar lumber. Lumber designated STK is selected from mill run for the tight knots in each piece, as differentiated from lumber that may contain loose knots or knotholes.

selenitic cement (selenitic lime) A type of lime cement that has had its hardening properties improved by the addition of 5% to 10% plaster of paris.

self-ballasted lamp An arc-discharge lamp with a special built-in filament coil that controls both starting and operating currents.

self-closing fire door A fire door equipped with a closing device.

self-desiccation The removal of free water by chemical reaction so as to leave insufficient water to cover the solid surfaces and to cause a decrease in the relative humidity of the system. The term is applied to an effect occurring in sealed concretes, mortars, and pastes.

self-drilling screw A fastener that does not require a predrilled hole.

self-faced stone An undressed stone such as a flagstone.

self-furring Metal lath or welded-wire fabric formed in the manufacturing process to include means by which the material is held away from the supporting surface, thus creating a space for *keying* of the insulating concrete, plaster, or stucco.

self-furring nail A nail with a flat head and a washer or spacer in the shank, used for fastening reinforcing wire mesh and spacing it from the nailing member.

self-furring nail

self-healing A redundancy in a system or a network (such as a backup power supply) that allows operations to continue in the event of failures in certain components.

self-leveling sealant A sealant, such as one used on concrete joints or cracks, with properties such that the simple force of gravity will level it.

self-sealing paint A type of paint that can be applied over a surface of inconsistent porosity to seal it and still dry to a uniform color and sheen.

self-spacing tile Ceramic tile with protuberances on the sides that space the tiles for grout joints.

self-stressing concrete (chemically prestressed concrete) Expansive-cement concrete (mortar or grout) in which expansion, if restrained, induces persistent compressive stresses in the concrete.

self-tapping screw A screw designed to cut its own threads in a predrilled hole.

self-test An automatic diagnostic procedure.

selvage (selvedge) **1.** An edge or edging that differs from the main part of a fabric or granule-surfaced roll roofing. **2.** The finished lengthwise edge of woven carpet that will not unravel or require binding.

selvage joint In roofing, a lapped joint mortised with mineral-surfaced cap sheets. A small part of the longitudinal edge of the

sheet below contains no mineral surfacing so as to improve the bond between the lapped top sheet surface and the bituminous adhesive.

semicircular arch A round arch whose intrados is a complete semicircle.

semicircular arch

semicircular dome A dome constructed in the shape of a half-sphere.

semiconductor 1. An electric conducting material, with resistivity in the range between metals and insulators, such as germanium or silicon. **2.** Miniature electric devices manufactured from semiconductor materials.

semidetached dwelling One of a pair of dwellings with a party wall between them.

semidetached house One of a pair of houses with a party wall between them.

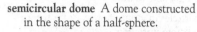

semidetached house

semidirect lighting Lighting from luminaires that direct 60° to 90° of the emitted light downward and the balance upward.

semidome A half-dome equal to one-fourth of a hollow sphere, as might be found above a semicircular niche or apse.

semiflexible joint In reinforced concrete construction, a connection in which the reinforcement is arranged to permit some rotation of the joint.

semigloss 1. The degree of surface reflectance midway between glass and eggshell. **2.** Paints and coatings displaying these properties.

semi-indirect lighting Lighting from luminaires that direct 60° to 90° of their emitted light upward and the balance downward.

semirigid frame A type of structural framework construction in which some flexibility is allowed at the joints of columns and beams.

sensible heat Heat that alters the temperature of a material without causing a change in state in that material.

sensor A device designed to detect an abnormal ambient condition, such as smoke or high temperature, and to sound an alarm or operate a device.

separate-application adhesive An adhesive consisting of two parts, each part being applied to a different surface. The surfaces are brought together to form a joint.

separation 1. The tendency of coarse aggregate to separate from the concrete and accumulate at one side as concrete passes from the unconfined ends of chutes, conveyor belts, or similar arrangements. **2.** The tendency of the solids to separate from the water by gravitational settlement.

septic system A wastewater treatment system installed at the site of a home. Systems typically include a septic tank in which waste is biologically processed and a drain field that allows remaining liquids to be absorbed by the soil.

septic tank A watertight receptacle that receives the discharge from a sewage system, or part thereof. They are constructed to separate solids from the liquid, digesting organic matter during a period of detention, and discharging the clarified liquids.

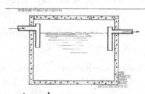

septic tank

sequence of operations An accounting of a system's procedures for start-up and shut-down, response to varying conditions, and certain scheduled operations.

sequence-stressing loss In posttensioning, the elastic loss in a stressed tendon resulting from the shortening of the member when additional tendons are stressed.

serial digital alarm communicator transmitter A transmitter that accepts output from devices such as a fire alarm control.

serial distribution A group of absorption trenches, seepage pits, or seepage beds arranged in a series so total effective absorption area of one is used before flow enters the next.

sericite A white clay material used in masonry as a filler in grouting cement. Also used as a paint extender.

series circuit A circuit supplying electrical energy to a number of devices connected so that the same current passes through each device.

series wiring A wiring configuration in which electrical current travels along one conductor or wire. Unlike parallel wiring, series wiring is seldom used.

serpentine wall Garden wall that follows a curved ogee path resembling a serpent.

server room A space (often designed with additional/backup power and cooling supply and protective systems) that houses computer servers and communications equipment.

service box Within a building, a metal box located at the point where the electric service conductors enter the building.

service conductors Those conductors (wires) that carry electrical current from the street mains, or transformers, to the service equipment of the building being supplied.

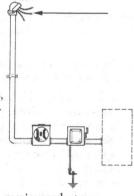

service conductors

service dead load The dead weight supported by a structural member.

service drop The overhead conductors that connect the electrical supply or communication line to the building being served.

service ell (street ell) A fitting for threaded pipe with a 45° or 90° bend, a male thread on one end, and a female thread on the other.

service entrance The location where an electrical supply line enters a building.

service entrance switch The circuit breaker or switch, with fuses and accessories, located near the point of entrance of supply conductors to a building and intended to be the main control and cut-off for the electrical supply to that building.

service lateral Those underground electrical service conductors (wires) between the street main, including any risers at a pole or other structure or from transformers, and the point of initial connection with the service entrance conductors in a terminal box or other enclosure either inside or outside a wall of the building being served. In the absence of such a box or enclosure, the service lateral connects the street main and the point at which the service conductors enter into the building.

S

service live load The live load specified by the general building code or bridge specification, or the actual nonpermanent load applied to a structural system.

service panel The main circuit breaker panel or fuse box where the electric service lead ties into the building's circuits.

service period In lighting, the number of hours per day that natural daylight provides a specified amount of illumination, often expressed as a monthly average.

service pipe 1. The water or gas pipe that leads from a supply source, usually public distribution mains in the street, to the particular building(s) being served. **2.** The pipe or conduit through which underground service conductors are run from the outside supply wires to the customer's property.

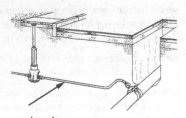

service pipe

service road A road or drive in a complex that is intended for vehicles making deliveries or collecting waste.

service systems The heating, ventilating, air-conditioning, water, and electric distribution systems in a building.

service tee A tee fitting used for threaded pipe and having one end threaded on the outside, and the other end and the branch threaded on the inside.

service temperature The highest temperature at which a material can be used without compromising its strength or other properties.

service valve In a piping system, any valve that isolates a device or apparatus from the rest of the system.

set 1. The condition reached by a cement paste, mortar, or concrete when it has lost plasticity to an arbitrary degree usually measured in terms of resistance to penetration or deformation. Initial set refers to first stiffening; final set refers to attainment of significant rigidity. **2.** To transform a resin or adhesive from its initial liquid or plastic state to a hardened state by physical or chemical action, such as condensation, polymerization, oxidation, vulcanization, gelation, hydration, or the evaporation of volatile ingredients. **3.** To drive a nail so far that its head is below the surface into which it has been driven. **4.** The overhang given to the points of sawteeth resulting in a kerf slightly wider than the saw to facilitate sawing motion.

setback The minimum distance required by code or ordinance between a building and a property line or other reference.

setback thermostat Programmable thermostat that can be set to different temperatures at different times of the day or week.

set point The desired value of the controlled variable (i.e., temperature, pressure).

setting bed The mortar subsurface to which terrazzo is applied.

setting block In glazing, a small block of wood, lead, neoprene, or other suitable material placed under the bottom edge of a light or panel to support it within the frame and prevent it from settling down onto the lower rabbet or channel.

setting shrinkage A reduction in volume of concrete prior to the final set of cement, caused by settling of the solids and by the decrease in volume due to the chemical combination of water with cement.

setting temperature The temperature to which an adhesive or resin must be subjected in order for setting to occur.

setting up That point in the initial drying of a paint or other liquid coating at which it is no longer able to flow.

settlement Sinking of solid particles in grout, mortar, or fresh concrete, after placement and before initial set.

settling The lowering in elevation of sections of pavement or structures due to their mass, the loads imposed on them, or shrinkage or displacement of the support.

settling basin An enlargement or basin within a water conduit that provides for the settling of suspended matter, such as sand; and is usually equipped with some means of removing the accumulated material.

set up 1. The stationing of a surveying instrument, such as a transit.

sewage Any liquid home waste containing animal or vegetable matter in suspension or solution. Sewage may include chemicals in solution; ground, surface, or storm water may be added as it is admitted to or passes through the sewers.

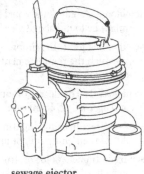

sewage ejector A plumbing device used to raise sewage to a higher elevation.

sewage gas The mixture of gases, odors, and vapors, sometimes including poisonous and combustible gases, found in a sewer.

sewage ejector

sewage treatment plant Structures and appurtenances that receive raw sewage and bring about a reduction in organic and bacterial content of the waste so as to render it less dangerous and less odorous.

sewer Generally, an underground conduit in which waste matter is carried in a liquid medium.

sewerage The entire works required to collect, treat, and dispose of sewage, including the sewer system, pumping stations, and treatment plant.

sewer appurtenances Manholes, sewer inlets, and other devices, constructions, or accessories related to a sewer system but exclusive of the actual pipe or conduit.

sewer brick Low-absorption, abrasive-resistant brick intended for use in drainage structures.

sewer lateral or side sewer Part of a sanitary sewer system that delivers the wastewater from a home to the main sewer lines.

sewer stub or sewer tap The point at which a home's sewer line joins the municipal sewer system.

s-green A description of lumber surfaced while at a moisture content of more than 19%.

shading coefficient The measure of the solar heat coming through the glass of a window. A low shading coefficient means reduced air conditioning expense.

shaft 1. That portion of a column between the base and the capital. **2.** An elevator well. **3.** A pit dug from the ground surface to a tunnel to furnish access and ventilation. **4.** Any enclosed vertical space in a building used for utilities or ventilation. **5.** Any cylindrical rod connecting moving parts in a machine.

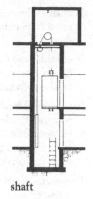
shaft

shaft wall Fire-resistant wall that isolates elevators, stairwells and vertical mechanical chases in high-rise construction. This wall must withstand the fluctuating (positive and negative) air-pressure loads created by elevators or air distribution systems.

shake 1. Roofing or sidewall material produced from wood, usually cedar, with at least one surface having a grain split face. **2.** A crack in lumber due to natural causes.

shake

shallow trench system A drainage system that can be used for dispersal of gray water for irrigation.

shank 1. The main body of a nail, screw, bolt, or similar fastener extending between the head and the point. **2.** The usually metal part of a drill or other tool that connects the working head to the handle.

shank hanger Device used to attach metal gutters to a structure by fastening to sheathing or rafters.

shape 1. A solid section, other than flat product, rod, or wire sections, furnished in straight lengths and usually made by extrusion; but sometimes fabricated by drawing. **2.** A wrought product that is long in relation to the dimensions of its cross section, which is of a form other than that of sheet, plate, rod, bar, tube, or wire. **3.** To work on a piece of wood or other material to make it conform to a predetermined desired or required pattern, or to render from its surface a specific texture or degree of smoothness.

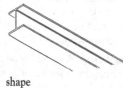
shape

shaper 1. A woodworking machine with a vertically revolving cutter for cutting irregular outlines, moldings, etc., in wood placed on a table below the cutter. **2.** A metalworking machine similar to a planer, except that the cutting tool is moved back and forth across the surface.

sharp sand Coarse sand made up of particles of angular shape.

shaving A very thin slice of wood removed in dressing, and used in some types of panels.

shear 1. An internal force tangential to the plane on which it acts. **2.** The relative displacement of adjacent planes in a single member. **3.** To cut metal with two opposing passing blades or with one blade passing a fixed edge.

shear block Face nailing of plywood to wall studs to prevent the wall from sliding and falling.

shear braces A bracing system, usually using metal brackets or straps, that eliminates most structural wall sheathing.

shear connector 1. A welded stud, spiral bar, short length of channel, or any other similar connector that resists horizontal shear between components of a composite beam. **2.** A timber connector.

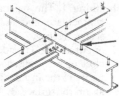
shear connector

shear diaphragms Members in a structure utilized to resist shear forces, such as those caused by wind load. *See also* **shear wall**.

shear failure (failure by rupture) Failure in which movement caused by shearing stresses in a soil or rock mass is of sufficient magnitude to destroy or seriously endanger a structure.

shearhead Assembled unit in the top of the columns of flat slab or flat plate construction to transmit loads from slab to column.

shearing force The algebraic sum of all the tangential forces acting on either side of the section at a particular location in a flexural member.

shearing machine An apparatus having a movable blade that passes a fixed cutting edge to cut metal.

shear legs A hoisting device from two or more poles fastened together near their apex, from which a pulley is hung to lift heavy loads.

shear plate 1. A shear-resisting plate used to reinforce the web of a steel beam. **2.** In heavy timber construction, a round steel plate usually inserted in the face of a timber to provide shear resistance in joints between wood and nonwood.

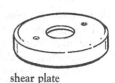
shear plate

shear-plate connector A type of timber connector employed in wood-to-wood or wood-to-steel applications.

shear reinforcement Reinforcement designed to resist shear or diagonal tension stresses.

shear slide A type of landslide where a section of earth slides away from the material beneath it in a single consolidated mass.

shear splice A type of splice designed to distribute the shear between the two members that it joins.

shear strain (shearing strain) The angular displacement or deformation of a member caused by a force perpendicular to its length.

shear strength The maximum shearing stress that a material or structural member is capable of developing, based on the original area of cross section.

shear stress The shear-producing force per unit area of cross section, usually expressed in pounds per square inch.

shear stud A short unthreaded bolt welded to the top flange of a steel beam. The shear studs are embedded in a concrete slab to form a composite beam and stud.

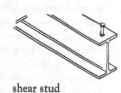
shear stud

shear wall (shearwall) A wall portion of a structural frame intended to resist lateral forces, such as earthquake, wind, and blast, acting in the plane or parallel to the plane of the wall.

sheath An enclosure in which posttensioning tendons are encased to prevent bonding during concrete placement.

sheathed cable Electric cable protected by nonconductive covering, such as vinyl.

S

281

sheathing

sheathing **1.** The material forming the contact face of forms. Also called *lagging* or *sheeting*. **2.** Plywood, waferboard, oriented strand board, or lumber used to close up side walls, floors, or roofs preparatory to the installation of finish materials on the surface. **3.** The first covering of exterior studs or rafters by boards, plywood, or particleboard.

sheathing

sheathing felt Roofing felt that has been saturated or impregnated, usually with some type of bitumen of low softening point (100° to 160°F).

sheathing paper *See* **building paper**.

sheave **1.** The grooved wheel of a pulley or block. **2.** The entire assembly over which a rope or cable is passed, including not only the pulley wheel but also its shaft bearings and side plates.

sheave block A pulley with a housing and bail.

she bolt A type of form tie and spreader bolt in which the end fastenings are threaded into the end of the bolt, eliminating cones and reducing the size of holes left in the concrete surface.

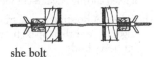
she bolt

shed dormer A dormer window having vertical framing projecting from a sloping roof, and an eave line parallel to the eave line of the principal roof. A shed dormer is designed to provide more space under a roof than a gabled dormer would provide.

shed dormer

sheepsfoot roller A powered or towed earth-tamping or compacting machine consisting of a large drum with projecting studs or feet with enlarged outer ends.

sheet A thin piece of material, such as glass, veneer, plywood, or rolled metal.

sheet asphalt A plant-mixed asphalt paving material containing sand that has passed through a 10-mesh sieve, and some type of mineral filler.

sheet glass (common window glass) Flat glass made by continuous drawing.

sheeting **1.** Structural wood panel, usually OSB or plywood, installed over studs, floor joists, or rafters/trusses. **2.** Planks used to line the sides of an excavation, such as for shoring and bracing. **3.** Sheet piling. **4.** A form of plastic in which the thickness is very small in proportion to length or width, and in which the plastic is present as a continuous phase throughout, with or without filler.

sheeting

sheeting driver An air hammer attachment that fits onto the ends of planks to allow their being driven without splintering.

sheeting jacks Push-type turnbuckles used to support sheeting in a trench.

sheet lath Heavier and stiffer than expanded-metal lath, it is fabricated by punching geometrical perforations in copper alloy steel sheet.

sheet lead Lead that has been cold-rolled into a sheet and whose designation is determined by the weight of one square foot of the finished product.

sheet metal Metal, usually galvanized steel but also aluminum, copper, and stainless steel, that has been rolled to any given thickness between 0.06″ and 0.249″ and cut into rectangular (usually 4′ × 8′) sections, which are then used in the fabrication of such items as ductwork, pipe, and gutters.

sheet-metal screw Usually a self-tapping screw, either round, flat, or pan-headed, threaded from tip to head, and used without a nut for fastening lapped sheet metal and some other materials. When equipped with a pointed tip, no tap hole is required.

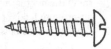
sheet-metal screw

sheet-metal work The fabrication, installation, and/or final product, such as the ductwork of a heating or cooling system, as performed or produced by a worker skilled in that trade.

sheetpiling A barrier or diaphragm formed of sheet piles that is used to prevent the movement of soil or keep out water during excavation and construction. Sheet piles are constructed of timber, sheet steel, or concrete.

Shelby tube A sampling device used to collect soil samples during a drilling operation.

shelf **1.** Any horizontally mounted board, slab, or other flat-surfaced device upon which objects can be stored, supported, or displayed. **2.** A ledge, as of rock, of a setback.

shelf angles Structural angles with holes or slots in one leg for bolting to the structure to support brickwork, stone, or terra cotta.

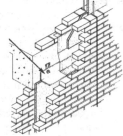
shelf angles

shelf cleat A wood (or other) strip that supports a shelf along one edge.

shelf life Maximum duration that a material can be stored and still remain in a usable condition.

shelf rest (shelf pin, shelf support) A type of angle bracket through whose vertical portion a pin is passed and inserted into one of several position-adjusting holes such as in a wall, or cabinet.

shelf standard A metal or wood shelf support with slots that allow adjustable placement of brackets to support shelving.

shell **1.** Structural framework. **2.** In stressed-skin construction, the outer skin applied over the frame members. **3.** Any hollow construction when accomplished with a very thin curved plate or slab.

shellac A transparent coating produced by dissolving lac, a resinous secretion of the lac bug, in denatured alcohol.

shell aggregate A type of aggregate made up of fine sands and the shells of mollusks, such as clams, oysters, and scallops.

shell construction **1.** A type of reinforced concrete construction in which thin curved slabs are primary elements. **2.** Construction in which a curved exterior surface has been obtained by using shaped steel and hardboard or curved plywood panels.

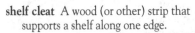

shell pile A load-supporting structural element that consists of a steel shell embedded in the ground and filled with concrete.

sherardizing A type of galvanizing process that is mainly used for small items such as fittings and fasteners. Sherardizing is the diffusion coating of zinc on steel, and is accomplished by submerging items in hot zinc dust.

shield In excavation, a short, hollow section mounted at the heading end of a jacked pipe or tunnel with a front-leaning plane to deter soil inflow.

shielded conductor An insulated electric conductor enclosed in a metal sheath or envelope.

shielded conductor

shielded metal-arc welding An arc welding process wherein coalescence is produced by heating with an arc between a covered metal electrode and the work. Shielding is obtained from decomposition of the electrode covering. Pressure is not used, and filler metal is obtained from the electrode.

shielded pair Two insulated wires in a cable wrapped with metallic braid or foil to prevent interference caused by external electric and magnetic fields and facilitate a clean transmission.

shielding (screening) 1. The erection of an interfering barrier (usually made of metal) to prevent electric, magnetic, or electromagnetic fields from escaping or entering an enclosed area. 2. Any device that blocks, diffuses, or redirects light, such as a baffle or lens.

shim A strip of metal, wood, or other material used to set base plates or structural members at the proper level for placement of grout, or to maintain the elongation in some types of posttensioning anchorages.

shingle 1. A roof-covering unit made of asphalt, wood, slate, asbestos, cement, or other material cut into stock sizes and applied on sloping roofs in an overlapping pattern. 2. A thin piece of material, such as wood, cement, asbestos, or plastic, used as an exterior wall finish over sheathing.

shingle

shingle lap A type of lap joint whereby the thinner of two tapered surfaces is lapped over the thicker.

shingle molding An exterior molding used at the joint of roof shingles and the fascia.

shingle nail A galvanized steel, aluminum alloy, or coated steel nail, often having a threaded shank. There are two types of shingle nails: one for wood shingles and one for asphalt shingles, the latter having a larger head.

shingle tile A flat clay tile used primarily for roofing and laid so as to overlap.

shinglewood Another name for western red cedar.

shingling hatchet (claw hatchet) Similar to a lath hammer, this tool consists of a hammer head, a hatchet, and a notch or nail claw.

shiplap Lumber that has been worked to make a rabbeted joint on each edge so that pieces may be fitted together snugly for increased strength and stability

shivering (peeling) The splintering that critical compressive stresses can precipitate in fired glazes or other ceramic coatings.

shiplap

shock load The impact load of material such as aggregate or concrete as it is released or dumped during placement.

shoe 1. Any piece of timber, metal, or stone receiving the lower end of virtually any member. Also called a *soleplate*. 2. A metal device protecting the foot, or point, of a pile. 3. A metal plate used at the base of an arch or truss to resist horizontal thrust 4. A support for a bulldozer blade or other digging edge to prevent cutting down. 5. A cleanup device following the buckets of a ditching machine. 6. A short section used at the base of a downspout to direct the flow of water away from a wall.

shoe

shoe molding A base shoe used at the bottom of a baseboard to cover the space between the finished flooring and the baseboard.

shooting board A jig to hold a board while its edge is being squared or beveled.

shooting plane A light plane used to square or bevel the edge of a board with a shooting board.

shop coat A surface coating applied in a fabricating shop.

shop drawings Drawings created by a contractor, subcontractor, vendor, manufacturer, or other entity that illustrate construction, materials, dimensions, installation, and other pertinent information for the incorporation of an element or item into the construction.

shop lumber (factory lumber) Lumber graded, in a shop or factory, according to the number of pieces, of designed size and quality, into which it may be cut.

shop painting The application of paint, usually one coat, to metals in the shop before shipment to a job site.

shop rivet A rivet driven in the shop.

shop steward A union official elected to represent members of a particular trade or department. Position responsibilities include soliciting new members, collecting dues, and initial negotiations for grievances.

shore (post, prop, strut, tom) A temporary support for formwork and fresh concrete or for recently built structures that have not developed full design strength.

shore hardness number A number representing the relative hardness of materials. It is the height of rebound observed when a standard hammer strikes the material being tested.

shore head A wood or metal horizontal member placed on and fastened to a vertical shoring member.

shore (post, prop, strut, tom)

S

shoring **1.** Props or posts of timber or other material in compression; used for the temporary support of excavations, formwork, or unsafe structures. **2.** The process of erecting shores.

shoring

shoring, horizontal A metal or wood load-carrying strut, beam, or trussed section used to carry a shoring load from one bearing point, column, frame, post, or wall to another. Horizontal shoring may be adjustable.

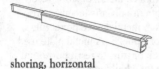

shoring, horizontal

shoring layout A drawing prepared prior to erection showing the arrangement of equipment for shoring.

short circuit An accidental electric connection of relatively low resistance between two points of different potential in an electric circuit, causing a high current flow between the two points.

short-cycling The turning off, followed by immediate restarting of a compressor or furnace.

short grain Wood in which part of the grain runs diagonally to the length of the piece and that is subject to failure under load.

short-length **1.** A piece of stock lumber usually less than 8′ (2.44 m) long. **2.** A piece of sawn hardwood usually less than 6′ (1.83 m) long (Brit).

short nipple A pipe nipple with a short unthreaded length between the threaded ends.

short nipple

short-oil varnish A varnish having a low vehicle content, containing less than 15 gallons of oil per 100 pounds of resin.

short ton A unit of measurement of weight in the English system equal to 2,000 pounds.

shotblasting A process similar to sandblasting, except that metal shot is used. Often used to remove oil and grime from old pavement surface.

shotcrete **1.** Mortar or concrete pneumatically projected at high velocity onto a surface. Also known as *air-blown mortar*. **2.** Pneumatically applied mortar or concrete, sprayed mortar, and gunned concrete.

shotcrete gun A pneumatic device used to deliver shotcrete or to move concrete.

shot saw A power tool that shoots steel pellets to cut and finish building stone.

shoulder **1.** An unintentional offset in a formed concrete surface usually caused by bulging or movement of formwork. **2.** The enlarged section of a bolt or screw shank just below the head.

shoulder bolt In construction, a fastener that attaches wall and roof paneling to framing members.

shoved joint A vertical joint in brickwork made by laying the brick in a bed of mortar and shoving the brick toward the last brick laid.

shower bath (shower, shower stall) The compartment and plumbing provided for bathing by overhead spray.

shower-bath drain A floor drain in a shower bath.

shower head A nozzle used to spray water in a shower bath.

shower mixer A valve in a shower bath used to mix hot and cold water supplies to obtain a desired water temperature.

shower bath
(shower, shower stall)

shower pan A pan of concrete, terrazzo, concrete and tile, or metal used as a floor in a shower bath.

shower room Part of the worker decontamination enclosure system. Usually located between the equipment room and the clean room, acts an as airlock between contaminated and noncontaminated areas. Contains hot and cold tap water suitable for showering during decontamination.

shower stall door A door in a shower partition.

show rafter A rafter, often ornamental, that is visible below a cornice.

shrinkage Volume decrease caused by drying and/or chemical changes, such as of concrete or wood.

shower stall door

shrinkage-compensating A characteristic of grout, mortar, or concrete made using an expansive cement in which volume increase, if restrained, induces compressive stresses intended to approximately offset the tendency of drying shrinkage to induce tensile stresses.

shrinkage cracking The cracking of a structure or member due to failure in tension caused by external or internal restraints from carbonation and/or reduction in moisture content.

shrinkage factor The percentage of reduction in volume of a given soil from bank to compacted state.

shrinkage limit The water content at which a reduction in water content will not cause a decrease in volume of the soil mass but an increase in water will increase the volume.

shrinkage loss Reduction of stress in prestressing steel resulting from shrinkage of concrete.

shrinkage reinforcement Reinforcement designed to resist shrinkage stresses in concrete.

shrink-mixed concrete Ready-mixed concrete mixed partially in a stationary mixer and then mixed in a truck mixer.

shrunk joint A joint made by placing a piece of heated pipe over the ends of two cool pipes and allowing it to contract.

shunt An electrical device with a low resistance or impedance connected in parallel across another electrical device to divert current from the second device.

shutter A movable cover or screen used to cover an opening, especially a window opening.

shutter butt A narrow hinge used on shutters and small doors.

shutter

shutting shoe A receptacle of metal or stone set in the paving or ground beneath a gate with two leaves to receive the vertical bolt.

SIA format A standard format developed for alarm system communications interfaces.

siamese connection A wye connection on the outside of a building with two inlet connections, used by the fire department to supply water to a sprinkler and/or standpipe system.

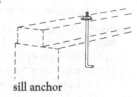

sick building syndrome (SBS) A condition where a significant proportion of the occupants of a building experience symptoms that are not related to a specific disease and that usually disappear when they leave the building. These symptoms are usually caused by indoor pollutants such as volatile organic compounds.

siamese connection

sidecasting Piling soil beside the excavation.

side-dump loader An earth loader with its bucket mounted on a pivot so it can be dumped to either side as well as in front.

side-entrance manhole A deep manhole with the access shaft built into the side of the inspection chamber.

sidehill A slope crossing the line of work.

sidehill cut An excavation for a highway through a sidehill, leaving a bank on one side only.

side lap The distance that a piece of material, such as steel roof deck, overlaps an adjacent piece.

side light A fixed frame of glass beside a window or door.

sidelighting A form of daylighting that provides natural, directional light.

side mounting Typical mounting detail for posts and rails to a fascia when floor mounting is not an option.

side post One of a pair of posts in a roof truss, each set at an equal distance from the center of the truss.

side vent A vent connected to the drain pipe at an angle of 45° or less from vertical.

sidewalk door A cellar door opening directly onto a sidewalk. The door is flush with the sidewalk when closed.

sidewalk door

sidewalk elevator 1. An elevator opening onto a sidewalk. 2. An elevator platform without a cab that rises to a level flush with a sidewalk.

sidewall The exterior wall of a building.

side yard The open space between a side of a dwelling and the property line.

siding (weatherboarding) Lumber or panel products intended for use as the exterior wall covering on a house or other building.

siding (weatherboarding)

sieve A metallic plate or sheet, a woven wire cloth, or other similar device, with regularly spaced apertures of uniform size, mounted in a suitable frame or holder, for use in separating material according to size. In mechanical analysis, an apparatus with square openings is a sieve; one with circular apertures is a screen.

sieve analysis The determination of the proportions of particles within certain size ranges in a granular material by separation on sieves of different size openings.

sieve number A number used to designate the size of a sieve, usually the approximate number of openings per inch. The number is applied to sieves with openings smaller than ¼″ (6.3 mm).

sight glass A glass tube used to indicate the liquid level in boilers, tanks, etc.

sight rail A series of rails set with a surveying instrument, and used to check the vertical alignment of a pipe in a trench.

signal Aggregate waves that are transmitted or received.

signal sash fastener A sash fastener beyond reach of a person. The fastener consists of a catch operated by a ring moved by a fitting on a long pole.

silica Silicon dioxide (SiO_2).

silica brick A refractory brick made from quartzite containing approximately 96% silica with alumina and lime.

silica flour (silica powder) Very finely divided silica; a siliceous binder cement that reacts with lime under autoclave curing conditions. The flour is prepared by grinding silica, such as quartz, to a fine powder.

silica gel (synthetic silica) A drying agent made from a form of silica.

silicate paint A paint utilizing sodium silicate as the binding agent.

silicon A metallic element used mainly as an alloying agent but used in pure form in electrical rectifiers.

silicon bronze A copper alloy used in hardware and other applications where a high resistance to oxidation is desirable.

silicon carbide An artificial product (SiC), granules of which may be embedded in concrete surfaces to increase resistance to wear or as a means of reducing skidding or slipping on stair treads or pavement. Silicon carbide is also used as an abrasive in saws and drills for cutting concrete and masonry.

silicone A resin, characterized by water-repellent properties, in which the main polymer chain consists of alternating silicon and oxygen atoms, with carbon-containing side groups (free radicals). Silicones may be used in caulking or coating compounds or admixtures for concrete.

silicone-carbide paper A tough, black, water-resistant sandpaper used in wet sanding and finishing.

silicone paint A heat- and chemical-resistant paint used on chimneys, stoves, and heaters, that requires heat to cure.

sill 1. The horizontal member of the bottom of a window or exterior door frame. 2. As applied to general construction, the lowest member of the frame of the structure, resting on the foundation and supporting the frame.

sill

sill anchor An anchor bolt used to fasten a sill to its foundation.

sill cock A water faucet located on the exterior of a building roughly at the top of the sill. A sill cock is usually threaded to provide a connection for a hose.

sill anchor

S

sill course A course of brickwork at a windowsill, usually projected to shed water.

sill plate or mudsill **1.** A horizontal framing member installed on a structure's foundation. **2.** The bottom framing member of an interior wall frame.

sill seal Insulation (fiberglass or foam) installed between the foundation wall and wood sill plate to fill gaps.

sill course

silo **1.** A tower-like structure, usually cylindrical, used to store items such as grain, coal, or minerals. **2.** A structure built in the ground to house a military missile.

silt (inorganic silt, rock flour) **1.** A granular material resulting from the disintegration of rock, with grains largely passing a No. 200 (75-micrometer) sieve. **2.** Such particles in the range from 2 to 50 micrometers in diameter.

siltation test A test to measure the impurities present in a sand sample.

silt box A steel box in a catch basin that can be removed to clean out the silt.

silt trap A trap that catches waterborne soil before it can enter a drainage system and obstruct flow.

silver solder A solder containing silver and having a high melting point, used for high-strength solder joints.

simple beam A beam without restraint or continuity at its supports.

simple span The support-to-support distance of a simple beam.

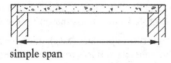

simple span

simulated masonry Products manufactured from a variety of materials including cement, minerals, epoxy, and fiberglass and used to simulate natural stone.

single-acting door Any door mounted so that it swings in only one direction.

single-acting hammer A pile-driving hammer using compressed air or steam to raise the ram to the height from which it is dropped.

single-cleat ladder A ladder made of two side rails with single treads between them.

single-cut file A file having serrations cut in one direction only.

single-duct system An air-conditioning system using one duct to serve a number of different areas.

single-hubpipe A pipe with a hub at one end only.

single-hubpipe

single-hung window A window with a movable and a fixed sash that is vertically hung.

single-package refrigeration system A factory-assembled and tested refrigeration system shipped in one section. No refrigerant-containing parts are connected in the field.

single-ply roofing A membrane roofing system that uses polymer-based synthetic material in flexible sheet or liquid form as an alternative to traditional built-up roofing. Made from a variety of bitumens, polymers,

single-ply roofing

fillers, plasticizers, stabilizers and other additives, single-ply roofing products are commonly applied over an insulating material and secured with ballast, mechanical fasteners or an adhesive.

single-point adjustable suspension scaffold A manually or power-operated platform supported by a single wire attached to its frame; used for light work only.

single-pole scaffold A platform resting on putlogs or crossbeams supported on ledger beams and posts on the outer side and by the wall on the inner side.

single-pole switch An electric switch with one movable and one fixed contact.

single prime contractor A constructor acting alone to fulfill the contractor's responsibility under the contract for construction.

single-roller catch A catch consisting of a spring-loaded roller on the door and a strike plate on the jamb. The door catches automatically and is opened by pushing or pulling.

single span Any structure without intermediate support.

single-stage absorption In an HVAC system, absorption chillers with one generator to evaporate refrigerant (water) from the solution.

single-stage curing An autoclave curing process in which precast concrete products are put on metal pallets for autoclaving and remain there until stacked for delivery or yard storage.

single-throw switch An electric switch opened or closed by the operation of a single pair of contacts.

single-wythe wall A stone, brick, or concrete wall that is one masonry unit thick.

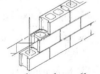

single-wythe wall

sink A plumbing fixture consisting of a water supply, a basin, and a drain connection.

sinker nail A wire nail with a small, slightly depressed head, driven below the surface of a wood member as a hidden fastener.

sink

sinter The process of forming a material by heating powder to a temperature just below its melting point so that it fuses together.

sintering grate A grate on which material is sintered.

siphonage The removal of fluid from a device, such as a trap, caused by suction produced by fluid flow.

siphon break **1.** During a toilet flush, the point when air is reintroduced into the trapway, ending the siphoning or pulling action that empties the bowl. **2.** A small groove in a surface that breaks capillary action and diverts water.

siphon trap An S-shaped plumbing trap containing water as a gas seal.

S-iron A retaining plate at each end of a tie rod used with a turnbuckle to tie two masonry walls together.

siphon trap

sisal A fiber from the leaves of the sisal plant; used in making rope and cord, and as reinforcing in plaster.

sistering The reinforcement of a structural member by nailing or attaching a stronger piece to a weaker one.

site The location of the project geographically, usually defined by legal boundaries.

site analysis services Services, as described in a schedule of designated services, that are necessary to determine limitations of the site and the corresponding project requirements.

site audit A review of activities at a construction site to identify inappropriate material and wasteful management practices.

site built The construction of a structure at the site where it is to remain.

site characterization In an environmental remediation operation, the process of defining and analyzing the contaminants and media of a site to determine the nature and extent of contaminants present.

site closeout The final phase in an environmental remediation effort that occurs after it is determined that site poses no future significant health risk.

site conditions clause A contract clause that typically requires the contractor to, at a minimum, inspect and understand the grade level and above-ground job site conditions prior to the start of any construction activities. The clause may also obligate the contractor to take full responsibility for the site's subsurface conditions.

site drainage 1. An underground system of piping carrying rainwater or other wastes to a public sewer. **2.** The water so drained.

site drawings Also called *civil drawings*, these illustrate a structure's relationship to the property, including various engineering improvements to the site, such as the sanitary system, utilities, paving walks, curbing, and so forth.

site investigation A complete examination, investigation, and testing of surface and subsurface soil and conditions. The report resulting from the investigation is used in design of the structure.

site plan A plan of the area of a proposed construction operation, including the building outline, parking, work areas, and/or property lines.

site preservation Minimizing the effect of a new building on the building site (environment).

site safety officer The person responsible for establishing the appropriate health and safety equipment to be used by workers at a hazardous waste site.

sitzbath A bathtub, designed to support a person in the sitting position, used for therapeutic treatment.

size To bring a piece to specified dimensions.

sized lumber Lumber uniformly manufactured to net surfaced sizes. Sized lumber may be rough, surfaced, or partly surfaced on one or more faces.

sizing (size) The application of diluted glue or adhesive to hardwood veneer to prepare the wood for application of a standard concentration of glue. Sizing reduces the amount of standard glue that will be absorbed. Sizing is often used when woods of different densities are glued together.

skeleton construction A type of construction in which loads are carried on frames made of beams and columns. Walls of upper floors are supported on the frames.

skeleton steps A simple stairway consisting of treads and a support carriage, but no risers.

skew back (skewback) A sloping surface against which the end of an arch rests, such as a concrete thrust block supporting the thrust of an arch bridge.

skew chisel A woodworking chisel with a sharp cutting edge of oblique angle and beveled sides.

skew corbel A specially shaped stone at the bottom of a stone gable forming an abutment for the coping, eave gutters, or wall cornices.

skewed Forming an oblique angle with a main center line.

skid resistance A measure of the frictional characteristics of a surface.

skim coat A thin coat of plaster, usually either the finish coat or the leveling coat.

skewed

skin 1. The materials, such as steel, aluminum and/or glass that make up a curtain wall. **2.** The outer veneer or ply of a lamination or built-up piece. **3.** The thin face of a hollow-core door. **4.** A tough layer formed on the surface of paint in a container.

skin friction The friction between soil and a structure, such as a retaining wall, or between soil and a pile.

skirt (apron) The border or section of molding under a window stool.

skirting block 1. A corner block where a base and vertical framing meet. **2.** A concealed block to which a baseboard is attached.

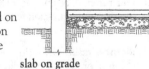

skylight A glazed opening in a roof to admit light.

skylight

slab A flat, horizontal (or nearly horizontal) molded layer of plain or reinforced concrete, usually of uniform but sometimes of variable thickness, positioned either on the ground or supported by beams, columns, walls, or other framework.

slab bar bolster A wire form used to support reinforcing while placing concrete.

slab form The formwork used in placing a concrete slab.

slabjacking A process used to correct concrete slabs that have sunk. Holes are drilled through the slab and materials are forced through the holes to raise the slab.

slab on grade A concrete slab placed on grade, sometimes having insulation board or an impervious membrane beneath it.

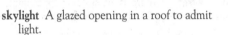

slab on grade

slab-on-grade construction A type of construction in which the floor is a concrete slab poured after plumbing and other equipment is installed.

slack The amount allowed for contingency in an estimate.

slack-rope switch A safety device that shuts off electric power to the drive of an elevator if the supporting cables become slack.

slag block A concrete masonry unit with blast furnace slag as the coarse aggregate.

slag concrete Portland cement concrete with blast furnace slag as the coarse aggregate.

slag sand A sandlike material made by crushing and grading blast furnace slag.

slake 1. To add water to quicklime to make putty. **2.** To crumble or disintegrate on exposure to moist air.

S

slamming strip An inlay along the edge of the lockstile of a flush wood door.

slant A sewer pipe connecting a house sewer to a common sewer.

slasher saw A set of circular saws operated in combination for quick cross-cutting of lengths of wood before chipping or grinding into pulp fibers, or as fuel.

slat A thin, narrow strip of wood, metal, or plastic.

slate A fine-grained metamorphic rock possessing a well-developed fissility (slaty cleavage) usually not parallel to the bedding planes of the rock.

slate batten A batten fastened across rafters in order to support slates on a roof.

slate boarding Close boarding used to support slates or tiles.

slate roll (slate ridge) A cylindrical rod formed from slate, having a V-shaped notch cut on its underside, and used to form the ridge on a roof.

slating 1. The installation of slates on a roof or wall. **2.** The slate shingles on a roof or wall, taken collectively.

slating

slave A mechanism under the control of a similar mechanism. A point that responds to a trigger. In fire safety systems, for example, if a fire alarm causes a siren to go off, the siren is the slave.

sled runner A hand tool with a curved or right-angled blade that is used to create uniform masonry joints.

sleeper One of many strips of wood fastened to the top of a concrete slab to support a wood floor

sleeper clip A metal clip used to fasten sleepers to a concrete subfloor.

sleeper joist A joist supported on sleepers.

sleeper wall Any short wall that supports floor joists.

sleeve See pipe sleeve.

slender beam A beam that if loaded to failure without lateral bracing of the compression flange would fail by buckling rather than in flexure.

slenderness ratio The ratio of effective length or height of a wall, column, or pier to the radius of gyration. The ratio is used as a means of assessing the stability of the element.

slewing Rotating the jib of a crane so the load moves through a horizontal arc.

sliced veneer Veneer sliced from a face of a squared-off log.

sliding fire door A fire door hung from an overhead track. The track may be sloped for direct gravity operation or horizontal with operation by weights, cables, and pulleys.

sliding sash Any window that moves horizontally in grooves.

slimline lamp A type of instant-starting fluorescent lamp with a single pin base.

sliding sash

slip Movement occurring between steel reinforcement and concrete in stressed, reinforced concrete indicating anchorage breakdown.

slip form (slipform, sliding form) A form that is pulled or raised as concrete is placed. The form may move in a generally horizontal direction to lay concrete evenly for highway paving or on slopes and inverts of canals, tunnels, and siphons; or vertically to form walls, bins, or silos.

slip joint 1. A vertical joint between an old and a new brick wall, made by cutting a slot in the old wall and filling it with brick projecting from the new wall. **2.** A joint in plumbing in which one pipe slips within another and the seal is made by a pressure device fitting over the joint, often threaded to the larger pipe.

slip-joint conduit Electric conduit connected by short couplings that slip over the ends of the conduit.

slip-joint conduit

slip-joint pliers Pliers having jaws connected by a rivet through a slot in one part, allowing two settings for the jaws.

slip match A veneer pattern that creates a repeat effect by laying similarly grained wood sheets alongside each other.

slip-resistant tile Ceramic tiles with abrasive particles or grooves in the surface.

slip stone A handheld, coarse sharpening stone used to hone cutting and woodworking tools.

slope 1. The angle of repose at which a soil material will stand without moving. **2.** The slope of a roof, ramp or paving expressed as one unit of rise to one unit of run.

slope map A map displaying the topography of an area, along with a discussion of topographic features.

slope ratio Relation of the horizontal projection of a surface to its rise. For example, 2′ horizontal to 1′ rise is shown as 2:1 or 2 to 1.

slop sink A deep sink set low on a wall, used to clean mops and to empty and clean pails.

slot outlet An air-supply outlet with a length-to-width ratio greater than 10:1.

slop sink

slot weld A weld between two members made by welding within a slot in one of the members.

sloughing Subsidence of shotcrete, due generally to excessive water in mixture.

slow-burning A misleading term implying a material is fire-safe. The term must be related to a particular test for interpretation.

slow-burning construction Heavy timber construction with large flat surfaces, as opposed to joisted construction.

slow-curing asphalt Liquid asphalt made of asphalt cement and oils of low volatility.

slow-evaporating solvent A solvent with a high boiling point, used to lengthen, and thus improve, the drying time of paint.

sludge 1. The semi-liquid, settled solids from treated sewage. **2.** Waste material composed of wet fines produced from grinding a terrazzo floor. **3.** Accumulated solids in the wash water reservoir of paint spray booths.

slugging The pulsating and intermittent flow of shotcrete material due to improper use of delivery equipment and materials.

sluice A steep, narrow passage for water. See also **sluicing**.

sluicing Moving soil by use of rapidly flowing water, for excavation or particle grading.

slump A measure of consistency of freshly mixed concrete, mortar, or stucco equal to the subsidence measured to the nearest ¼″ (6 mm) of the molded specimen immediately after removal of the slump cone.

slump block A concrete masonry unit intentionally removed early from a mold so that it slumps slightly.

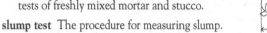
slump block

slump cone A mold in the form of the lateral surface of the *frustum* of a cone with a base diameter of 8″ (203 mm), top diameter 4″ (102 mm), and height 12″ (305 mm), used to fabricate a specimen of freshly mixed concrete for the slump test. A cone 6″ (152 mm) high is used for tests of freshly mixed mortar and stucco.

slump cone

slump test The procedure for measuring slump.

slurry A mixture of water and any finely divided insoluble material, such as Portland cement, slag, or clay in suspension.

slurry seal machine A self-propelled machine used for delivering a mixture such as an asphalt emulsion slurry seal.

slurry wall A vertical trench used to contain, capture, or redirect groundwater flow in the vicinity of a environmental remediation site.

smart actuator A device that combines control and sensor capabilities.

smart building *See* **intelligent building**.

smart escalator An escalator that has controls, such as automatic slowing or stopping of the system when it is not being used.

smart glass Glazing that can adjust for opacity and light and heat transmission.

smart lighting Lighting designed for energy efficiency. May include high efficiency fixtures, daylighting and automatic controls that make adjustments based on conditions such as occupancy.

smart time scheduling A control system for lighting and HVAC based on time settings rather than occupancy sensors.

smart window Any window that uses removable coatings or a light-sensitive suspended particle device to allow for greater control of solar gain.

smelting The process of melting a metal, typically in a blast furnace, or fusing it to separate it from impurities.

smoke 1. A suspension in air of particles that are usually solid. **2.** Carbon or soot particles less than 0.1 micron in size, resulting from incomplete combustion, such as of oil, wood, or coal.

smoke and fire vent A vent cover, installed on a roof, that opens automatically when activated by a heat-sensitive device, such as a fusible link.

smoke chamber The transition portion of a chimney between the fireplace and the flue.

smoke damper A damper arranged to close and stop flow automatically when smoke is detected.

smoke-developed rating A relative number indicating the smoke produced when the surface

smoke chamber

of a material burns, as measured during an ASTM E-119 flame spread test.

smoke shelf A concave shelf at the back of a smoke chamber to redirect downdrafts up the chimney.

smoke test A test using a nontoxic, visible smoke to determine the routes taken by air currents and/or to detect leaks.

smoke zone or smoke control zone 1. A building space with compartmentalization and pressurization that can control smoke. **2.** An HVAC zone where smoke is detected.

smoother bar A heavy, rod-like attachment to excavation equipment that breaks up clumps of soil disturbed during excavation operations.

smoothing iron A handheld, hot iron with a long handle, used for smoothing asphaltic pavement joints.

smoothing plane A small, fine carpenter's plane used for finishing.

smooth-surfaced roofing A built-up roofing membrane, the top surface of which is either hot-mopped asphalt, an asphalt emulsion of a cutback coating, or an inorganic top felt.

smooth-surfaced roofing

snake 1. A long, resilient wire used by electricians in running wires through conduit. The snake is pushed through and then used to pull the wires. **2.** A flexible metal wire used to clear clogged plumbing fixtures.

snap cutter One of a variety of hand tools used to cut tile, cast iron pipe, and fittings.

snap cutter

snapping line A layout line made by stretching a chalked line across a surface and snapping the line.

snap switch A manually operated switch used to control low-power indoor circuits.

snap tie A proprietary concrete wall-form tie, the end of which can be twisted or snapped off after the forms have been removed.

snifter valve A valve that allows air into a tank or piping system automatically through the use of a mechanical device or by other external means.

snips *See* **tin snips**.

snow load The live load allowed by local code, used to design roofs in areas subject to snowfall.

soakaway An excavated pit designed to receive excess surface water for gradual draining.

soaking period In high-pressure and low-pressure steam curing, the time during which the live steam supply to the kiln or autoclave is shut off and the concrete products are exposed to the residual heat and moisture.

soap A brick or tile of normal face dimensions but with a nominal thickness of 2″.

soapstone A soft rock containing a high proportion of talc; used for such items as sinks, bench tops, and carved ornaments.

S

socket **1.** British term for the enlarged end of bell-and-spigot pipe. **2.** A mechanical device for supporting a lamp or plug fuse and completing the electric circuit.

socket plug A plug for the end of an interior-threaded pipe fitting consisting of a threaded piece with a recess into which a tool is placed to turn it.

socket weld A pipe joint made by use of a socket weld fitting that has a female end or socket for insertion of the pipe to be welded.

socket wrench A box wrench with a recessed socket at the end of a shank that fits over the head of a nut or bolt for tightening or removing the nut or bolt.

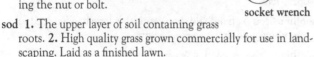

socket wrench

sod **1.** The upper layer of soil containing grass roots. **2.** High quality grass grown commercially for use in landscaping. Laid as a finished lawn.

soda-acid fire extinguisher A fire extinguisher that discharges water under pressure. The pressure is produced by mixing soda and acid to generate carbon dioxide.

sodium silicate A material that is used to waterproof concrete and mortar, to encapsulate asbestos, and to insulate.

sodium silicate adhesive A noncombustible, inorganic, high-strength adhesive with unsatisfactory weather resistance.

sodium vapor lamp An electric discharge lamp that produces light when sodium vapor is activated by current passing between two electrodes. Commonly used in street lights, they can take three to four minutes to warm up to full light output.

soffit The underside of a part or member of a structure, such as a beam, stairway, or arch.

soffit board A board that forms the soffit of a cornice.

soffit

soffit bracket A bracket used to mount an exposed exterior door closer to a door frame head or transom bar.

softening point The temperature at which bitumen softens or melts; used as an index of fluidity.

soft solder A solder used to join metals in welding and brazing applications. It is typically a mixture of tin and lead, although soft solder of pure lead is used in some industries to join copper.

soffit bracket

soft water Water with a low mineral content.

softwood **1.** A general term referring to any of a variety of trees having narrow, needle-like or scale-like leaves, usually coniferous. **2.** The wood from such trees. The term has nothing to do with the actual softness of the wood; some softwoods are harder than certain of the hardwood species.

soft glass Glass susceptible to thermal shock because it has a high coefficient of thermal expansion and a low softening point.

soft-mud brick Brick produced by molding, often by hand, relatively wet clay (20% to 30% moisture). When the inside of the mold is sanded to prevent sticking of the clay, the product is sand-struck brick. When the molds are wetted to prevent sticking, the product is water-struck brick.

soft rot Decay in wood, where the residue is chiefly cellulose.

soil A generic term for unconsolidated natural surface material above bedrock.

soil absorption system A disposal system, such as an absorption trench, seepage bog, or seepage pit, that utilizes the soil for subsequent absorption of treated sewage.

soil-cement Soil, Portland cement, and water mixed and compacted in place to make a hard surface for sidewalks, pool linings, and reservoirs, or for a base course for roads.

soil class A classification of soil by particle size, used by the U.S. Department of Agriculture: (a) gravel, (b) sand, (c) clay, (d) loam, (e) loam with some sand, (f) silt-loam, and (g) clay-loam.

soil classification test A series of tests combined with sensory observations used to classify a soil. The tests may include such aspects as grain size, distribution, plasticity index, liquid limit, and density.

soil creep The slow movement of a mass of soil down a slope, caused by gravity and aggravated by pore water.

soil drain A horizontal soil pipe.

soil hydraulic conductivity A measure of the rate at which water will flow through a soil matrix.

soil mechanics The application of the laws and principles of mechanics and hydraulics to engineering problems dealing with soil as a building material.

soil pipe A pipe that conveys the discharge from water closets or similar fixtures to the sanitary sewer system.

soil profile A vertical section through a site showing the nature and sequence of layers of soil.

soil sample A representative sample of soil from a specific location or elevation of a construction site, usually extracted to determine bearing capacity.

soil stabilizer **1.** A machine that mixes in-place soil and an added stabilizer, such as cement or lime, in order to stiffen the soil. **2.** A chemical added to soil to stiffen it and increase the stability of a soil mass.

soil stack A vertical soil pipe that carries the discharge from water closet fixtures.

soil vapor extraction A process that removes volatile organic compounds from a soil matrix through vacuum extraction of the air in the soil.

solar altitude A vertical measurement, between 0 and 90 degrees, used in solar analysis and design.

solar azimuth A horizontal measurement used for solar analysis and design.

solar cells Components in solar energy collection systems, they come in four basic types: (a) single-crystalline, (b) multicrystalline, (c) amorphous thin film, and (d) hybrid.

solar collector Any device intended to collect solar radiation and convert it to energy.

solar constant The average rate at which radiation energy from the sun is received at the surface of the earth, equal to 430 Btus per hour per square foot (1.94 cal. per min. per sq. cm.).

solar energy The radiant energy from the sun.

S

solar flat plate collector A solar collector in the shape of a flat plate, consisting either of a series of photovoltaic cells or a sandwich panel made up of a black surface, a film of circulating water or air, and a transparent cover.

solar flat plate collector

solar gain Heat that enters a building through windows as solar energy. Considered a "passive" heat source, as opposed to those sources that artificially generate heat.

solar heat exchanger A means of transferring the heat in storage in a solar heating system to the area to be heated.

solar heat gain coefficient (SHGC) The measure of the solar radiation admitted through a window, including radiation directly transmitted and that absorbed and subsequently released inward.

solar heating system An assembly of components, including collectors, heat exchangers, piping, storage system, controls, and supplemental heat source, used to provide heat and/or hot water to a building, with the sun as the main source of energy.

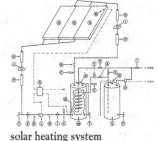
solar heating system

solar hot-water system A system that uses a collector, typically mounted on a roof, to absorb heat from sunlight.

solar insolation The amount of solar energy available at a given time in a given area.

solarium A room or porch with a great deal of glass and located for maximum exposure to the sun.

solarium

solar orientation The alignment of a building relative to the sun, initially set either for maximum or minimum heat gain, depending on the local climate.

solar pin That part of a building's heat load produced by solar radiation striking the building or passing through windows.

solar radiation Energy from the sun, which can be absorbed and converted into heat energy, reflected or transmitted.

solar screen 1. An openwork or louvered panel of a building positioned to act as a sun shade. 2. A perforated wall used as a sun shade.

solar storage Fluid and/or rocks used to hold some of the heat energy collected by a solar heat collector.

solder 1. An alloy, usually lead-tin, with a melting point below 800°F (427°C), used to join metals or seal joints. 2. The process of joining metals or sealing joints using solder and heat.

soldered joint A gas-tight pipe joint made by applying solder to a heated joint.

soldering gun A tool with a pistol grip and a small electrically heated bit that reaches operating temperatures rapidly, used to solder electrical components.

soldering nipple A nipple with one end threaded and the other plain. The plain end is used for a soldered joint and the threaded end for a mechanical joint.

soldier 1. A vertical wale used to strengthen or align formwork or excavations. 2. A masonry unit set on end so its long, narrow face is vertical on the face of the wall.

soldier beam A rolled-steel section driven into the ground to support a horizontally sheeted earth bank.

soldier course A course of brick units set on end with the long, narrow face vertical on the wall face.

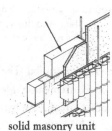

soldier pile 1. In an excavation, a vertical member that supports horizontal sheeting and is supported by struts across the excavation and by embedment below the excavation. 2. A vertical member used to support formwork and held in place by struts, bolts, or wires.

soldier course

solenoid An electromagnetic coil used to activate a mechanical device or switch.

solenoid valve A valve opened by a plunger in which movement is controlled by an electromagnet.

soleplate 1. A solepiece or shoe that serves as a base for studs in a core of solid wood or mineral composition, as opposed to a partition. 2. A plate welded or bolted to the underside of a plate girder that bears on a pad.

solid core 1. The inner layers of a plywood panel that contain no open irregularities, such as gaps or open knotholes, and in which the grain runs perpendicular to the outer plies. Solid core is primarily used as underlayment for resilient floor covering. 2. A flush door, used in entries and as fire-resistant doors in which particleboard or wood blocks completely fill the area between the door skins.

solid-core door A door having a core of solid wood or mineral composition.

solid frame The frame of a door or window having jambs, head, and sill made of solid pieces of timber rather than parts.

solid glass door A door in which the glass essentially provides all the structural strength.

solid loading Filling a drill hole with explosive, except for a stemming space at the top.

solid masonry unit 1. A masonry unit whose minimum net cross-sectional area parallel to its bearing surface is 75% or more of its gross cross-sectional area. 2. A masonry unit having holes less than 3/4″ (2 cm) wide or less than 0.75 sq. in. (5 sq cm) passing through, or having frogs that do not exceed 20% of its volume. In either type, up to three handling holes, not exceeding 5 sq. in. (32.5 sq cm) each, are allowed. The total area of all through holes may not exceed 25% of the gross area.

solid masonry unit

solid masonry wall A wall built of solid masonry units with all joints filled with mortar and no hollow wythes.

solid modeling The general type of geometric modeling where the elements being

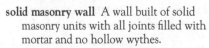
solid masonry wall

S

modeled and operated on are closed and bounded, enclosing a volume. Solid modeling can represent solid shapes but also is misnamed because it can also represent the shapes of voids, such as a room.

solid mopping The application of hot asphalt over an entire roof surface.

solid punch A steel rod used to drive bolts out of holes.

solid rock Rock that cannot be moved or processed without being blasted.

solids content 1. The portion of a protective coating material that remains on the surface after drying. Sometimes referred to as nonvolatiles. 2. In an adhesive, the percentage of weight of the nonvolatile matter.

solid-state welding A welding process in which coalescence takes place at temperatures below the melting point of the metals being joined and without use of a brazing filler metal. Sometimes this is accomplished through the use of pressure.

solid-web steel joist A steel truss or light beam having a solid web; usually cold-formed from sheet steel and having a channel shape.

solvent A liquid in which another substance may be dissolved.

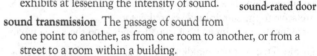

solid-web steel joist

solvent adhesive An adhesive having a volatile organic liquid as a vehicle.

solvent extraction An ex situ remediation process in which contaminated sediment, soil or sludge is mixed with a solvent to separate the contaminant from its existing matrix.

solvent welding A process where a chemical liquid solvent is used to soften and join together two or more thermoplastic materials.

sonic pile driver 1. A vibrating pile-driving hammer. Puts a pile into its resonant vibration range, causing it to slide down into the soil. 2. A machine used to drive piles or sheet piling into soil using a head vibrated at a high frequency, usually less than 6,000 times per minute.

Sonotube® A product consisting of a preformed casing made of laminated, waxed paper; used to form cylindrical piers or columns.

Sonotube®

sound A vibratory disturbance, with the frequency in the approximate range between 20 to 20,000 cycles per second, capable of being detected by a human ear

sound absorption The process of dissipating sound energy.

sound attenuation Measures taken to soundproof a wall or subfloor.

sound attenuation batt Insulation designed to reduce noise transmission.

sound attenuation fire batt (SAFB) Insulation that provides acoustical and thermal control, as well as fire resistance.

sound damping The use of fibrous sound-absorbing material in a partition to reduce sound transmission. Damping in floor/ceiling construction has a wider application for impact sound than for airborne sound.

sound deadening board A board with good sound absorption qualities; used in sound-control.

sound insulation (acoustic insulation) The use of materials and assemblies to reduce sound transmission from one area to another or within an area

sound isolation Attenuating sound transmission between adjacent spaces.

sound knot A dead knot in wood that is undecayed, at least as hard as the surrounding wood, and held firmly in place.

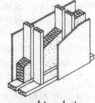

sound insulation
(acoustic insulation)

sound leak Opening in a partition that allows air (and sound) to pass through. Examples include small holes in a wall, openings for electrical boxes and plumbing, and cracks around doors.

sound masking Provision of white noise to provide privacy or reduce distractions.

soundness The freedom of a solid from cracks, flaws, fissures, or variations from an accepted standard.

sound pressure The change in pressure resulting from vibration in the audible frequency range. Conversational speech at close range produces a sound pressure of about one dyne per sq. cm.

sound pressure level (SPL) Expressed in decibels, the SPL is 20 times the logarithm to the base 10 of the ratio of the sound pressure to a reference pressure of 20 micropascals.

soundproofing 1. The design and construction of a building or unit to reduce sound transmission. 2. The materials and assemblies used in a building or unit to reduce sound transmission.

sound-rated door A door constructed to provide greater sound attenuation than that provided by a normal door; usually carrying a rating in terms of its sound transmission class (STC).

sound reduction factor The effectiveness, measured in decibels, that a building assembly exhibits at lessening the intensity of sound.

sound-rated door

sound transmission The passage of sound from one point to another, as from one room to another, or from a street to a room within a building.

sound transmission class (STC) A single number indicating the sound insulation value of a partition, floor-ceiling assembly, door, or window, as derived from a curve of insulation value as a function of frequency. The higher the number is, the greater the insulation value.

sound transmission loss The reduction in sound when it passes through a partition, wall, or ceiling.

southern yellow pine A species group composed primarily of loblolly, longleaf, shortleaf, and slash pines. Various subspecies also are included in the group.

space frame Any three-dimensional structural frame capable of transmitting loads in the three dimensions to supports. A space frame is usually an interconnected system of trusses or rigid frames.

space heater A small heating unit, usually equipped with a fan, intended to supply heat to a room or portion of a room. The source of heat energy may be electricity or a fluid fuel.

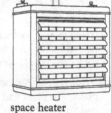

space heater

S

space planning The process and act of laying out and designing space for a tenant's needs.

spacer 1. A device that maintains reinforcement in proper position, or keeps wall forms a given distance apart before and during concreting. 2. A small block of wood or other material placed during installation on the edges of a pane of glass to center it in the channel and maintain uniform width of sealant beads to prevent excessive sealant distortion.

spacing criterion The maximum distance between interior light fixtures that will ensure a uniform illumination on a work plane.

spacing factor (power spacing factor) An index related to the maximum distance of any point in a cement paste or in the cement paste fraction of mortar or concrete from the periphery of an air void.

spackle (speckling, sparkling) A paste, or a dry mixture blended with water to form a paste; used to fill holes and cracks in plaster, wallboard, or wood.

spade A sturdy digging tool having a thick handle and a flat blade that can be pressed into the ground with a foot.

spading Consolidating mortar or concrete by repeatedly inserting and withdrawing a flat, spadelike tool.

spall A fragment usually in the shape of a flake, detached from a larger mass by a blow through the action of weather, by pressure, or by expansion within the larger mass. A small spall involves a roughly circular depression not greater than 20 mm in depth nor 150 mm in any dimension; a large spall may be roughly circular or oval or, in some cases, elongated.

span The horizontal distance between supports.

spandrel (spandril) That part of a wall between the head of a window and the sill of the window above it.

spandrel beam A beam in the perimeter of a building spanning columns, and usually supporting floor or roof loads.

spandrel glass Opaque glass used in curtain wall construction to conceal structural elements.

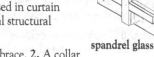
spandrel glass

spanner 1. A horizontal cross brace. 2. A collar beam.

spar 1. A common rafter. 2. A heavy, round timber. 3. A bar used as a gate latch.

sparge pipe A perforated pipe used to distribute flushing water for a urinal.

spark arrester (bonnet) A device at the top of a chimney or stack to catch sparks, embers, or other ignited material over a given size.

spar varnish A varnish with superior weather-resisting qualities; used on exterior wood.

spat A protective sheet, usually stainless steel, installed at the bottom of a door frame.

spatter dash (spatterdash) A rich mixture of Portland cement and coarse sand, thrown onto a background by a trowel, scoop, or other appliance so as to form a thin, coarse-textured, continuous coating. As a preliminary treatment, before rendering, it assists bonding of the undercoat to the background, improves resistance to rain penetration, and evens out the suction of variable backgrounds.

spec Abbreviation for "on speculation," used when a building is constructed before a buyer or tenant is obtained.

special assessment A charge imposed by a government on a particular class of properties to defray the cost of a specific improvement or service; presumably of benefit to the public but of special benefit to the owners of the charged properties.

special hazards insurance Insurance coverage for damage such as sprinkler leakage, water damage, collapse, and coverage for materials in transit (or off-site). The damage must be caused by additional perils and risks not covered in standard property insurance.

special matrix terrazzo A finish flooring consisting of colored aggregate and organic matrix.

special-quality brick A brick that is durable when used under conditions of exposure.

specialty drawings Graphic representations of the unique requirements of various spaces' special uses (such as kitchens, libraries, retail spaces, and home theater systems). Specialty drawings outline the coordination among other building systems, most commonly the mechanical and electrical systems.

special use permit Permission granted to a landowner to use property in a way that would not normally be allowed in a zoning district.

species A category of biological classification. A species is a class of individuals having common attributes and designated by a common name.

specific adhesion A bond created by chemical means, rather than mechanical.

specification A detailed and exact statement of particulars, especially a statement prescribing materials, dimensions, and workmanship for something to be built or installed.

specifications or specs Documents that define the qualitative requirements for products, materials, and workmanship upon which the contract for construction is based.

specific gravity 1. The ratio of the mass of a unit volume of a material at a stated temperature to the mass of the same volume of gas-free distilled water at a stated temperature. 2. The ratio of the density of one substance to another when used as the standard. Water is the standard for determining specific gravity of solids and liquids; hydrogen is used for gases.

specific gravity, bulk The ratio of the mass in air of a unit volume of a permeable material, including both permeable and impermeable voids normal to the material, at a stated temperature to the mass in air of equal density of an equal volume of gas-free distilled water at a stated temperature.

specific gravity factor The ratio of the weight of aggregates, including all moisture, as introduced into the mixer to the effective volume displaced by the aggregates.

specific heat A quantity that describes the ability of a body to absorb heat and increase its temperature. The specific heat of a substance is the ratio of the amount of heat that must be added to a unit mass to raise its temperature through one degree to the amount of heat that is required to raise the temperature of an equal mass of water one degree.

specific volume The volume of a given weight of a material, typically a gas, expressed as cubic feet per pound at specific conditions. Simply put, the reciprocal of density.

S

293

specifier One who writes or prepares specifications.

spectral power distribution The distribution of radiant power, usually expressed in watts per nanometer, with respect to wavelength.

spectrophotometer An instrument for measuring the intensity of radiant energy of desired frequencies absorbed by atoms of molecules. Substances are analyzed by converting the absorbed energy to electrical signals, proportional to the intensity of radiation.

speculative builder A contractor who develops, constructs, and then sells (or leases) a property.

speed square A triangular-shaped tool used to mark perpendicular and angled lines.

spigot 1. The end of a pipe that fits into the bell, or upset, end of another pipe to form a joint after caulking. **2.** A faucet.

spike A heavy nail, sometimes having a square cross section, and 3″ to 12″ (7.6 to 30.5 cm) long.

spike-and-ferrule installation A type of gutter installation in which the gutter is secured by long nails in metal sleeves.

spiked grill fastener A multi-tooth connector made of steel that is used to fasten heavy timbers. Commonly used in roof truss and flooring construction.

spile 1. A peg or plug of wood. **2.** A spout for directing the flow of sap from sugar maples. **3.** A heavy wooden stake.

spillway A passage to convey overflow water from a dam or similar structure.

spindle 1. A small axle on which an object turns. **2.** The bar in a lock that connects the knob(s) or handle(s) with the latching mechanism. **3.** A short, turned-wood part, such as a baluster.

spiral A continuously wound length of reinforcing steel in the form of a cylindrical helix.

spirally reinforced column A column in which the vertical bars are enveloped by spiral reinforcement, (i.e., closely spaced, continuous hooping).

spiral reinforcement Continuously wound reinforcement in the form of a cylindrical helix.

spiral spacer Reinforcement that keeps spirals a uniform distance apart.

spiral stair A flight of stairs whose treads wind around a central newel in a spiral or helix shape. *See also* **circular stair**.

spire Any long, slender, pointed construction on top of a building. A spire is often a narrow, octagonal pyramid set on a short, square tower.

spirit level A device used to set an instrument to true horizontal or true vertical, consisting of a glass tube nearly filled with liquid so a traveling air bubble is formed. Cheap levels are bent concave; better levels are ground internally to an overall concave shape.

spit The depth of one hand-shovel blade.

splash block A small masonry block set in the ground beneath a rain gutter to receive roof drainage and prevent puddling or soil erosion.

spiral reinforcement

spiral stair

splashboard 1. A board placed on a wall at a sink to protect the wall. **2.** A weather molding placed on the bottom of an outside door. **3.** A board set against a wall at a scaffold to protect the wall.

splashboard

splash brush A brush used to apply water to a finish coat of plaster while it is being troweled smooth.

splash lap That part of the overlap of a seam in sheet-metal roofing that extends onto the flat surface of the next sheet.

splay 1. To form with an oblique angle, or to bevel. **2.** To spread out or extend.

splay brick (cant brick) A brick with one side beveled at about 45°.

splayed edge A bevel across the full thickness of a piece of wood.

splayed hangers Hangers installed at an angle rather than perpendicular to a support grid or channel.

splayed joint A joint between the ends of two adjacent members in which the ends of each are beveled to form an overlap.

splice Connection of two similar materials to another by lapping, welding, gluing, mechanical couplers, or other means.

splice plate A plate laid over a joint and fastened to the pieces being joined to provide stiffness.

splice

spline (false tongue, feather, slip feather, slip tongue) A piece of metal or wood used to join two pieces of wood, such as decking, together.

spline joint A joint formed by inserting a spline into slots formed in the two pieces to be joined.

splinter 1. A small, thin, sharp piece of wood broken from a larger piece. **2.** To split or break something into splinters.

split 1. A crack extending completely through a piece of wood or veneer. **2.** A tear in a built-up membrane resulting from tensile stresses. **3.** A masonry unit one-half the height of a standard unit.

split astragal Two pieces of molding attached to the meeting edges of two leaves of a pair of swinging doors, allowing both leaves to be active.

split astragal

split-batch charging A method of charging a mixer in which the solid ingredients do not all enter the mixer together. Cement or different sizes of aggregate may be added separately.

split-bolt connector A device that holds electrical wires in close contact with a simple assembly. Wires are fed through the device and held in place by a nut tightened on a threaded shaft.

split-conductor cable A cable in which each conductor consists of two or more insulated wires normally in parallel.

split face An exposed, rough face of a masonry unit or stone created by splitting rather than forming or sawing the face.

split-face block Concrete masonry units with one or more faces produced by purposeful fracturing of the unit to provide architectural effects in masonry wall construction.

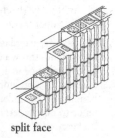
split face

split fitting A section of electric conduit that is split longitudinally. After conductors have been placed in one half, the unit is assembled and secured with screws.

split frame (split jamb) A door frame with the jambs split in two or more pieces to allow the use of a pocket-type sliding door.

split hanger A two-piece pipe hanger that is tightened around a pipe.

split jamb A two-piece doorjamb that can be adjusted to accommodate varying wall thicknesses.

split level A type of house in which the floor levels on one side of the house are one-half story above or below those on the other side.

split level

split-phase motor An induction motor that typically has two windings, a main and an auxiliary. While noisier, more expensive, and harder to start than three-phase motors, single-phase motors are excellent for medium duty applications and where stops and starts are somewhat frequent.

split pin A pin or spike that spreads during insertion, or that can be split after insertion.

split ribbed block Concrete masonry units featuring a ribbed face with ribs that have been slit off, providing a rough texture.

split-ring connector A timber connector consisting of a metal ring set in circular grooves in two pieces; the assembly being held by bolts.

split-ring connector

split-spoon sampler A sampling device used to collect soil samples during well drilling.

splitter damper A single blade damper hinged at one end, installed to divert air from a main duct into a branch duct.

splitting tensile test (diametral compression test) A test for tensile strength in which a cylindrical specimen is loaded to failure in diametral compression applied along the entire length.

spoil Dirt or rock excavated and removed from its original location but not used for embankment or fill.

spokeshave A carpenter's tool for shaping curved edges, consisting of two handles and used as a drawing knife or planing tool.

spokeshave

sponge rubber Expanded rubber having interconnected cells, used as a resilient padding and as thermal insulation.

spoon 1. A small, steel plasterer's tool used in finishing moldings. **2.** A recovery tool used in soil sampling.

spot In gypsum construction, to treat fastener heads with joint compound material. As the fastener sets below the surface of the board, an indentation is formed, allowing a recess, which is filled by spotting.

spot elevation A point on a map or plan with its existing or proposed elevation noted.

spot ground A piece of wood attached to a plaster base and used as a gauge for thickness of applied plaster.

spotter 1. The person who directs a truck driver into a loading or unloading position. **2.** The horizontal framework between the machinery deck of a pile driver and the leads.

spotting 1. Directing trucks for loading or unloading. **2.** Spots of adhesive material used to fasten a veneer to its backing. **3.** A defect in a painted surface consisting of spots of a different color, shade, or gloss than the rest.

spot-weld A small circular weld between two metal pieces made by applying heat and pressure.

spray bar A pipe with ports used to apply liquid asphalt to a road surface.

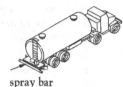

spray bar

spray booth An enclosed or partly enclosed area used for the spray-painting of objects, usually equipped with a waterfall system to catch overspray and/or a filtered air supply and exhaust system.

spray drying A method of evaporating the liquid from a solution by spraying it into a heated gas.

sprayed acoustical plaster An acoustical plaster applied with a special spray gun. The plaster usually has a rough surface and may be perforated with hand tools before hardening.

sprayed fireproofing An insulating material sprayed directly onto structural members with or without wire mesh reinforcing to provide a fire endurance rating.

sprayed fireproofing

spray-in-place insulation An insulation material applied with a gun or sprayer to surfaces or cavities. Examples include loose cellulose fiberfill and polyurethane foam.

spray painting Applying paint, lacquer, or a similar material with a special tool activated by air pressure.

spray pond A component of a cooling system that cools water for use by forcing it upward through spray nozzles and collecting it in a pool.

spreader 1. A piece of wood or metal used to hold the sides of a form apart until the concrete is placed. **2.** A brace between two wales. **3.** A device for spreading gravel or crushed stone for a pavement base course. **4.** A stiffening member used to keep door or window frames in proper alignment during shipment and installation. **5.** An additive that increases the surface area over which a given volume of liquid will spread when poured on a solid or other liquid.

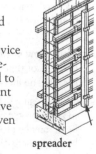

spreader

spreader box A piece of heavy construction equipment that connects to the back of a vehicle and spreads rock, asphalt, concrete, or other materials.

spread footing A generally rectangular prism of concrete, larger in lateral dimensions than the column or wall it supports; used to distribute the load of a column or wall to the subgrade.

spread footing

S

spreading rate **1.** The number of square yards of surface covered by a gallon of paint. **2.** The rate at which bitumen or other material is applied to the surface of a roof.

spread-of-flame test A fire test for roof coverings in which a specified flame source is applied to a sample while a specified air stream is directed at the sample. The sample may be mounted horizontally or at an angle, depending on the intended end use of the product.

spreadsheet Large, wide sheet with many columns that is used to tabulate all estimates and sub-bids when putting a bid together.

spring **1.** An elastic body or shape, such as a spirally wound metal coil, that stores energy by distorting and imparts that energy when it returns to its original shape. **2.** The line or surface from which an arch rises.

spring-bolt (cabinet lock) A spring-loaded, beveled bolt that self-latches when a door or drawer is closed.

spring brace Temporary brace used to adjust wood frame walls for straightness and plumb.

spring buffer An assembly containing a spring that is designed to absorb and dissipate kinetic energy, such as that from a descending elevator car or counterweight.

spring clamp Any clamp with pressure exerted by a spring. One variety is used to hold parts during gluing; another is used as a temporary electric connection.

springer (skewback, summer) **1.** The stone from which an arch springs. **2.** The bottom stone of the coping of a gable. **3.** The rib of a groined vault.

spring hinge A hinge with one or more springs mounted in its barrel to return a door to the closed position. The hinge may be single-acting or double-acting for a swing door.

spring hinge

springing line The horizontal line connecting the points from which an arch or arches rise.

sprinklered Said of an area of a building that is protected from fire by an automatic sprinkler system.

sprinklered

sprinkler flow switch A fire sprinkler line sensor that detects water flow and notifies the fire alarm control panel.

sprinkler head A distribution nozzle used in a fire-protection sprinkler system. The head may be closed by a plug held in place by a heat-sensitive device.

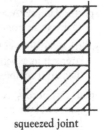
sprinkler head

sprocket (cocking piece, sprocket) A wedge-shaped piece of wood attached to the upper side of a rafter at the eave to form a break in the roof line.

spruce-pine-fir (SPF) Canadian woods of similar characteristics that have been grouped for production and marketing. The SPF species have moderate strength, are worked easily, take paint readily, and hold nails well. They are white to pale yellow.

sprung molding A molding that has its interior corner beveled off to better fit a right-angle joint.

spud **1.** A hand tool used to strip bark from logs. **2.** A sharp, narrow bar or spade used for removing gravel and roofing from a built-up roof.

spud vibrator A vibrator used for consolidating concrete, having a vibrating casing or head that is inserted into freshly placed concrete.

spud vibrator

spun concrete Concrete compacted by centrifugal action, such as in the manufacture of pipes and utility poles.

spun lining The smooth bituminous lining in a corrugated metal pipe formed by spinning the pipe around its axis.

spur **1.** An appendage to a supporting member such as a buttress, shore, or prop. **2.** A decorative stone base that makes the transition from a round column to a square or polygonal plinth. **3.** A carpenter's tool with a sharp point used for cutting veneer. **4.** A rock ridge left projecting from a side wall after a blast. **5.** A short length of railroad track; usually parallel to a main track and used for loading, unloading, or storage.

square **1.** A quantity of shingles, shakes, or other roofing or siding materials sufficient to cover 100 square feet when applied in a standard manner; the basic sales units of shingles and shakes.

square edge In context of acoustical tile, a square-edge is not beveled and creates a hairline joint when installed.

square-framed Joinery framing with all angles of stiles, rails, and mountings cut square.

square roof A roof with two sloping surfaces, each at an angle of 45° from the horizontal.

square up To trim a timber or piece of wood, using a plane, so that its cross section is rectangular.

squaring Aligning or constructing an item or assembly so that all angles are 90°.

squeegee **1.** Window cleaning tool. **2.** Fine gravel installed for grading a floor prior to placement of concrete).

squeezed joint A joint formed by coating surfaces of two pieces with cement and squeezing them together.

squinch A small arch built across an interior corner of a room to support a superimposed load, such as a dome or the spire of a tower.

squeezed joint

squint A small oblique opening in an interior wall of a medieval church to allow a view of the high altar from the aisles.

stab To roughen bricks in preparation for a plaster coat.

stability A measure of the ability of a structure to withstand overturning, sliding, buckling, or collapsing.

stabilization Temporary structural and weather protection used to maintain the essential form and structural integrity of the existing property before restoration begins.

stabilizer A substance that makes a solution or suspension more stable, usually keeping particles from precipitating.

stack **1.** A chimney. **2.** A vertical structure containing one or more flues for the discharge of hot gas. **3.** A vertical supply duct in a warm, air-heating system. **4.** Any vertical plumbing pipe, such as soil pipe, waste pipe, vent, or leader pipe. **5.** A collection of vertical plumbing pipes. **6.** A tier of shelves for books. **7.** To place trusses on framed walls where they will be installed.

S

stack bond (stacked bond) A masonry pattern bond in which all vertical and horizontal joints are continuous and aligned.

stack bond (stacked bond)

stackhead A vertical duct that discharges exhaust air into the atmosphere at high velocity.

stack vent 1. The extension of a soil or waste stack above the highest horizontal drain or fixture connected to the stack. **2.** A device installed through a built-up roof covering to allow entrapped water vapor to escape from the insulation.

stadia rod (stadia) A graduated surveyor's rod used with a transit or similar instrument to determine distances. An observed intercept on the rod, as defined by two lines in the reticle of a telescope, is converted into the distance between the instrument and the rod by use of similar triangles.

staff bead A thick exterior molding placed between a door or window and a masonry wall.

staff man (rodman) A person who sets the leveling rod for a surveyor in leveling or stadia work.

stage grouting Sequential grouting of a hole in separate steps or stages, in lieu of grouting the entire length at once.

staggered Descriptive of fasteners, joints, or members arranged in two or more rows so that the beginning of each row is offset from the adjacent one.

staggered course A course, of shingles or tiles for instance, where the butts do not form a continuous horizontal line.

staggered riveting Rows of rivets installed in a staggered pattern.

staggered-stud partition A partition made of two rows of studs with alternating studs supporting opposite faces of the partition and each stud making contact with only one wall. Staggered-stud partitions often have an interwoven fiberglass blanket to improve the sound-insulation value of a partition.

staging (scaffolding) 1. A temporary working platform against or within a building for construction, repairs, or demolition. **2.** A temporary working platform supported by the temporary timbers in a trench.

staging (scaffolding)

stain 1. Color in a dissolving vehicle. When spread on wood or similar material, the stain penetrates and gives color to the material. **2.** A discoloration in the surface of a material, such as wood or plastic.

stainless steel Any of a number of steels alloyed with chromium and nickel. Depending on the alloy, the metal may possess good corrosion resistance, high heat tolerance, or high strength.

stair 1. A single step. **2.** A series of steps or flights of steps connected by landings, used for passage from one level to another.

stair carriage or stringer Support framing members for stair treads, usually a 2″ × 12″ notched plank. Also known as a "rough horse."

staircase 1. A single flight or multiple flights of stairs including supports, frameworks, and handrails. **2.** The structure containing one or more flights of stairs.

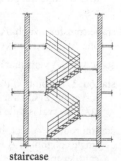

staircase

stair headroom The least clear vertical distance measured from a nosing of a tread to an overhead obstruction.

stair landing A level platform installed at the point where stairs change direction, at the top of a flight of stairs, or between flights.

stair rise The height of a stair (vertical distance between treads).

stairwell A vertical shaft enclosing a stair.

stairwell

stake A short, pointed piece of wood or metal driven into the ground as a marker or an anchor.

staking out The process of driving stakes for batter boards to locate the limits of an excavation, or to identify the boundaries of a parcel of land.

stalactite A downward-pointing deposit formed as an accretion of mineral matter produced by evaporation of dripping water from the surface of rock or of concrete; commonly shaped like an icicle.

stalagmite An upward-pointing deposit formed as an accretion of mineral matter produced by evaporation of dripping water, projecting from the surface of rock or concrete, commonly conical in shape.

stamp A seal required on drawings for commercial projects that contains the architect's or engineer's name and registration number. The individual's signature is usually required over the stamp.

stanchion 1. A vertical post or prop supporting a roof, window, etc. **2.** An upright bar or post, as in a window, screen, or railing.

stanchion

standard 1. A grade of lumber suitable for general construction and characterized by generally good strength and serviceability. In light framing rules, the standard grade applies to lumber that is 2″ to 4″ thick and 2″ to 4″ wide. It falls between the construction and utility grades. **2.** General recognition and conformity to established practice.

standard absorption trench An absorption trench 12″ to 36″ (30 to 90 cm) wide, containing 12″ (30 cm) of clean, coarse aggregate and a distribution pipe covered with a minimum of 12″ (30 cm) of earth.

standard air Air with a density of 0.075 lb per cu. ft. (0.0012 gm per cc) that is close to air at 68°F (20°C) dry bulb and 50% relative humidity at a barometric pressure of 29.9″ (76 cm) of mercury.

standard and better A mix of lumber grades suitable for general construction. The *and better* signifies that a portion of the lumber is actually of a higher grade than standard (but not necessarily of the highest grade). The proportion of higher grades included is a factor in determining market value.

standard atmosphere A pressure equal to 14.7 lb per sq. in. (1.01×10 dynes per sq. cm).

standard brick Brick with the dimensions 2¼″ × 3¾″ × 8″. Standard modular brick measures 2⅔″ × 4″ × 8″.

standard deviation A statistic, used as a measure of dispersion in a distribution, that is equal to the square root of the arithmetic average of the squares of the deviations from the mean.

S

standard hook A hook at the end of a reinforcing bar made in accordance with a standard.

standard knot Any knot in wood 1½″ (3.8 cm) or less in diameter.

standard matched Tongue-and-groove lumber with the tongue and groove offset, rather than centered as in center-matched lumber.

standard net assignable area The area of a project that can be rented to an occupant.

standard penetration resistance (Proctor penetration resistance) The load required to produce a standard penetration of a standard needle into a soil sample at a standard rate.

standard penetration test A test to estimate the degree of compactness of soil in place by counting the number of blows required to drive a standard sampling spoon 1′ (0.3 m) using a 140-pound (64-kg) weight falling 30″ (0.8 m).

standard sand Ottawa sand accurately graded to pass a U.S. Standard No. 20 (850-micrometer) sieve and be retained on a U.S. Standard No. 30 (600-micrometer) sieve, for use in the testing of cements.

standard temperature and pressure (STP) The standard conditions used to measure properties of matter. Standard temperature is 32°F, 0°C or 273° Kelvin—the freezing point of water. Standard pressure is the pressure exerted by a column of mercury 760 mm high, referred to as 1% atmosphere.

standard tolerance A generally accepted tolerance for a specific product.

standard wire gauge The legal standard wire gauge in Great Britain and Canada.

standby lighting Lighting provided to supply illumination in the event of failure of the lighting system so that normal activities can continue.

standing finish Those items of interior finish that are permanent and fixed, as distinguished from such items as doors and movable windows.

standing leaf An inactive leaf of a double-door bolted in the closed position.

standing seam A seam, in sheet metal and roofing, made by turning up two adjacent edges and folding the upstanding parts over on themselves.

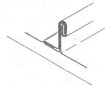

standing seam

standing waste A vertical overflow pipe connected to piping at the bottom of a water tank to control the height of storage.

standpipe 1. A pipe or tank connected to a water system and used to absorb surges that can occur. **2.** A pipe or tank used to store water for emergency use, such as for fire fighting.

standpipe system A system of tanks, pumps, fire department connections, piping, hose connections, connections to an automatic sprinkler system, and an adequate supply of water used in fire protection.

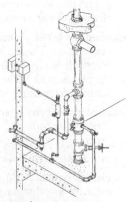

standpipe system

staple A double-pointed, U-shaped piece of metal used to attach wire mesh, insulation batts, building paper, etc.; usually driven with a staple gun.

staple gun (stapler) A spring-driven gun used to drive staples used for fastening materials such as building paper and batt insulation.

staple hammer A hand tool that holds a magazine of staples and drives a staple when the face strikes a surface.

stapling flange A flange on the edges of a faced insulation batt or blanket that is used to staple the insulation to studwork.

starter 1. A device used with a ballast to start an electric-discharge lamp. **2.** An electric controller used to accelerate a motor from rest to running speed and to stop the motor.

starter board A 6″ or 8″ board used at the eave of a roof to provide a solid nailing surface for the first courses. A starter board is also used in reroofing to replace the old shingles at the eaves.

starter piece The piece of pipe that starts the branch line. It extends from the cross main or riser nipple to the first head for a fire protection system.

starter strip (starting strip) A strip of composition roofing material applied along the eaves before the first row of shingles is laid.

starting newel The first support post for a handrail at the bottom of a stairway.

starved box A variable air volume box that has a less-than-desired flow despite a fully open damper.

starved joint A poorly bonded glue joint resulting from the use of too little glue.

statement of probable construction cost A cost estimate prepared by the design professional during each of the design phases for the owner's use.

statement of work A written statement describing the procurement of architectural and engineering services, including preliminary or schematic design, design development, and construction document preparation.

static head The static pressure of a fluid expressed as the height of a column of the fluid that the pressure could support.

static pressure 1. The pressure exerted by a fluid on a surface at rest with respect to the fluid. **2.** The pressure a fan must supply to overcome resistance to airflow in an air distribution system.

statics That branch of mechanics dealing with forces acting on bodies at rest. Statics is the basis of structural engineering.

static test 1. A test that subjects a curtain wall to a pressure differential equal to a specified wind pressure. **2.** A test that simulates the flow of water over a curtain wall during a hurricane. **3.** A test to determine the pressure exerted by a fluid while the system is at rest.

static transfer switch A device that transfers an electrical load from one AC power source to another without interruption.

static vent A vent that does not have a fan.

station 1. A point on the Earth's surface that can be determined by surveying. **2.** On a survey traverse, particularly a roadway, every 100′ interval is called a station.

station roof 1. A roof shaped like an umbrella and supported by a single column. **2.** A long roof supported on a row of columns and cantilevering off one or both sides of the column line.

S

station yards of haul The product of the number of cubic yards in a haul and the number of 100′ (30.48-m) stations through which it is hauled.

statute A law or enactment of the various federal, state, and local rule-making bodies such as Congress, state legislatures, and local planning and zoning boards.

statute of frauds A statute specifying that certain kinds of contracts, such as for the sale or lease of real property, are unenforceable unless signed and in writing, or unless there is a written memorandum of terms signed by the party to be charged. Statute varies by state.

statute of limitations Provision of law establishing a certain time limit from an occurrence during which a judgment may be sought from a court of law.

statutory bond A bond of which the content and form are established by statute.

statutory requirements Requirements that are embodied in the law.

staunching piece A gap left between adjacent concrete bays until the concrete has cured and shrunk, when it is then filled with more concrete.

stave 1. A narrow strip of curved wood used in the construction of barrels, buckets, or tanks. **2.** A ladder rung.

stay Anything that stiffens a frame or stabilizes some structural component.

stay bolt A long, metal rod with a threaded end and used as a stay.

stay rod Any rod-shaped tie used temporarily or permanently to prevent spreading of the parts it connects.

stay

steam box An enclosure for steam-curing concrete products.

steam chest A container in which wood is steamed for bending or forming.

steam cleaner A machine that provides pressurized steam to a nozzle for the purpose of cleaning grease or dirt from a surface. Detergents or chemicals are sometimes added.

steam curing Curing of concrete or mortar in water vapor at atmospheric or higher pressures and at temperatures between about 100°F and 420°F (40°C and 215°C).

steam-curing cycle 1. The time interval between the start of the temperature-rise period and the end of the soaking period or the cooling-off period. **2.** A schedule of the time and temperature of periods that make up the cycle.

steam grid humidifier An air duct humidifier that introduces steam into the air through perforated pipes.

steam hammer A pile hammer activated by steam pressure.

steam heating system A heating system in which heat is transferred from a boiler or other source, through pipes, to a heat exchanger. The steam can be above, at, or below atmospheric pressure.

steaming A process in which logs are heated with steam or hot water in special vats prior to peeling them into veneer. Steaming results in smoother veneer and improved recovery from the log. A similar process is used to prepare wood for bending or shaping.

steam injection In an environmental remediation operation, the process of injecting steam into soil to aid in the extraction of soil vapor or groundwater.

steam pipe Any pipe for the conveyance of steam.

steam stripping 1. A process of removing volatile contaminants from soil or water by passing steam through the soil or water. **2.** A method of removing wallpaper.

steam traps A device that discharges condensate air and prevents the passage of steam.

steam vat (steam chest) A container used in steaming to soften logs or flitches before peeling or slicing veneers.

steel Any of a number of alloys of iron and carbon, with small amounts of other metals added to achieve special properties. The alloys are generally hard, strong, durable, and malleable.

steel domes 19″ × 19″ or 30″ × 30″ metal forms used in two-way joist construction of cast-in-place concrete slabs.

steel erector A contractor who undertakes to place, plumb, and secure a steel structure.

steel-frame construction Construction in which steel columns, girders, and beams comprise the structural supporting elements.

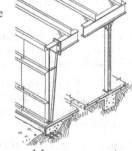

steel-frame construction

steel inspection An inspection prior to concrete placement to ensure that steel rebar, window bucks, and other components are properly installed in compliance with the approved foundation plan.

steel sheet Cold-formed sheet or strip steel shaped as a structural member for the purpose of carrying the live and dead loads in lightweight concrete roof construction.

steel sheet piling Interlocking rolled-steel sections driven vertically into the ground to serve as sheeting in an excavation or to cut off the flow of groundwater.

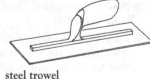

steel sheet piling

steel stair fill Concrete that is poured into metal pans to form stair treads and landings.

steel stud anchor A steel clip fastened to a door frame and used to secure the frame to a steel stud.

steel trowel A smooth concrete finish obtained with a steel trowel.

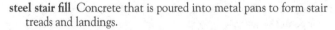

steel trowel

steening Brickwork laid without mortar.

steering brake A brake to slow or stop one side of a tractor.

steering clutch A clutch that can disconnect power from one side of a tractor.

stem bars Bars used in the wall section of a cantilevered retaining wall or in the webs of a box. When a cantilevered retaining wall and its footing are considered as an integral unit, the wall is often referred to as the stem of the unit.

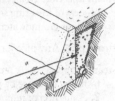
stem bars

stemming A suitable inert, incombustible material, such as rock bits or other packing, or device used to confine or separate explosives in a drill hole, or to cover explosives in mudcapping.

stenciling A decorative painting technique using hand-cut or pre-manufactured cut-out shapes.

step brazing A method of brazing in which filler metals of successively lower brazing temperatures are used to prevent softening of previously brazed joints.

step down transformer An electric transformer with a lower voltage at the secondary winding terminals than at the primary winding terminals.

step joint 1. A notched joint used to connect two wood members meeting at an angle, such as a tie beam and a rafter. 2. A joint between two rails with different heights and/or cross sections.

stepped flashing Flashing at the intersection of a sloped roof and a wall or chimney. The upper edge of each sheet is stepped with relation to adjacent sheets to maintain a safe distance from the sloped roof surface.

stepped footing 1. A wall footing with horizontal steps to accommodate a sloping grade or bearing stratum. 2. A column or wall footing composed of two or more steps on top of one another to distribute the load.

stepped foundation A foundation constructed in a series of steps that approximate the slope of the bearing stratum. The purpose is to avoid horizontal force vectors that might cause sliding.

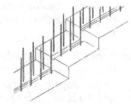
stepped foundation

stepping Lumber designed to be used for stair treads. Stepping is vertical-grained and is customarily shipped kiln-dried, surfaced three sides, and bull-nosed on one edge. Besides the "C & Better" and "D" grades of solid wood, stepping is also made from particleboard for use where it will be covered.

stepping off Laying off the required length of a rafter by use of a carpenter's square.

step-plank Any hardwood lumber used to make steps, usually 1¼" to 2" (3.2 to 5.1 cm) thick.

step soldering A method of soldering in which solders with successively lower application temperatures are used to prevent the disturbing of previously soldered joints.

step-taper pile A pile formed of straight-sided, thin-walled segments of spirally corrugated steel that, once driven into the ground, are filled with concrete.

step-up transformer An electric device that used to convert a power supply to a higher voltage circuit.

stereobate The foundation, substructure, or platform upon which a building is constructed.

stereo photogrammetry The science of measuring from two photographs taken from known positions in relation to the building or structure. Using known dimensions of parts of the structure, other dimensions can then be determined by scaling.

stereotomy The practice of cutting stone to specified forms and then placing it.

sterling A grade of Idaho white pine boards, equivalent to #2 common in other species.

stick 1. Any long, slender piece of wood. 2. A waxed paper cartridge containing an explosive, usually 1⅛" × 8" (3 × 20 cm). 3. A rigid bar fastened to the bucket and hinged at the boom of a power shovel or a backhoe.

stick built A term describing frame houses assembled piece-by-piece from lumber delivered to the site with little or no previous assembly into components. The more typical type of residential construction is stick built.

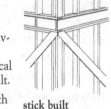
stick built

sticker An additive that increases the strength with which water-soluble materials attach to solid surfaces.

sticky cement Finished cement that develops low or zero flowability during or after storage in silos, or after transportation in bulk containers, hopper-bottom cars, etc. Sticky cement may be caused by: (a) interlocking of particles, (b) mechanical compaction, (c) electrostatic attraction between particles, or (d) moisture.

stiffened compression element A compression element that has been stiffened on its weak axis in order to resist buckling on that axis.

stiffener 1. A bar, angle, channel, or other shape attached to a metal plate or sheet to increase its resistance to buckling. 2. Internal reinforcement, usually light-gauge channels, for a hollow metal door.

stiff leg derrick An erection derrick with a vertical mast shorter than its boom. The mast is tied from its peak to a triangular structural frame at its base by rigid steel members.

stiffness Resistance to deformation.

stiffness factor A measure of the stiffness of a structural member. For a prismatic member, the stiffness factor is equal to the ratio of the product of the moment of inertia of the cross section and the modulus of elasticity for the material to the length of the member.

stile 1. The vertical members forming the outside framework of a door or window. 2. A roofing tile with an S-shaped cross section.

stilt 1. A post used to raise a structure above ground or water level. 2. A compression member placed above or below a similar member to gain additional height.

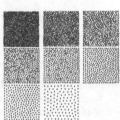
stile

stinger A slang term for a welding rod.

stipple To make dots or short dashes on a surface as a decorative effect.

stippled finish A dotted or pebbly textured finish on the surface coat of paint, plaster, or porcelain enamel, induced by punching the unset surface with a stiff brush.

stipple

S

stipulated sum agreement A contractual agreement in which a fixed amount is established for performance of the work.

stirrup 1. A reinforcement used to resist shear and diagonal tension stresses in a concrete structural member. **2.** A steel bar bent into a U or box shape and installed perpendicular to, or at an angle to the longitudinal reinforcement, and properly anchored. **3.** Lateral reinforcement formed of individual units, open or closed, or of continuously wound reinforcement. The term *stirrups* is usually applied to lateral reinforcement in flexural members and the term *ties* to lateral reinforcement in vertical compression members.

stitched veneer Veneer sheets composed of random width pieces of veneer sewn together with heavy thread. Several stitch lines run across the width of each piece of stitched veneer. In veneer production, the random width pieces are run through a large sewing machine and reclipped in desired (usually standard 48″ to 54″) sizes. Stitched veneer is usually used in the core plies of plywood.

stitch rivet One of a number of rivets installed at regular intervals and in a straight line to connect two parts and provide lateral stiffness.

stitch welding The union of two or more parts with a line of short, equally spaced welds.

stock 1. Material or devices readily available from suppliers. **2.** The body or handle of a tool. **3.** A frame to hold a die when cutting external threads on a pipe. **4.** The total value of the equity in a corporation.

stock brush A brush used to apply water to base coat or finish coat plaster to improve workability.

stock lumber Lumber cut to standard sizes and readily available from suppliers.

stock millwork Millwork manufactured in standard shapes and sizes and readily available from suppliers.

stockpile Material stored for later use, such as topsoil.

stock size A standard size of an item that is readily available from suppliers.

stoichiometric combustion A controlled combustion where fuel is reacted with the exact amount of oxygen required to oxidize all carbon, hydrogen, and sulfur in the fuel to carbon dioxide, water, and sulfur dioxide.

stone 1. Individual blocks of rock processed by shaping, cutting, or sizing. For use in masonry work. **2.** Fragments of rock excavated, usually by blasting, from natural deposits and further processed by recrushing and sizing. For use as aggregate. **3.** A carborundum or other natural or artificial hone used to sharpen cutting edges of tools.

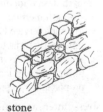
stone

stone dust Pulverized stone used in the construction of walkways or other stable surfaces. The dust is mixed with soil and compacted or used with gravel to fill spaces between irregular stones. Stone dust is a byproduct of stone-crushing operations.

stonemason A craftsman skilled in constructing stonemasonry, including any preparation at the site.

stone sand Fine aggregate resulting from the mechanical crushing and processing of rock.

stone seal An asphaltic wearing surface made by rolling aggregate into a heavy application of bituminous material. Application is usually repeated several times.

stone veneer, thin Stone cut 2″ thick or less and applied to an interior or exterior building surface in a non-load-bearing manner. Granite, marble, limestone, slate and travertine are most commonly used.

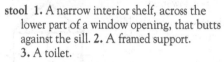
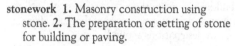
stonework

stonework 1. Masonry construction using stone. **2.** The preparation or setting of stone for building or paving.

stool 1. A narrow interior shelf, across the lower part of a window opening, that butts against the sill. **2.** A framed support. **3.** A toilet.

stool cap The molding that sits on the windowsill and extends into a room to form a kind of shelf.

stool

stop 1. On doors, the molding on the inside of the doorjamb that causes the door to stop in its closed position, preventing it from swinging through. On windows, the molding that covers the inside face of the jamb. **2.** A type of molding nailed to the face of a door frame to prevent the door from swinging through. A stop is also used to hold the bottom sash of a double-hung window in place. **3.** A valve used to shut off water supply to a fixture.

stop-and-waste cock A stopcock with a drain plug in the valve. When the valve is closed, water in the downstream piping can be drained through the plug. A stop-and-waste cock is often used at connections to outdoor faucets so the exterior piping can be drained for the winter.

stop box A cast-iron pipe installed vertically underground and covered with a lid. It provides access (with a tool on a pole) to a water cut-off valve for the home.

stopcock A valve to shut off the flow of fluid in a branch of a distribution system in a building.

stop glazing Either the lip at the back of a rabbet or the molding applied at the front. These serve to hold a light in a sash or frame with the help of spacers.

stopped miter A miter joint used with members of different thickness.

stop valve Any valve in a piping distribution system that is used to stop flow.

stop work order or stop order An order issued by the owner's representative to stop work on a project. Reasons for the order include failure to conform to specifications, unsatisfied liens, labor disputes, and inclement weather.

storage life (shelf life) The length of time that a product, such as a package adhesive or sealant, can be stored at a specified temperature range and remain usable.

storefront sash An assembly of light metal members that form a frame for a fixed-glass storefront.

storm clip A clip on the exterior of a glazing bar that holds the pane in place.

storefront sash

S

storm drain A drain used to convey rain water, subsurface water, condensate, or similar discharge, but not sewage or industrial waste.

storm sewer A sewer used for conveying rainwater and/or similar discharges, but not sewage or industrial waste, to a point of disposal.

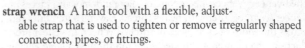

storm drain

story That part of a building between the upper surface of a floor and the upper surface of the floor above. Building codes differ in designations applied when a part of a story is below grade

story-and-a-half The designation of a building in which the second story rooms have low headroom at the eaves.

story-and-a-half

story pole A pole used to gauge height (in stories) during masonry construction.

story post An upright post that helps support a beam on which a floor rests.

story rod A wooden rod of a length equal to one story height, sometimes divided in parts each equal to one riser in a stair, for use in stair construction.

straightedge (rod) A rigid, straight piece of wood or metal used to strike off or screed a concrete surface to proper grade, or to check the flatness of a finished grade.

straight grain A piece of wood in which the principal grains run parallel to its length.

straight joint 1. A continuous joint in a wood floor formed by the butt ends of parallel boards. 2. An edge joint between two parallel timbers. 3. Vertical joints in masonry that form a continuous straight line.

straight-joint tile A term describing single-lap tiles designed to be laid so that the edges of each course run in a straight line from eave to ridge.

straight-line theory An assumption in reinforced-concrete analysis according to which the strains and stresses in a member under flexure vary in proportion to the distance from the neutral axis.

strain Deformation of a material resulting from external loading. The measurement for strain is the change in length per unit of length.

strain gauge A sensitive electrical or mechanical instrument used to measure strain in loaded members or objects, usually for research or development.

straining beam (straining piece, strutting piece) 1. A horizontal strut in a truss, placed above the tie beam or the bottom of the rafters, usually mid-height. 2. The chord between the upper ends of the posts in a queen post truss.

strake A single row of clapboard siding.

strand A prestressing tendon composed of a number of wires twisted about a center wire or core.

stranded wire A group of fine wires used as a single electric conductor.

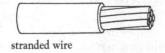

stranded wire

S-trap An S-shaped trap that provides a water seal in a waste or soil pipe. Used most commonly at sinks and lavatories.

strap A metal plate fastened across the intersection of two or more timbers.

strap bolt A bolt in which the middle portion of its shank is flattened, so the unit can be bent in a U-shape.

straphanger A hanger made from a thin, narrow strip of material.

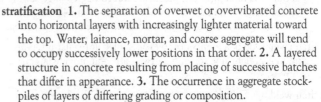

S-trap

strap wrench A hand tool with a flexible, adjustable strap that is used to tighten or remove irregularly shaped connectors, pipes, or fittings.

stratification 1. The separation of overwet or overvibrated concrete into horizontal layers with increasingly lighter material toward the top. Water, laitance, mortar, and coarse aggregate will tend to occupy successively lower positions in that order. 2. A layered structure in concrete resulting from placing of successive batches that differ in appearance. 3. The occurrence in aggregate stockpiles of layers of differing grading or composition.

stratified rock Rock that has been formed by compaction, cementation, or crystallization of successive beds of deposited material.

stratum A bed or layer of rock or soil.

straw bale A low-embodied-energy farming by-product that can be used to build highly insulated (R-50), fire- and termite-resistant walls. Paired with stucco exterior and drywall interior, straw bale is a plentiful green construction alternative to lumber.

streamline flow Fluid flow in which the velocity at every point is equal in magnitude and direction.

street elbow (service ell, street ell) A pipe elbow with male threads on one end and female threads on the other.

street floor That floor of a building nearest to street level. According to some building codes, the street floor is a floor level not more than 21″ above or 12″ below grade level at the main entrance.

street lighting luminaire A complete lighting unit intended to be set on a pole or post with a bracket.

strength design method (ultimate strength method) A design method in which service loads are increased sufficiently by factors, often referred to as load factors, to obtain the ultimate design load. The structure or structural element is then proportioned to provide the desired ultimate strength.

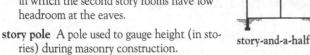

street lighting luminaire

stress Intensity of internal force (i.e., force per unit area) exerted by either of two adjacent parts of a body on the other across an imagined plane of separation. When the forces are parallel to the plane, the stress is called *shear stress;* when the forces are normal to the plane, the stress is called *normal stress;* when the normal stress is directed toward the part on which it acts it is called *compressive stress;* when it is directed away from the part on which it acts it is called *tensile stress.*

stress corrosion cracking Cracking of metal that occurs under the combined influence of certain corrosive environments and applied or residual stresses.

S

stress crack An external or internal crack in a plastic as a result of applied tensile stresses.

stressed-skin construction Construction in which a thin material on the surface of a building is used to carry loads.

stressed-skin panel A panel assembled by fastening plywood or similar sheets over a frame or core, on both sides. The result is a composite structural member.

stress-graded lumber Lumber graded for strength according to the rate of growth, slope of grain, and the number of knots, shakes, and other defects.

stressing end In prestressed concrete, the end of the tendon from which the load is applied when tendons are stressed from one end only.

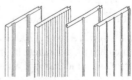

stressing end

stress relaxation Stress loss developed from strain when a constant length is maintained under stress.

stress-relief heat treatment A process used to relieve residual stresses that involves the uniform heating of a material or structure to a high temperature.

stress-strain diagram A diagram for a particular material with values of stress plotted against corresponding values of strain, usually with stress plotted as the ordinate.

stretcher A masonry unit laid with its length parallel with the face of the wall.

stretcher bond (running bond, stretching bond) A masonry bond with all courses laid as stretchers and with the vertical joint of one course falling midway between the joints of the courses above and below.

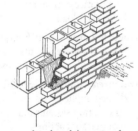

stretcher bond (running bond, stretching bond)

striated Having parallel grooves in the face, as in a fluted column.

striated face The face of a plywood panel that has been given closely spaced, shallow grooves to provide a vertical pattern.

striated face

strike 1. In masonry, to cut off the excess mortar at the face of a joint with a trowel stroke. 2. To remove formwork. 3. A work stoppage by a body of workers. 4. A metal plate installed on a door frame where a latch or dead bolt engages.

strike breaker One who takes the place of a worker who has gone on strike.

strike off (strikeoff) To remove concrete in excess of that which is required to fill the form evenly or bring the surface to grade, performed with a straightedged piece of wood or metal by means of a forward sawing movement or by a power-operated tool appropriate for this purpose.

strike plate (strike, striking plate) A plate or box, mounted in a jamb, with a hole or recess shaped to receive and hold a bolt or latch from a lock on the door.

striker A slightly beveled metal plate attached to a strike plate to guide a door latch to its socket.

strike reinforcement A metal piece welded inside a hollow metal frame to which the strike plate is attached and which also serves to strengthen the frame.

striking The removal of temporary supports from a structure.

strike plate
(strike,
striking plate)

stringcourse (belt course) A horizontal band of masonry, usually narrower than the other courses, which may be flush or projecting, and plain or ornamented.

stringer or string 1. A secondary flexural member parallel to the longitudinal axis of a bridge or other structure. 2. A horizontal timber used to support joists or other cross members in floors or ceilings. 3. In stairs, the framing member that supports the treads.

stringer bead A continuous bead of weld metal made by moving the electrode in a direction parallel to the bead without much transverse oscillation.

strip 1. Board lumber 1″ in nominal thickness and less than 4″ in width, frequently the product of ripping a wider piece of lumber. The most common sizes are 1″ × 2″ and 1″ × 3″. 2. To remove formwork or molds. 3. To remove an old finish with paint removers. 4. To damage the threads on a nut or bolt.

strip core (blackboard) A composite board whose core is made of strips of wood, laid loose or glued together. Veneer is glued to both sides of the core with its grain at right angles to the grain of the core pieces.

strip flooring Flooring made of wood strips that are pieced together in a tongue-and-groove assembly.

strip mopping A method of applying hot bitumen to a roof deck in parallel strips.

stripper 1. A liquid compound formulated to remove coatings by chemical and/or solvent action. 2. A drywall tool with a serrated wheel for cutting wallboard.

stripping felt A narrow strip of roofing felt used to cover a flange of metal flashing.

stripping shovel A mobile power shovel with a long boom for ditching and removal of bulk material.

strip pouring A quick and economical way to place concrete slabs, allowing for a continuous pour with control points cut after the concrete has set.

stroke 1. A run of clapboard on the side of a house. 2. A row of steel plates in a steel chimney.

strongback A frame attached to the back of a form to stiffen or reinforce it during concrete placing or handling operations.

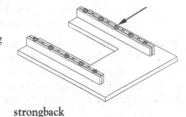

struck joint A masonry joint in which excess mortar is removed by a stroke of a trowel.

strongback

struck tools Tools that perform work by placing the sharp end (called the *working* end) of the tool on a surface and striking the other end (the *struck* end) with a hammer. The working end of a struck tool must be kept sharp. The face on the struck end must be kept smooth and even to prevent off-center blows. Examples of struck tools include chisels, punches, and wedges.

structural A term applied to those members in a structure that carry an imposed load in addition to their own weight.

structural analysis The determination of stresses in members in a structure due to imposed loads from gravity, wind, earthquake, thermal effects, etc.

structural clay facing tile A structural clay tile with a finished ceramic face.

structural clay tile A hollow masonry unit molded from clay, shale, fireclay, or a mixture of such materials.

structural drawings Graphic representations of the members, assemblies, and systems that will transmit live and dead loads of the structure to the earth.

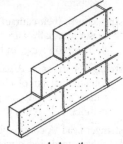
structural clay tile

structural engineering That branch of engineering concerned with the design of the load-supporting members of a structure.

structural failure 1. The inability of a structure or structural member to perform its intended function, perhaps caused by collapse or excessive deformation. **2.** A marked increase in strain without an increase in load.

structural frame All the members of a building or other structure used to transfer imposed loads to the ground.

structural glued-laminated timber A wooden structural member made from selected boards strongly glued together.

structural glued-laminated timber

structural insulated panel (SIP) A rigid panel, often constructed of polyurethane foam insulation between a facing material such as foil or oriented strand board, that exhibits both strength and thermal insulating properties.

structural joists and planks Lumber 2″, 3″, or 4″ thick and 6″ or wider, graded for its strength properties. Such planks are used primarily for joists in residential construction and graded, in descending order, select structural, #1, #2, and #3. The #1 and #2 grades are usually marketed in combination as #2 & Better.

structural light framing A category of dimension lumber up to 4″ in width that provides higher bending strength ratios for use in engineered applications, such as roof trusses. The lumber is often referred to by its fiber strength class, such as 175f for #1 & Better Douglas Fir, or as stress-rated stock.

structural lumber Any lumber with nominal dimensions of 2″ or more in thickness and 4″ or more in width and that is intended for use where working stresses are required. The working stress is based on the strength of the piece and the use for which it is intended, such as beams, stringers, joists, planks, posts, and girders.

structural plate With reference to drainage structures, heavily corrugated steel plate, usually curved, that are bolted together to form large pipes, arches, and other drainage structures.

structural shape A hot-rolled or cold-formed steel member of standardized cross section and strength, generally used in a structural frame. Common structural shapes are angle irons, channels, tees, H-sections, and wide flange beams.

structural steel Steel rolled in a variety of shapes and manufactured for use as load-bearing structural members.

structural steel fastener Any fastener used to connect structural steel members to each other, or to supporting elements, or with concrete to make a composite section.

structural tee 1. A structural shape made by cutting a wide-flange beam or I-beam in half. **2.** A hot-rolled steel member shaped like the letter T.

structural timber Structural lumber with a nominal dimension of 5″ or more on each side, used mainly as posts or columns.

structural veneer Veneers used in the construction of structural plywood panels, as opposed to decorative veneers.

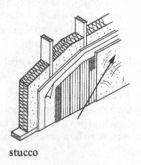
stuctural steel

structure A combination of units fabricated and interconnected in accordance with a design and intended to support vertical and horizontal loads.

structure-borne sound Sound energy imparted directly to and transmitted by the elements of a structure. Plumbing noises traveling through pipes are a good example.

structured cabling system (SCS) A cabling system that supports several applications, such as video, voice, data, and building automation systems functions, reducing the overall cost and space required for cabling.

structure height The vertical distance from grade to the top of the structure.

stub A short projecting element.

stub mortise A mortise that does not project entirely through the piece in which it is cut.

stub tenon A short tenon cut to fit a stub mortise.

stub wall A low wall, usually 4″ to 8″ (100 mm to 200 mm) high, placed monolithically with a concrete floor or other members to provide for control and attachment of wall forms.

stucco A cement plaster used to cover exterior wall surfaces; usually applied over a wood or metal lath base.

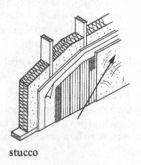
stucco

stud **1.** A vertical member of appropriate size (2″ × 4″ to 4″ × 10″) (or 50 mm × 100 mm to 100 mm × 250 mm) and spacing (16″ to 30″) (or 400 mm to 750 mm) to support sheathing or concrete forms. **2.** A framing member, usually cut to a precise length at the mill, designed to be used in framing building walls with little or no trimming before it is set in place. Studs are most often 2″ × 4″, but 2″ × 3″, 2″ × 6″ and other sizes are also included in the stud category. Studs may be of wood, steel, or composite material. **3.** A bolt having one end firmly anchored.

stud anchor A floor anchor for studs in a wall.

stud bolt A bolt firmly anchored in, and projecting from, a structure, such as a concrete pad; used to secure another member, as in bolting a sill plate to concrete.

stud bolt

stud crimper A hand tool used to speed metal stud construction. The crimper joins vertical studs to horizontal runners with a piercing action.

stud driver A device for driving a hardened steel fastener into concrete or other hard material, consisting of a hand-held driver that positions the fastener. A blow on the head of the driver forces the fastener into the material.

stud grade A grade of framing lumber under the National Grading Rule established by the American Lumber Standards Committee. Lumber of this grade has strength and stiffness values that make it suitable for use as a vertical member of a wall, including use in load-bearing walls.

stud gun A powered stud driver.

stud opening A rough opening in wood framing.

stud partition A partition in which studs are used as the structural base. Wallboard is usually applied over the studs.

stud partition

stud shoe A metal bracket used as reinforcing for a vertical stud.

stud welding Attaching a special metal shear stud into a steel member by resistance welding. A special gun is used to hold the stud and provide electric current for the welding process. Shear studs are used for composite construction in which steel and concrete form a composite beam.

study **1.** A preliminary sketch. **2.** The programming phase of facility planning that precedes building design.

stuff Sawn timber.

styrene butadiene rubber (SBR) or SB latex A synthetic rubber material widely used for applications including wiring insulation and for conventional carpet backing. SBR contains potentially toxic chemicals.

Styrofoam® The brand name of a lightweight extruded polystyrene product noted for its excellent thermal insulation performance.

sub Contraction of subcontractor.

subbase **1.** A layer in a pavement system between the subgrade and base course or between the subgrade and the concrete pavement. **2.** The bottom front strip or molding of a baseboard.

subbase

subbasement **1.** A level, or levels, of a building below the basement. **2.** A story immediately below a basement.

subbid A bid offered by a subcontractor.

subbidder (subbidder) A person or entity who has a direct contract with the contractor for a portion of the work at the site.

subcontract An agreement between a prime contractor and a contractor (specializing in a particular trade) for the completion of a portion of the work for which the prime contractor is responsible.

subcontractor One under contract to a prime contractor by subcontract for completion of a portion of the work for which the prime contractor is responsible.

subcooling Cooling a refrigerant below its saturated condensing temperature.

subdrain A pipe with perforations or open joints that has been buried in a trench backfilled with pervious soil for the purpose of intercepting groundwater or seepage.

subfeeder An electric feeder that originates at a distribution center other than the main distribution centers and supplies one or more branch-circuit distribution centers.

subfloor (blind floor, counterfloor) A rough floor laid on floor joists and serving as a base for the finish floor. A subfloor may also be used as a structural diaphragm to resist lateral loads.

subfloor (blind floor, counterfloor)

subflooring Plywood sheets or construction grade lumber used to construct a subfloor.

subflorescence Salt crystallization inside masonry units, which can cause damage to the units' internal structure. It may occur when the mortar surrounding the masonry units has a lower rate of permeability than the units themselves.

subframe (rough buck, subbuck) **1.** A structural frame of wood members or channel-shaped metal members that support the finish frame of a door or window. **2.** A framework supporting wall siding or panels.

subgrade **1.** The soil prepared and compacted to support a structure or a pavement system. **2.** The elevation of the bottom of a trench in which a sewer or pipeline is laid.

submetering Measuring utilities used by individual tenants in a building.

submittal A sample, manufacturer's data, shop drawing, or other such item submitted to the owner or the design professional by the contractor for the purpose of approval or other action, usually a requirement of the contract documents.

submit the bid Deliver a bid to a sponsor.

S

sub-purlin (subpurlin) A light structural section used as a secondary structural member, and used in lightweight concrete roof construction to support the formboards over which the lightweight concrete is placed.

sub-purlin

subrail A component of a stair rail system that runs down the stairs from the top of the string and upon which the balusters are mounted.

subsealing The placing of a waterproofing material under an existing pavement to waterproof the pavement and to fill voids in the subsoil.

subsidence Settlement over a large area as opposed to settlement of a single structure.

subsill 1. An additional sill fitted to the outside of a window as a stop for screens and to increase the shedding distance. 2. A rough doorsill fixed to the groundsill.

subsoil The bed or stratum of soil lying immediately below the surface soil and which is usually devoid of humus or organic matter.

subsoil drain A drain installed to collect subsurface or seepage water and convey it to a point of disposal.

substantial completion The condition of the work when the project is substantially complete, and ready for owner acceptance and occupancy. Any items remaining to be completed should, at this point, be duly noted or stipulated in writing.

substantial performance A party's performance of most of its contractual obligations, which entitles it to payment of at least a portion of the contract price and precludes a termination for default.

substation An assembly of equipment, including switches, circuit breakers, buses, and transformers, for switching power from one voltage or system to another.

substitution A product, material, or piece of equipment offered in place of that specified.

substrate An underlying material that supports or is bonded to another material on its surface.

substrate failure A surface failure in a concrete wall at a joint, caused by a sealant of high tensile strength that tears weak concrete or mortar from the face of the joint.

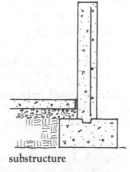

substructure The foundation of a building that supports the superstructure.

substructure

subsurface contamination The presence of hazardous materials in the soil or groundwater under a site.

subsurface investigation (geotechnical investigation) The sampling and laboratory testing process (including soil borings) to establish subsurface profiles, relative strengths, compressibility, and other characteristics of strata deep enough to affect project design.

subsurface sand filter A sewage filtering system consisting of a number of lines of perforated pipe or drain tile surrounded by coarse aggregate, an intermediate layer of sand as filtering material, and a system of underdrains to carry off the filtered liquid.

subsurface sewage disposal system A system for treating and disposing of domestic sewage, usually from a single residence, by means of a septic tank and a soil absorption system.

successful bid A bid that is accepted by a sponsor for award of contract; a low bid (assuming that there will be an award of contract).

suction 1. The absorption of water from a plaster finish coat by the base coat, gypsum block, or gypsum lath, functioning to increase bond and promote better adhesion to the base. 2. The adhesion of mortar to brick. 3. A vacuum effect created when wind load on a building creates a load in an outward direction.

suction pump A pump that draws water from a reservoir at a level lower than the pump, or from a pipe operating at a lower hydraulic gradient than the pump.

sulfate attack A chemical and/or physical reaction between sulfates usually in soil or groundwater and concrete or mortar, primarily with calcium aluminate hydrates in the cement-paste matrix, often causing deterioration.

sulfate resistance The ability of concrete or mortar to withstand sulfate attack.

sulfur cement A cement of clay or a similar substance, usually with additives such as sulfur, metallic oxides, silica, or carbon used for sealing joints and coatings in high-temperature areas such as furnace fire boxes.

summer 1. A horizontal beam supporting floor joists or a wall of a superstructure. 2. Any heavy timber that serves as a bearing surface. 3. A lintel of a door or window. 4. A stone set on a column as a support for construction above, such as a base for a column-supported arch.

sump 1. A pit, tank, or basin that receives sewage, liquid waste, seepage, or overflow water and is located below the normal grade of the disposal system. A sump must be emptied by mechanical means. 2. A depression in a roof deck at a drain.

sump pump A small pump used to remove accumulated waste or liquid from a sump.

sun-bearing angle, or relative solar azimuth Used in solar analysis and design, the solar azimuth angle relative to the direction a building surface faces.

sunken joint A small, narrow depression on the face of a piece of plywood. Sunken joints occur over joint gaps in the core plies of plywood.

sunk fillet A fillet formed by a groove in a flat surface.

sun shade Feature on the exterior of a building that provides shade.

sunspace Interior building space where temperatures are permitted to vary beyond a limited range of "comfort" conditions. When the mass overheats with warmth from the sun, the additional stored energy is retrained to help warm the building when outdoor temperatures cool off.

sun-tempered buildings Buildings designed using standard construction methods, but oriented optimally on the site and featuring carefully designed windows to reduce heating load.

super Contraction of superintendent.

superabsorbent materials Materials, often granular, that absorb liquid several times their own weight. An example is humus, which maintains moisture in soil.

S

superheated steam Steam at a temperature higher than the saturation temperature corresponding to the pressure.

superheaters Devices that add heat to saturated fluids.

superimposed drainage 1. A natural drainage system developed by erosion and having little relation to the area's geological structure. 2. A man-made drainage system developed against the existing geological structure.

superintendent The contractor's representative who is responsible for field supervision, coordination, completion of the work, and the prevention of accidents.

superplasticizer Concrete admixture that adds to the concrete's workability despite low water-cement ratio.

superstructure 1. The part of a building or other structure above the foundation. 2. The part of a bridge above the beam seats or the spring line of an arch.

supervision Direction of work performed by the contractor's (or others) workers on site, as specifically defined by the contract.

supervisory control and data acquisition (SCADA system) A computer system used in industry to monitor and control facility operations and status. The automation capabilities of a SCADA system provide early detection of developing problems and improve efficiency in acquiring data and generating reports.

supplemental agreement A change to an existing contract accomplished by the mutual action of the parties.

supplemental authorization A written agreement that authorizes a modification to a service contract.

supplemental general conditions Written modifications to the general conditions that become part of the contract documents.

supplemental instructions to bidders Written modifications to the instructions to bidders that become part of the bidding requirements.

supplemental services Services described in the schedule of designated services that are outside the normal range of services, including renderings, energy studies, value analyses, project promotion, and expert testimony.

supplied air respirator (SAR) A breathing apparatus, connected by a long air hose either to a source of compressed air or to an air pump.

supplier A person or entity who supplies materials or equipment from off the site for the work, including specially fabricated work.

supply air The conditioned air delivered to a space or spaces.

supply air outlets Air terminals (such as grilles and diffusers) for the discharge of supply air.

supply fixture unit A measure of the probable demand on a water supply by a particular type of plumbing fixture. The value depends on the volume of water supplied, the average duration of a single use, and the number of uses per unit time.

supply grille A grille through which conditioned air is delivered to a space.

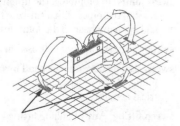
supply grille

supply line In a hot water heating system, the pipe that carries water from the water heater to individual heating units.

supply mains 1. The pipes that bring water, gas, or other elements into a building. 2. The pipes through which the heating or cooling fluid flows from the source of heating or cooling to the laterals or risers leading to heating or cooling units.

supply system The connected ducts, plenums, and fittings through which conditioned air is transferred from a heat exchanger to the space or spaces to be conditioned.

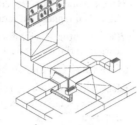
supply system

supreme A grade in Idaho white pine equivalent to B & Better 1&2 clear, the highest grade of select lumber.

surcharge load A force or combination of charges acting on the upper level of a retaining wall.

surety An individual or company that provides a bond or pledge to guarantee that another individual or company will perform in accordance with the terms of an agreement or contract.

surety bond A bond or pledge made by an individual or company that guarantees another individual's or company's performance according to a contract's terms.

surface-active agent An additive to a concrete mix to reduce the surface tension of the mixing water and facilitate wetting, penetrating, emulsifying, dispersing, solubilizing, foaming, and frothing of other additives.

surface bonding A method of bonding masonry that involves first laying masonry units with no mortar or cement between the blocks, and then coating the surface with a fiber-reinforced mortar.

surface burning characteristic A rating of interior and surface finish material providing indexes for flame spread and smoke developed.

surface course The top course of asphalt pavement; the wearing course.

surface drain A drain that eliminates water that pools around the foundation of a home. Also referred to as a superficial peripheral drain, it consists of a trench sloped away from the foundation, lined with plastic and filled with crushed rock.

surfaced sizes Sized lumber may be rough, surfaced, or partly surfaced on one or more faces.

surface hinge A hinge, often ornamental, mounted on a face of a door rather than the edge.

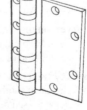
surface hinge

surface impoundment A lagoon or pond designed to hold waste materials and prevent their escape to the environment.

surface metal raceway Any surface-mounted furrow or channel constructed of metal to house and protect electrical conductors.

surface moisture (free water, surface water) Free water retained on surfaces of aggregate particles and considered to be part of the mixing water in concrete, as distinguished from absorbed moisture.

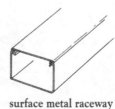
surface metal raceway

surface planer A machine used to plane and smooth the surface of materials such as wood, stone, or metal.

surface retarder A retarder applied to a form, or to the surface of freshly mixed concrete to delay setting of the cement in order to facilitate construction joint cleanup or to facilitate production of exposed aggregate finish.

surface tension That property, due to molecular forces, in the surface film of all liquids that tends to prevent the liquid from spreading.

surface vibrator A vibrator used for consolidating concrete by application to the top surface of a mass of freshly mixed concrete.

surface voids Cavities visible on the surface of a solid.

surface wiring switch An electric switch intended for mounting on a surface, such as a wall post, with most of the body of the switch exposed.

surfacing weld A deposit of weld metal on the surface of a material to provide desired dimensions or properties.

surfactant Chemicals that change the properties of surfaces with which they come in contact. In construction, surfactants are used as wetting or spreading agents and to facilitate the mixture of normally incompatible substances. They also affect the emulsification characteristics and alter the behavior (the dispersion, suspension, or precipitation) of pesticides in water. In environmental remediation, a surfactant is used to solubilize contaminants and mobilize the highly contaminated fines material.

surge A transient voltage effect that causes a sudden, undesirable rise or fall in current.

surge arrester A protective electrical device used to absorb overvoltage surges so that circuitry and equipment are not damaged.

surge pile or bin Stored aggregate or asphaltic concrete awaiting later use.

surge tank A tank in a water supply system used to absorb water during a sharp pressure rise or to supply water in a sudden pressure drop.

survey 1. A topographic or boundary mapping of a job site. 2. Taking measurements of an existing building. 3. The examination of the physical or chemical characteristics of the site.

surveying The measurement of distances, elevations, or angles of the earth's surface, including natural and man-made features; usually for the preparation of a plan or map.

surveyor An engineer or technician skilled in surveying.

survey stakes Small pieces of wood, usually 1″ × 2″ or 2″ × 2″ that have been cut and pointed for driving into the ground to mark a survey line or some boundary of construction.

survey traverse A sequence of lengths and angles between points on the earth established and measured by a surveyor. A survey traverse is used as a reference in making a detailed survey.

suspended acoustical ceiling A ceiling designed to be sound absorbent and to be hung from the structural slab or beams in an area.

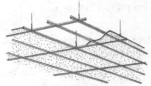

suspended acoustical ceiling

suspended ceiling (dropped ceiling) A finished ceiling suspended from a framework below the structural framework.

suspended ceiling (dropped ceiling)

suspended metal lath A system in which metal lath is attached to a light framing system that, in turn, is hung from a structural slab or beam.

suspended span A span supported by two cantilevers or a cantilever and a column or pier, used mainly in bridges and some roofs.

suspended-type furnace A unit furnace designed to be suspended from a floor or roof and to supply heated air through ducts to areas other than that in which the furnace is located.

suspension bridge A bridge with a deck suspended from cables that are typically raised on towers.

suspension bridge

suspension of work A situation in which the contractor must stop performance of the contract work.

suspension roof A roof system supported by cables attached to a frame.

sustainable design Selecting technologies and materials and using them in a manner that avoids depleting natural resources. Considers material durability and sustainability, which involves not only the energy and environmental costs to replace the materials, but availability and rate of regeneration.

swage 1. A tool or die used to bend or form cold metal. 2. A tool used to set the teeth of a saw by bending each one to the proper transverse angle.

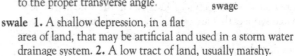

swage

swale 1. A shallow depression, in a flat area of land, that may be artificial and used in a storm water drainage system. 2. A low tract of land, usually marshy.

swan-neck 1. The connector between a gutter outlet and the downspout. 2. The curved portion of the handrail of a stair connecting the rail to a newel post.

swaybrace A diagonal brace used to resist wind or other lateral forces.

sweating 1. A soldering technique for joining metal parts. 2. A gloss that develops on a dull, or matte, paint or varnish film and is caused by rubbing the dry film. 3. A collection of condensation moisture on a surface that is below the dew point of the air.

sweat joints In plumbing, the union of two copper pipes made by coating the pipes with a tin-based solder, fitting them together, and applying heat to the joint, fusing the pipes together.

swedge bolt An anchor bolt used in concrete or mortar.

sweep 1. The curvature or bend in a log, pole, or piling; classified as a defect. 2. A bend in an electrical conduit.

sweep fitting Any plumbing fitting with a large radius curve.

sweep fitting

S

sweep strip A flexible weather-stripping used on the top and bottom edges of a revolving door.

swell factor The ratio of the volume of loose excavation material to the volume of the same material in place.

swelling A volume increase caused by wetting and/or chemical changes.

swift A reel or turntable on which pre-stressing tendons are placed to facilitate handling and placing.

swing 1. The movement of a door on hinges on a pivot. 2. The rotation of a power shovel on its base.

swing check valve A check valve with a hinged gate to permit fluid to flow in one direction only; used mostly where fluid velocities are low.

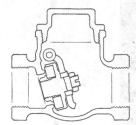

swing check valve

swinging latch bolt A latch bolt hinged to a lock or door and operated by swinging rather than sliding.

swinging scaffold A scaffold suspended from a roof by hooks, and a block and tackle.

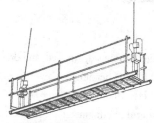

swinging scaffold

swing leaf 1. The active leaf in a double door. 2. A hinged sash in a casement window.

swing saw (pendulum saw) A power saw for ripping, deep cutting, or cross cutting that is mounted on a frame and pushed down into the material to be cut.

swing-up garage door A rigid overhead door that opens as one unit.

swirl finish (sweat finish) A nonskid texture imparted to a concrete surface during final troweling by keeping the trowel flat and using a rotary motion.

switch A device used to open, close, or change the connection of an electric circuit.

switch-and-receptacle An electrical outlet controlled by its own on-off switch. When the switch is turned off, the flow of electricity to the plug is terminated.

switchboard A large panel, frame, or assembly with switches, overcurrent and other protective devices, fuses, and instruments mounted on the face and/or back. Switchboards are usually accessible from front or rear and are not intended to be mounted in cabinets.

switchboard

switchgear Any switching and interrupting devices combined with associated control, regulating, metering, and protective devices, used primarily in connection with the generation, transmission, distribution, and conversion of electric power.

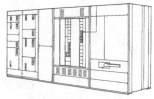

switchgear

switch plate A flush plate used to cover an electric switch.

swivel joint A special pipe fitting designed to be pressure-tight under continuous or intermittent movement of the equipment to which it is connected.

swivel spindle A spindle in a door latch unit with a joint in the middle of its length to allow one knob to be fixed by a stop; works while the other knob operates the latch.

synchronous motor A constant-speed electric motor, usually direct current. The operational speed is equal to the frequency of supply voltage divided by one-half the number of cycles or windings on the machine.

syneresis The contraction of a gel, usually evidenced by the separation from the gel of small amounts of liquid; a process possibly significant in the bleeding and cracking of fresh Portland cement mixtures.

synergy Interaction of substances, materials or systems that produces effects or results not caused by any of the individual substances or entities on its own.

synthetic gypsum A chemical product consisting primarily of calcium sulfate dehydrate derived primarily from an industrial process.

synthetic resin Any of a large number of products similar to natural resins and made either by polymerization or condensation or modifying a natural material.

synthetic rubber A chemically manufactured elastomer, rubber-like in its degree of elasticity.

system effect factor A pressure loss factor that recognizes the effect of fan inlet and outlet restrictions that influence fan performance.

Système International D'unités or International System of Units The contemporary metric system of international standard measures.

systems A process of combining prefabricated assemblies, components, and parts into single assembled units utilizing industrialized production, assembly, and other methods.

S

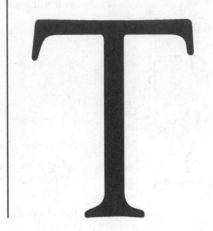

Abbreviations

t temperature, time, ton

T tee, township, true, thermostat

Ta tantalum

TAD thoroughly air dried

TAT turnaround time

tb turnbuckle

TB through bolt

TBE threaded both ends

TBR timber

T&C threaded and coupled

TC terra-cotta

TCE trichloroethylene

TCLP Toxicity Characteristic Leaching Procedure

TDD (TTY) Telecommunications Devices for the Deaf (teletypewriter)

Te tellurium

TE table of equipment, trailing edge

tech technical

TEL telephone

TEM Total Energy Management

temp temperature, temporary

TEMP temperature

TER terrazzo

TEU technical escort unit

tf tar felt

T&G, T and G tongue-and-groove

TG&B tongued, grooved, and beaded

tg&d tongued, grooved, and dressed

TH true heading

THK thick

therm thermometer

THERMO thermostat

thou thousand

thp thrust horsepower

THRU through

Ti titanium

TL transmission loss

tlr trailer

TM technical manual

tn ton, town, train

TN true north

tnpk, tpk turnpike

TNT trinitrotoluene

to take-off (estimate)

TOC total organic carbon

TOE thread one end

TOL tolerance

T-O-L Thread-O-Let

tonn tonnage

topog, topo topography

TOT total

tp tar paper

TPH total petroleum hydrocarbon

tps townships

tr tread

trans transom

TRANS transformer

transp transportation

trf tuned radio frequency

trib tributary

trib ar tributary area

ts tensile strength

TSCA Toxic Substance Control Act

TSD treatment, storage, and disposal

TSDF treatment, storage, and disposal facility

TSS total suspended solids

T&T truck and trailer

TTY text telephone (teletypewriter)

TU trade union, transmission unit

TUB tubing

TV terminal velocity

twp township

TYP typical

Definitions

tabby A type of concrete consisting of a mixture of lime and water, with shells, gravel, or stones used to make blocks for masonry.

tabia Compacted soil mixed together with lime and small stones.

tabled joint A joint in stonemasonry in which a projection cut into a stone fits into a channel cut into the stone below.

table joint Any of various joints in which the fitted surfaces are parallel to the edges of the pieces being joined, with a vertical break in the middle. This break is termed the table. Such joints are used in lengthening structural members.

tachometer A device for measuring rotation speed. Usually refers to revolutions per minute (RPM) of a revolving shaft.

tachymeter An instrument used by surveyors that determines distance, direction and elevation variation with one observation.

tack 1. A short, sharp-pointed nail with a large head used in laying linoleum and carpets. 2. A strip of metal, usually lead or copper, used to secure edges of sheet metal in roofing. 3. The property of an adhesive that allows it to form a strong bond as soon as the parts and adhesive are placed in contact. 4. To glue, weld, or otherwise fasten in spots.

tack coat A coat of emulsion that enhances the bonding of two layers of asphalt.

tack dry The state of an adhesive at which it will adhere to itself although it seems dry to the touch.

tack-free dry The point where paint or varnish is no longer sticky.

tackle A mechanism for moving materials, including raising or lowering them on a rope and pulley block.

tackless strip A metal or wood strip, with many small upstanding hooks, fastened to a floor or stair. A carpet is stretched beyond the strip. Hooks engage the backing and hold the carpet in place.

tack rag A rag treated with slow-drying or nondrying varnish or resin, and used to clean dirt and foreign matter from the surfaces of articles before they are painted.

tack rivet A rivet used to hold work steady during riveting.

tack weld 1. A temporary weld to position parts. 2. One of a series of short welds used where a continuous weld is unnecessary.

tag A strip of sheet metal folded on itself and used as a wedge to hold metal flashing in a masonry joint.

tagline 1. A line that runs from a crane boom to a clamshell bucket and keeps the bucket firmly in position during operations. 2. A safety line used by workers performing a job at high elevations or in other dangerous locations where a fall could be injurious. 3. Rope attached to a heavy member, such as a truss or beam, or a piece of equipment being lifted. The rope is used by a worker on the ground to control the item as it swings up.

tail 1. The bottom part of a slate shingle. 2. The part of the roof rafter that extends beyond the wall line. 3. The section of a stone step that extends into a wall. **tail**

tail bay 1. The area between a wall and the nearest column line. 2. In a framed floor or roof, the area between an end wall and the nearest girder.

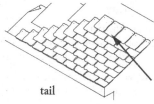

tail

tail beam A short beam or joist with one end set in a wall and the other supported by a header.

tail cut 1. A cut in the lower end of a rafter where it overhands the wall; sometimes ornamental. 2. The seat cut at the lower end of a rafter.

tail in 1. To secure or fasten one end of a timber, such as a floor joist, at a wall. 2. To secure one end or edge of a projecting unit of masonry, such as a cornice.

tailing iron A steel member, built into a wall, to take the upward thrust of a cantilevered member that projects from the wall.

tailings 1. Stones left on a screen used to grade material, as in a crushing operation. 2. The waste material or residue of a product.

tailpiece 1. A subordinate joist, rafter, or the like supported by a header joist at one end and a wall or sill at the other. 2. A handle on the bar end of a two-man power saw.

take-off (takeoff, quantity takeoff) 1. The process in which detailed lists are compiled, based on drawings and specifications, of all the material and equipment necessary to construct a project. The cost estimator uses this list to calculate how much it will cost to build the project. 2. The activity of determining quantities from drawings and specifications. 3. The actual quantity lists.

take-up The process of plaster substrate absorbing water from plaster during installation.

take-up block A pulley block rigged so that its weight or a spring will prevent slack in the lines passing through it.

talc A mineral with a greasy or soapy feel. It is very soft, having the composition $Mg_3Si_4O_{10}(OH)_2$.

talus 1. A slope formed of rock debris at the bottom of a steep slope. 2. The slope of a wall.

tambour door A sliding door made of thin folding panels on a flexible backing.

tamp To press a loose material such as soil or fresh concrete into a firm, compact mass by pounding repeatedly on it.

T

tamper **1.** An implement used to consolidate concrete or mortar in molds or forms. **2.** A hand-operated device for compacting flooring topping or other unformed concrete by impact from the dropped device in preparation for strikeoff and finishing. Contact surface often consists of a screen or a grid of bars to force coarse aggregates below the surface that prevents interference with floating or trowelling.

tandem **1.** A pair of moving parts in which one part follows the action of the other, such as tandem rollers. **2.** A vehicle made up of two units attached to one another, as the cab and trailer of a truck.

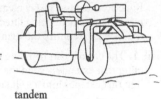

tandem

tandem wiring (master-slave wiring) A method of wiring two or more luminaires with one ballast. It saves on installation costs, and electricity.

tang The narrow extending tongue or prong on an object by which it is affixed to another piece, such as the tongue that secures a chisel to a handle.

tangent A straight line or curve that touches another curve at a single point without crossing the curve.

tangential In reference to lumber, it refers to sawing the truck or branch in a longitudinal fashion, perpendicular to the growth rings. Flat-grained lumber is produced by this method.

tank farm One contiguous site that may contain multiple tank fields, typically used for defining and classifying underground storage tanks and aboveground storage tanks.

tank field One continuous cluster of tanks.

tankless water heater (instantaneous heater) A water heater system that relys on instantaneous increases of power to heat water as it is demanded. It has only a mini-assembly of heater coils 36" high or less, and provides an endless amount of hot water. It stores only the water within the coils, and is consequently energy efficient.

tannin A soluble compound found in all woods that can cause staining.

tap **1.** A connection to a water supply. **2.** A faucet. **3.** A tool used to cut internal threads. **4.** To cut internal threads in a nut or hole.

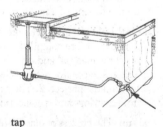

tap

tap bolt A machine bolt that screws into a hole in material without requiring a nut or internal threads. A fully threaded bolt.

tap borer A hand tool used to bore tapered holes, as in a lead pipe for a connection.

tape **1.** A flexible measuring strip of fabric or steel marked off with lines similar to the scale of a carpenter's rule, usually contained in a case to allow rewinding or retracting after use. **2.** One of a number of adhesive-backed fabric or paper strips of assorted width used for various purposes within construction systems.

tape balance A balancing device on a sash that counterbalances

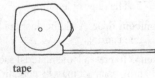

tape

the weight of the sash by force of a metal tape coiled around a spring-loaded reel.

tape correction A correction that is figured into a distance measured by a tape in order to compensate for errors resulting from the condition of the tape or the manner in which it was handled.

tape joint A flat joint that is reinforced with tape and sealed with joint compound to provide additional strength.

tape measure (tapeline) A steel strip used by builders and surveyors to measure distances, usually graded in feet, tenths, and hundredths of a foot for use in surveying and engineering. Or, graded in feet, inches, and fractions of an inch for use in the building trades.

taper A gradual diminution of thickness, diameter, or width in an object.

tapered edge strip A tapered insulation strip used to elevate roofing components at the roof's perimeter and penetrations.

tapered insulation Preformed rigid insulation used on roofs at drains or roof intersections requiring a slight change in elevations at the surface.

tapered tenon A tenon decreasing in width from the root to the projecting end.

tapershank The tapered end of a tool that fits into a socket.

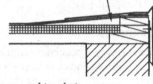

tapered insulation

tapersplit Shakes produced by using a mallet and froe, a sharp steel blade, to obtain split faces on both sides. The taper of the shake is a result of reversing the block and splitting the shake from a different end after each split.

taper thread A screw thread developed on a frustum of a cone, used in piping systems to ensure a tight joint, also used on some fasteners such as a screw or plug for a hole with worn threads.

taper tie A tapered steel rod used to hold concrete forms in place. When the forms are being stripped, the tie is removed by hitting the small diameter end and forcing it out through the hole created by the larger end.

tapia An adobe-like building material primarily made from earth or clay.

taping compound (embedding compound) A compound specifically formulated and manufactured for use in embedding of joint reinforcing tape at gypsum board joints.

taping knives Drywall finishing knives that range in width from 4" to 24". The smaller knives (4" to 6") are utilized for taping, spotting fasteners, taping angles and finishing. The wider blade knives are used in finishing. All have square corners.

taping strip **1.** A strip of roofing felt laid over the joint between adjacent precast concrete units before roofing operations. **2.** A strip of tape used to cover the joint between adjacent roof insulation boards.

tapped fitting Any fitting that has one or more tapped internal threads to receive threaded pipe.

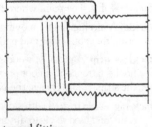

tapped fitting

T

tapped tee A cast-iron soil pipe tee with a tapped outlet.

tar A dark, glutinous oil distilled from coal, peat, shale, and resinous woods, used as surface binder in road construction and as a coating in roof installation.

tar-and-gravel roofing Built-up roofing made up of gravel or sand, poured over a heavy coating of coal-tar pitch applied to an underlayer of felt.

tar-and-gravel roofing

tar cement Heavier grades of tar for use in construction and maintenance of bituminous concrete pavements.

tare The weight of a rail car, truck, or other conveyance when empty. The tare is deducted from the gross weight of the vehicle and its contents to determine the weight of the freight in figuring freight rates.

target date Date imposed on an activity or project by the user or client that constrains or otherwise modifies the network analysis. There are two types: target start dates, and target finish dates.*

target hardening Prohibiting entry or access to a facility by installing security hardware such as keyed or tamper-proof window locks, deadbolts for doors, interior door hinges, and rock-guard shields for windows.

target leveling rod A leveling rod fitted with a target to facilitate setting or reading.

target schedule A schedule devised or selected as an objective measure against which actual performance can be gauged.*

tarpaulin A waterproof cloth, generally used in large sheets to cover and protect construction materials or other goods stored outdoors.

task 1. A cohesive, individual unit of work that is part of the total work needed to accomplish a project. **2.** Well-defined component of project work; a discrete work item. There are usually multiple tasks for one activity.*

task lighting Lighting directed to a specific work surface or area.

T-bar A light-gauge, T-shaped member used to support panels in a suspended acoustical ceiling.

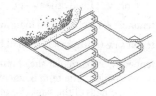

T-bar

TBE Threaded both ends. Term used when specifying or ordering cut measures of pipe.

T-beam A beam composed of a stem and a flange in the form of a T; usually of reinforced concrete or rolled metal.

T&C Threaded and coupled; an ordering designation for threaded pipe.

tearoff (ripoff) The stripping off of all roof coverings down to the wood.

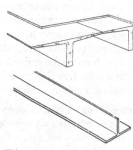

T-beam

teazel An angle post used in the construction of a timber-frame building.

teco A metal strap used to secure roof rafters and trusses to the top horizontal wall plate. Sometimes referred to as a hurricane clip.

tee 1. An elaborately turned finial in the shape of an umbrella; used as finishing ornamental on pagodas, stupas, and topes. **2.** A metal structural member with a T-shaped cross-section. **3.** A pipe fitting that has a side port at right angles to the run.

tee handle A T-shaped handle used in place of a doorknob to operate the bolt on a door lock.

tee iron 1. A piece of flat, heavy sheet metal, shaped into a T and equipped with drilled, countersunk holes; used to reinforce joints in wood construction. **2.** A steel T-beam section.

tee joint A joint between two members that intersect at right angles to form a T-shaped joint.

tee nut A fastening device that consists of an internally threaded cylinder and a disk with prongs around the perimeter of the bottom. The prongs are driven into wood to hold the nut steady while a bolt is inserted.

telegraphing Grain or other defects showing through on a smooth plywood panel.

telephone exchange The switching center for interconnecting telephone lines that terminate at the central station switch.

telescope To insert or slide one piece inside another.

telescoping boom A boom that can be extended as a telescope from within itself, eliminating the need to manually add sections to lengthen its reach.

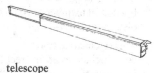

telescope

telltale Any device designed to indicate movement of formwork or a point along the length of a pile under load.

tellurometer An electronic instrument that measures distance by transmitting microwaves to an object and determining how fast they bounce back. The time measurement is then converted to feet and inches.

temper 1. To bring clay mortar or plaster to the proper consistency by moistening and mixing. **2.** To make stronger through processes of heating and cooling. In metals, stress is relieved without affecting hardness by heating to a temperature of 500°F. Other processes use 1,000°F to heat materials that are then allowed to cool naturally in order to soften them and reduce brittleness. **3.** The process of treating a wood surface with liquid resins or oils to increase resistance to water damage.

temperature, curing The temperature to which an assembly with adhesives is subjected in order to cure the adhesive.

temperature differential The difference in temperature between two spaces within a building, or between the indoor and outdoor temperature. Temperature differential causes natural convection currents and air to migrate through cracks and open doors, windows, or other means of egress.

temperature reinforcement Reinforcement designed to carry stresses resulting from temperature changes.

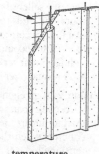

temperature rise The increase of temperature caused by absorption of heat or internal generation of heat, as by hydration of cement in concrete.

temperature reinforcement

temperature rise period The time interval during which the temperature of a concrete product rises at a controlled rate to the desired maximum in autoclave or atmospheric-pressure steam curing.

temperature stress Stress in a structure or a member due to changes or differentials in temperature in the structure or member.

tempered air Air that is warmed before being distributed.

tempered glass Glass that is prestressed by heating and then rapidly cooled, a process that makes it two to four times stronger than ordinary glass. It shatters into pebbles instead of slivers.

tempered water Warm water between 85°F and 110°F.

tempering The addition of water and mixing of concrete or mortar as necessary to bring it to the desired consistency during the prescribed mixing period. For mixed concrete, this will include any addition of water as may be necessary to bring the load to the correct slump on arrival at the work site, but not after a period of waiting to discharge the concrete.

template (templet) 1. A thin plate or board frame used as a guide in positioning or spacing form parts, reinforcement, or anchors. **2.** A full-size mold, pattern, or frame shaped to serve as a guide in forming or testing contour or shape. **3.** A member of stone, metal or wood that is embedded in a wall in order to support a horizontal member or beam to distribute the load.

template (templet)

temporary construction cost Includes costs of erecting, operating, and dismantling nonpermanent facilities, such as offices, workshops, etc., and providing associated services such as utilities.*

temporary shoring Shoring installed to support a structure while it is being built, and removed when construction is finished.

temporary stress A stress that may be produced in a precast concrete member or a precast concrete component during fabrication or erection, or in cast-in-place concrete structures due to construction or test loadings.

temporary shoring

tenant's improvements Improvements made to a house or parcel of land by a tenant at his own expense that commonly become a part of the property and cannot be removed by a departing tenant unless the owner gives his consent.

tender 1. An offer showing a willingness to buy or sell at a specific price and under specific conditions. **2.** In futures, the act on the part of a seller of a contract of giving notice to the clearinghouse that he intends to deliver the physical commodity in satisfaction of a futures contract. **3.** A hooktender. **4.** A railroad car attached to a steam locomotive for carrying fuel and water, or a ship providing similar services in a fleet. **5.** A formal offer of a bid.

tendon A steel element such as a wire, cable, bar, rod, or strand used to impart prestress to concrete when the element is tensioned.

tendon profile The path or trajectory of the prestressing tendon.

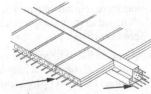
tendon

tenon A projecting, tongue-like part of a wood member designed to be inserted into a slot or mortise of another member to form a mortise and tenon joint.

tenon-and-slot mortise A glued wood joint formed by a tenon and mortise. Usually the two pieces join at a right angle to each other.

tenoner A machine used to form tenons.

tenon saw (mitre saw) A small backsaw used for cutting tenons.

tenpenny A particular size of nail, generally 3″ long with 69 of them to a pound. The term is believed to have originated in England, where nails of the same size once sold for ten pence per hundred.

tensile strain Elongation of a material that has been subjected to tension by pulling or stretching.

tensile strength Maximum unit stress that a material is capable of resisting under axial tensile loading, based on the cross-sectional area of the specimen before loading. Tensile strength is often used as a measure of fastener strength; it is the load required to pull open the fastener.

tensile stress Stress resulting from tension.

tensile test A test in which a sample is subjected to increasing longitudinal pulling stress until the material fractures.

tension The state or condition imposed on a material or structural member by pulling or stretching.

tension member A tie or other structural member subjected to tension.

tension reinforcement Reinforcement designed to carry tensile stresses such as those in the bottom of a simple beam.

tension rod A rod in a truss or structure that connects opposing components and prevents them from spreading.

tension wood Defective wood usually cut from the upper side of hardwood branches or leaning trunks. The processed lumber exhibits high longitudinal shrinkage, causing warping and splitting.

teratogen A term used in labeling hazardous construction or other materials that adversely affect the human reproductive system.

teredo A marine worm that damages untreated wood by boring holes in it.

terminal 1. An element attached to the end of a conductor or to a piece of electric equipment to serve as a connection for an external conductor. **2.** A decorative element forming the end of an item of construction. **3.** A point of departure or arrival such as a railway or airport terminal.

terminal

terminal box A box, on a piece of electrical equipment, that contains leads from the equipment, ready for connection to a power source. The box is usually provided with a removable cover.

terminal landing The top and bottom elevator landing. (Courtesy of National Elevator Industry, Inc.)

terminal unit 1. A unit, at the end of a duct in an air-conditioning system, through which air is delivered to the conditioned space. **2.** Devices located near the conditioned space that regulate the temperature and/or volume of supply air to the space.

terminal velocity The average speed of an airstream in an air-conditioning system as it reaches the end of its throw, used as an indicator of comfort level and degree of draftiness.

termites Insects that destroy wood by eating the wood fiber. Termites are social insects that exist in most parts of the United States, but they are most destructive in the coastal states and in the Southwest. Termites can enter wood through the ground or above the ground, although the subterranean type is most common in the United States. They eat the softer springwood first and prefer sapwood over heartwood.

termite shield A sheet of metal used on a foundation wall or pier as a projecting shield to prevent the passage of termites from the ground to a structure.

terms of payment Defines a specific time schedule for payment of goods and services and usually forms the basis for any contract price adjustments on those contracts that are subject to escalation.*

terne metal An alloy of lead, composed of up to 20% tin.

terneplate Sheet steel coated with an alloy of lead and tin, used chiefly in roofing.

terrace 1. An embankment with a level top surface and stabilized side slopes. It may be used for agriculture or paved and/or planted for recreational use. **2.** A platform or paved embankment adjoining a building and used for recreational purposes.

terrace

terra-cotta Units of hard, unglazed fired clay, used for ornamental masonry.

terrazzo A type of flooring material made from marble or other stone chips set in Portland cement and polished when dry.

tertiary beam A beam that transfers its load to a secondary beam.

tertiary treatment An advanced stage of the wastewater treatment process, employing methods such as ion exchange, carbon absorption, reverse osmosis, and demineralization of residual solids.

terrazzo

tessera A small square of glass, stone, tile, or marble used in geometric and figurative mosaic work in pavements or floors.

test A trial, examination, observation, or evaluation used as a means of measuring a physical or chemical characteristic of a material, or a physical characteristic of a structural element or a structure.

test cylinder A sample of a concrete mix, cast in a standard cylindrical shape, cured under controlled or job conditions and used to determine the compressive strength of the mix after a specified time interval.

test method The testing procedures used to determine whether a specific product conforms with a particular standard.

test piling A foundation piling that is installed on the site of a proposed construction project and used to conduct load tests to determine the size and quantity of pilings needed for the actual structures.

test pit An excavation made to examine the subsurface conditions on a potential construction site. Samples are taken at specified elevations for lab analysis.

test plug A device that contains water test pressure in drainage pipes as part of a process to check plumbing systems for leakage.

test pressure Air pressure or water pressure used in plumbing to test pipes and fittings for strength and water-tightness.

test tee A pipe tee that is inserted into a drainage system to test it for leaks by subjecting it to water pressure.

texture 1. The pattern or configuration apparent in an exposed surface, as of concrete or mortar, including roughness, streaking, striation, or departure from flatness. **2.** In referring to lumber, it is the grain of the wood.

texture brick A rough, fired clay brick, often multicolored, commonly used for facing work.

texture

texture cracking Cracking tensile failure, caused by temperature drop in members subjected to external restraints, or temperature differential in members subjected to internal restraints.

texturing The process of producing a special texture on unhardened or hardened concrete.

T&G joint See **tongue-and-groove joint**.

thatch A roof covering made of straw, red palm leaves, or similar material fastened together to shed water and provide thermal insulation.

T-head 1. In precast framing, a segment of girder crossing the top of an interior column. **2.** The top of a shore formed with a braced horizontal member projecting on two sides and forming a T-shaped assembly.

theodolite An instrument used in precise surveying consisting of an alidade that is equipped with a telescope, a leveling device, and an accurately graduated horizontal circle. A theodolite may also have an accurately graduated vertical circle.

therm A quantity of heat equal to 100,000 Btus.

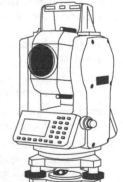
theodolite

thermal break (thermal barrier) An element of low conductivity placed between two conductive materials to limit heat flow; for use in metal windows or curtain walls that are to be used in cold climates.

thermal bridge The interruption of a layer of thermal insulation by another material with high thermal conductivity (such as metal).

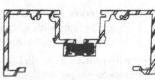
thermal break (thermal barrier)

thermal coefficient of expansion A measure of the expected amount of expansion of a material when heated.

315

thermal conductance The rate at which heat flows from one surface of a material to the other, usually measured or specified at the rate over a unit area and under a unit temperature differential.

thermal conduction The process of heat transfer through a material by internal molecular action.

thermal conductivity A property of a homogeneous body measured by the ratio of the steady state heat flux, time rate of heat flow per unit area, to the temperature gradient, temperature difference per unit length of heat flow path, in the direction perpendicular to the area.

thermal cutout An overcurrent protective device containing a heater element and a renewable fusible member that opens the circuit. The thermal cutout is not designed to interrupt short-circuit currents.

thermal desorption A process that removes organic contaminants (as vapors or condensed liquids) from soils, sludges, and other solid media, which may then be destroyed in a permitted incinerator or used as supplemental fuel.

thermal diffusivity Thermal conductivity divided by the product of specific heat and unit weight. The term is an index of the facility with which a material undergoes temperature change.

thermal expansion The change in length or volume experienced by a material or mass when subjected to a change in temperature.

thermal factor A factor applied in doing lighting calculations that compensates for changes in the light given off by a fluorescent lamp when the ballast/lamp configuration varies from that used in photometric tests.

thermal insulating cement A dry mixture of cement and granular, flaky, fibrous, or powdery materials of low conductivity; when mixed with water and applied to a surface, it dries to provide an insulating covering.

thermal insulation (heat insulation) A material that provides a high resistance to heat flow. Examples are foamed plastics, mineral or glass fibers, cork, and foamed glass. The material is used in the form of blankets, boards, blocks, and poured or granular fill.

thermal lag The time required to add or remove heat from a mass before it reaches the design set point temperature.

thermal limit switch Automatically disconnects the electric supply if air flow through a heater is reduced below a safe level.

thermal mass cooling A method suited to climates with wide diurnal temperature swings. It involves running cool nighttime air across a large indoor building mass, such as a slab. The cool thermal mass then absorbs heat during the day.

thermal movement Changes in dimension of masonry or concrete resulting from fluctuations of temperature over time.

thermal oxidation A process that heats a gas stream to a sufficiently high temperature with adequate residence time to oxidize the hydrocarbons to carbon dioxide and water.

thermally protected A label appearing on the nameplate of a motor or motor-compressor that identifies it as being provided with a thermal protector. *See also* **thermal protector**.

thermal protector A protective device consisting of one or more sensing elements and an external control device to guard against overheating that can result from overload or failure to start.

thermal radiation The transmission of heat through electromagnetic waves from a warm surface to a cooler one.

thermal regenerator A device within a heating duct that is used to recover excess heat and send it to other areas.

thermal resistance (R value) The resistance of a material to the transmission of heat.

thermal shock The subjection of a material or body, such as partially hardened concrete, to a rapid change in temperature that may be expected to have a potentially deleterious effect.

thermal storage capacity A material's capacity to store the sun's heat for later use.

thermal stress (temperature stress) Stress induced in an object or structural member by restraint against movement required to accommodate temperature changes.

thermal transmittance (U-value) The measure of the rate of heat flow per unit area under steady conditions from the fluid on the warm side of a barrier to the fluid on the cold side, per unit temperature difference between the fluids.

thermal unit A unit of heat energy, usually the British thermal unit (Btu) in the English system or the calorie in the metric system.

thermal valve A valve with an activating element that responds to temperature or rate of temperature change.

thermite welding A welding process that uses heated liquid metal and slag resulting from the ignition of a mixture of ferric oxide and aluminum particles.

thermocouple A device for measuring temperature that has two junctions of two dissimilar metals. When the two junctions are at different temperatures, the junction connected to an instrument, such as a voltmeter, generates voltage.

thermoelectric Conversion of electricity to heat or heat to electricity.

thermography Process used to photograph heat being emitted from a structure. The photographs are in color, and the rate of heat loss related to particular colors.

thermomechanical pulping (TMP) A process in which wood chips are heated and softened by steam before being ground into fibers.

thermometer well A specially designed enclosure connected into the piping system and into which a thermometer can be placed to measure the temperature of a fluid in the piping.

thermopiles A support system used in permafrost that relies on temperature differences and natural convection for removing heat from the ground to keep it frozen or refrozen.

thermoplastic Becoming soft when heated and hard when cooled.

thermosetting Becoming rigid by chemical reaction and not remeltable.

thermostat An electric switch controlled by an element that responds to temperature; used in heating and/or cooling systems.

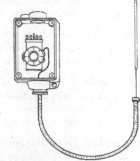

thermostatic expansion valve A valve regulating the flow of volatile   **thermostat**

refrigerant to a cooling unit, activated by changes in cooling unit pressure and superheat of the refrigerant leaving the cooling unit.

thermostatic trap A steam trap using a thermally actuated element to expand and close a discharge port when a designed amount of steam flows through it, and to contract and allow condensate to flow through as the temperature drops; usually used on steam radiators.

thermosyphon system A solar-powered water heating system that requires no pump or electricity to run.

thick-lift asphalt pavement A paving process in which the asphalt course is placed in one or more lifts of 4″ or more compacted thickness.

thimble 1. A protective sleeve in a part intended to hold an item supported by, or passing through, the part. **2.** A protective sleeve of metal in the wall of a chimney used to hold the end of a stovepipe or smoke pipe.

thin butt A defect in western red cedar shakes in which the butt end of the shake fails to meet minimum thickness requirements.

thin edge A defect in shakes in which the thickness of 24″ shakes, within 10″ of the butt, is less than half the minimum specified thickness.

T-hinge (tee-hinge) A T-shaped, surface-mounted hinge. The crossarm of the T is fixed to a door frame or post while the length of the T supports a door or gate.

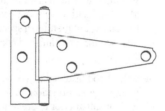

T-hinge (tee-hinge)

thinner Any volatile liquid used to lower the viscosity of a paint, adhesive, or other like material.

thin-set Descriptive of bonding materials for tile that are applied in a layer approximately ⅛″ (3mm) thick.

thin-shell concrete Reinforced or prestressed concrete used to form a large shell. The thickness of the concrete is small relative to the span of the shell.

thin-wall conduit Electric conduit with a wall thickness that will not support threads. Sections are joined by couplings held in place by setscrews.

third clear (Factory Select, #3 clear) The highest grade of shop lumber. This grade will yield a high percentage of cuttings, but represents a small portion of a mill's total shop production.

third-party commissioning Independent assessment of systems to ensure that their installation and operation meets design specifications and that they are as efficient as possible.

thixotropy The property of a material that enables it to stiffen quickly while standing, but acquire lower viscosity with mechanical agitation, such as certain gels. Material having this property is called thixotropic.

thoroughly air-dried (TAD) Lumber that is air-dried sufficiently to meet the grading rule requirements for dry lumber.

thread A ridge, of uniform cross section, following a helix on the external or internal surface of a cylinder.

thread-cutting screw Screw that cuts threads in metal and joins metal sections.

threaded anchorage An anchoring device that is provided with threads to facilitate attaching the jacking device and to effect the anchorage.

threaded insert A device forced into a predrilled hole to form screw threads.

threaded joint A mechanical joint between threaded pipes.

threaded stud A fastening device with one pointed end driven into a material, such as concrete, and the other end is threaded and extends above the surface for the attachment of structural members.

threads, classes Loose tolerance threads are Class 1. Threads with Class 2 listing are usually stainless steel with normal commercial tolerances. Class 3 are made to stricter tolerances and fit much more tightly. When letter "A" is added to a class, it indicates external threads, or screws. Letter "B" means internal threads or nuts.

three-coat work The application of three coats of plaster: scratch coat, brown coat, and finish coat.

three-hinged arch An arch designed in two parts that are pin-connected, or hinged, to each other and to their supports. Such arches are designed for convenience of transportation and erection.

three-phase line A conductor installation that is able to carry heavy loads of electricity.

three-phase motor Induction motor in which the voltage varies from phase to phase by 120°.

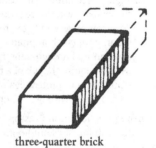

three-point lock An assembly that latches the active leaf or a pair of doors at three points.

three-quarter brick A brick that has a length equal to approximately three-quarters that of a normal brick.

three-quarter brick

three-quarter turn Descriptive of a stair that turns through 270° in its progress from top to bottom.

three-way strap A metal strap used to tie three members of a wood truss together at a joint.

three-way switch An electric switch used to control lights from two different points, as from two different ends of a hallway.

three-way valve A valve with two outlets and one inlet or two inlets and one outlet.

three-wire system A system of electric power supply consisting of three conductors, one of which, the neutral wire, is maintained at a potential midway between the other two.

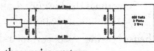

three-wire system

T

threshold 1. A shaped strip on the floor between the jambs of a door; used to separate different types of flooring, or to provide weather protection at an exterior door. **2.** The level of lighting or volume of illumination that permits an object to be seen a specified percentage of the time with specified accuracy.

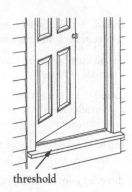

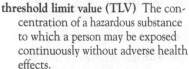
threshold

threshold limit value (TLV) The concentration of a hazardous substance to which a person may be exposed continuously without adverse health effects.

thriebeam Often used as bridge railing, it is steel plate (usually galvanized) with lengthwise raised areas or corrugations.

throat 1. A groove cut in the underside of an exterior projecting piece, such as a sill or coping, to prevent water running back across the underside to a wall. **2.** The thinnest part supporting a stair stringer. **3.** The thinnest area of concrete stairs—between the intersection of the back of the tread and bottom of the riser and the underside of the soffit. **4.** The thinnest part of a weld. **5.** The opening in a wood plane hand tool where the shavings pass through.

through bolt A bolt that passes completely through the members it connects.

through bond The transverse bond formed by masonry units extending through a wall.

through lintel A lintel having thickness equal to that of the wall in which it is placed.

through-penetration An opening through a fire-resistive partition or floor/ceiling assembly to provide for an item (such as piping) to pass through it. Through-penetrations usually require the use of a firestop system to protect against the spread of fire through the opening.

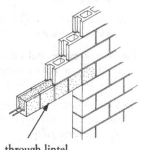
through lintel

through-penetration firestop A system for sealing through-penetrations in fire-resistant floors, walls, and ceilings.

through tenon A tenon extending completely through the part in which the mortise is cut.

through-wall flashing A flashing that extends completely through a wall, as at a parapet.

throw 1. The distance a latch bolt extends. **2.** The horizontal or vertical distance an airstream travels after leaving an outlet until its velocity is reduced to a specific value. **3.** The effective distance between a fixture and the area being illuminated.

throwout bearing A bearing that rides on the clutch jackshaft, that carries the engage-and-disengage mechanism.

thrust 1. The amount of force or push exerted by or on a structure. Sometimes the horizontal component of that force. **2.** In an arch, the resultant force normal to any cross-section of the arch.

thrust bearing A support for a shaft that is designed to resist its end thrust.

thrust block Additional support introduced at the point of intersection between a brace and its support.

thrust washer A washer that holds a rotating part in place and keeps it from moving sideways relative to its bearings.

thumb nail bead Refers to quarter-round molding commonly found on the inside edge of rails and stiles on glazed or raised panel doors.

thumbscrew A screw that has a head that is either curled or flattened so it can be turned with a thumb and fingers.

thumbturn A small lever attached to a bolt lock that operates the lock much like a key by turning it.

tie 1. A member or device that keeps two separate parts together, i.e., tie beam. **2.** Loop of reinforcing bars encircling the longitudinal steel in columns. **3.** A tensile unit adapted to holding concrete forms secure against the lateral pressure of unhardened concrete, with or without provision for spacing the forms a definite distance apart, and with or without provision for removal of metal to a specified distance from the finished concrete surface.

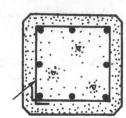
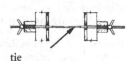
tie

tieback A rod fastened to a deadman, a rigid foundation, or a rock or soil anchor to prevent lateral movement of formwork, sheet pile walls, retaining walls, or bulkheads.

tie bar 1. Bar at right angles to, and tied to, minimum reinforcement to keep it in place. **2.** Bar extending across a construction joint.

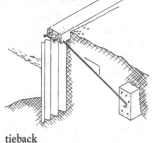
tieback

tie beam 1. A concrete beam that connects individual pile caps or spread footings. **2.** A horizontal timber that connects the lower end of two opposite rafters to prevent spreading.

tie plate A plate used to join two components or parallel parts of a built-up structural-steel member.

tie wall A wall built perpendicular to a spandrel wall for lateral stability.

tie wire 1. A wire used to hold forms together so they will not spread when filled with concrete. **2.** A single-strand wire used to tie reinforcing in place or metal lath to a column.

tige The shaft of a column, from the base moldings to the capital.

tight buildings Buildings that are designed to minimize infiltration air in order to reduce heating and cooling costs.

tight knot A knot in a piece of lumber that is sound and poses no detriment to the proper use of the wood member.

tight sheathing 1. Tongue-and-groove or matched boards nailed to rafters or studs that may run at an angle to provide stiffness to

T

the roof or wall. **2.** Excavation sheathing with the vertical planks interlocked for use in saturated soils.

tile A thin rectangular unit used as a finish for walls, floors or roofs, such as ceramic tile, structural clay tile, asphalt tile, cork tile, resilient tile, and roofing tile.

tile-and-a-half tile A roof tile that is the same length as the other tiles on a roof, but 1½ times as wide.

tileboard **1.** A wallboard with a factory-applied facing that is hard, glossy, and decorated to simulate tile. **2.** A square or rectangular board of compressed wood or vegetable fibers, used for ceiling or wall facings.

tile creasing A water-shedding barrier at the top of a brick wall consisting of two courses of tile that project beyond both faces of the wall.

tile field A system of distribution tile normally associated with a septic tank and leaching field system.

tile field

tile pin A pin that is passed through a roofing tile into the sheathing underneath to hold tile in place.

tile tie A section of heavy wire used to help secure roof tile.

till (glacial till, boulder clay) An unstratified glacial deposit consisting of compacted pockets of clay, silt, sand, gravel, and boulders, usually having good bearing capacity.

tilting fillet (cant strip, doubling piece) A thin wedge placed under the eave course of shingles or tiles to shed water more effectively.

tilting level A surveyor's level with a bubble mounted on the telescope and a provision for slight tilting of the telescope and level. The upright axis of the unit does not need to be vertical, but the level and telescope must be precisely aligned.

tilting mixer A small mixer for concrete or mortar that is emptied by tilting the mixer about a horizontal pivot.

tilting mixer

tilt-up (tilt-up construction) A method of concrete construction in which members are cast horizontally at a location adjacent to their eventual position and tilted into place after removal of forms.

tilt-up

timber **1.** Square-sawn lumber having a minimum nominal dimension of 5″ in the United States, or approximately equal cross-dimension greater than 4″ × 4½″ or 10 cm × 11 cm in Britain. **2.** Any heavy wood beam used for shoring or bracing.

timber connector One of a variety of metal connectors used in conjunction with bolts to form connections of timbers. Usually the bolt holds the timbers together while the connector prevents slippage.

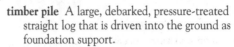

timber connector

timber-framed building A building that has timbers for above-ground structural elements (except foundations).

timber pile A large, debarked, pressure-treated straight log that is driven into the ground as foundation support.

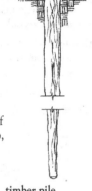

time-delay fuse A fuse in an electric circuit that takes more than 12 seconds to open at 200% load.

time-dependent deformation Combined effects of autogenous volume change, contraction, creep, expansion, shrinkage, and swelling occurring during an appreciable period of time, not synonymous with inelastic behavior or volume change.

timber pile

timekeeper A representative of the contractor who keeps records of hours worked by employees of the contractor and allocates the hours to various parts of the work.

time line Schedule line showing key dates and planned events.*

timely completion Completion of the specified work, or an agreed upon portion of the work, within the required time limits.

time of completion A specific date stated in the construction contract for substantial completion of the work.

time of haul In production of ready-mixed concrete, the period from first contact between mixing water and cement until completion of discharge of the freshly mixed concrete.

time sensitive costs Costs defined by the duration for which they are needed. The longer the particular item (such as a dumpster) is on the job site, the higher the cost.

time system A system of clocks and control devices, with or without a master timepiece, that will display current time at various locations and may include devices to program other systems, such as bells.

time-temperature curve Rate at which the temperature increases in a fire-testing furnace. Developed by ASTM, NFPA and UL, this curve is adhered to in all fire-resistive testing.

time value of money **1.** The time-dependent value of money stemming both from changes in the purchasing power of money (i.e., inflation or deflation), and from the real earning potential of alternative investments over time. **2.** The cumulative effect of elapsed time on the money value of an event, based on the earning power of equivalent invested funds. **3.** The expected interest rate that capital should or will earn.*

time weighted average (TWA) The average exposure of a person during eight hours to an airborne chemical hazard.

tin A lustrous, white, malleable metal with a low melting point, highly resistant to corrosion; used to make alloys and solder, and to coat sheet metal. When it is added to brass alloys, it provides greater strength and hardness, and also provides corrosion resistance from salt water.

T

tin-cap A small, flat metal washer used under roofing nails.

tin-clad fire door A door of two or three plywood planks, core covered with metal sheets and constructed in accordance with specifications of labeling authorities.

tingle A flexible metal clip designed to hold a sheet of metal or glass.

tinker's dam Small dam of sand or other non-combustible material that contains the spread of molten solder.

tinning Coating metal with a tin alloy for corrosion protection or as a presoldering procedure.

tinplate Sheet steel or iron that has been coated with tin as protection against corrosion.

tin roofing A roof covering of tinplate or terneplate.

tin-clad fire door

tin roofing

tin saw A saw used for cutting brick kerfs.

tin snips Strong shears with a blunt nose, used to cut sheet metal.

tint A light color made by diluting a color with white.

tip up A downspout extension that directs water away from the building.

titanium A lightweight, silver-gray metal that is strong and quite expensive. It is often added to other metals. Its properties include corrosion resistance against chlorides, acids and salt water.

title sheet The first page of a set of blueprints that usually includes location, name, a plot or sketch of the project, designer's name, and a list of the prints.

tobacco juicing Term given to the brown residue that comes out of asphalt products during weathering. Typically, it is seen on asphalt roofing shingles after periods of sun and rain in the first year exposed to the weather. It often disappears after that time.

tobermorite 1. A mineral found in northern Ireland and elsewhere, having the formula $5CaO \cdot 6SiO_2 \cdot 5H_2O$. **2.** Ca:4H O, the artificial product tobermorite, G, of Brunauer, a hydrated calcium silicate having CaO/SiO ratio in the range 1.39 to 1.75. It forms minute layered crystals that constitute the principal cementing medium in Portland cement concrete, a mineral with 5 mols of lime to 6 mols of silica, usually occurring in plate-like crystals that are easily synthesized at steam pressures of about 100 psi and higher. Tobermorite is the binder in several properly autoclaved products.

tobermorite gel The binder of concrete cured moist or in atmospheric-pressure steam. A lime-rich gel-like solid containing 1.5 to 2.0 mols of lime per mol of silica.

toe 1. Any projection from the base of a construction or object to give it increased bearing and stability. **2.** That part of the base of a retaining wall that projects beyond the face away from the retained material. **3.** The lower portion of the lock stile. **4.** The junction between the base metal and the face of a filled weld. **5.** To drive a nail at an oblique angle. **6.** That portion of sheeting below the excavated material. **7.** The part of a blasting hole furthest from the face.

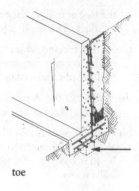

toe

toeboard A vertical barrier at floor level erected along exposed edges of a floor opening, wall opening, platform, runway, or ramp to prevent falls of materials.

toehold A batten or board temporarily nailed to a sloping roof as a footing for workers.

toe joint A joint between a horizontal timber and another angle from the horizontal, as between a rafter and a plate.

toenail 1. To drive a nail at an angle. **2.** A nail driven diagonally from a stud into another framing member.

toenailing Fastening a piece of lumber by driving nails obliquely to the surface. Alternate nails may be opposing to increase holding power.

toeplate A metal bar fastened to an outer edge of grating or the rear of a tread, and projecting above the surface to form a lip. *See also* **kickplate**.

toe wall A low wall built at the bottom of an embankment for greater stability.

toggle bolt A bolt and nut assembly used to fasten objects to a hollow wall or a wall accessible from only one side. The nut has pivoted wings that close against a spring when the nut end of the assembly is pushed through a hole and is open on the other side.

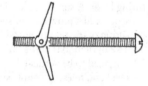

toggle bolt

toggle switch A lever-actuated snap switch.

toilet enclosure A compartment placed around a water closet for privacy.

toilet partition One of the panels forming a toilet enclosure.

tolerance 1. The permitted variation from a given dimension or quantity. **2.** The range of variation permitted in maintaining a specified dimension. **3.** A permitted variation from location or alignment.

toilet enclosure

tom *See* **shore**.

ton 1. A measure of weight equal to 2,000 lbs. or 907.2 kg. **2.** In cooling systems, a measurement of chiller size equal to 12,000 Btus of heat removal per hour.

T

tongue One edge of a piece of lumber that has been rabbeted from opposite faces, leaving a projection intended to fit into a groove cut into another board.

tongue and groove **1.** Lumber machined to have a groove on one side and a protruding tongue on the other, so that pieces will fit snugly together, with the tongue of one fitting into the groove of the other. **2.** A type of lumber or precast concrete pile having mated projecting and grooved edges to provide a tight fit, abbreviated "T&G."

tongue and groove

tongue-and-groove joint (T&G joint) A joint made by fitting a structural, cast-in-place surface for metal. The joint may also be welded. For plastic or wood pieces, the joint may be glued.

tongue-and-lip joint (tongue joint) A type of tongue-and-groove joint, except the tongue is wedge-shaped and the groove is tapered to receive it.

tongued miter A miter joint incorporating a tongue.

ton of refrigeration A measure of refrigerating effect equal to 12,000 Btu per hour.

tooled finish (tooled surface) A stone finish with the surface having 2 to 12 concave grooves per inch.

tooled joint A masonry joint in which the mortar has been shaped or worked before it sets.

tooling time The period of time after applying sealant in a joint during which it is possible to tool.

tooth A fine texture in a paint film provided by pigments or by abrasives used in sanding, providing a base for adhesion of a second coat.

toothed plate (bulldog plate, toothed gusset) A punched metal plate in which the punched metal protrudes from one side, forming teeth. Toothed plates are used for timber connections.

toothed ring A toothed ring that serves as a timber connector, generally used in the manufacture of large member wood trusses.

toothing Cutting or chipping out courses in old work as a bond for new work.

toothing plane A plane with a serrated blade mounted nearly vertically, used to roughen a surface prior to application of a veneer.

top beam A collar beam.

top car clearance The clearance between the top of an elevator car, or crosshead if provided, and the lowest overhead obstruction when the car is level with the top terminal landing.

top chord The upper section of a truss.

top coat The final coat in a paint system.

top cut The vertical cut at the top of a rafter.

top form Form required on the upper or outer surface of a sloping slab or thin shell.

top hung-in window An awning window that pivots at the top, and the bottom swings in.

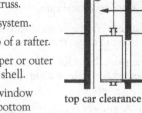

top car clearance

top lap The shortest distance between the lower edge of an overlapping roof shingle and the uppermost edge of the one in the course directly below.

topographic survey A review process and record of information regarding the surface conditions of a proposed construction site. The resulting drawing normally uses contour lines to convey the surface height or depth relative to a given level.

topping **1.** A layer of concrete or mortar placed to form a floor surface on a concrete base. **2.** A structural, cast-in-place surface for precast floor and roof systems. **3.** The mixture of marble chips and matrix that, when properly processed, produces a terrazzo surface.

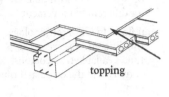

topping

topping compound A compound specifically formulated and manufactured for use over taping or all-purpose compounds to provide a smooth, level surface for applying decoration.

top plate A member on top of a stud wall on which joists rest to support an additional floor or to form a ceiling.

top plate

topsoil The surface layer of soil, usually containing organic matter, a mixture of particle sizes, and some animal life.

torch brazing A brazing process in which the heat is supplied by a torch.

torching The application of a lime mortar under the up-slope edges of roof tiles or slates. In full torching, the exposed underside of slates between battens is mortared.

torch soldering Soldering, with a gas flame supplying the required heat.

torpedo level Hand level, which may have a laser, that is used to level and plumb short members.

torque **1.** The force used to rotate something. **2.** Turning or twisting energy measured as the product of a force and a lever arm. **3.** That which tends to produce rotation.

torque converter A hydraulic coupling that introduces slippage to increase torque.

torque viscometer An apparatus used for measuring the consistency of slurries in which the energy required to rotate a device suspended in a rotating cup is proportional to viscosity.

torque wrench A tool used to turn nuts, bolts, and other similarly threaded fasteners. Unlike other types of wrenches, torque wrenches have a built-in device that measures the amount of torque (the turning or twisting force) applied to a nut or bolt. The device may be an audible signal, a scale built into the handle, or a dial, calibrated scale, or light. Torque wrenches limit the maximum amount of force that can be used, thereby preventing damage to the fastener and the material to which it is fastened.

torsel A piece of timber, steel, or stone placed under one end of a beam or joist to distribute its load.

torsion The twisting of a structural member by two equal and opposite torques.

T

torsion strength The measure of resistance of a material to twisting about an axis. For example, the amount of force needed to twist a fastener apart.

torus roll A joint in sheet metal or lead roofing at the intersection of two surfaces with different slopes. The joint is formed to allow for differential movement.

total float In CPM terminology, the difference between the time available to accomplish an activity and the estimated time required.

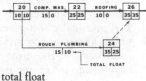

total float

total harmonic distortion (THD) A measure of the distortion of sinusoidal waves by harmonics. Fluorescent fixtures reflect harmonics back into the electrical supply system.

total heat The sum of both latent and sensible heat.

total pressure (impact pressure) The sum of velocity pressure and static pressure in a moving liquid. Measurement is expressed in inches of water.

total rise of a roof The vertical distance between the plate and the ridge of a roof.

total run The distance covered by a rafter including any overhang.

total suspended solids (TTS) A measure of the solid or particulate content of a liquid.

touch sanding A light surface sanding.

tower crane A crane with a fixed vertical mast that is topped by a rotating boom and equipped with a winch for hoisting and lowering loads. The winch can be moved along the boom so that any location within the diameter of the boom can be reached.

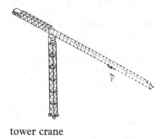

tower crane

tower equalizing lines Pipes installed to directly connect adjacent cooling tower basins to maintain a common basin water level.

toxic Causing an adverse health effect.

toxic agent A substance that causes an adverse health effect in humans.

TPH test A test used to determine the total quantity of petroleum hydrocarbons in a sample of soil or water.

T-plate A flat metal plate in the shape of a T, used to join or strengthen a right angle joint made between two timbers.

trabeated system A system of building construction using beams or lintels supported by columns.

trabeation (entablature) A type of construction utilizing horizontal members supported by columns. Trabeation is a common method of construction in Japan.

track 1. A light gauge U-shaped metal member attached to a floor and used to anchor studs for a partition. **2.** A U-shaped member attached to a floor,

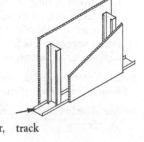

track

ceiling, door or window header; used as a guide for a sliding or folding partition, door, or curtain. **3.** A pair of special structural shapes with fastenings or ties for a craneway, movable wall, or railroad. **4.** An electrical raceway that allows the placement and use of a variety of types of luminaires along it.

tracking solar collection systems Solar collection systems that are typically pole-mounted to the ground and that track the east-west movement of the sun for maximum energy collection.

track lighting A series of movable lights arranged in a line and located on a strip on the ceiling or mounted flush. Typically used for accent or spot lighting.

track roller In a crawler machine, the small wheels that are under the track frame and that rest on the track.

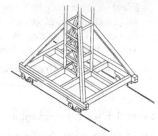

track roller

traction The friction developed between a body and a surface on which it moves relative to the total amount of driving force exerted.

traction elevator A car and counterweight attached to opposite ends of a hoist rope, which is moved by a traction machine. Used in buildings with more than five floors. (Courtesy of National Elevator Industry, Inc.)

traction machine An electric machine in which the friction between the hoist ropes and the machine sheave is used to move the elevator car. (Courtesy of National Elevator Industry, Inc.)

traction steel Steel with strength of 180,000 to 190,000 pounds per square inch that is used to manufacture wire rope.

tractor A vehicle on tracks or wheels used for towing or operating equipment.

trade (craft) 1. The business or work in which one engages regularly and which may require manual skill. **2.** A group representing a particular occupation or craft.

trade discount A dealer discount offered to the contractor by the supplier representing the difference between list price and the actual charge for goods or services.

trade stacking Having more subcontractor or prime contractor work crews performing different types of construction work in the same area than is efficient for the flow of the work.

traffic 1. The total number of messages handled by a communications channel in a given period, expressed in hundred call seconds (CCS) or other units. **2.** Number of vehicles within a specific space and time.

traffic cone A pliable and highly visible cone used as a temporary marker to direct traffic away from a work area.

traffic paint Paint formulated to withstand vehicular traffic, and to be highly visible at night; used to mark traffic lanes and pedestrian crossings.

train A string of connected or unconnected vehicles or mobile equipment, such as a paving train, which consists of mobile machines to lay the various courses of a pavement.

trajectory of prestressing force The path along which the prestress is effective in a structure or member. It is coincident with the

center of gravity of the tendons for simple flexural members and statically indeterminate members that are prestressed with concordant tendons; but, it is not coincident with the center of gravity of the tendons of a statically indeterminate structure that is prestressed with nonconcordant tendons.

transducer A substance or device that converts input energy into output energy of a different form, such as a photoelectric cell.

transfer bond In pretensioning, the bond stress resulting from the transfer of stress from the tendon to the concrete.

transfer column A column in a multistory framed building that is not continuous to the building foundation. At some floors the column is supported by a girder or girders, and its load transferred to adjacent columns.

transfer girder A girder that supports a transfer column.

transfer grille A grille or pair of grilles that allow air to move from one space to another, installed in locations such as a wall or door.

transfer medium The material that carries heat from a solar collector to a living area or storage medium.

transfer strength In prestressed concrete, the concrete strength required before stress is transferred from the stressing mechanism to the concrete.

transfer switch A mechanism designed to switch an electrical conductor from one circuit to another without interrupting the current flow.

transformed section A hypothetical section of one material arranged so as to have the same elastic properties as a section of two materials.

transformer An electric device with two or more coupled windings, with and without a magnetic core, for introducing mutual coupling between circuits; generally used to convert a power supply at one voltage to another voltage.

transformer

transformer bank Two or more transformers that are situated in the same enclosure, such as in a transformer vault.

transistor A semiconductor device with three terminals that amplifies electrical power.

transit A surveyor's instrument used to measure or lay out horizontal or vertical angles, or measure distance or difference in elevation.

transit-and-stadia survey A type of survey in which angles are measured with a transit and distances are measured with a transit and a stadia rod.

Transite® A well-known brand name of high-strength cement made from Portland cement and non-asbestos fibers. It is sold in panels and as preformed items and pipe. Formerly made of asbestos-reinforced cement.

transit

transition strip A strip of wood, metal, or vinyl that covers adjacent areas of flooring to ensure a smooth transition.

transit line Any line of a survey traverse projected by use of a transit or similar instrument.

transit mix (transit-mixed concrete) Concrete that is wholly or mainly mixed in a truck mixer, usually while in transit to the job site.

translucent Descriptive of a material that transmits light, but diffuses it sufficiently that an object cannot be seen clearly through the material.

transmission 1. A gear set or similar device that permits changes in the speed/power ratio and/or direction of rotation. 2. The process by which incident light leaves an object or surface from a side other than that which receives the light. 3. The transportation of bulk electricty from the generator to the substation.

transmission length (transfer length) The distance at the end of a pretensioned tendon necessary for the bond stress to develop the maximum tendon stress.

transmission line Any electric line operating at 69,000 or more volts.

transmission loss (TL) The decrease in energy during transmission from one surface of a medium to another, such as, through a panel or wall.

transmittance The capacity of glass to transmit solar energy—in visible light, as well as ultraviolet and infrared ranges. Transmittance is measured in percentages of each type of light.

transom A glazed or solid panel over a door or window, usually hinged and used for ventilation. The transom and bar may be removable for passage of large objects.

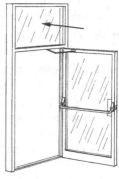
transom

transom bar 1. A horizontal member separating a transom from the door or window. 2. Any intermediate horizontal member of the frame of a door or window.

transverse cracks Cracks that develop at right angles to the long direction of the member.

transverse joint A joint parallel to the intermediate dimension of a structure.

transverse load A load applied at right angles to the longitudinal axis of a structural member, such as a wind load.

transverse prestress Prestress that is applied at right angles to the principal axis of a member.

transverse reinforcement Reinforcement at right angles to the principal axis of a member.

transverse rib A rib in vaulting that spans the nave, cross aisle, or aisle, at right angles to the longitudinal axis of the area spanned.

transverse section A section of a building taken at right angles to its longest dimension.

transverse shear A shearing action or force perpendicular to the main axis of a member.

trap 1. A plumbing fixture so constructed that, when installed in a system, a water seal will form and prevent backflow of air or gas, but permit free flow of liquids. 2. A removable section of stage floor.

trap

trap coil A coil used in the electromagnet that activates a trip.

trapdoor A door set in a floor, ceiling, or roof.

trapeze hanger A horizontal rigid member used to suspend pipes from two rods in instances where longitudinal movement is anticipated.

traprock (trap) Any of various fine-grained, dense, dark-colored igneous rocks, typically basalt or diabase.

trap seal (deep seal) The head (vertical distance between the crown weir and the dip of a trap fixture), expressed linearly resisting back pressure. The height of the water in a toilet bowl, it prevents sewer gases from entering the building.

trapdoor

trapway (passageway) The channel in a toilet that connects the bowl to the waste outlet. The trapway is where siphonic action occurs, and its size is measured in terms of the largest ball that can pass through it.

trash rack A grid of metal bars placed in front of a water inlet to collect larger solids transported by the water.

trave 1. A beam or timber crossing a building. **2.** A panel in a ceiling delineated by beams or timbers.

travel (rise) The vertical distance between the bottom landing of an elevator or escalator and the top landing.

traveler An inverted, U-shaped structure usually mounted on tracks that permit it to move from one location to another to facilitate the construction of an arch, bridge, or building.

travelers Electrical wires between three-way and/or four-way switches.

traveling cable A cable made of electric conductors, that connects an elevator or dumbwaiter car with a fixed electric outlet in the hoistway.

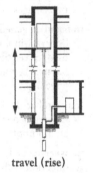

travel (rise)

traveling crane A tower crane mounted on tires, crawlers, or rails.

traverse 1. To plane across the grain of wood. **2.** A structural crosspiece, such as a transom bar. **3.** A gallery or loft crossing a building. **4.** A barrier to allow passage by an official or dignitary, but discouraging unauthorized passage.

traverse closure The calculated line that closes a traverse; the resultant error of measurements, sometimes referred to as "linear error of closure."

travertine A variety of limestone deposited by running water, usually stratified; used for interior walls and for floors.

tray ceiling A horizontal ceiling constructed part way up the slope of a gabled roof.

tray ceiling

tread 1. The horizontal part of a stair. Historically, treads have been made from 5/4″ × 12″ vertical-grain lumber called stepping. In recent years, however, most stepping has been made from particleboard that is given a bullnosed edge. **2.** High-friction coating on a pulley for the purpose of reducing slippage of the belt.

tread length The length of a tread measured perpendicular to the travel line of the stair.

tread plate A fabricated metal tread with a slip-resistant surface.

tread return The projection of a tread beyond the stringer of an open stair.

tread run The horizontal distance between nosings of stair treads.

tread width The horizontal distance from the nosing of a stair tread to the riser above the tread. The tread run plus nosing.

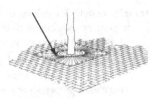

tread width

treated wood Wood that has been treated with a wood preservative.

tree-dozer An attachment for a tractor or bulldozer consisting of metal bars and a cutting blade, used to clear bushes and small trees.

tree grate A metal grating set around a tree and flush with a pavement.

tree grate

trellis A latticework of wood or metal, usually used to support vines.

tremie A pipe or tube through which concrete is deposited under water, having at its upper end a hopper for filling and a bail for moving the assemblage.

tremie concrete Subaqueous concrete placed by means of a tremie.

tremie seal The depth to which the discharge end of the pipe is kept embedded in the fresh concrete that is being placed. The seal is a layer of concrete placed in a cofferdam for the purpose of preventing the intrusion of water when the cofferdam is dewatered.

trench box (trench shield) Box-shaped sheathing made of wood or steel, permanently braced across a trench for excavation and pipe laying. The trench-box unit is pulled along the trench as excavation and pipe laying proceed. It prevents sidewall cave-in during trenching.

trench brace

trench brace A normally adjustable device, used as crossbracing to support sheeting in a trench.

trench drain A cast-in-place or preformed concrete trench usually covered with a grate that serves as both a drain and a collection point for runoff water or other liquid.

T

trench duct A trough with removable covers through which electric power and control cables are run. It can be a metal unit that is set in concrete or formed in a concrete slab. The top of the covers are level with the floor.

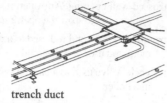
trench duct

trenched footing (neat excavation) A type of footing in which a trench has been excavated to the exact dimensions of the desired footing, and the concrete poured directly into the trench. This type of footing requires no formwork.

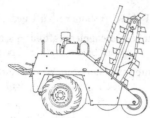

trench excavator A self-propelled machine with a side-mounted shovel or chain of buckets, used to excavate trenches.

trench excavator

trench header duct A preformed duct laid in concrete slabs to provide raceways for wiring and communication lines.

trench jack A hydraulic or screw jack used as a cross brace in a trench bracing system.

trestle ladder A self-supporting, portable ladder that is often paired with another to form a portable scaffold system.

trial batch A batch of concrete prepared to establish or check proportions of the constituents.

trial pit A small pit dug to investigate the soil, sometimes dug to bedrock or other dense material.

triangular truss A light wood roof truss used for short spans.

triangulation A method of surveying over long distances by establishing a network of triangles. Most sides in the network are computed from a known side that may be calculated, and two measured angles. Lengths are measured periodically as a check.

triaxial compression test A test whereby a specimen is subjected to a confining hydrostatic pressure, and then loaded axially to failure.

triaxial test A test whereby a specimen is subjected simultaneously to lateral and axial loads.

tricalcium aluminate A compound having the composition $3\text{CaO} \cdot \text{Al}_2\text{O}_3$, abbreviated C_3A. Its mineral phase name is *celite*.

tricalcium silicate A compound having the composition $3\text{CaO} \cdot \text{SiO}_2$ abbreviated C3S. An impure form of it (alite or celite) is a main constituent of Portland cement. *See also* **alite**.

trig 1. Bricks laid in the center of a wall to prevent sagging and keep it plumb. **2.** Metal clip used to hold a mason's line.

trigger The condition or event that initiates or causes an interlocking process to occur.

trilateration A method of surveying similar to triangulation, except that distances are measured by electronic instruments, and the two angles are calculated. Quadrilaterals are used in addition to triangles.

trim Millwork, primarily moldings and/or trim to finish off and cover joints around window and door openings.

trim hardware Decorative finish hardware that is functional or used to operate functional hardware.

trimmer (header) 1. A short beam that supports one or more joists or beams at an opening in the floor. **2.** A beam or joist inserted in a floor on the long side of a stair opening and supporting a header.

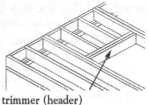
trimmer (header)

trimming joist A joist parallel to the common joists, but of larger cross-section, possibly two pieces nailed together, that support a trimmer.

trimming rafter A rafter supporting the end of a header. *See also* **trimming joist**.

trimstone (trim) Decorative masonry members on a structure built or faced largely with other masonry; includes sills, jambs, lintels, coping, cornices, and quoins.

trip (release catch) A device that releases a mechanism, such as a pawl.

trip coil A control device that utilizes a solenoid to open a circuit breaker.

triple glazing Energy-saving windows composed of three panes of glass sandwiched together with air in between. Gases such as argon or others may be used to increase efficiency. *See* **three-phase motor**.

triplex cable Electrical cable composed of three conductor cables twisted together.

tripod A three-legged, adjustable stand for an instrument.

trivet A low support for a surveying instrument used where a tripod cannot be accommodated.

troffer A long recessed lighting unit that is usually installed flush with the ceiling.

trolley beam An exposed steel beam on the underside of a structure, used to support a trolley crane.

trolley hoist Lifting apparatus that features a hoist that moves along an I-beam that is supported above the work area.

Trombe wall A sun-facing wall made of stone, concrete, or other thermal mass material that collects heat during daylight hours and releases it to the space behind the wall over a 24-hour period.

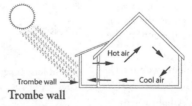
Trombe wall

trough A channel used to contain electric power or control cables.

trowel 1. A flat, broad-blade, steel hand tool used in the final stages of finishing operations to impart a relatively smooth surface to concrete floors and other unformed concrete surfaces. **2.** Also a flat, triangular-blade tool used for applying mortar to masonry.

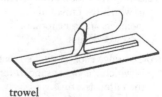
trowel

T

trowel finish The smooth finish surface produced by troweling.

troweling machine A motor-driven device used to trowel concrete. It functions via orbiting steel trowels on radial arms that rotate on a vertical shaft.

truck crane A crane mounted on a wheeled vehicle.

truck mixer A concrete mixer suitable for mounting on a truck chassis and capable of mixing concrete in transit.

truck crane

truck zoning device A freight elevator mechanism that allows the car to be maneuvered within a limited distance above a landing while the door and hoistway door are open.

true wood Heartwood.

trunk 1. The main wood shaft of a tree. 2. The shaft portion of a column. 3. Descriptive of the main body of a system, as a sewer trunk line. 4. Transmission channel that runs between two central office or switching devices, connecting exchanges to the main telephone network.

trunk sewer A main sewer that receives flow from many tributaries, covering a large area.

trunnion A pivot consisting of two cylinders or pins projecting from the body of the pivoted object.

truscan order One of the five classical orders featuring columns in which the bottom one-third is straight and the top two-thirds are tapered.

truss 1. A structural component composed of a combination of members, usually in a triangular arrangement, to form a rigid framework; often used to support a roof. 2. An escalator frame.

truss

trussed beam A beam, usually composed of timber that is reinforced by a center post beneath the beam and two rods running from the bottom of the post to the ends of the beam.

trussed joist (open-web joist) A joist in the form of a truss.

trussed joist (open-web joist)

trussed purlin A lightweight trussed beam used as a purlin.

trussed rafter roof A roof system in which the cross-framing members are some form of light wood truss.

trussed ridge roof A pitched roof in which the upper support for the rafters is a truss.

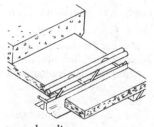
trussed purlin

trussed-wall opening Any opening in a framed structure with a truss system used to span the opening.

truss rod 1. A metal rod used as a tension member in a truss. 2. A metal rod used as a diagonal tie.

try cock Valve used as a backup system of determining water level in a boiler.

try square A normally graduated square, the legs of which are at 90°.

T-shore A shore with a T-head.

tube-and-coupler scaffold A scaffold system using tubes for posts, bearers, braces, and ties. Special couplers connect the parts.

tube axial fans Single-width airfoil wheel fans arranged in a cylinder to discharge air radially against the inside of the cylinder.

tub trap The U-shaped section of a bathtub drainpipe designed to hold a water seal to prevent sewer gasses from entering.

tubular fiber form Long tubes of compressed paper used as concrete forms. Often used to pour footings.

tubular lock A type of bored lock that encloses the bolt within a tube.

tubular scaffolding Scaffolding manufactured from galvanized steel or aluminum tube and connected by clamps.

tubular-welded-frame scaffold A scaffold system using prefabricated welded sections that serve as posts and horizontal bearers. The prefabricated sections are braced laterally with tubes and bars.

tubular-welded-frame scaffold

tuck A recess in a horizontal masonry joint formed by raking out mortar in preparation for tuck pointing.

tuck pointing A method of refinishing old mortar joints. The loose mortar is dug out and the tuck is filled with fine mortar that is left projecting slightly or with a fillet of putty or lime.

tuff, volcanic tuff A highly porous, low-density rock comprising volcanic particles sometimes in stone applications or as a thermal insulation material.

tufted carpet Carpet manufactured by pile yarn punched through carpet backing material that has been previously woven. The pile is cut after.

tumbler The mechanism in a lock holding the bolt until operated by a key.

tumbler switch A lever-operated electric snap switch.

tung oil An oil that dries to a clear finish and resists water, it is often used as a wood finish.

tungsten-halogen lamp (quartz lamp) An incandescent lamp that consists of a tungsten filament, a gas containing halogens, and an envelope of a high-temperature resistant material, such as quartz. The lamp is small compared with lamps of similar wattage.

tungsten steel A very hard, heat-resistant carbon steel containing tungsten.

tunneling Excavating a horizontal hole greater than six feet in diameter.

tunnel test Refers to an ASTM standard test of a material's surface-burning characteristics.

turbidimeter A device for measuring the particle-size distribution of a finely divided material by taking successive measurements of a suspension in a fluid.

turbidimeter fineness The fineness of a material such as soil or Portland cement, usually expressed as total surface area in square centimeters per gram, as determined with a turbidimeter. *See also* **Wagner fineness.**

turbidity The clarity of water, measured by light transmission through a water sample.

turbine Any of various machines that convert the kinetic energy of a moving fluid to mechanical energy. The turbine is often used for driving an electric generator.

turnbuckle A device for adjusting the length of a rod or cable, consisting of a right screw and a left screw coupled by a link.

turnbuckle

turned bolt A machine bolt, usually with a hexagonal head, the shank of which is finished to a close tolerance.

turned work Pieces of stone or woodwork having a circular cross-section, such as posts and balusters. Turned work is usually cut on a lathe.

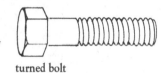

turned bolt

turning Shaping objects by use of cutting tools while the piece to be shaped is rotated on a lathe.

turning gouge Hand tool used to rough cut wood on a lathe.

turning piece A wood template used by a mason to form a small arch that does not require centering.

turning vane One of several curved fins that are placed inside ductwork at a change in direction in order to promote better airflow.

turn-of-the-nut method A method used to pretension high-strength bolts. The nut is turned from the snug-tight position by either a man using his full strength on a wrench, or a few hits with an impact wrench.

turn piece A small knob used to control a deadbolt from the inside of a door, usually crescent- or oval-shaped for gripping with thumb or fingers.

turnstile A barrier rotating on a vertical axis, usually allowing movement in one direction only and admitting one person at a time. A turnstile is sometimes coin- or token-operated.

turpentine A thin, volatile oil obtained by steam distillation from the wood or exuded resin of certain pine trees. Turpentine was once widely used as a paint thinner and solvent, but now is replaced by solvents derived from petroleum.

tusk tenon A tenon with stepped reinforcing at the lower side.

twin cable A cable consisting of two parallel, insulated conductors fastened side-by-side through the insulation, or by common wrapping.

twin-filament lamp An incandescent lamp with two filaments that are wired independently, used as double-function lamps as in automobile stop lights or as three-level wattage lamps.

twin pug (pugmill) A mill with two blade-supporting shafts running in opposite directions.

twist The distortion from a flat plane caused by the turning or winding of the edges of a board.

twist drill A drill with one or more helical cutting grooves, used to drill holes in metal, wood, and plastic.

twisted pair Two single, insulated wires that have been twisted together to reduce the likelihood of interference to and from other wire pairs.

twisted pair

two-by-four A commonly used piece of timber with nominal dimensions 2″ thick by 4″ wide, i.e., 5 cm thick by 4 cm wide.

two-coat work The application of two coats of plaster: a base coat followed by a finish coat.

two-core block A concrete masonry unit with two hollow cells.

two-four-one (2–4–1) Structural wood panels, at least 1⅛″ thick, designed for single-floor applications over joists spaced 48″ apart and also used as roof

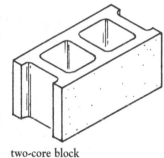

two-core block

sheathing in heavy timber construction. The term is synonymous with APA-Rated Sturd-I-Floor and is a registered trade name of the American Plywood Association.

two-handed saw A large saw with two handles intended to be worked by two people; used for felling trees or cross cutting logs, largely replaced by chain saws.

two-part line A single strand, rope, or cable that is doubled back around a sheave to double the capacity.

two-stage absorption A refrigeration system in which the hot refrigerant (water) vapor travels to a second generator. There, upon condensing, it supplies heat for further refrigerant vaporization from the absorbant of intermediate concentration that flows from the first generator.

two-stage curing A process whereby concrete products are cured in low-pressure steam, stacked, and then autoclaved.

two-way joist construction Floor or roof construction in which the floor or roof is supported on two mutually perpendicular systems of parallel joists.

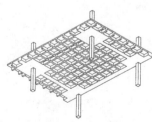

two-way joist construction

two-way reinforcement (two-way system) A system of reinforcement. Bars, rods, or wires are placed at right angles to each other in a slab, and are intended to resist stresses due to bending of the slab in two directions.

T

two-way slab A reinforced concrete slab in which the main reinforcing runs in two directions, parallel to the length and width of the panel.

Type I & IA Duty Rating Safety ratings for ladders based on load limits. Type I is rated for 250 lbs. Type IA is rated for 300 lbs.

type S fuse A mechanism designed to fit into a screw shell socket to determine whether the fuse is blown. Available in 15, 20, and 30 amperes.

Type X A fire-resistant type of gypsum wallboard.

Tyvek® house wrap Brand name for a weather protector building wrap. In addition to adding insulation, it works to keep moisture from penetrating walls while also allowing the building to "breathe."

T

U

Abbreviations

u unit

U uranium

UBC Uniform Building Code

UCI uniform construction index

UCS unconfined compressive strength

UDC universal decimal classification

U/E unedged

UF, UGND underground feeder

Ug micrograms

ugnd underground

UHF ultra-high frequency

UL Underwriters' Laboratories, Inc.

Uld unloading

ult ultimate

unassm unassembled

unfin unfinished

unglz unglazed

unhtd unheated

unidir unidirectional

unins. uninsurable

unmtd unmounted

uns unsymmetrical

up upper

UPS uninterruptible power supply, United Parcel Service

ur, UR urinal

URD underground residential distribution

USBM U.S. Bureau of Mines

USBR U.S. Bureau of Reclamation

USG United States gauge

USP United States Primed

UST underground storage tank

Util utility

UTMCD Uniform Traffic Manual for Control Devices

UTP unshielded twisted pair

UV Ultraviolet

USGBC U.S. Green Building Council

Definitions

U-bolt A bolt formed in the shape of the letter U, with threads on the ends to accommodate nuts.

U factor A measure of thermal transmission.

U-bolt

UL label A seal of certification attached by Underwriters' Laboratories, Inc. to building materials, electrical wiring and components, storage vessels, and other devices, attesting that the item has been rated according to performance tests on such products, is from a production lot that made use of materials and processes identical to those of comparable items that have passed fire, electrical hazard, and other safety tests, and is subject to the UL reexamination service.

UL label

ultimate design resisting moment The theoretical, applied bending moment that will cause failure in a reinforced concrete member through yield in the tensile reinforcing steel or crushing of concrete.

ultimate load 1. The maximum load a structure can bear before its failure due to buckling of column members or failure of a component. **2.** The load at which a unit or structure fails.

ultimate set The final degree of firmness obtained by a plastic compound after curing.

ultimate shear stress The stress at a section loaded to its maximum in shear.

ultimate strength The maximum resistance to load that a member or structure is capable of developing before failure occurs. With reference to cross sections of members, the largest moment, axial force, torsion, or shear a material can sustain without failure.

ultimate wiring capacity The average load per unit of area required to cause the rupture and subsequent failure of a supporting mass.

ultrasonic soldering A soldering process, usually performed without a flux, wherein unwanted surface films are removed from the base metal by transmitting high frequency sounds through molten solder, resulting in a more effective welding of the base metal with the solder.

ultrasonic testing A nondestructive method used to test the structural integrity of a pile or a metal element.

ultrasonic welding A process in solid-state welding wherein the metal parts are held together under pressure and joined by applying high frequency sound waves to their surface.

ultraviolet Literally, beyond violet in the color spectrum. A minor component of the sun's radiation, it is also produced artificially in arc lamps.

unbalanced partition A wall built with two layers of drywall on one side and a single layer on the other. Such an imbalance occurs when fire or sound insulation is just needed on one side.

unbalancing A technique used in the pricing process to allocate estimated costs to accounts whose definitions do not fully reflect the nature of the cost being allocated. The purpose of unbalancing is to achieve a desired business result such as improved cash flow. For example, a disproportionate amount of overhead costs may be allocated in a contract bid to early project activities so that early income is maximized.*

unbonded member A posttensioned, prestressed concrete element in which tensioning force is applied against end anchorages only, with tendons free to move within the elements.

unbonded posttensioning Posttensioning in which the tendons are not grouted after stressing.

unbonded tendon A tendon not bonded to the concrete section.

unbound base Material without a binder added.

unbraced frame A structural frame, resistant to the lateral load carried, due to the ability of its members and connections to withstand bending and shear stresses without additional diagonal bracing, K-bracing, or other extra supporting devices.

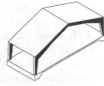

unbraced frame

unbraced length The greatest length between points on a compression member not restrained against lateral movement by beams, slabs, or bracing.

unbraced length of column Distance between adequate lateral supports.

uncased pile An unlined concrete pile, often reinforced, that has been poured directly into a bored hole in the ground.

uncertainty 1. The total range of events that may happen and produce risks (including both threats and opportunities) affecting a project. (Uncertainty = Threats + Opportunities.) **2.** All events, both positive and negative whose probabilities of occurrence are neither 0% nor 100%. Uncertainty is a distinct characteristic of the project environment.*

unclassified excavation An excavation priced with a single unit cost, regardless of whether the excavator finds loose fill or embedded rock. Bidders on an unclassified excavation assume the risk for whatever appears.

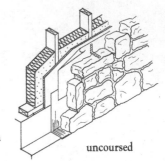

unconsolidated Not compacted.

uncoursed Descriptive of irregularly placed masonry, which is not laid in courses with continuous horizontal joints, but in a seemingly random pattern.

uncoursed

underbed In terrazzo floor construction, the base mortar, usually horizontal, into which strips are embedded and on which terrazzo topping is applied.

undercloak 1. The part of a lower sheet in sheet metal roofing that serves as a seam. **2.** A course of tiles or slate used in roofing to provide an underlayer for the first course installed at the eaves.

undercoat 1. A coat of paint that improves the seal of wood or of a previous coat of paint, and provides a superior adhesive base for the top coat. **2.** A paint used as a base for enamel. **3.** A colored primer paint.

unconfined compressive strength (UCS) A measure of soil bearing capacity, determined by a variety of tests.

undercured Descriptive of concrete, paint, sealant, or other substances applied in wet or elastic form, which have not had time to harden properly because of unsuitable environmental conditions.

undercut To cut away a lower portion of architectural stonework, creating a projection above it that functions as a drip.

undercut door A door with greater than normal clearance at the floor to give more ventilation to an area.

underdrain A drain installed in porous fill under a slab to drain off ground water.

underfloor heating A heating system that uses embedded pipes or electric heating cables beneath a finished floor so that heat radiates steadily from floor level.

underfloor raceway A raceway installed beneath a finished floor to distribute electric conductors and communications wiring. Suitable for use in a concrete floor.

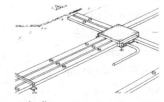

underfloor raceway

underground facilities All pipelines, conduits, ducts, cables, wires, utility access-ways, vaults, tanks, tunnels or other such facilities or attachments, and any encasements containing such facilities that have been installed underground to furnish any of the

following services or materials: electricity, gases, steam, liquid petroleum products, telephone or other communications, cable television, sewage and drainage removal, traffic or other control systems or water.*

underground feeder (UF) cable Cable rated for external, underground use.

underground piping Piping that has been or will be laid beneath the surface of the ground.

underground storage tank (UST) A tank system that has at least 10% of its volume underground.

underlay A material, such as asphaltic felt, which isolates a roof covering from the deck.

underlayment Structural wood panels designed to be used under finished flooring to provide a smooth surface for the finish material.

underlining felt The material, usually a No. 15 felt, applied to a wood roof deck before shingles are laid.

underpinning The construction of new substructure support beneath a column or a wall, without removing the superstructure, in order to increase the load capacity or return it to its former design limits.

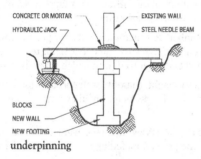
underpinning

underrun **1.** Opposite of overrun; same as pickup; the (lesser) difference between the actual cost and the estimate. **2.** Decrease in estimated quantity.

undersanded With respect to concrete, containing an insufficient proportion of fine aggregate to produce optimum properties in the fresh mixture, especially workability and finishing characteristics.

underslung car frame The car frame of an elevator with fastening sheaves for hoisting cables attached at or below its platform.

Underwriters' Laboratories, Inc. (UL) A private, nonprofit organization that tests, inspects, classifies, and rates devices and components to ensure that manufacturers comply with various UL standards.

undisturbed sample A sample taken from soil in such a manner that the soil structure is deformed as little as possible.

undressed Descriptive of lumber products that have not been surfaced.

uneven grain Wood grain showing a distinct difference in appearance between springwood and summerwood. Examples are ring-porous hardwoods such as oak, and softwoods such as yellowpine that have soft springwood and hard, dense summerwood.

uneven grain

unfaced insulation Board, batt, or blanket insulation without a vapor barrier.

unglazed tile A hard ceramic tile of homogeneous composition throughout, deriving its color or texture from the materials used and the method of manufacture. The unglazed tile is used for floors or walls.

uniformat A registered trademark system for organizing a cost estimate that is based on building systems.

uniform angle sander A drywall finishing tool with a right-angled sanding device that can be adjusted to fit corner angles. The tool enables a worker to sand both sides of a corner simultaneously.

Uniform Building Code (UBC) One of several national building codes used to promote building safety, health and public welfare. Published by the International Conference of Building Officials (ICBO), it is used throughout the western U.S. states.

Uniform Construction Index The forerunner of the MasterFormat adopted by the Construction Specifications Institute in 1978. This system divides technical data and all related accounting, specifying, and tracking functions into 16 divisions.

Uniform Fire Code A model law devoted to ensuring practices consistent with protecting life and property from fire, explosion, and hazardous materials and conditions.

uniform grading A particle-size distribution of aggregate in which pan fractions are approximately uniform with no one size or group of sizes dominating.

uniformity coefficient A coefficient related to the size distribution of a granular material, obtained by dividing the size of the sieve of which 60% of the sample weight passes by the size of the sieve of which 10% of the sample weight passes.

uniform load A load distributed uniformly over a structure or a portion of a structure.

uninterruptible power supply An electrical supply system that will continue to provide current during a power outage.

union **1.** A confederation of individuals who share the same trade or similar trades and who have joined together for a common purpose. **2.** A pipe fitting used to join two pipes without turning either pipe, consisting of a collar piece that is slipped on one pipe, and a shoulder that is threaded or soldered on that pipe and against which the collar piece bears. Unions allow dismantling a fitting without disturbing the pipe.

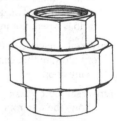
union

union elbow A pipe elbow outfitted with a union coupling at one end that makes it possible for the coupling end to connect with the end of a pipe without turning or disturbing the pipe.

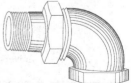
union elbow

union tee A pipe tee with a union-type joint on one end.

unitary air-conditioner A fabricated assembly of equipment to move, clean, cool, dehumidify, and sometimes heat the air, consisting of a fan, cooling coil, compressor, and condenser.

U

unit cost The cost per unit of measurement.

unit hours Work hours per unit of production.*

united inches The sum of the length and width of a piece of rectangular glass, each in inches.

unit heater A factory-assembled heating unit consisting of a housing, a heating element, a fan and motor, and a directional outlet.

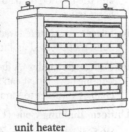
unit heater

unitized Wood products securely gathered into large standard packages or units, usually fastened with steel straps and often covered by tough paper or plastic.

unit lock A preassembled lock.

unit price 1. Current and accurate cost of materials, equipment, and labor used to develop a unit price estimate. **2.** The sum stated in a project bid representing the price per unit for materials and/or services.

unit price contract A construction contract in which payment is based on the work done and an agreed on unit price. The unit price contract is usually used only where quantities can be accurately measured.

unit-type vent One of several relatively small openings on the roof of a structure, equipped with a metal frame and housing as well as manual or automatic hinged dampers that are opened in case of fire.

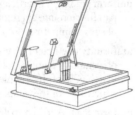
unit-type vent

unit ventilator A unit with operable air inlets, and often with heating and/or cooling coils, that conveys outdoor air into an interior room.

unit water content 1. The quantity of water per unit volume of freshly mixed concrete, often expressed as pounds or gallons per cubic yard. **2.** The quantity of water on which the water-cement ratio is based, not including water absorbed by the aggregate.

universal Descriptive of a door lock, door closer, or similar piece of hardware that can be used on either a left-hand or right-hand swing door.

universal joint A coupling between two shafts set at an angle that allows them to rotate and swivel.

universal joint

universal motor A motor that can operate on either alternating or direct current and is usually less than one horsepower.

unkeyed Loose plaster that has become detached from the lath, requiring additional repair other than canvassing.

unloader A control for an electric-motor-driven compressor. The unloader controls the pressure head of the compressor and allows the motor to be started at low torque by disconnecting one or more cylinders during the initial period of operation.

UN number A classification code assigned to a particular material by the Department of Transportation.

unprotected corner Corner of a slab with no adequate provision for load transfer, so that the corner must carry over 80% of the load.

unrestrained member A structural member that is allowed to rotate freely about its supports.

UN number

unseasoned Lumber that has not been dried to a specified moisture content before surfacing. The American Softwood Lumber Standard defines unseasoned lumber as having a moisture content above 19%.

unsound Not firmly made, placed, or fixed, and thus subject to deterioration or disintegration during service exposure.

unsound plaster Hydrated lime, plaster, or mortar that contains particles that are unhydrated and may expand later, causing popping or pitting.

unstable soil Earth material, other than running, that because of its nature or the influence of related conditions, cannot be depended upon to remain in place without extra support, such as would be furnished by a system of shoring.

unstiffened member A structural member or portion thereof that must withstand compressive force, but is not reinforced in the direction perpendicular to that in which it bends most readily.

unsupported wall height Masonry wall construction limit for wall height set by local codes determined by ratio of wall thickness to height of wall.

updating The regular review, analysis, evaluation, and reporting of progress of the project, including recomputation of an estimate or schedule.*

uplift The upward load that wind or water pressure can create on a building.

upright A vertical length of stone or timber. Also any vertical structural member, such as a stanchion.

upset 1. To make an object or part of an object shorter and thicker by hammering on its end. **2.** A flaw in timber caused by a heavy blow or impact that splits fibers across the grain. **3.** In welding, an increase in volume at the point of the weld caused by applied pressure.

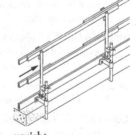
upright

upset welding A process of resistance-welding making use of both the pressure and heat generated by the flow of current as it passes through the resistance provided at the contact point of the surfaces being welded.

upstand (upturn) That portion of a flashing or roof covering that is run up a wall without being tucked in, and which is usually covered with stepped flashing.

upstanding beam A beam projecting above a concrete floor rather than concealed beneath it.

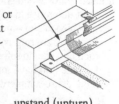
upstand (upturn)

U

upturned beam A concrete beam that extends above the slab it is designed to support.

urea-formaldehyde foam insulation (UFFI) Insulation used in wall cavities as an energy conservation measure during the 1970s that has relatively high indoor concentrations of formaldehyde.

urea resin adhesive A powder which is mixed with water before use and has high early strength and good heat resistance. Urea resin adhesive is not recommended for exposure to moisture or use in poorly fitted joints.

urea resin glue Urea-formaldehyde resin, an adhesive used in the manufacture of hardwood plywood and interior particleboard panel products. Urea is a soluble, crystalline material found in the urine of mammals, but it is also produced synthetically for the manufacture of plastics and adhesives.

urinal A plumbing fixture designed for the collection of urine and equipped with a water supply for flushing.

urinal

U-stirrup A rod shaped like a U used in reinforced concrete construction.

U-strap Column anchorage to a concrete base using a steel U-shaped anchor bolt embedded in concrete. A wall form tie made of heavy wire bent into a U-shape.

utility 1. A grade of softwood lumber used when a combination of strength and economy is desired. Utility grade is suitable for many uses in construction, but lacks the strength of Standard, the next highest grade in light framing, and is not allowed in some applications. 2. A grade of Idaho white pine boards, equivalent to #4 Common in other species. 3. A grade of fir veneer that allows white speck and more defects than are allowed in D grade. Utility grade veneer is not permitted in panels manufactured under Product Standard PS-1–83. 4. Gas, water, electrical, and sewer beyond 5' from a building.

utility easement areas Areas utility company personnel are legally authorized to access in order to service the lines.

utility and better (Util&Btr) A mixture of light framing lumber grades, the lowest being utility. The *"and better"* signifies that some percentage of the mixture is of a higher grade than utility, but not necessarily of the highest grade. In joist and plank grades, the corresponding term is *"#3&Btr."*

utility pole An outdoor pole installed by a utility company for the support of telephone, electric, and other cables.

utility sheet Metal sheeting that is mill-finished and cut into numerous widths and lengths for general use within the building construction industry.

utility transformer A transformer of primary and secondary wire coils that reduces voltage supplied by a utility to a facility.

utility vent A pipe that helps provide an air supply within a drainage system or fixture to prevent siphonage. The vent rises above the highest water level of the fixture, and turns downward before connecting to the main vent.

utility window A hot-rolled steel window, generally inexpensive, equipped with a hopper light and a fixed light and used principally in garages, shops, and basements.

U-trap A running trap, built in the shape of a U, that forms a seal against the passage of gasses in a pipe, while allowing liquid to flow freely.

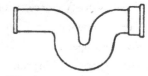

U-trap

U-tube (manometer) A U-shaped glass tube filled with water or mercury and used to measure pressure by liquid displacement.

U-value (thermal transmittance) The time rate of heat flow per unit area between fluids on the warm side and cold side of a barrier, calculated in accordance with the difference in unit temperature between the two test fluids.

U-tube (manometer)

U

333

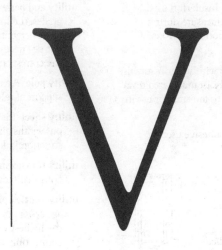

Abbreviations

V volt, valve, V-groove, vertical shear

V1S vee one side

VA volt amperes

VAC volts alternating current

vac vacuum

val value, valuation

van vanity

VAP vapor

VAR visual-aural range, volt-ampere reactive

var variation, varnished

VAT vinyl-asbestos tile

VAV variable air volume

VC veneer core

VCP visual comfort probability

VCT vinyl composition tile

VD vapor density

VDC volts direct current

vel velocity

ven veneer

vent, VENT ventilator, ventilate

ver version

verm vermiculite

vert, VERT vertical

VF video frequency, vinyl-faced

VG vertical grain

vhf very high frequency

VHO very high output

Vib vibrating, vibrator

vic vicinity

VIF verify in field

vil village

vis visibility, visual

VIT vitreous, vitrified

Vitr vitrification

vit ch vitreous china

vj V-joint

VLDPE very low density polyethylene

vlf very low frequency

VLF vertical linear foot

VLR very long range

voc volatile organic compound

vol, VOL volume

vou voussoirs

VP vent pipe

VRP vinyl reinforced polyester

VS versus, vent stack, vapor seal

VT vacuum tube, variable time

VTR vent through roof

VU volume unit

Definitions

vacuum 1. A space that contains no matter. **2.** A space with reduced air or water pressure.

vacuum breaker A pipe or fixture element in a water supply system that prevents siphon or suction action and the backflow that can result.

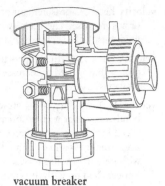

vacuum breaker

vacuum circuit breaker An electrical circuit breaker with its surge protection components housed in a vacuum.

vacuum concrete Concrete from which water and entrapped air are extracted by a vacuum process before hardening occurs.

vacuum saturation A process for increasing the amount of filling of the pores in a porous material, such as lightweight aggregate, with a fluid, such as water, by subjecting the porous material to reduced pressure in the presence of the fluid.

vacuum tank In a hot water heating system, a container that holds condensate drawn into the boiler by a vacuum pump.

vadose zone The layer of unsaturated soil above the groundwater level.

valance lighting Lighting from sources that are concealed and shielded by a board or panel at the wall-ceiling intersection. This lighting may be directed either upward or downward.

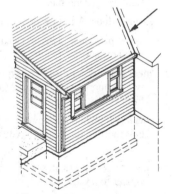

valance lighting

valley The place where two planes of a roof meet at a downward, or V, angle.

valley board A board nailed to a valley rafter of a roof to accommodate the metal gutter.

valley flashing Sheet metal with which the valley of a roof is lined.

valley

valley gutter The exposed open gutter in the valley of a roof, constructed with sloping sides.

valley jack A rafter in a roof system that is cut shorter than a common rafter and connects the ridge and a valley rafter.

valley rafter In a roof frame, the rafter that follows the line of the valley and connects the ridge to the wall plate along the line where the two inclined, perpendicular sides of the roof meet.

valley roof A pitched roof having one or more valleys.

valley shingle A specially cut shingle designated for placement next to a valley with grains that run parallel to the valley.

valley tile A trough-shaped roofing tile used at a roof valley to create a channel for water runoff.

value 1. The utility of an object or service or its worth consisting of the power of purchasing other objects or services. **2.** The relative lightness or darkness of a color.

value engineering (VE) An analysis and comparison of cost versus value of building materials, equipment, and systems. VE considers the initial cost of construction, coupled with the estimated cost of maintenance, energy use, life expectancy, and replacement cost.

valve Device that regulates a liquid or gas flowing through piping.

valve flutter Any undesirable movement of a valve during operation.

valve motor An electric or pneumatic motor that operates a valve within an air-conditioning system, regulating it at a location remote from the unit.

valve plug A valve component that moves against a stationary seat to control pressure or flow of a fluid.

valve seat The stationary portion of the valve which, when in contact with the movable portion, stops the flow.

valve seat

valve stem A threaded, spindle-like valve component that connects the valve plug to an external handle, gearset, or actuator.

vaneaxial fan A fan with a disk-type wheel within a cylinder and a set of air-guided vanes on the wheel. The fan is either belt-driven or connected directly to a motor.

vane test, vane shear test A test used to evaluate soil strength and consistency. After a hole is drilled into the earth, a spinning four-bladed shaft is inserted and the soil's resistance to the shaft's vanes is measured.

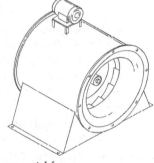

vaneaxial fan

vanity A cabinet and lavatory combination for a bathroom.

van stone A type of joint made by using loose flanges on lengths of pipe whose ends are lapped over to give a bearing surface for the flange.

vapor barrier Material used to prevent the passage of vapor or moisture into a structure or another material, thus preventing condensation within them.

vapor barrier

vapor heating system A system of steam heating functioning at or near atmospheric pressure, wherein the condensed liquid is returned to the boiler by gravity.

vapor lock The formation of vapor in a pipe carrying liquids, thereby preventing normal flow.

vapor migration Penetration of vapor through walls or roofs, caused by the vapor pressure differential between the inside and the outside of a structure.

vapor pressure A component of atmospheric pressure caused by the presence of vapor. Vapor pressure is expressed in inches, centimeters, or millimeters of height of a column of mercury or water.

V

vapor retarder A layer of material used to protect roof insulation and moisture-sensitive roofing membranes from the damaging effects of internal humidity.

vapor retarder felt A felt material containing asphalt and used for sealing.

variable air volume (VAV) An air distribution system capable of automatically delivering a reduced volume of constant temperature cool air to satisfy the reduced cooling load of individual zones.

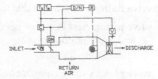

variable air volume (VAV)

variable costs Those costs that are a function of production, e.g., raw materials costs, byproduct credits, and those processing costs that vary with plant output such as utilities, catalysts and chemical, packaging, and labor for batch operations.*

variable inlet vanes VIV Vaned dampers installed at the inlet to a fan to produce a spin to the air entering the fan, which reduces the fan performance.*

variable-volume air system An air-conditioning system that automatically regulates the quantity of air supplied to each controlled area according to the needs of the different zones, with preset minimum and maximum values based on the load in each area.

variegated Descriptive of a material or surface having streak-marks, or patches of a different color or colors.

varnish A clear, colorless substance used chiefly on wood to provide a hard, glossy, protective film. Varnish is manufactured from resinous products dissolved in oil, alcohol, or a number of volatile liquids.

varnish stain A colored transparent varnish with a lesser power of penetration than a true stain.

varved clay Sedimentary soil with alternating layers of clay and silt or fine sand that display contrasting colors as they dry, formed during the differing sedimentation conditions in various seasons of the year.

vault **1.** An enclosure built above or below ground large enough to accommodate human entry. Vaults are used to install, operate, and maintain electrical cables and equipment. **2.** A masonry structure with an arched ceiling. **3.** A room used for storage of valuable records and/or computer tapes that is of fire-resistant construction, has safe electric components, and has a controlled atmosphere.

V-bank filter box An air-handling unit section in which the filters are arranged in a V configuration in order to provide a greater filter surface.

V-beam sheeting Corrugated sheeting formed with flat, V-angled surfaces instead of a rippled or curved surface.

V-brick Brick that is vertically perforated.

V-cut **1.** Descriptive of a style of lettering carved into stone in acute, triangular cuts. **2.** A V-shaped saw cut or incision in wood.

vehicle **1.** In painting, a substance, such as oil, in which pigments are mixed for application. **2.** Any device for conveyance.

vein A thin layer or deposit of one material in another, usually in an approximate plane.

veining The lines that appear in soft bitumens as the material ages.

velocity head A measurement of the velocity of fluid through a pipe or watercourse. The velocity head is equal to the height that the fluid must fall to achieve that same velocity.

velocity riser A vertical pipe with a smaller diameter than adjacent piping that is used to increase the flow velocity of the material conveyed in the pipeline.

veneer **1.** A masonry facing attached to the backup, but not so bonded as to act with it under load. **2.** Wood peeled, sawn, or sliced into sheets of a given constant thickness and combined with glue to produce plywood. Veneers laid up with the grain direction of adjoining sheets at right angles produce plywood of great stiffness and strength, while those laid up with grains running parallel produce flexible plywood most often used in furniture and cabinetry construction.

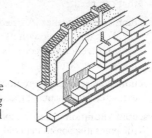

veneer

veneer adhesives Any of several basic substances used in the gluing of veneers to produce plywood. These include blood glue, soybean glue, and phenolic resins. Other adhesives made from urea, resorcinol, polyvinyl, and melamine are sometimes used in edge gluing, patching, and scarfing.

veneer base Gypsum lath sheeting, ordinarily cut in widths of four feet and in varying lengths and thicknesses, with a core of gypsum and a special paper facing that is receptive to veneer plaster.

veneered construction Construction of wood, reinforced concrete, or steel, faced with a thin layer of another material such as structural glass or marble.

veneered door A hollow or solid core door with veneer faces.

veneered plywood Plywood faced with a decorative veneer, usually wood or plastic.

veneer plaster A mill-mixed gypsum plaster made up of one or two components and desired for its bond, strength, and ease of installation.

veneer saw A hand or power saw with a fine-toothed blade used to cut wood veneers.

veneer tie A wall tie designed to hold a veneer to a wall.

veneer wall A wall with a facing that is attached to, but not bonded to, the wall.

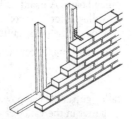

veneer wall

venetian blind **1.** A blind made of thin slats mounted so as to overlap when closed and provide spaces for admitting light and/or air when open. The mounting usually consists of strips of webbing. The unit is operated by cords. **2.** Adjustable exterior slatted shutters.

venetian mosaic A type of terrazzo topping in which large chips of stone are incorporated.

Venetian plaster A wall plaster applied with a trowel in several layers and often coated with a polished waxed finish.

venetian blind

venetian window *See* **palladian window.**

vent **1.** A pipe built into a drainage system to provide air circulation, thus preventing siphonage and back pressure from affecting the function of the trap seals. **2.** A stack through which smoke, ashes, vapors, and other airborne impurities are discharged from an enclosed space to the outside atmosphere. **3.** Any opening serving as an outlet or inlet for air.

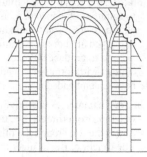
venetian window

vent cap Piece at the top of a ventilation pipe.

vent connector A metal pipe connecting the exhaust of a burner to a chimney.

vented form A concrete form that retains the solid constituents of concrete and permits the escape of water and air.

ventilating brick A brick with holes in it for the passage of air.

ventilating jack A sheet metal hood over the inlet to a vent pipe to direct flow into the pipe.

ventilation A natural or mechanical process by which air is introduced to or removed from a space, with or without heating, cooling, or purification treatment.

ventilator **1.** A device or opening in a room or building through which fresh air enters the enclosure and stale air is expelled. **2.** A pivoted sash or framework, outfitted with hinged panes of glass that may be opened without opening the sash.

ventilator

ventilator frame An assembly designed to accommodate a pivoted sash, with two rails and two stiles into which the operable panes are set.

vent pipe A small-diameter pipe used in concrete construction to permit the escape of air in a structure being concreted or grouted.

vent sash A small, operable light usually hinged on its upper edge in a window, which may be swung open to allow some ventilation without opening the entire sash.

vent screed A plaster screed with a vented portion, used in roof overhang soffits.

vent sash

vent stack main vent A vertical vent pipe whose functions are to provide air circulation to or from any part of a building drainage system and to protect its trap seals from siphonage.

vent system **1.** Piping that provides a flow of air to or from a drainage system to protect trap seals from siphonage or back pressure. **2.** A chimney or vent combined with a vent connector to form a clear passageway for expulsion of vent gases from gas-burning equipment to the outside air.

venturi 1 A short tube with a constriction used to measure liquid or gas velocity by the differential pressure as the liquid or gas flows.

2. A constricted throat in an air passage of a carburetor used to mix fuel and air.

veranda A covered porch or balcony along the outside of a building, intended for leisure.

verbal quotation A contractor's record of a subcontractor's cost proposal for designated work. Quotes given by phone should be followed by a written proposal.

verdigris A bluish green patina that naturally forms on copper, bronze, or brass over time as it is exposed to atmospheric conditions.

verge 1. The edge that projects over the gable of a roof. **2.** A small shaft of a column employed for ornamental effect. **3.** The unpaved section of a road right-of-way.

verge fillet A strip of wood that is fastened to the roof battens of a gable to cover the upper edges of the gable wall.

vermiculite A group name for certain clay minerals, hydrous silicates or aluminum, magnesium, and iron that have been expanded by heat. Vermiculite is used for lightweight aggregate in concrete and as a loose fill for thermal insulating applications.

vermiculite concrete Concrete in which the aggregate consists of exfoliated vermiculite.

vermiculite plaster A fire-retardant plaster covering for steel beams, concrete slabs, and other heavy construction materials, constituted with an aggregate of very fine exfoliated vermiculite.

Vermont slate A type of slate from Vermont used in paving and roofing.

vertical bar An upright muntin.

vertical bond Masonry bond with one brick or block placed directly on top of the next.

vertical broken joint A tiling pattern that offsets each vertical row by half its length.

vertical exit A stairway, ramp, fire escape, escalator, or any other route of movement that serves as an exit from the floors of a building above or below street level.

vertical-fiber brick A paving brick cut with wire during the manufacturing process, and laid on pavement with the wire-cut side exposed to view.

vertical firing Mechanical arrangement of gas, oil, or coal burners in a furnace so that fuel is vertically discharged up from burners below or down from burners on top.

vertical force The load on a structure from a force acting vertically.

V

vertically pivoted window A window having a sash that pivots on a vertical axis near its center so the outside of the glass can be conveniently cleaned.

vertical pipe Any pipe or fitting that makes an angle of 45° or less with the vertical.

vertical sand drain A boring through clay or silt filled with sand or gravel to allow soil to drain. Most effective when bored to permeable soil below.

vertical siding A type of exterior wall cladding consisting of wide matched boards.

vertical siding

vertical sliding window A window with one or more sashes that move only in a vertical direction and make use of friction or a ratchet device to remain in an open position.

vertical slip form A form that is jacked vertically during construction of a concrete structure. Movement may be continuous with placing or intermittent with horizontal joints.

vertical sliding window

vertical spring-pivot hinge A spring hinge that is mortised into the heel of a door, which in turn is fastened with pivots to its door head and the floor.

vertical transportation An elevator, escalator, or lift.

vertical tray conveyors A vertical conveying system designed to carry trays or boxes.

very high output fluorescent lamp (VHO lamp) A rapid-start fluorescent lamp designed to operate on a very high current, providing a light flux per unit length of lamp higher than that obtained from a high output fluorescent lamp.

vestibule An anteroom or foyer leading to a larger space.

V-groove Any of several longitudinal cuts made on the faces of pieces of lumber or plywood. Edges are sometimes chamfered to create a V where pieces are placed edge to edge. A V may also be machined the length of the piece to provide decoration. A V pattern also may be used to form tongue-and-groove connections on hardwood or plywood.

vibrated concrete Concrete compacted by vibration during and after placing.

vibrating roller A towed or self-propelled roller with a motor-driven vibrating mechanism.

vibrating screed A machine designed to act as a vibrator while leveling freshly placed concrete.

vibration Energetic agitation of freshly mixed concrete during placement by mechanical devices, either pneumatic or electric, that create vibratory impulses of moderately high frequency that assist in evenly distributing and consolidating the concrete in the formwork.

vibration, external The process of attaching vibrating devices at strategic positions to concrete forms. Particularly applicable to the manufacture of precast items and for vibration of tunnel-lining forms. In the manufacture of concrete products, external vibration or impact may be applied to a casting table.

vibration, internal The process of inserting one or more vibrating elements into concrete at selected locations. Generally applicable to cast-in-place construction.

vibration isolator A flexible support for any form of vibrating or reverberating machinery, piping, or ductwork, serving to reduce the vibrations that are carried to the remainder of the building structure.

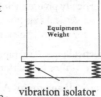

vibration isolator

vibration limit That time at which fresh concrete has hardened sufficiently to prevent its becoming mobile when subjected to vibration.

vibrator An oscillating machine used to agitate fresh concrete so as to eliminate gross voids, including entrapped air, but not entrained air, and to produce intimate contact with form surfaces and between embedded materials.

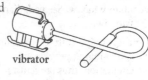

vibrator

vibratory hammer A pile-driving hammer normally weighing from three to five tons that with rapid vibrations causes the soil around the pile to change to a liquid state. The weight of the hammer then slips the pile down through the soil.

vibratory plate A piece of construction machinery that is used to compact soil, pavement underlayment materials, or pavement.

vibroflotation A process used to stabilize and otherwise improve soil for building by compacting loose granular soils.

victaulic pipe Pipe with fittings that allow for movement along joints while remaining watertight.

view sash A picture window with panes of glass divided by muntins.

vinyl A thermoplastic compound made from polymerized vinyl chloride, vinylide chloride, or vinyl acetate. Vinyl is typically tough, flexible, and shiny.

vinyl-clad Doors and windows that have a vinyl covering applied over all exposed surfaces that make up the frame and jamb.

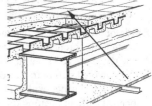

vinyl composition tile A floor tile similar to vinyl-asbestos floor tile except the asbestos has been replaced by glass fiber reinforcing.

vinyl composition tile

vinyl overlay An overlay applied to panel products and moldings and usually printed with a color and grain, needing no further finishing.

vinyl siding Building siding most often consisting of rigid polyvinyl chloride (PVC) used to cover exterior walls.

vinyl tile A vinyl floor tile similar to vinyl-asbestos floor tile, but that does not contain mineral or other fibers.

virgin-growth lumber Wood products harvested from mature trees, with attributes such as a tighter grain, superior hardness, dimensional stability, and durability.

vis, vice, vise 1. A spiral staircase, generally of stone, with steps winding around a central shaft. 2. A screw stair.

viscometer Instrument for determining viscosity of slurries, mortars, or concretes.

viscosity The degree to which a fluid resists flow under an applied force.

viscous filter An air-cleaning filter that employs a surface covered with viscous oil or fluid, to which dirt particles and other airborne impurities cling as the air passes through.

V

vise A gripping tool with adjustable jaws controlled by a lever or screw, used to clamp an object firmly in place while work is being done on it.

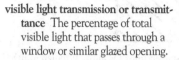

vise

visible light transmission or transmittance The percentage of total visible light that passes through a window or similar glazed opening.

visible transmittance VT A window rating that includes the amount of visible light that is transmitted. VT ranges between 0 and 1, and the higher rating indicates more light transmitted.

vision panel Glazed opening installed in a door as specified by plans and specifications.

visqueen Plastic sheeting, typically 4 mil or 6 mil in thickness.

visual comfort probability VCP A system that evaluates the visual comfort of a lighting system by rating its direct discomfort glare.

vitreous In describing glazed products, a degree of impermeability in a material characterized by low water absorption. Vitreous is less than 3% absorption in floor and wall tile and low voltage electrical porcelain, and less than 0.3% absorption in other materials.

vision panel

vitreous china Material with a glass-like finish used for plumbing fixtures.

vitrification 1. A process of melting soil or wastes to a liquid form, and then cooling the liquid to a hard, glass-like material. **2.** The fusion of grains in clay products under high kiln temperatures, resulting in closure of pores and an impervious material. **3.** A process that transforms the chemical and physical characteristics of hazardous waste so that the treated residues contain hazardous material immobilized in a vitreous mass.

vitrified brick Glazed brick that is impervious to moisture and highly resistant to chemical corrosion.

vitrified-clay pipe Glazed earthenware pipe favored for use in sewage and drainage systems because it is impervious to water and resistant to chemical corrosion.

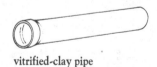

vitrified-clay pipe

V-joint *See* **V-shaped joint.**

void-cement ratio Volumetric ratio of air plus net mixing water to cement in a concrete or mortar mixture.

void ratio In a mass of granular material, the ratio of the volume of voids to the volume of solid particles.

voids The air spaces between particles in granular material such as sand, or a paste such as mortar.

void-solid ratio The ratio of the sum of the areas of window and door openings to the gross area of an exterior wall of a building.

volatile organic compounds Regulated compounds that lead to the formation of ozone and smog and may be hazardous to humans. VOCs contain carbon molecules and are volatile enough to evaporate from material surfaces into indoor air at normal temperatures.

volatility The tendency of a substance to vaporize, or "escape," from the liquid phase or from the surface of a solid to the vapor phase.

volt The unit of voltage or potential difference equal to the voltage between two points of a conducting wire carrying a constant current of one ampere, when the power dissipated between the points is one watt.

voltage A measurement of electrical potential for a building's circuits.

voltage control Device that maintains a constant, uniform voltage in electrical circuits.

voltage drop The difference in voltage between any two points in an electric circuit.

voltage, nominal A value assigned to a circuit or system for classification purposes. Operation may actually be possible within a range, and the nominal voltage is often the midpoint of that range.

voltage regulator A control device within an electrical system that automatically keeps the voltage supply constant despite variable line voltage at the point of input.

voltage relay A voltage monitoring device that detects changes in levels of current and controls the flow of electricity accordingly.

voltage-to-ground 1. The voltage between a conductor in a grounded electric circuit and the point of the circuit that is grounded. **2.** The maximum voltage between a conductor in an ungrounded electric circuit and another conductor.

volt-ampere The product obtained by multiplying one volt times one ampere, equivalent to one watt in direct-current circuits and to one unit of apparent power in alternating current circuits.

voltmeter An instrument used to measure the voltage drop between any two points in an electric circuit.

voltmeter

volume batching The measuring of the constituent materials for mortar or concrete by volume.

volume damper Device installed in a duct system circuit to add resistance to the circuit for air volume balancing.

volume method of estimating cost A cost estimating method determined by multiplying an estimated cost per unit of volume by the volume of the building.

volumetric test 1. A test to determine the integrity of an underground storage tank through review and comparison of tank volume. **2.** A test to determine the air content of fresh concrete.

volute 1. The spiral-turned handrail detail at the first tread on some stairs. **2.** The spiral treatment that appears at the top of a column, usually Ionic.

vomitory An opening dividing a bank of seats to provide access to and egress from an aisle.

V

voussoir A wedge-shaped masonry unit in an arch or vault, the converging sides of which are cut along radii of the vault or arch.

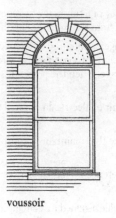

voussoir

V-shaped joint (V-joint, V-tooled joint) **1.** A V-shaped horizontal joint in mortar, formed with a steel jointing tool, that serves to resist rainwater penetration. **2.** Two adjacent wood boards, with beveled or chamfered edges, that form a joint in the same plane.

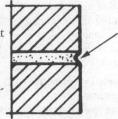

V-shaped joint (V-joint, V-tooled joint)

V-tool A gauge with a V-shaped cutting edge.

vulcanization A chemical process in which a rubber compound becomes less plastic, more resistant to swelling by organic liquids, and more elastic, and its elastic properties are extended over a greater temperature range.

V

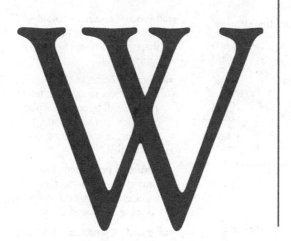

Abbreviations

w water, watt, weight, wicket, wide, width, work, with

W watt, west, western, width

w/ with

WA with average

WAF wiring around frame

WB welded base, water ballast, waybill

WBT wet-bulb temperature

WC water closet

WCLIB West Coast Lumber Inspection Bureau

wd wood, window

Wdr wider

WF wide flange, white fir

wfl waffle

wg wing, wire gauge

wh watt-hour

WH water heater

WHP water horsepower

whr watt-hour

whse, WHSE warehouse

WI wrought iron

WK week, work

wm wattmeter

W/M weight or measurement

WM wire mesh

w/o water-in-oil, without

WP waterproof, weatherproof, white phosphorus

wpc watts per candle

w proof waterproofing

wrc western red cedar

wrt wrought

WS weather strip

wsct/wains wainscoting

wt, Wt weight

WT water table, watertight

ww white wash, white woods

WWM welded wire mesh

WWPA Western Wood Products Association

Definitions

waferboard A panel product made of discrete wafers of wood bound together by resin, heat, and pressure. Waferboard can be made of timber species, such as aspen, that are not suitable for lumber or plywood manufacture.

waffle slab A reinforced concrete slab with equally spaced ribs parallel to the sides, having a waffle appearance from below.

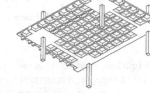

waffle slab

wage rate Hourly rate of pay on which the employee's paycheck is calculated, agreed upon between the employer and employee.

Wagner fineness The fineness of Portland cement, expressed as total surface area in square centimeters per gram, determined by the Wagner turbidimeter apparatus and procedure.

wagon drill A movable rig for positioning and holding a pneumatic drill, consisting of a mast with a carrier for the drill and a supporting wheeled carriage.

wagon-head dormer A dormer with a semicircular roof.

W

wainscot The lower portion of an interior wall whose surface differs from that of the upper wall.

wainscot cap The finish molding at the top of wainscot.

waist The narrowest section of a column or similar building element.

wale (waler, whaler) 1. Timber placed horizontally across a structure to strengthen it. 2. Horizontal bracing used to stiffen concrete form construction and to hold studs in place.

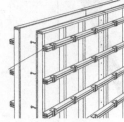

wale (waler, whaler)

walk-in box A refrigerator or freezer large enough for one or more persons to enter to deposit or retrieve food.

walking edger A concrete finishing tool mounted on a long handle and used to produce a radius at the edge of a concrete slab.

wall A vertical element used primarily to enclose or separate spaces.

wall anchor A steel strap used to attach a back-up surface to the masonry fascia.

wall beam 1. A special form of reinforced concrete framing, used for some apartment buildings, in which walls between apartments are used as deep, transverse beams supporting one edge of a floor slab below and one above. The deep beams alternate between floors and between columns. 2. A header bolted to a wall and used to support joists or beams.

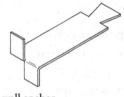

wall anchor

wall-bearing construction A structural system where the weight of the floors and roof are carried directly by the masonry walls rather than the structural framing system.

wallboard A manufactured sheet material used to cover large areas. Wallboards are made from many items, including wood fibers and gypsum. In North America, the most common is gypsum board, a gypsum-based panel bound by sheets of heavy paper. It is used to cover interior walls and ceilings in place of wet plaster.

wall box (beam box, wall frame) 1. A bracket fixed to a wall to support a structural member. 2. A metal box set in a wall to house an electric switch or receptacle.

wall bracket A bracket fastened to a wall and used to support a structural member, pipe, an electric insulator, an electric fixture, or a section of scaffolding.

wall clamp A brace or tie used to connect two parallel walls to each other.

wall column A steel or concrete column fully or partly embedded in a wall.

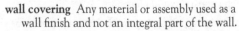

wall bracket

wall covering Any material or assembly used as a wall finish and not an integral part of the wall.

wall crane A crane with a horizontal arm supported from a wall or columns of a building. The arm may support a trolley.

wall expansion A break designed into a wall to allow room for building materials to expand and contract with thermal conditions.

wall footing A strip of reinforced concrete that is wider than the wall it supports and thus distributes the wall load over a wider area.

wall form A retainer or mold erected to give the necessary shape, support, and finish to a concrete wall.

wall furring Strips of wood or shaped sheet metal attached to a rough wall to provide a plane on which lath and plaster, paneling, or wainscoting may be installed.

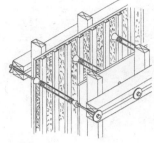

wall form

wall grille A perforated plate, casting, punched sheet, or frame used to conceal an opening, radiator, or the like, while allowing a passage for air.

wall guard A protective, resilient strip attached to a wall to protect the surface from carts, transporters, or other movable conveyors.

wall hanger A stirrup or bracket fixed to a wall and used to support one end of a horizontal member.

wall height The vertical distance from the top of a wall to its support, such as a foundation or support beam.

wall hook 1. A special large nail or hook used as a beam anchor or used on a beam plate. 2. A hook or bracket fixed in a masonry wall to hold downspouts, lightning rods, etc.

wall outlet An electric outlet mounted in a wall with a decorative cover.

wall plate The top plate in construction, placed on top of studs and bearing the joists of the next floor above.

wall puller A simple hand tool used to help position a stud wall or similar construction.

wall rib A longitudinal rib against an exterior wall of a vaulting compartment.

wall shaft A small column, supported on a corbel or bracket that appears to support a rib of a vault.

wall sign 1. A sign mounted on, or fastened to, a wall. 2. A sign attached to the exterior wall of a building and projecting not more than a code-defined distance.

wall spacer A metal tie used to hold forms in position until concrete has set.

wall tie A metal strip or wire used to tie masonry wythes together, or tie a masonry veneer to a wood or concrete frame or wall.

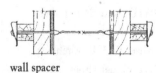

wall spacer

wall tile A glazed tile used in a wall facing.

wall-wash luminaire A luminaire located adjacent to or on a wall with most of its light directed onto the wall.

wane A defective edge of a board due to remaining bark or a beveled end. A wane is usually caused by sawing too near the surface of the log.

ward 1. A baffle in a lock to prevent use of an unauthorized key. 2. A division of a hospital or jail.

warehouse set The partial hydration of cement stored for a time and exposed to atmospheric moisture, or mechanical compaction occurring during storage. A partially hardened, unopened bag of cement.

warm-air heating system A heating system in which warm air is distributed through a single register or series of ducts. Circulation may be by convection (gravity system) or by a fan in the ductwork (forced system).

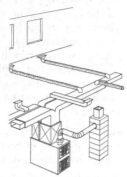

warm-air heating system

warning pipe An overflow pipe with an outlet located so that discharge can be readily observed.

warping A deviation of a slab or wall surface from its original shape, caused by temperature and/or moisture differentials within the slab or wall.

warping joint A longitudinal or transverse joint, with bonded steel or tie bars passing through it, with the sole function of permitting warping of pavement slabs when moisture and temperature differentials occur in the pavement.

warranty A premise made by a seller or contractor responsible for work performed under a contract that the work performed is fit for the purpose intended and is free from structural, electrical, mechanical and other defects.

warren truss (warren girder) A truss consisting of horizontal top and bottom chords, separated by sloping members, and without vertical pieces.

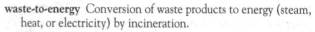

warren truss (warren girder)

wash The top surface of a building element, such as a coping or sill, intended to shed water.

washable Descriptive of a material that may be washed repeatedly without a noticeable deterioration in appearance or function.

wash boring A boring made in the earth by a rotary drill that uses water to stabilize the sides of the hole.

wash coat A thin, almost transparent coat of paint applied to a surface as a sealer or stain.

wash cut A beveled cut in a stone windowsill that diverts water away from the window and the structure.

washer 1. A flat ring of rubber, plastic, or fibrous material, used as a seal in a faucet or valve or to minimize leakage, as in a threaded connection. **2.** A flat ring of steel that may be split, toothed, or embossed; used in threaded connections to distribute loads, span large openings, relieve friction, or prevent loosening.

washer

wash primer Any thin paint that promotes adhesion of the subsequent coat.

waste The difference between the quantity that must be included in an estimate and the quantity that is actually to be installed, added to the quantity as a percentage. Applicable to materials, waste should not be applied to labor or confused with productivity.

waste and overflow fitting A common bathtub fixture that provides one outlet for a tub drain and another in case water rises above a safe level.

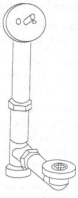

waste and overflow fitting

waste-disposal unit An electrically operated device for grinding waste food and disposing of it through the plumbing drainage pipes.

waste heat recovery An economical measure that involves capturing the heat produced by a process in a cyclic process to serve another need, such as preheating combustion air.

waste management Activities undertaken to minimize, contain, control, store, transport, treat, or dispose of waste material.

waste pipe A pipe to convey discharge from plumbing pipes.

waste-to-energy Conversion of waste products to energy (steam, heat, or electricity) by incineration.

wastewater Water that has been used in a process and contains contaminants.

wasting (dabbing) Splitting of excess stone with a hand tool so that the resulting surfaces of block are nearly flat.

water-absorption The amount of water absorbed by a material under specified test conditions. Commonly expressed as a weight percent of the test specimen.

waterborne preservative Preservative salts dissolved in water and transferred to the wood during the treating process.

water-cement ratio The ratio of the amount of water, exclusive only of that absorbed by the aggregates, to the amount of cement in a concrete or mortar mixture. The ratio is preferably stated as a decimal by weight.

water closet (WC, flushable closet, toilet) A plumbing fixture used to receive human wastes and flush them to a waste pipe.

water crack A fine crack in a coat of plaster, caused by applying a coat over a previous coat that had not dried sufficiently or by using plaster with a water content that is too high.

water closet (WC, flushable closet, toilet)

water gauge A manometer filled with water.

water-gel explosive A wide variety of explosives, all containing water and ammonium nitrate, used for blasting. The explosives are sensitized with an explosive, such as TNT or smokeless powder, with metals such as aluminum, or with other fuels. They may be premixed at a plant or site-mixed immediately prior to loading in a blast hole.

water hammer A loud thumping noise in a water service line due to the surge of suddenly checked water.

water level A simple device used to see that surfaces are level. Most often a long (up to 50'), thin plastic tube filled with water.

water-level control A control on a boiler used to maintain the water at a safe level.

W

water lime A hydraulic lime or cement that will set under water.

waterline The highest water level in a cistern or flush tank to which the shutoff should be adjusted.

water main A main supply pipe in a water system providing water for public or community use.

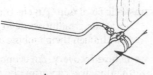
water main

water outlet 1. Any opening or end of pipe used to discharge water to a plumbing fixture, boiler, or other device. **2.** An opening through which water is discharged to the atmosphere.

waterproof adhesive An adhesive that, when properly cured, is not affected by water.

"waterproofed" cement Cement interground with a water-repellent material such as calcium stearate.

waterproofing Any of a number of materials applied to various surfaces, e.g., a building foundation, to prevent the infiltration of water.

waterproofing

water reactive A substance that generates heat or gas in the presence of water.

water-reducing agent A material that either increases workability of freshly mixed mortar or concrete without increasing water content, or maintains workability with a reduced amount of water; the effect being due to factors other than air entrainment.

water-repellent cement A hydraulic cement having a water-repellent agent added during the process of manufacture, with the intention of resisting the absorption of water by the concrete or mortar.

water-repellent paper Gypsum board paper surfacing that has been formulated or treated to resist water penetration.

water retentivity The property of mortar that prevents rapid loss of mix water by absorption.

water ring A perforated manifold in the nozzle of dry-mix shotcrete equipment through which water is added to the materials.

water-saving toilet A toilet that uses between 1.6 and 3.5 gallons per flush.

water seal The water in a trap acting as a seal against the passage of gases.

water-service pipe The part of a water-service main owned by the water department.

watershed The area of land that drains naturally into a stream or complex of streams.

water softener A device to remove calcium and magnesium salts from a water supply, usually by ion exchange.

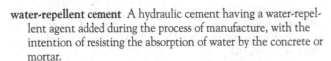

waterstop A thin sheet of metal, rubber, plastic, or other material inserted across a joint to obstruct the seeping of water through the joint.

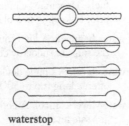
waterstop

water-supply stub A vertical supply pipe that supplies fixtures and that reaches less than one story in height.

water-supply system 1. The system that supplies water throughout a building, including the service pipe(s), distribution and connecting pipes, fittings, and control valves. **2.** The system that supplies water to units in a community, including reservoirs, tunnels, and pipelines.

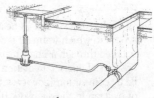
water-supply system

water table 1. The top surface of groundwater. **2.** A horizontal, sloped ledge on an exterior wall with a drip molding to prevent water from running down the outside of the wall below.

water tap (faucet) An outlet valve for water.

watertight A joint, enclosure, or assembly that is designed to prohibit the entry of water.

water tube boiler A boiler in which water circulates through the tubes, while combustion gases surround the tubes.

water tap (faucet)

water vapor diffusion The movement of water vapor through permeable materials to areas of different water vapor pressure.

water vapor transmission The rate of water vapor flow, under steady specified conditions, through a unit area of a material, between its two parallel surfaces. Metric unit of measurement is 1 g/24 h. × m^2 × mm Hg.

watt A unit of power equal to the power dissipated in an electric circuit in which a potential difference of 1 volt causes a current flow of 1 ampere. The power required to do work at the rate of 1 joule per second.

watt-hour A unit of work equivalent to the power of 1 watt operating for 1 hour, which is equal to 3,600 joules.

wattmeter A calibrated instrument that measures electric power in watts.

wave front The surface of the wave sphere created when sound waves radiate from the source in all directions, forming a spherical shape.

wavelength The distance between two successive points in a sound or light wave that are characterized by the same phase of oscillation.

wax A material obtained from vegetable, mineral, and animal matter that is soluble in organic solvents, and solid at room temperature. Wax is applied in a liquid or paste form on wood and metal surfaces to provide gloss, and to protect the surface.

wearing course A topping or surface treatment to increase the resistance of a concrete pavement or slab to abrasion.

weather 1. The length of shingle or tile that is exposed, as measured along the slope of a roof. **2.** To deteriorate or discolor when exposed to the weather. **3.** To slope a surface for the purpose of shedding rainwater.

weatherboard **1.** Boards or siding lapped to shed water. **2.** A projecting member attached to the bottom rail of an external door to divert water from the sill or threshold.

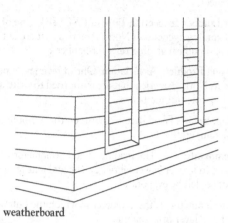

weatherboard

weathered **1.** Descriptive of a surface exposed long enough that distinct weathering has taken place. **2.** Having an upper surface that is sloped to shed water.

weathering **1.** Changes in color, texture, strength, chemical composition, or other properties of a natural or artificial material due to the action of the weather. **2.** The mechanical or chemical disintegration and discoloration of the surface of wood, caused by exposure to light, the action of dust and sand carried by the wind, and the alternate shrinking and swelling of the surface fibers due to continual variation in temperature and moisture content brought on by changes in the weather.

weather resistance The ability of a material or coating to resist weathering.

weatherseal channel A channel installed, flanges downward, on the top of an exterior door.

weatherstrip A strip of wood, metal, felt, plastic, or other material applied at an exterior door or window to seal or cover the joint made by the door or window with the sill, casings, or threshold.

weather-struck joint (weathered joint) A horizontal masonry joint sloped from a point inside the face of the wall to the face of the wall so the joint will shed water.

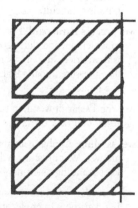

weather-struck joint (weathered joint)

weaving The alternate lapping of shingles on opposite surfaces when two adjacent roofs intersect.

web **1.** That part of a beam or truss between the flanges or chords, used mainly in resisting shear stresses. **2.** The walls connecting the face shells of a hollow concrete masonry unit.

web crippling Local failure of the web of a structural member caused by a concentrated load.

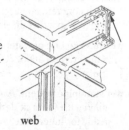

web

web member In a truss, a secondary structural member that joins the top and bottom chords.

web plate A steel plate forming the web of a built-up girder truss, or beam.

web reinforcement Reinforcement placed in a concrete member to resist shear and diagonal tension.

web splice A splice joining two web plates.

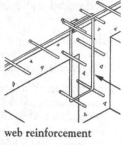

web reinforcement

web stiffener **1.** A vertical steel shape attached to the web of a structural member and used to prevent web buckling. **2.** Blocking added to the web of a composite wood structural member to prevent web buckling.

wedge A piece of wood or metal tapering to a thin edge, used to adjust elevation or tighten formwork.

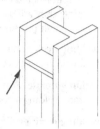

web stiffener

wedge anchor A wedging device used in the anchorage of a tendon in posttensioned, prestressed concrete.

weep cut A groove cut in the underside of a horizontal board or masonry unit that extends from an exterior wall to keep water from moving back toward the wall.

weep hole (weeps) **1.** A small hole in a wall or window member to allow accumulated water to drain. The water may be from condensation and/or surface penetration. **2.** A small hole in a retaining wall located near the lower ground surface. The hole drains the soil behind the wall and prevents build-up of water pressure on the wall.

weighted average An average in which certain entries, usually at the extremes of the range of values, are manipulated so a number is obtained that is believed to be more representative of the true mean than that obtained by straight averaging.

weir A structure across a ditch or stream used for measuring, or for diverting the flow of water.

weir head The depth of water above the top of a flat weir, or above the bottom of a notch in a notched weir.

weld **1.** To build up or fasten together, as with cements or solvents. **2.** To fasten two pieces of metal together by heating them until there is a fusing of material, either with or without a filler metal.

welded butt splice A reinforcing bar splice made by welding the butted ends.

welded cover plate A cover plate welded to the flange of a structural member, usually to increase strength in the middle length of a span.

welded reinforcement Concrete-reinforcing steel joined by welding, and most often used in columns to extend vertical bars.

welded-wire fabric (welded-wire mesh) A series of longitudinal and transverse

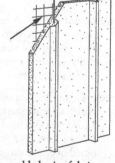

welded-wire fabric (welded-wire mesh)

W

wires of various gauges, arranged at right angles to each other and welded at all points of intersection; used for concrete slab reinforcement.

welding cables The pair of cables supplying electric energy for use in welding. One lead connects a welding machine with an electrode; the other lead connects the machine with the work.

welding fittings Wrought steel elbows, tees, reducers, saddles, and the like, beveled for butt welding to pipe. Forged fittings with hubs or with ends counter-bored for fillet welding to pipe; they are used for small pipe sizes and high pressures.

welding fitting

welding neck flange A flange with a long neck that is beveled for butt-welding to pipe.

welding nozzle A stub-pipe that is shop-welded to a vessel to facilitate welding a connecting pipe in the field.

welding rectifier Equipment used in arc welding to convert AC power to DC power.

welding rod Filler metal in the form of wire or rod used in gas welding and brazing and in some arc welding processes.

welding screw A screw with lugs or weld projections on the top or underside of the head to facilitate attaching the screw to a metal part by resistance welding.

weldment Any assembly made by welding parts together.

weld metal That part of a weld that was melted while welding.

weld nut A nut designed to be attached to a part by resistance welding.

well (wellhole) 1. Any enclosed space of considerable height, such as an air shaft or the space around which a stair winds. 2. A correction device for ground water. 3. A wall around a tree trunk to hold back soil. 4. A slot in a machine or device into which a part fits.

well drill A rotary drill mounted on a truck and used to drill for water.

well-graded aggregate Aggregate with a particle-size distribution producing maximum density, or minimum void space.

well point A perforated pipe surrounded by pervious soil to permit the pumping of groundwater.

well-point system A series of well points connected to a header and used to drain an area or to control groundwater seepage into an excavation.

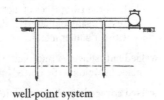

well-point system

welt 1. A seam in sheet metal, formed by folding over the edges of two sheets, interlocking the folded portions, and flattening the formed seam. 2. A strip of wood fastened over a seam, joint, or shaped piece fastened over an angle for reinforcing.

welt

welting strip A strip of sheet metal at the intersection of the roof and a vertical surface. One edge of the strip is fastened to the roof, and the other edge is bent to lock with the lower edge of a vertical sheet.

West Coast Lumber Inspection Bureau (WCLIB) One of seven regional grading agencies in North America that are authorized to write and publish grading rules for lumber.

Western Wood Products Association One of seven regional grading agencies in North America that are authorized to write and publish grading rules for lumber.

wet-bulb depression The difference between the dry-bulb and the wet-bulb temperatures.

wet-bulb temperature 1. The reading on a thermometer whose bulb is enclosed in a layer of wet fabric. 2. The ambient temperature of an object cooled by evaporation.

wet-bulb thermometer A thermometer in which the bulb is enclosed in a layer of wet fabric.

wet construction Any construction using materials that are placed or applied in other than a dry condition, as in concreting or plastering.

wet mix Concrete containing too much water, immediately evidenced by a runny consistency.

wet-mix shotcrete Shotcrete in which all ingredients, including mixing water, are mixed before introduction into the delivery hose. The shotcrete may be pneumatically conveyed or moved by displacement.

wet process In the manufacture of cement, the process in which the raw materials are ground, blended, mixed, and pumped while mixed with water. The wet process is chosen where raw materials are extremely wet and sticky, making drying before crushing and grinding difficult.

wet process hardboard Hardboard manufactured by a process in which a wood slurry is combined with a resin binder; and is then dried, first on a screen and later under pressure. The pressure imparts high density to the board and also sets the binder.

wet rot Decay of wood, in the presence of moisture and warmth, as a result of attack by fungi.

wet sanding Applying an abrasive to a surface while one or both are wet. Wet finishing reduces dust and keeps sanding machinery cleaner.

wet screeds Concrete strips placed beforehand at the proper elevation to act as height guides when pouring a concrete slab.

wet screening Screening to remove from fresh concrete any aggregate particles larger than a certain size.

wet sieving Use of water during sieving of a material on a No. 200 or (even finer) No. 325 sieve.

wet standpipe system A standpipe system filled with water at design pressure, ready for immediate use.

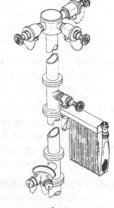

wet standpipe system

wet strength The measured strength of an adhesive joint after it has been submerged in a liquid.

wettability The degree to which a surface will absorb a liquid poured upon it.

wettest stable consistency The condition of maximum water content at which cement grout or mortar will adhere to a vertical surface without sloughing.

wetting 1. The soaking of asbestos materials to diminish the level of air contamination they give off when disturbed. The water is mixed with saturants and surfactants before it is applied to the asbestos. **2.** Coating a base metal with filler metal prior to soldering or brazing.

wetting agent A substance capable of lowering the surface tension of liquids, facilitating the wetting of solid surfaces and permitting the penetration of liquids into the capillaries.

wet-use adhesive Adhesives used in glue-laminated timber that will perform satisfactorily under a wide variety of uses, as well as exposure to weather and dry atmosphere.

wheelabration The application of steel abrasives in the form of grit or shot to clean a surface.

wheelbarrow A hand cart with one wheel in front plus two legs and two handles in back; used to move materials short distances.

wheel ditcher A trench digger consisting of a vehicle equipped with a large wheel on which buckets are mounted.

wheel excavator (continuous excavator) 1. A self-propelled unit for excavating a bank with buckets mounted on wheels. The earth is deposited on a conveyor belt for loading up hauling units. **2.** Any power-driven machine used in excavation that moves on wheels, rather than crawlers.

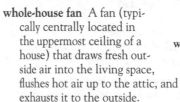

wheel excavator
(continuous excavator)

wheel trencher A trencher using a rotating wheel made of two ring plates with buckets every 20° to 30° around the circumference.

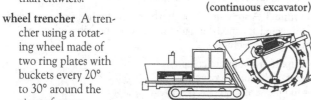

wheel trencher

whetstone A piece of natural or manufactured abrasive stone used to sharpen cutting tools.

whip A slang term that describes the vertical movement of improperly secured underlayment.

white coat (finish coat) A lime-putty plaster coat with a troweled finish.

white lead Basic lead carbonate, white in color; used as a pigment in exterior paints, ceramics, and putty.

white Portland cement Portland cement made from materials with low iron content, used to produce concrete or mortar that is white in color.

white roof systems Roofing systems that reflect heat and have proven to reduce the net cooling requirements of a facility by as much as 20%.

white rot A type of decay in wood caused by a fungus that leaves a white deposit.

whitewash Water added to quicklime or slaked lime, whiting, and glue and applied like paint, leaving a white coating.

whiting Calcium carbonate pigment, used as an extender in paint, putty, and whitewash.

Whitney strain diagram The assumed strain distribution of a reinforced concrete member in the yield state that is used in ultimate strength design.

whole-brick wall A brick wall having a thickness equal to the length of one brick.

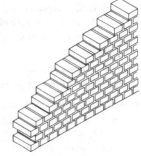

whole-brick wall

whole-building Integration of a building's systems to maximize sustainable and/or economic functioning by considering many factors including use of energy and other resources, building materials, site preservation, and indoor air quality so that a structure can run at its maximum efficiency; enhance user health, comfort, and productivity; and have the least impact on the environment.

whole design solution A design approach that considers and integrates all building systems, starting with the early design stages—for optimum efficiency and sustainability in the structure. Also holistic, integrated or whole building design.

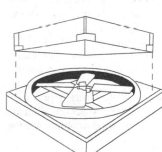

whole-house fan A fan (typically centrally located in the uppermost ceiling of a house) that draws fresh outside air into the living space, flushes hot air up to the attic, and exhausts it to the outside.

whole-house fan

wicking The tendency of water or any liquid to rise above an established level to saturate a porous solid.

wide-flange beam A hot-rolled steel beam having a cross section resembling an H and having wider flanges than an I-beam.

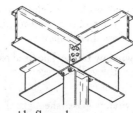

wide-flange beam

wide-throw hinge A rectangular hinge having extra-wide leaves.

Williot diagram A graphical method used to determine deflections of trusses.

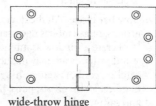

wide-throw hinge

winch A stationary hoisting machine having a rotating drum around which a cable, rope, or chain is wrapped.

windage loss Water removed by circulating air, as in a cooling tower.

wind brace A brace provided in a frame to support the frame against wind loads.

wind column A vertical support to a wall system that is designed to help the system withstand wind loads.

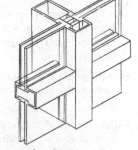

winch

wind drift The horizontal deflection of a structure as a result of wind.

wind electric system A single turbine, smaller than the utility-scale models, but much more efficient than the old-fashioned windmill, producing clean, affordable electricity for a rural home, farm, or business.

winder A tread that is used where a stair turns a horizontal angle, or in a spiral stair; shaped like a wedge or truncated wedge.

windlass A device for lifting heavy objects, usually consisting of a horizontal drum turned by a lever or crank. A cable runs from the object over a pulley to the drum where a cable is wound or unwound to move the object.

wind load The horizontal load used in the design of a structure to account for the effects of wind.

window 1. A normally glazed opening in an external wall to admit light and, in buildings without central air-conditioning, air. 2. An assembly consisting of a window frame, glazing, and necessary appurtenances. 3. A small opening in a wall, partition, or enclosure for transactions, such as a ticket window or information window.

window

window cleaner's anchor An anchor, fastened near a window on an outside wall of a multistory building, to which a person washing the window from the outside can attach a safety belt.

window cleaner's platform A platform suspended from a trolley on the roof of a multistory building used for outside window-washing and maintenance.

window frame The fixed part of a window assembly attached to the wall and receiving the sash or casement and necessary hardware.

window glass (sheet glass) A soda-lime-silica glass made in continuous sheets of varying thickness and cut to size as required.

window hardware All the devices, fittings, or assemblies necessary to operate a window as intended. Window hardware may include catches, cords, fasteners, hinges, handles, locks, pivots, pulls, pulleys, and sash weights.

window frame

window lift (sash lift) A handle fastened to the lower sash of a sliding window for use in moving the sash up or down.

window of time analysis A method of measuring labor productivity losses on a specific project by comparing labor productivity achieved during a period of time when work activities are subject to disruptive events and conditions, versus productivity during a period of time without disruption.

window schedule A tabulation, usually on a drawing, listing all windows on a project; and indicating sizes, number of lights, type of sash and frame, and hardware required.

window spring bolt A spring bolt, used on a sash that is not counter-balanced, that holds the sash in a desired position.

window tinting Film applied to windows to reduce the amount of solar heat transmission through the glass by increasing the solar reflection (not necessarily visible reflection) and solar absorption of the glass.

window trim The finished casing around a window.

window wall An exterior curtain wall using a frame containing windows that may be fixed or operable. The glazing may be clear, tinted, and/or opaque.

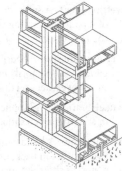

window wall

window well A well dug outside of a below-grade window, commonly a basement window, to allow it to operate.

wind power Energy from wind, usually collected by wind turbines.

wind pressure The pressure produced when wind blows against a surface.

windrow 1. A ridge of loose soil, such as that produced by the spill off of a grader blade. 2. A row of leaves or snow heaped up by the wind.

wind stop 1. A weatherstrip used around a door or window. 2. A strip of wood or metal that covers the joint where a sash or casement meets a stile. 3. Any wood or metal strip used to cover a crack in an exterior wall of a building.

wind tunnel A structure through which a controlled stream of wind is directed at a model in order to study the probable effects of wind on a structure.

wind uplift The upward component of the force produced as wind blows around or across a structure or an object.

wing 1. A section or addition extending out from the main part of a building. 2. The off-stage space at a side of a stage. 3. One of the four leaves of a revolving door.

wing

wing nut A nut provided with wing-like projections to facilitate turning by fingers and thumb.

wing pile A bearing pile (post), usually of concrete, widened at the upper portion to form part of a sheet pile wall.

wing screw A screw provided with wing-like projections to facilitate turning by fingers and thumb. A wing screw is often used to secure windows against unauthorized entry.

W

wing wall 1. A short section of wall at an angle to a bridge abutment, used as a retaining wall and to stabilize the abutment. **2.** A short section of wall used to guide a stream into an opening, such as at a culvert or bridge.

wing wall

wiped joint A wiped lead joint connecting two lead pipes.

wire A drawn metal strand or filament.

wire cloth A stiff fabric of woven wire, usually having a larger mesh than insect screen material, and used as reinforcing in plaster, in sieves, and as a leaf catcher in gutters. The number of openings per square inch designates the fineness of the mesh.

wire, cold-drawn Wire made from the rods hot-rolled from billets and then cold-drawn through dies.

wire comb A tool with long wire teeth, used to scratch a coat of wet plaster to increase the bond of a subsequent coat.

wire-cut brick Brick that has been extruded to shape and then cut to length with wire before firing.

wire gauge 1. The diameter of a wire as defined by several different systems. Ordinarily, the thicker the wire is, the smaller the gauge number is. **2.** A device for measuring the thickness of a wire; usually consists of a metal sheet with standard-sized notches on one or more edges.

wire glass Sheet glass with wire mesh embedded in the glass to prevent shattering.

wire holder An electrical insulator having a screw or bolt for fastening the insulator to a support.

wire lath A welded-wire mesh used as a base for plaster.

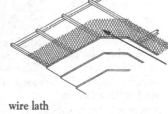

wire lath

wire nail A very thin nail with a small head, often used as a finishing nail.

wire nut A connector for two or more electric conductors, made in the form of a plastic cap with an internal spring-thread. It is turned over the parallel or twisted ends of the conductor wires.

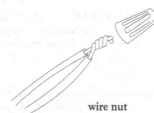

wire nut

wire rope A rope made of twisted steel strands laid around a central core.

wire saw A machine for sawing stone using a rapidly moving continuous wire carrying a slurry of sand or other abrasive material.

wireway A raceway with a removable cover for access to wires inside.

wire winding Application of high tensile wire, wound under tension by machines, around concrete circular or dome structures to provide tension reinforcing.

wiring Installing or connecting electrical components. Also refers to systems of wires for electrical work, such as lighting, switchboards, and so forth.

wiring box A box used in interior electric wiring at each junction point, outlet, or switch, that serves as protection for electric connections and as a mounting for fixtures or switches.

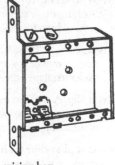

wiring box

witness corner A marker set on a property line near to, but not at, a corner. Its relation to the corner is recorded.

wobble-wheel roller A pneumatic compactor consisting of a weighted bed mounted on a series of wheels that are loosely assembled, and allowed to work the surface of a soil.

wood block floor A finished floor consisting of rectangular blocks of a tough wood such as oak, set in mastic with the end grain exposed, usually over concrete slab. A wood block floor is used where very heavy traffic and heavy loads are expected.

wood-cement concrete A Portland cement concrete using sawdust and wood chips as aggregate.

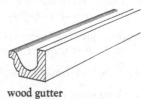

wood-cement concrete

wood-cement particle (WCP) board A high-density board manufactured in Europe for use on exteriors, or where fire resistance is needed. Wood particles are combined with Portland cement, or another mineral, as a binder.

wood-fibered plaster A calcined gypsum plaster containing shredded or ground wood fiber added during manufacture.

wood filler A liquid or paste compound used to fill pores or checks in wood before finishing.

wood fire-retardant treatment The impregnation of wood or wood products with special solutions to reduce the flame spread of the finished product.

wood firestop Wood pieces that are placed between framing members in a wall to slow the spread of fire.

wood flooring Flooring consisting of dressed and matched boards.

wood flour A finely ground, dried, wood powder that is used in plastic wood as an extender. It is also employed in some glues, and in the molding of plastics.

wood-frame construction A type of construction where floors and roofs, as well as exterior and other bearing walls are of wood, rather than masonry.

wood gutter A gutter, under the eaves of a roof, made from a solid piece of wood or built-up boards.

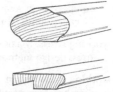

wood gutter

wood lath Narrow strips of wood used as a base for plaster.

wood molding Wood strips that are factory-shaped to commercially available patterns.

wood oil 1. Tung oil used in the manufacture of varnish, putty, etc. **2.** An oleoresin used for caulking and waterproofing.

wood molding

W

wood preservative Any chemical preservative for wood, applied by washing-on or pressure impregnating. Products used include creosote, sodium fluoride, copper sulfate, and tar or pitch.

wood screw A threaded fastener that cuts its way into wood when turned under axial pressure.

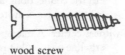

wood screw

wood slip A wooden nailing strip fixed in a masonry or concrete wall as a means of attaching wood trim or furring strips.

wood stud anchor (nailing anchor) A metal clip attached to the inside of a door frame and used to secure the frame to a wood stud partition.

wood treatment Treatment of wood with a preservative to prevent or retard its decay.

woodwork Work produced by carpenters and woodworkers, especially the finished work.

woodwork

work 1. All labor and materials required to complete a project in accordance with the contract documents. **2.** The product of a force times the distance traveled.

workability That property of freshly mixed concrete or mortar that determines the ease and homogeneity with which it can be mixed, placed, compacted, and finished.

work breakdown structure (WBS) A hierarchical breakdown of a project that contains successive levels of detail. Provides a way to incorporate project details as they become available without having to prepare an entirely new estimate or budget at each new level.

work capacity The greatest volume in number and/or size of construction projects that a contractor can manage efficiently without increasing the overhead costs.

workflow The sequences of task-related communication among people (normally the project team) to accomplish sequences of tasks and the needed data flows to support those sequences.

working (movement) The alternate swelling and shrinking of wood caused by changes in its moisture content induced by variations in the humidity of its environment.

working life The length of time a liquid resin or adhesive remains useful after the ingredients have been mixed.

working load Forces normally imposed on a member in service.

working stage A section of an assembly room or auditorium partially cut off from the audience section by a proscenium wall. It is equipped with some or all of the following: scenery loft, gridiron, fly gallery, and lighting equipment.

working stress Maximum permissible design stress using working stress design methods.

working stress design A method of proportioning structures or members for prescribed working loads at stresses well below the ultimate, and assuming linear distribution of flexural stresses.

work light 1. A light in a theater used to provide illumination for rehearsing, scene shifting, or other work onstage or backstage. **2.** A lamp in a protective cage, and having a long, heavy, flexible cord; used to provide temporary illumination in work areas.

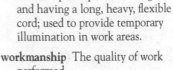

work light

workmanship The quality of work performed.

work package The grouping of tasks or scopes of work during the procurement phase that can be contracted together for coordination and economy of scale.

work plane In lighting, the plane at which work is usually done and at which the level of illumination is specified and measured.

worksheet The paper on which the calculations supporting the final estimate are recorded.

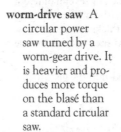

worm-drive saw A circular power saw turned by a worm-gear drive. It is heavier and produces more torque on the blasé than a standard circular saw.

worm-drive saw

woven valley A roof valley where shingles or other roofing materials from both sides of the valley are woven together as they are applied by overlapping alternate courses.

woven valley

woven-wire fabric A prefabricated steel reinforcement for concrete composed of cold-drawn steel wires mechanically twisted together to form hexagonally shaped openings.

wrap-around hinge An offset hinge attached to the rabbeted rail of a door or window, providing increased surface area for mounting.

wrap-around hinge

wrapping A method of applying narrow strips of veneer around a curved surface, such as a piece of furniture.

wreath A short, curved portion of a stair handrail that joins the newel post to the rail.

wrecking ball A heavy steel ball or concrete mass on a heavy chain or cable swung by a crane to demolish parts of a structure.

wrecking strip A small piece or panel fitted into a formwork assembly in such a way that it can be easily removed ahead of main panels or forms, making it easier to strip those major form components.

wrench A hand tool consisting of a handle and a jaw at one end; used to turn or hold a bolt, nut, pipe, or fitting. The jaw may be shaped for a specific-sized object or may be adjustable.

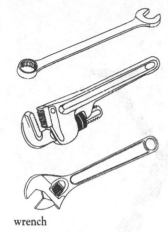

wrinkling (crinkling, reveling) 1. The distortion in a paint film that appears as ripples and may be deliberately induced as a decorative effect or accidentally caused by drying conditions or paint applied too thickly. **2.** The rippling or crinkling of the

wrench

surface of an adhesive, usually not affecting performance. **3.** The rippling of an area of veneer, caused by lack of contact with the adhesive.

wrought Descriptive of metals or metalwork shaped by hammering with tools.

wrought iron 1. Nearly pure ductile iron with a very small percentage of silica throughout. Once the surface iron decomposes, the silica surfacing prevents further oxidation. Wrought iron is no longer commercially available. **2.** A number of easily welded or wrought irons with low impurity content used for water pipes, tank plates, or forged work.

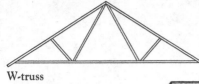

W-truss

wrought nail A nail wrought by hand, often having a head with a decorative pattern.

wrought timber Wood that has been planed on one or more surfaces.

wye

W-truss A wood roof truss with the web members in the shape of a "W."

wye A pipe fitting with a side outlet that is any angle other than 90° to the main run or axis.

wythe (leaf) Each continuous vertical section of a wall one masonry unit in thickness.

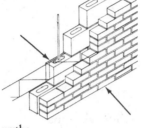

wythe

Abbreviations

X experimental

XBAR crossbar

XEPS extruded expanded polystyrene

XH, X HVY extra heavy

XL extra large

xr without rights

X STR extra strong

xw without warrants

XXH double extra heavy

Definitions

X-brace (cross brace) A paired set of sway braces.

xeriscaping Landscaping featuring native, drought-tolerant, well-

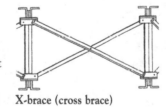

X-brace (cross brace)

adapted plant species, especially in dry climates, to avoid the need for irrigation. Xeriscaping typically calls for less fertilizer and fewer pest control measures than traditional landscapes.

xonolite Calcium silicate monohydrate (C_3S_2H), a natural mineral that is readily synthesized at 150°C to 350°C under saturated steam pressure. Xonolite is a constituent of sand-lime masonry units.

x-ray diffraction 1. The diffraction of x-rays by substances having a regular arrangement of atoms. **2.** A phenomenon used to identify substances having such structure.

x-ray fluorescence Characteristic secondary radiation emitted by an element as a result of excitation by x-rays, used to yield a chemical analysis of a sample.

x-ray protection Lead encased in sheetrock or plaster to prevent the escape of radiation from the encased room.

xylol An aromatic hydrocarbon distilled from coal tar, used as a solvent for paints and varnishes.

X

Y

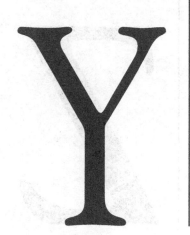

Abbreviations

y yard

Y yttrium, wye, Y-branch

yd yard

yp yellow pine

YP yield point

YR year

YS yield strength

Definitions

yakima pine *See* **ponderosa pine**.

yard **1.** A unit of length in the English system equal to 3′. **2.** A measure of concrete. One cubic yard = 3′ × 3′ × 3′ in volume, or 27 cubic feet. **3.** A term applied to that part of a plot not occupied by the building or driveway.

yardage **1.** An amount of excavated material equal to the volume in cubic yards. **2.** An area of surface (two dimensions) measured in square yards.

yard drain A drain in a pavement or earth surface, used to drain surface water.

yard lumber (general building construction lumber) Lumber graded according to its size, length, and intended use, stockpiled in a lumber yard.

Y-branch (wye branch) A plumbing system branch, in the shape of a Y.

yelm A bundle of reeds or straw used as thatching material for a roof.

Y-fitting (wye fitting) A pipe fitting in the shape of a Y. One arm is usually at 45° to the main fitting and may be of reduced size.

yield **1.** The volume of freshly mixed concrete produced from a known quantity of

Y-fitting

ingredients. **2.** The total weight of ingredients divided by the unit weight of the freshly mixed concrete. **3.** The number of product units, such as block, produced per bag of cement or per batch of concrete. **4.** A rate of return as calculated by the profit earned on an investment over a specified period of time.

yield point **1.** The point at which a stressed material begins to exhibit plastic properties. **2.** The point beyond which the material will not return to its original length.

yield strength The stress, less than the maximum attainable, at which the ratio of stress to strain has dropped well below its value at low stresses, or at which a material exhibits a specified limiting deviation from the usual proportionality of stress to strain.

Y-level A surveyor's level with the telescope and level supported by Y-shaped fittings. The telescope can be removed and reversed in order to average out internal sources of error.

yoke **1.** A tie or clamping device around column forms or over wall or footing forms to keep them from spreading as a result of lateral pressure of fresh concrete. **2.** Part of a structural assembly for slipforming that keeps the forms from spreading and transfers form loads to the jacks. **3.** A collar for supporting pipe. **4.** A soil pipe fitting in the shape of a Y. **5.** Pipe assembly used to install a water meter.

yoke

yoke vent **1.** A pipe connecting upward from a soil or waste stack to a vent stack for venting. **2.** A vertical or 45° relief vent of the continuous waste and vent type formed by the extension of an upright wye-branch inlet of the horizontal branch to the stack. Referred to as a *dual yoke vent* when two horizontal branches are thus vented by the same relief vent.

Young's modulus *See* **modulus of elasticity**.

Y-shaped fitting Plumbing fitting in the shape of a Y. Also called a wye.

Z

Abbreviations

z zero, zone

Z modulus of section

ZI zone of interior

Zn azimuth, zinc

Definitions

Z-bar A Z-shaped member that is used as a main runner in some types of acoustical ceiling.

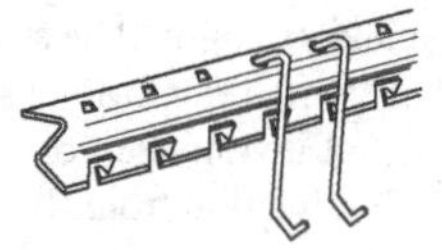

Z-bar flashing Z-shaped flashing installed above horizontal door or window trim to keep rainwater from seeping into a structure.

zee A light-gauge member with a Z-like cross section. The flanges of the Z are approximately at right angles to the web.

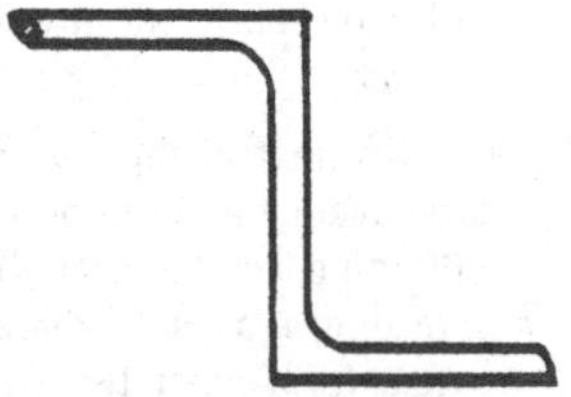

zeolite A group of hydrous aluminum silicate minerals or similar synthetic compounds used in water-softening equipment.

zero lot line A form of cluster housing development in which individual dwelling units are placed on separately-platted lots with very small yards and in some there is no side yard. They may be attached to one another, but not necessarily.

zero-slump concrete Concrete of stiff or extremely dry consistency showing no measurable slump after removal of the slump cone.

zero-water urinal An energy-efficient, wall-mounted urinal that uses no running water, other than for occasional servicing to clean the unit.

Z-bar

zee

z flashing Z-shaped metal flashing applied between panels of plywood siding to shed water.

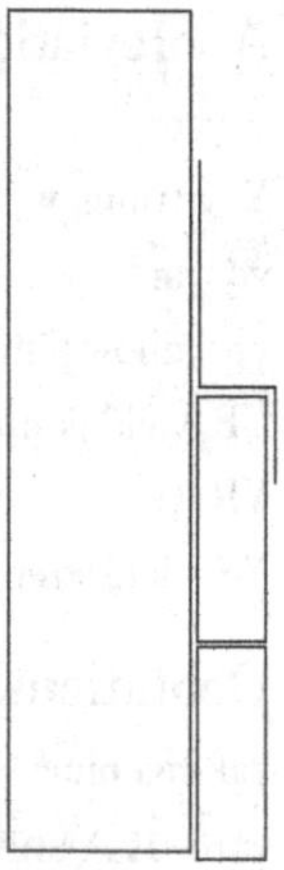

z-furring channel A z-shaped, galvanized steel track used to mechanically attach blanket or rigid plastic foam insulation to interiors of exterior walls. Also used to attach drywall to concrete, brick, concrete block, and tile.

zigzag rule A folding ruler made of pieces connected by a pivot.

zinc A metallic element used for galvanizing steel sheet and steel or iron castings, as an alloy in various metals, as an oxide for white paint pigment, and as a sacrificial element in a cathodic protection system. Characterized by its brittleness in ordinary temperatures.

zinc dust Zinc ground to a fine powder and used as a pigment in priming paints, especially those for use on galvanized surfaces.

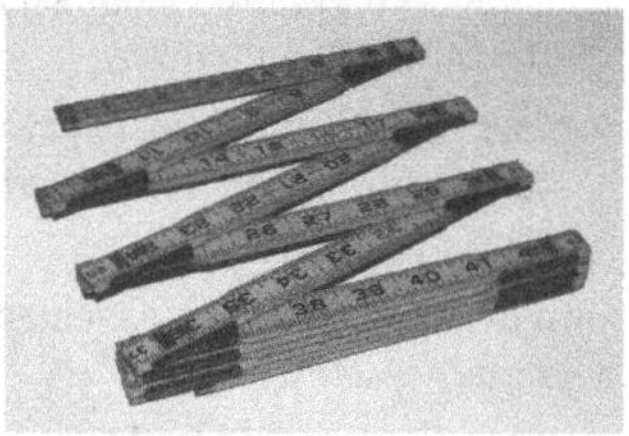

zinc oxide (zinc white) A pigment used in paints to provide durability, color retention, and hardness, and to improve sag resistance.

z flashing

zigzag rule

zone 1. A space or group of spaces in a building with common control of heating and cooling. 2. A form of public control over land use. 3. An area covered by a sprinkler system.

zoned building design Separating areas of a facility by establishing a series of clearly discernible zones to control access.

zone valve A control device for the flow of water or steam to areas, or zones, of a building.

Appendix: Symbols

Lighting Outlets

○	Ceiling surface incandescent fixture
⊢○	Wall surface incandescent fixture
Ⓡ	Ceiling recess incandescent fixture
⊢Ⓡ	Recess incandescent fixture
A ○ 3b	Standard designation for all lighting fixtures; A= fixture type, 3 = circuit number, b = switch control
Ⓑ	Ceiling blanked outlet
⊢Ⓑ	Wall blanked outlet
Ⓔ	Ceiling electrical outlet
⊢Ⓔ	Wall electrical outlet
Ⓙ	Ceiling junction box
⊢Ⓙ	Wall junction box
Ⓛ PS	Ceiling lamp holder with pull switch
⊢Ⓛ PS	Wall lamp holder with pull switch
Ⓛ	Ceiling outlet controlled by low voltage switching when relay is installed in outlet box
⊢Ⓛ	Wall outlet—same as above
◇	Outlet box with extension ring
EX →	Exit sign with arrow as indicated
▭O▭	Surface fluorescent fixture
▭O▭ P	Pendant fluorescent fixture
▭OR▭	Recessed fluorescent fixture
⊔O⊔	Wall surface fluorescent fixture
╾═╾	Channel mounted fluorescent fixture
▭O▭▭▭	Surface or pendant continuous row fluorescent fixtures

OR▭▭	Recessed continuous row fluorescent fixtures
●	Incandescent fixture on emergency circuit
▭•▭	Fluorescent fixture on emergency circuit

Receptacle Outlets

⊢⊖	Single receptacle outlet
⊢⊖	Duplex receptacle outlet
X ⊢⊖	Duplex receptacle outlet × indicates above counter max height = 42″ or above counter
⊢⊖	Weather proof receptacle outlet
⊢⊕	Triplex receptacle outlet
⊢⊕	Quadruplex receptacle outlet
⊢⊖	Duplex receptacle outlet—split wired
⊢⊕	Triplex receptacle outlet—split wired
⊢△	Single special purpose receptacle outlet
⊢△	Duplex special purpose receptacle outlet
⊢⊖ R	Range outlet
⊢⬤ DW	Special purpose connection—dishwasher
⊢⊖ XP	Explosion-proof receptacle outlet. max. height = 36″ to [cl]
⊢⊕ X	Multi-outlet assembly
Ⓒ	Clock hanger receptacle
Ⓕ	Fan hanger receptacle
⊡⊖	Floor single receptacle outlet

Floor duplex receptacle outlet

 Floor special purpose outlet

 Floor telephone outlet—public

 Floor telephone outlet—private

 Underfloor duct and junction box for triple, double, or single duct system as indicated by number of parallel lines

 Cellular floor header duct

Switch Outlets

S Single pole switch, max. height = 42″ to [CL]

S_2 Double pole switch

S_3 Three-way switch

S_4 Four-way switch

S_D Automatic door switch

S_K Key operated switch

S_P Switch and pilot lamp

S_{CB} Circuit breaker

S_{WCB} Weatherproof circuit breaker

S_{MC} Momentary contact switch

S_{RC} Remote control switch (receiver)

S_{WP} Weatherproof switch

S_F Fused switch

S_L Switch for low voltage switching system

S_{LM} Master switch for low voltage switching system

S_T Time switch

S_{TH} Thermal rated motor switch

S_{DM} Incandescent dimmer switch

S_{FDM} Fluorescent dimmer switch

$\vdash\!\ominus_S$ Switch and single rceptable

$\vdash\!\ominus_S$ Switch and double recpetacle

$\vdash\!\ominus_A$
S_A Special outlet circuits

Institutional, Commercial, and Industrial System Outlets

Nurses' call system devices—any type

Paging system devices—any type

Fire alarm system devices—any type

F   Fire alarm manual station—max. Height = 48″ to [CL]

F   Fire alarm horn with integral warning light

Fire alarm thermodetector, fixed temperature

S   Smoke detector

Fire alarm thermodetector, rate of rise

F   Fire alarm master box—max. Height per fire department

H   Magnetic door holder

ANN   Fire alarm annunciator

Staff register system—any type

Electrical clock system devices—any type

Public telephone system devices

Private telephone system devices

Watchman system devices

Sound system, L = speaker,
V = volume control

Other signal system devices:
CTV = television antenna;
DP = data processing

SC Signal central station

Telephone interconnection box

PE Pneumatic/electric switch

EP Electric/pneumatic switch

GP Operating room grounding plate

P 6 Patient ground point; 6 = number
of jacks

Panelboards

Flush mounted panelboard and cabinet

Surface mounted panelboard and cabinet

Lighting panel

Power panel

Heating panel

Controller (starter)

Externally operated disconnect switch

Bus Ducts and Wireways

T T T Trolley duct

B B B Busway (service, feeder, or plug-in)

C C C Cable through ladder or channel

W W W Wireway

J Bus duct junction box

Electrical Distribution or Lighting System, Aerial, Lightning Protection

Pole

Street light and bracket

Transformer

Primary circuit

Secondary circuit

Auxiliary system circuits

Down guy

Head guy

Sidewalk guy

Service weather head

Lightning rod

— L — Lightning protection system conductor

Residential Signaling System Outlets

Push button

Buzzer

Bell

Bell and buzzer combination

Annunciator

Outside telephone

Interconnecting telephone

Telephone switchboard

BT Bell ringing transformer

357

O   Electric door opener

M   Maid's signal plug

R   Radio antennal outlet

CH   Chime

TV   Television antenna outlet

T   Thermostat

Underground Electrical Distribution or Lighting System

M   Manhole

H   Handhole

TM   Transformer—manhole or vault

TP   Transformer pad

– – – –   Underground direct burial cable

   Underground duct line

   Street light standard fed from underground circuit

Panel Circuits and Miscellaneous

– – – –   Conduit concealed in floor or walls

– – – – –   Wiring exposed

   Home run to panelboard—number of arrows indiates number of circuits

   Home run to panelboard—two-wire circuit

   Home run to panelboard— number of slashes indicates number of wires (when more than two)

LS–L1,3,5   Home run to panelboard—"LS" indicates panel designation; L1, 3, 5, indicates circuit breaker no.

— C —   Clock circuit, conduit and wire

— E —   Emergency conduit and wiring

— T —   Telephone conduit and wiring

   Feeders

—o   Conduit turned up

—●   Conduit turned down

G   Generator

M   Motor

5   Motor—numeral indicates horsepowper

I   Instrument (specify)

T   Transformer

   Remote start/stop push button station

HTR   Remote start/stop push button station

   Electric heater wall unit

Common Abbreviations on Drawings

EWC   Electric water cooler

EDH   Electric duct heater

AFF   Above finished floor

UH   Unit heater

GFI   Ground fault interrupter

GFP   Ground fault protector

GFCB   Ground fault circuit breaker

EC   Empty conduit

WP   Weatherproof

VP   Vaporproof

HVAC

Ductwork Symbols

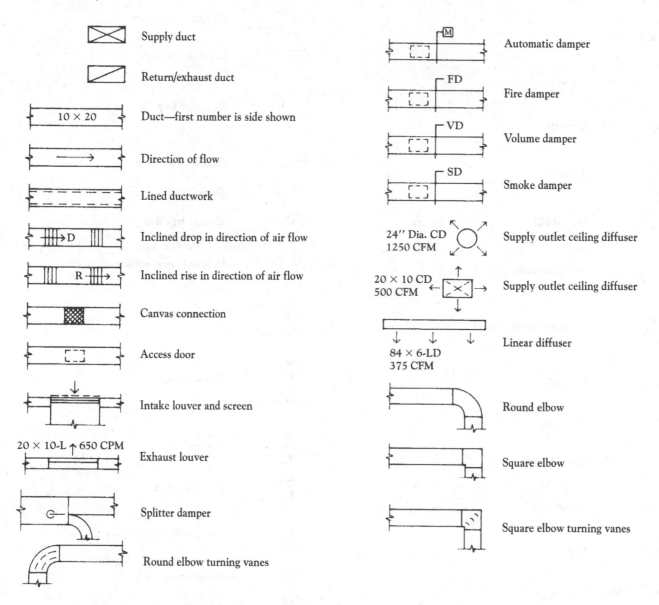

Supply duct

Return/exhaust duct

10 × 20   Duct—first number is side shown

Direction of flow

Lined ductwork

Inclined drop in direction of air flow

Inclined rise in direction of air flow

Canvas connection

Access door

Intake louver and screen

20 × 10-L ↑ 650 CPM   Exhaust louver

Splitter damper

Round elbow turning vanes

Automatic damper

FD   Fire damper

VD   Volume damper

SD   Smoke damper

24″ Dia. CD
1250 CFM   Supply outlet ceiling diffuser

20 × 10 CD
500 CFM   Supply outlet ceiling diffuser

84 × 6-LD
375 CFM   Linear diffuser

Round elbow

Square elbow

Square elbow turning vanes

Double duct air system

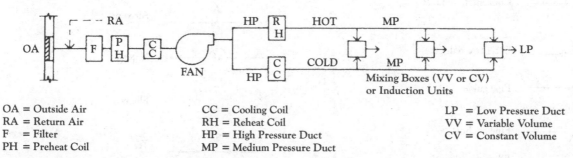

OA = Outside Air
RA = Return Air
F = Filter
PH = Preheat Coil

CC = Cooling Coil
RH = Reheat Coil
HP = High Pressure Duct
MP = Medium Pressure Duct

LP = Low Pressure Duct
VV = Variable Volume
CV = Constant Volume

Appendix: Symbols

Valves, Fittings, and Specialties

	Gate			Flow direction
	Globe		Up/Dn	Pipe pitch up or down
	Check			Expansion joint
	Butterfly			Expansion loop
	Solenoid			Flexible connnection
	Lock shield		T	Thermostat
	2-way automatic control			Thermostatic trap
	3-way automatic control		F & T	Float and thermostatic trap
	Gas cock			Thermometer
	Plug cock			Pressure gauge
	Flanged joint		FS	Flow switch
	Union		P	Pressure switch
	Cap			Pressure Reducing valve
	Strainer		H	Humidistat
	Concentric reducer		A	Aquastat
	Eccentric reducer			Air vent
	Pipe guide		M	Meter
	Pipe anchor			Elbow
	Elbow looking up			Tee
	Elbow looking down			

Equipment

HWS	Hot water heating supply
HWR	Hot water heating return
CHWS	Chilled water supply
CHWR	Chilled water return
D	Drain line
CW	City water
FOS	Fuel oil supply
FOR	Fuel oil return
FOV	Fuel oil vent
FOG	Fuel oil gauge line
PD	Pump discharge
	Low pressure condensate return
LPS	Low pressure steam
HPS	Medium pressure steam
MPS	High pressure steam
BD	Boiler blow-down

Piping Symbols

Valves, Fittings, and Specialties

Gate
Globe
Check
Butterfly
Solenoid
Lock shield
2-way automatic control
3-way automatic control
Gas cock

Plug cock
Flanged joint
Union
Cap
Strainer
Concentric reducer
Eccentric reducer
Pipe guide
Pipe anchor
Flow direction
Elbow looking up
Elbow looking down
Up/Dn	Pipe pitch up or down
Expansion joint
Expansion loop
Flexible connection
T	Thermostat
Thermostatic trap
F & T	Float and thermostatic trap
Thermometer
Presssure gauge
FS	Flow switch
P	Pressure switch
Pressure reducing valve
Temperature and pressure relief valve
H	Humidistat

A	Aquastat		——G——	Gas—low pressure
Air vent	Air vent		——MG——	Gas—medium pressure
M	Meter		——HG——	Gas—high pressure
	Hose bibb		——CA——	Compressed air
	Elbow		——V——	Vacuum
	Tee		——VC——	Vacuum cleaning
	"Y"		——N——	Nitrogen
	OS and Y gate		——N₂O——	Nitrous oxide
	Shock absorber		——O——	Oxygen
	House trap		——LOX——	Liquid oxygen
	"P" trap		——LPG——	Liquid petroleum gas
	Floor drain			

—— IW —— Indirect waste

– – –S– – – Sanitary below grade

——S—— Sanitary above grade

– – –ST– – – Storm below grade

——ST—— Storm above grade

– – – – – Vent

——CWV—— Combination waste and vent

– – –AW– – – Acid waste below grade

——AW—— Acid waste above grade

– – –AV– – – Acid vent

——– –—CW Cold water

——– –—HW Hot water

– – – – –HWC Hot water circulation

——DWS—— Drinking water supply

——DWR—— Drinking water return

Fire Protection Piping Symbols

——F—— Fire protection water supply

——WSP—— Wet standpipe

——DSP—— Dry standpipe

——CSP—— Combination standpipe

– – –SP– – – Automatic fire sprinkler

—o——o— Upright fire sprinkler heads

—●——●— Pendant fire sprinkler heads

Fire hydrant

Wall fire dept. connection

Sidewalk fire dept. connection

FHR
o–⊞⊞⊞ Fire hose rack

FHC Surface mounted fire hose cabinet

FHC Recessed fire hose cabinet

Plumbing Fixture Symbols

Baths

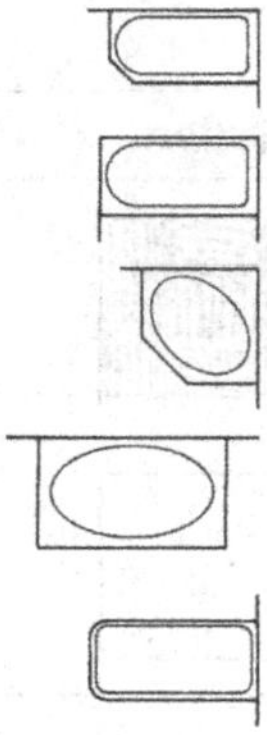

Corner

Recessed

Angle

Whirlpool

Institutional or island

Showers

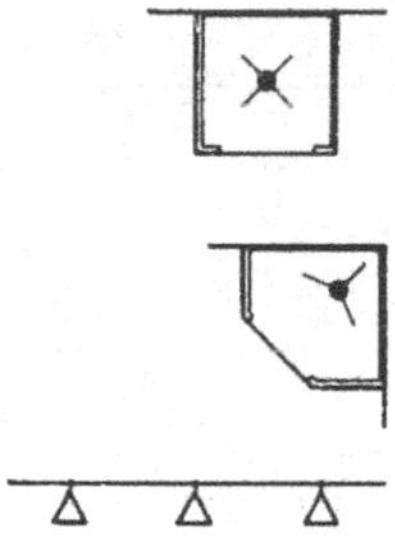

Stall

Corner stall

Wall gang

Water Closets

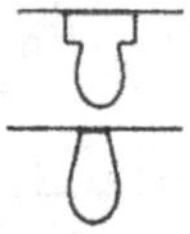

Tank

Flush valve

Bidet

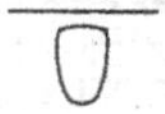

Bidet

Urinals

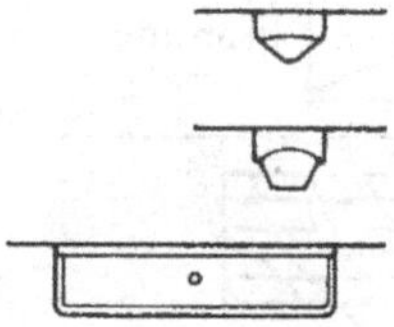

Wall

Stal[l]

Troug[h]

Lavatories

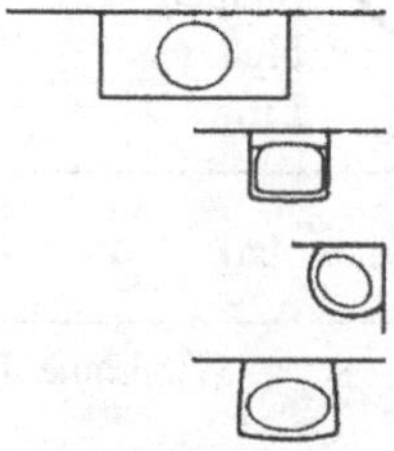

Vanity

Wall

Counter

Pedestal

Kitchen Sinks

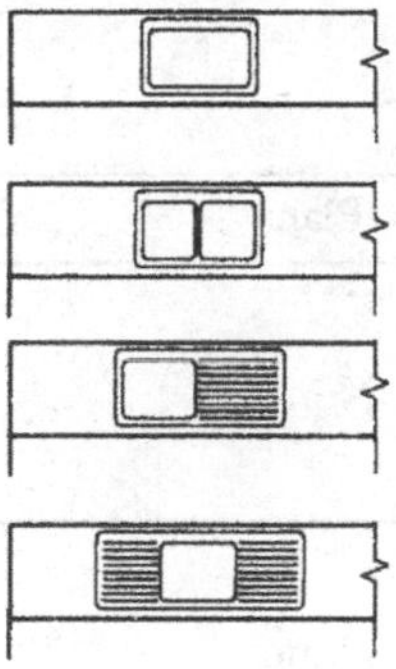

Single basin

Twin basin

Single drainboard

Double drainboard

Dishwasher

DW — Dishwasher

Drinking Fountains or Electric Water Coolers

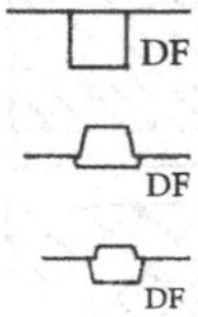

DF — Floor or wall

DF — Recesssed

DF — Semirecessed

Laundry Trays

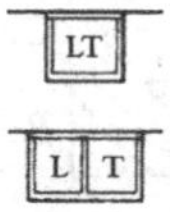

LT — Single

L T — Doubl[e]

Service Sinks

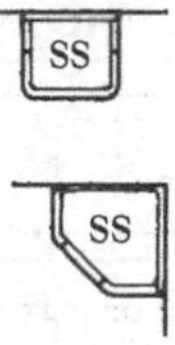

SS — Wall

SS — Floor

Wash Fountains

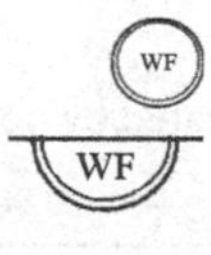

WF — Circular

WF — Semicircular

Hot Water

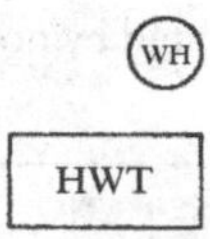

WH — Heater

HWT — Tank

Separators

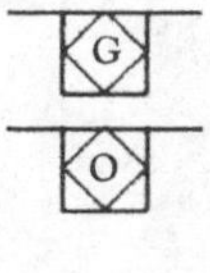

G — Gas

O — Oil

Material Indication Symbols

Table A-1   Material Indication Symbols

Material	Plan	Elevation	Section
Earth	None	None	
Porous fill	None	None	
Concrete			Same as plan
Brick			Same as plan
Concrete block			Same as plan
Stone	Cut stone / Rubble	Cut stone / Rubble	Same as plan
Steel	OR OR	None	
Framing wood	Wall or partition	Siding / Finish wood	Blocking / Framing
Plywood	Indicated by note	Indicated by note	
Sheet metal flashing	Indicated by note		Shaped Heavy Line
Batt insulation		None	Same as plan

(Continued)

Table A-1 Material Indication Symbols (Continued)

Material	Plan	Elevation	Section
Rigid insulation		None	Same as plan
Glass			Small scale Large scale
Gypsum wallboard			Same as plan
Acoustical tile		None	
Ceramic wall tile			Same as plan
Floor tile		None	

Table A-2 Window and Door Symbols

Type	Plan	Elevation
Double-hung windows		
Casement window		Indicates Window Hinge
Slider		
Exterior door		
Interior door		
Bifold door